National
Fire Protection
Association

Wildland Fire Fighter

Principles and Practice

SECOND EDITION

Joe Lowe
Jeff Pricher

JONES & BARTLETT
LEARNING

Jones & Bartlett Learning
World Headquarters
5 Wall Street
Burlington, MA 01803
978-443-5000
info@jblearning.com
www.jblearning.com

National Fire Protection Association
1 Batterymarch Park
Quincy, MA 02169-7471
www.NFPA.org

International Association of Fire Chiefs
4025 Fair Ridge Drive
Fairfax, VA 22033
www.IAFC.org

Jones & Bartlett Learning books and products are available through most bookstores and online booksellers. To contact Jones & Bartlett Learning directly, call 800-832-0034, fax 978-443-8000, or visit our website, www.jblearning.com.

Substantial discounts on bulk quantities of Jones & Bartlett Learning publications are available to corporations, professional associations, and other qualified organizations. For details and specific discount information, contact the special sales department at Jones & Bartlett Learning via the above contact information or send an email to specialsales@jblearning.com.

24015-3

Production Credits
General Manager and Executive Publisher: Kimberly Brophy
VP, Product Development: Christine Emerton
Senior Managing Editor: Donna Gridley
Executive Editor: Bill Larkin
Content Development Editor: Ashley Procum
Editorial Assistant: Alexander Belloli
VP, Sales: Phil Charland
Project Manager: Lori Mortimer
Project Specialist: John Coakley
Digital Products Manager: Jordan McKenzie
Digital Project Specialist: Angela Dooley
Director of Marketing Operations: Brian Rooney
Production Services Manager: Colleen Lamy
VP, Manufacturing and Inventory Control: Therese Connell
Composition: S4Carlisle Publishing Services
Cover and Text Design: Scott Moden
Senior Media Development Editor: Troy Liston
Rights Specialists: John Rusk and Liz Kincaid
Cover Image: Photo courtesy of Joe Lowe
Printing and Binding: LSC Communications

Library of Congress Cataloging-in-Publication Data
Names: Lowe, Joseph D., author. | Pricher, Jeffrey, author.
Title: Wildland fire fighter : principles and practices / Joseph Lowe, Jeffrey Pricher.
Description: Second edition. | Burlington, Massachusetts : Jones & Bartlett Learning, [2021] | Includes bibliographical references and index. |
Identifiers: LCCN 2020000278 | ISBN 9781284042115 (paperback)
Subjects: LCSH: Wildfire fighters--United States--Handbooks, manuals, etc. | Wildfires--United States--Prevention and control.
Classification: LCC SD421.3 .L69 2021 | DDC 634.9/6180973--dc23
LC record available at https://lccn.loc.gov/2020000278

6048

Printed in the United States of America
25 24 23 22 21   10 9 8 7 6 5 4 3 2

Courtesy of Joe Lowe.

Brief Contents

Contents

CHAPTER 15 Basic Firing Operations 327

Skill Drills

Preface

The second edition of *Wildland Fire Fighter: Principles and Practice,* was significantly updated to better serve wildland fire fighter instructors and students.

This second edition now meets and exceeds the following National Fire Protection Association (NFPA) and National Wildfire Coordinating Group (NWCG) standards:

- NFPA 1051, *Standard for Wildland Firefighting Personnel Professional Qualifications*, 2020 Edition (Chapters 4 and 5)
- NWCG S-190, *Introduction to Wildland Fire Behavior,* 2020 Edition
- NWCG S-130, *Firefighter Training,* 2008 Edition
- NWCG L-180, *Human Factors in the Wildland Fire Service,* 2014 Edition

The book now has two distinct sections: NFPA Wildland Fire Fighter I (NWCG Fire Fighter Type 2) and NFPA Wildland Fire Fighter II (NWCG Fire Fighter Type 1). The first eight chapters set the foundation for Wildland Fire Fighter I knowledge and understanding. These chapters include new and updated content coverage on wildfire history and wildfire in the 21st century, a sample case demonstrating incident development, wildland/urban interface considerations, hand tools, and much more.

The final seven chapters build upon the knowledge and skills outlined in the first eight chapters while introducing more advanced Wildland Fire Fighter II topics. These chapters include new and updated content coverage on human resources, radio communications, chainsaw operations, basic firing operations, and more.

Each of the four appendices correlates to one of the NFPA or NWCG standards mentioned above. The appendices break down the standards into digestible points and ties a chapter and page number to each job performance requirement or learning objective.

Whether you're following NFPA or NWCG teaching requirements, this new organization will allow you the flexibility to teach your Wildland Fire Fighter I and II course(s) exactly the way you wish.

About the Authors

Joe Lowe

Fire Chief Joseph Lowe is a 36-year veteran of the fire service. He came up through the ranks in the southern California fire service. After 25 years, he then took the position of Chief/Division Director for the State of South Dakota Wildland Fire Suppression Division where he served as Chief for 11 years before his retirement in 2012.

Chief Lowe was the National Type 2 Incident Commander for Rocky Mountain Area Team C from 2006 until 2012. He has also served as Chairman of the Rocky Mountain Coordinating Group (RMCG). The RMCG is established to promote the safe management of wildland fire and all risk incidents through interagency cooperation, communication, and coordination in the 5 state Rocky Mountain area.

The one accomplishment he is most proud of is the establishment of the Great Plains Interstate Fire Compact, which was formally established in 2007 by ratification of the 110th United States Congress. He was the author and its first compact administrator.

In 2006 he was part of a team of subject matter experts who did the revision to I-300/400 Incident Command courses at the National Interagency Fire Center.

Joe and his wife, Wendy, live in the Black Hills of South Dakota. Joe continues to consult and teach on wildland fire topics.

Jeff Pricher

Jeff Pricher is a 24-year veteran of the fire service currently working as a Division Chief managing a Special Operations program and serving as the Fire Marshal for two agencies in Oregon. He maintains credentialing in several areas in wildland, as a Paramedic, Fire Investigator, Instructor, Fire Officer and Part 107 credentialing with the Federal Aviation Administration. In addition, Jeff has served on red card committees, the NFPA 1051 committee, and is an associate member of the NWCG Interagency Fire Unmanned Aerial Systems subcommittee. He has participated in NWCG curriculum rewrites over the years as well as serving as a member on state level task forces related to wildland.

Jeff has been teaching wildland firefighting to new fire fighters since 1998 and has worked on handcrews, engines, and as a helicopter rappeller with the US Forest Service putting out fires in many states in the western US and Alaska. He currently serves on a Type 2 Interagency Incident Management Team in the Operations section.

Acknowledgments

The authors, the Jones & Bartlett Learning Public Safety Group, the National Fire Protection Association, and the International Association of Fire Chiefs would like to thank all of the advisors and reviewers that generously offered their time, expertise, and talent to the making of this second edition.

Authors

Joe Lowe
Chief/Division Director (Retired)
South Dakota Division of Wildland Fire Suppression
Rapid City, South Dakota

Jeff Pricher
Division Chief Fire Marshal/Special Operations
Scappoose Fire District
Columbia River Fire and Rescue
Scappoose, Oregon

Reviewers

Dena M. Ali, MPA
Captain
Raleigh Fire Department
Raleigh, North Carolina

Hudson Babler
Dallas Fire-Rescue Department
Dallas, Texas

Christopher Baker
Volunteer Advocate, Region IX Advocate Manager
Everyone Goes Home Program
National Fallen Firefighters Foundation
Clovis, California

Toby Ballard
Captain, Engine Boss
Missoula Rural Fire District
Missoula, Montana

Seth Barker, CTO, FO
Battalion Chief, Training Officer
Big Sky Fire Department
Big Sky, Montana

Lawrence Biehler
Fire Fighter, EMT
Wharton Fire Department
Wharton, New Jersey

Jovan Blake
Fire Fighter
San Francisco Fire Department
San Francisco, California

Pete K. Blakemore
Battalion Chief (Retired)
California Department of Forestry and Fire Protection
Sacramento, California
College of the Desert Fire Academy
Palm Desert, California

David Bray
Battalion Chief over Training
Las Vegas Fire and Rescue
Las Vegas, Nevada

William H. Campbell
Chief (Retired)
New York State Incident Management Team Program
Albany, New York

Ryan E. Carey, MSL
Rio Hondo College
Wildland Fire Technology
Whittier, California

Charlie Carpenter
Fire Chief
Millry Volunteer Fire and Rescue Department
Millry, Alabama

Peter J. Copeland
Midway Fire and Rescue
Pawleys Island, South Carolina

Edward T. Costa Jr., NREMT, BCA, NBFD
Fire Fighter
North Bend Fire Department
North Bend, Oregon

Brad Cronin, NRP, CFPS
Captain
Newport Fire Department
Newport, Rhode Island

Michael Destefano, MPA
Brevard County Fire Rescue
Rockledge, Florida

Vincent Dias
Fire Chief
Alamo West Fire Rescue
Alamogordo, New Mexico

Thomas Fentress
Fire Chief
Keener Township Volunteer Fire Department
Demotte, Indiana

Ken Fowler, MSL
Assistant Director
Louisiana State University
Fire Emergency Training Institute
Baton Rouge, Louisiana

Les Gandee
Instructor, Wildland Fire Fighter, Fire Fighter
West Side Volunteer Fire Department
Saint Albans, West Virginia
West Virginia Division of Forestry
Charleston, West Virginia
Florida Division of Forestry
Fort Myers, Florida

Anthony Gianantonio
Deputy Fire Chief
Palm Bay Fire Rescue
Palm Bay, Florida

Christopher L. Gilbert, PhD
Training Officer, Tactical Medic, Hazardous Materials Officer
Gainesville, Florida

Robert S. Goldenberg, MPA
Investigator, Inspector, Live Fire Instructor
Florida State Fire College
Ocala, Florida

Kevin Goodwin
Captain
Central Kitsap Fire and Rescue
Silverdale, Washington

Dale Hall
Captain
LeRoy-Rose Lake Fire Department
LeRoy, Michigan

Mike Holberton, BA
Captain
Columbia County Fire Rescue
Martinez, Georgia

Jeremy A. Keller, CWMS
Bellefontaine Fire and Emergency Medical Services
Bellefontaine, Ohio

Jason S. Kirk
St. Johns Emergency Services
St. Johns, Arizona

Jeremy Mathis, FO, GIFireE
Captain
Covington Fire Department
Covington, Georgia

Robert E. Mathis Jr.
Assistant Fire Chief
Spearfish Fire Department
Spearfish, South Dakota

Jason L.P. McMillan, EMT-P, FO1
Springfield Fire Department
Springfield, Illinois
Illinois Fire Service Institute
Champaign, Illinois

Jonathan Robert Morris
Lieutenant, AEMT, Instructor 1, TIFMAS Engine Boss
Burkburnett Fire Department
Burkburnett, Texas

Andrew Myhra
Captain
Columbia River Gorge National Scenic Area
United States Forest Service
Hood Rover, Oregon

John Newlin
Deputy Chief
Gainesboro Volunteer Fire and Rescue Company
Frederick County, Virginia

Nelson Ojeda, AIFireE
Fire Fighter, Paramedic
Tulsa Fire Department Tulsa, Oklahoma

Richard E. Presley
Lieutenant
French Broad Fire and Rescue Department
Alexander, North Carolina

Michael Rath
Fire Fighter, Paramedic
Lexington Fire Department
Lexington, Kentucky

William A. Raulerson
Florida Forest Service
DeLeon Springs, Florida

Roland Rose
Battalion Chief
Columbia River Gorge National Scenic Area
United States Forest Service
Hood Rover, Oregon

James Schiller
Ontario, Canada

Justin Schmollinger
Battalion Chief
Cal Fire Training Center
Ione, California

Gary Stevenson
Training Officer
Clark County Fire Department
Las Vegas, Nevada

David M. Wade, MAS, MICP
Cal Fire Training Center
Ione, California

Edward A. Wright, M.Ed
Lieutenant (Retired)
Kitsap County Fire District #18
Adjunct Instructor
U.S. Fire Administration National Fire Academy
Adjunct Instructor
Bates Technical College
Adjunct Instructor
Washington State Fire Marshall's Office

Gray Young
Chief of Training
South Bossier Fire
Elm Grove, Louisiana
Assistant Manager
Louisiana State University, Fire and Emergency Training Institute
Baton Rouge, Louisiana

Special Thanks from Joe Lowe

I want to thank all those mentors who have invested in me during my lengthy fire career. Their coaching helped me unlock my potential to see what I could become. I am grateful.

I want to thank the Orange County Fire Authority for sending me, early in my career, to the best wildland fire schools and National Wildfire Coordinating Group training opportunities. I want to applaud those in the organization's wildland stations, those working on hand crews, and the dozer operators, as they all share the same passion I have for wildland fire suppression.

A special thanks to the late South Dakota Governor Bill Janklow for hiring me to lead the South Dakota Division of Wildland Fire Suppression as Fire Chief. His support allowed us to get the equipment and resources needed to build the strongest and best-trained fire organization in a five-state region of the US.

The following agencies provided photographs and support for the book and I am grateful for their contributions. They are as follows:

- South Dakota Division of Wildland Fire Suppression
- Orange County Fire Authority
- California Department of Forestry and Fire Protection (Cal Fire)
- USDA Forest Service

Many thanks to the National Interagency Fire Center, Cal Fire, and South Dakota Wildland Fire Division for their support in the wildland fire community.

I also want to thank the members of the South Dakota Wildland Fire Division for their professionalism in fighting wildland fires. You are some of the best damned fire fighters in our nation.

Special Thanks from Jeff Pricher

Jeff Pricher would also like to thank his family for the continued support during the many hours put into this project.

There are several mentors and agencies that were a part of his formative years in wildland and public safety. Without them, he would not be able to share the information, skills, and knowledge submitted in this text. They are: Jim Howe, Darren Kennedy, Roland Rose, Jeff Dimke, Troy Corn, Tony Engel, Tom Taylor, Gil Dustin, Marc Crain, Renee Beams, Jim Trammell, Dennis Klein, Loretta Duke, Dave Jacobs, Devon Wells, Jack Deshong, Mike Greisen, Rod Altig, Jeff Walker, Pete Peterson, NW12, PNW2, Columbia River Gorge National Scenic Area (USFS), Malheur National Forest, Mount Hood Ranger District, Mt. Adams Ranger District, Snowmass Fire District, Cascade Locks Fire & EMS, and the Scappoose Fire District.

SECTION 1

Wildland Fire Fighter I

CHAPTER 1

Wildland Fire Fighter I

The Wildland Fire Service

KNOWLEDGE OBJECTIVES

After studying this chapter, you will be able to:

- Understand the history of wildland firefighting in the United States. (pp. 4–6)
- Describe fire's role in the natural world. (pp. 5–6)
- Understand wildfire in the 21st Century. (pp. 6–8)
- Describe the general prerequisite qualifications needed to become a wildland fire fighter. (**NFPA 1051: 4.1.1**, pp. 6–12)
- Outline the roles and responsibilities of a Wildland Fire Fighter I (NWCG Fire Fighter Type 2). (p. 8)
- Outline the roles and responsibilities of a Wildland Fire Fighter II (NWCG Fire Fighter Type 1). (pp. 8–9)
- Describe what a red card is and why it is important to keep it up-to-date. (pp. 10–11)
- Describe station life and daily routine of a wildland fire fighter. (pp. 10–11)
- Describe the daily routine of a wildland fire fighter. (pp. 11–12)
- Understand how to properly use reference materials. (pp. 12–17)

SKILLS OBJECTIVES

After studying this chapter, you will be able to perform the following skills:

- Use and reference the *Incident Response Pocket Guide* (IRPG). (pp. 12–13)
- Use and reference the Position Task Book (PTB). (pp. 14–15)
- Use and reference the Fire Danger Pocket Cards. (pp. 15–17)

Additional Standards

- **NWCG L-180 (NFES 2985)**, *Human Factors in the Wildland Fire Service*
- **NWCG S-190 (NFES 2902)**, *Introduction to Wildland Fire Behavior*
- **NWCG S-130 (NFES 2731)**, *Firefighter Training*

You Are the Wildland Fire Fighter

Today is your first day on the job. You are nervous, but excited. The time you have spent learning all of the required information has you feeling very confident you will be successful in this new realm of firefighting. The night before your first day of work, you prepared your fireline pack, had your go bag packed and ready to go, and reviewed all the components of engine company operations. In your mind, you are prepared. You arrive 15 minutes early and realize that most of your crew members are already there in physical training gear (PT gear). The engine captain is barking orders about the 10s and the 18s and laying into everyone about duty, respect, integrity, and the five communication responsibilities. As you hustle to get in line with everyone else, the following questions come to mind:

1. Where is my IRPG, and is it accessible right now for reference?
2. Why is the captain asking questions about duty, respect, integrity, and the five communication responsibilities?
3. This experience is very intense. Should I post this on my social media account to share with all of my friends?
4. Am I really prepared for this job? Is there anything else I could have done to prepare myself for this fire season?

Access Navigate for more practice activities.

Wildland Firefighting History

Throughout history, wildfires have been catalogued by scientists and historians for the integral part they have played within nature and human society. Some portions of this wildfire history have seen catastrophic destruction, yet, over time, humankind have learned to live with fire and its significance **TABLE 1-1**. Most recently, the destructiveness of fires has spurred local, state, and federal stakeholders as well as researchers to look more closely at problems arising from wildfires. Meanwhile, as this development continues, wildland fire fighters continue to be on the front lines protecting our communities and natural resources.

Learning from History

It is important to note how we can learn from history and use it to combat what we currently have control over in terms of today's resources, objectives, training, and education. The emphasis on training and education cannot be overstated. Your success as a wildland fire fighter will be based on your mentored experiences and those you gather when in the field on your own. Future experiences will need to have a substantial foundation, built from what we know and have learned from the past.

Recent research has shown the prevalence of fire during the past 1500 years. A 2018 study of paleoecology (study of past ecosystems) surmised that Native Americans used deliberate fires to manage landscapes. The team of researchers used pollen and charcoal records in addition to computer modeling to determine these conclusions. Another component of this research found that the old forests were more parklike, meaning there was less fuel buildup around the base of trees compared to the fuel-choked forests of today. Deliberate burning (prescribed fires) over time likely leads to healthy forests. Native Americans were identified as being essential stakeholders in forest ecology, and it is important that we incorporate their traditions into how we shape forest management today.

DID YOU KNOW?

Modern wildland firefighting can be dated back to a document created in 1919 by the second chief of the Forest Service, Henry Graves. In the document titled *A Policy of Forestry for the Nation*, under the heading Fire Protection, he established five objectives:

1. Need for a universal, effective service for suppressing and preventing wildfires.
2. Prompt fire detection and suppression through improvements to roads, trails, lookout stations, aircraft bases, etc.
3. Measures to reduce natural wildfire ignition.
4. Public education on the dangers of wildfires and the need for public cooperation.
5. Law enforcement involvement to punish those who start wildfires through carelessness or arson.

To some extent, the policies outlined above from the 1919 document are still in use today. Which of the above points do you think are still relevant?

TABLE 1-1 Historical Fires

Year	Fire	Location	Acres Burned	Fatalities
1845	"The Great Fire"	New York	1.5 million	Unknown
1871	Peshtigo Fire	Wisconsin	1.2 million	1200–2500
1871	The Great Michigan Fire	Michigan	2.5 million	250–300
1881	The Thumb Fire	Michigan	1 million	282
1910	The Great Fire of 1910	Idaho and Montana	3 million	87

Civilian Conservation Corps

The year 1933 was a significant year for wildland firefighting in the United States. On April 5, 1933, Executive Order 6101 was signed by President Franklin Delano Roosevelt to create the Civilian Conservation Corps (CCC). The CCC, as part of the New Deal economic relief program, established a voluntary unemployment public relief program to assist young men who needed to provide for their families during the Great Depression. With the creation of this program, 10 work classifications were authorized. One of them was "Forest Protection." During the history of the program, which lasted from 1933 to 1942, 3470 fire lookout towers and an estimated total of 97,000 miles of fire roads were built. In the course of this time period, the lookout towers aided in the detection and early notification of fires so that the resources necessary to control those fires could be requested. Access roads helped to facilitate the necessary logging operations but also afforded fire crews the ability to get to areas that would otherwise be inaccessible. Today's hotshots are direct descendants of the firefighting crews that were born in the CCC firefighting program.

Fighting Fire from the Air

In 1934, an Intermountain U.S. Forest Service (USFS) regional forester proposed a program that would deploy fire fighters from a plane to increase the success of initial attack on wildland fires **FIGURE 1-1**. This proposal highlighted the necessity of these fire fighters to be self-sufficient for days at a time. In 1939, the proposal became a working experiment that continues to this day. A couple of notable firsts with this program included the first jump on a fire in 1940 and the first female smokejumper in 1981.

FIGURE 1-1 A modern fixed-wing aircraft.
Courtesy of California Department of Forestry and Fire Protection.

Historical Changes of Wildland Fires

While fires in our forests, grasslands, and other wildlands have remained a constant, much has changed over the years. One of the biggest changes has been the frequency and ferocity of fires. Since the 1970s, it seems as though every year has been the worst fire year ever. There is some truth to this, but one of the biggest changes has been with regard to fire intensity and fire spread.

If we look back to the fires in the 19th century, the statistics highlight large-scale fires burning over a million acres every few years. As time has gone on, people have been migrating to and establishing communities within formerly remote forest and grassland areas. As a result, the number of fires has decreased, but the number of acres affected by fire has increased.

A large part of this can be attributed to longstanding forest policies that have called for the immediate suppression of all fires. The one variable to all

FIGURE 1-2 Ladder fuels.

of this has been that, more recently, an abundance of the fires have been human caused rather than naturally occurring.

All of this has led to forests and forested areas being overgrown with ladder fuels, which has resulted in situations that create high-intensity fires, instead of the lower-intensity ones that had historically been occurring **FIGURE 1-2**. Higher-intensity fires are harder to put out, take more manpower, and cost significantly more than previous fires. If we are to be successful in combatting future wildfires, we need to take a long hard look at the historical research and make use of the lessons learned over a century ago to make the best use of the advanced tools and knowledge we have today.

Wildfire in the 21st Century

The wildland firefighting landscape has changed in recent years due to changes in climate, population shifts, and patterns of development. These changes will continue to shape how we respond to fire activity.

Climate Change

Climate change has caused longer fire seasons, more intense fires, and unpredictable fire behavior, thus placing fire fighters at greater risk. In the foreseeable future, these fires will increase in intensity and destructive capability, causing significant effects on our economy and communities.

Recent evidence indicates that expansive, intense fires are migrating north and east and occurring in areas that have not historically been associated with significant wildfire activity. Droughts are occurring for longer periods of time in Western states and the Great Plains. In places where snow has been historically common, snowpacks are lower than in the past. Additionally, temperatures are increasing from the southern to the northern part of our country.

Commerce and migration patterns are the first to point to the effects of these changes in weather patterns. Individuals are shifting from cities to more suburban and rural areas for reasons other than the cost of living and price of homes. California winemakers, for example, seem to be moving north, where the climate is more hospitable.

Population Shifts

The **wildland/urban interface (WUI)**, is an area where undeveloped land, populated with vegetative fuels, meets with manmade structures **FIGURE 1-3**. As the population continues to grow, cities are getting increasingly more expensive to live in. These costs and the increasingly limited space in cities are forcing people into community outskirts near a WUI. Alternately, some people simply want a house with modern conveniences that is surrounded by the beauty of nature.

Developments in the WUI are sometimes allowed to be more dense than in the past, and, in some areas, building components are not required to be noncombustible. Development in the WUI puts fire fighters and residents at risk as a result of the close proximity of homes to one another.

FIGURE 1-3 Wildland/urban interface.

DID YOU KNOW?

Smokey Bear is a mascot that was created by the U.S. Forest Service in 1944 **FIGURE 1-4**. Smokey, a brown bear depicted with jeans, a hat, and sometimes a shovel, was created to serve as a public outreach icon, informing the public about wildfires and their dangers.

In 1950, Smokey came to life when an orphaned bear cub was found badly burned after a wildfire in New Mexico. The cub soon became a public interest story, and, during his treatment, many people across the United States wrote to the Forest Service asking about the cub's recovery. Upon recovery, the cub was transported to the National Zoo in Washington, DC, where he became an icon for conservation and wildfire prevention, a real-life Smokey. Smokey received so many gifts and letters during his time at the zoo that he was given his own zip code. Smokey died at the zoo in 1976. After his death, he was transported back to New Mexico and buried in Smokey Bear Historical Park. Today, Smokey's spirit is still alive more than ever.

Smokey's slogan to the public changed over the years, but the most well-known "Only YOU can prevent forest fires" in 1947 and then "Only you can prevent wildfires" from 2001 to the present day have stuck with the public. Smokey is still a public outreach figure for children and adults. Those looking to learn more can visit smokeybear.com.

FIGURE 1-4 Smokey Bear.

Technological Effects

As technology continues to improve, so does the technological world of wildland firefighting. Technological innovations have seen greater emphasis placed on tools, such as heat detecting satellites, geographic information systems (GIS), global positioning systems (GPS), enhanced communication systems, and real-time location for all fire fighters present at a fire.

Forthcoming technologies and tools promise enhancement to wildfire decision support software, injury databases, real-time mapping, smoke forecasting, and localized fire season predictions. These advances are a direct result of federal legislation signed into the Wildfire Management Technology Advancement Act in 2018.

Wildland Firefighting as a High-Reliability Organization

Despite the incredible power of technology, it is important to remember the human side of the wildfire community. Between 2002 and 2012, there was much emphasis from the wildland community to continue to foster a safety culture and be a leader that develops professionals.

One successful strategy was instituting the principles of high-reliability organizations (HROs) into the wildfire culture. An HRO is an organization that has succeeded in avoiding disasters in environments where disasters are common or expected. The concept is not new, and HROs have undergone several adaptations, based on the industry. In the early 1980s, researchers in California observed the operational cultures of U.S. Navy aircraft carriers, air traffic controllers, and the nuclear power industry. Based on the risk involved in these operational cultures, the researchers were looking for common factors that reduced risk and kept accidents to a minimum. Through the course of this research, the HRO concept was created.

Wildland firefighting as an HRO employs five characteristics:

- Preoccupation with failure. Share and report both large and small errors and mistakes. Small errors left unacknowledged could lead to a serious accident.
- Reluctance to oversimplify. Consider options other than the common practice. Utilize detailed checklists to ensure the job is done properly. It is not always beneficial to choose the quickest option.
- Sensitivity to operations. If you see errors being made, potential hazards, or something out of the ordinary, speak up!
- Commitment to resilience. Practice and perfect your skills. Commit to your team. Consider worst-case scenarios and determine action items for those scenarios.
- Deference to expertise. Rely on those with the most expertise, not those with seniority or the highest rank.

There are vast amounts of research conducted on this subject, but, for everyday fire fighters, understanding these five characteristics will allow you to successfully navigate the inevitable risk associated with wildland firefighting. More information on wildland firefighting as an HRO can be found in Chapter 1 of the *Wildland Fire Incident Management Field Guide*.

Roles and Responsibilities

The first step in understanding the organization of the fire service is to learn your roles and responsibilities as a **Wildland Fire Fighter I (NWCG Fire Fighter Type 2)** or **Wildland Fire Fighter II (NWCG Fire Fighter Type 1)**. As you progress through this text, you will learn what to do and how to do it so that you can take your place confidently among the ranks. This book discusses the training and performance qualifications for wildland fire fighters at these levels as outlined in the following standards:

- NFPA 1051, *Standard for Wildland Fire Fighter Personnel Professional Qualifications*
- NWCG S-190, *Introduction to Wildland Fire Behavior*
- NWCG S-130, *Firefighter Training*
- NWCG L-180, *Human Factors in the Wildland Fire Service*

The following sections describe the roles and responsibilities of a Wildland Fire Fighter I (NWCG Fire Fighter Type 2) and a Wildland Fire Fighter II (NWCG Fire Fighter Type 1). In some cases, the same or similar roles and responsibilities may appear under Wildland Fire Fighter I and Wildland Fire Fighter II. This enables fire fighters to learn the fundamentals of those concepts as a Wildland Fire Fighter I and increase their knowledge of the same concepts as a Wildland Fire Fighter II.

Wildland Fire Fighter I (NWCG Fire Fighter Type 2)

General roles and responsibilities for the Wildland Fire Fighter I (NWCG Fire Fighter Type 2) include the following:

- Understand the Wildland Fire Service.
- Understand and work within the Incident Command System (ICS)
- Understand fireline safety.
- Use and maintain personal protective equipment (PPE).
- Inspect and maintain hand tools and equipment.
- Understand the use for heavy ground and air equipment.
- Work as a member of a team.
- Assemble and prepare for response.
- Understand basic wildland fire behavior.
- Recognize hazards and unsafe situations.
- Construct and secure a fireline.
- Reduce the threat of fire exposure.
- Understand wildland/urban interface situations.
- Mop-up a fire area.
- Patrol a fire area.

Wildland Fire Fighter II (NWCG Fire Fighter Type 1)

General roles and responsibilities for the Wildland Fire Fighter II (NWCG Fire Fighter Type 1) include the following:

- Evaluate the readiness of crew members.
- Inspect and maintain PPE.
- Work as a member of a team.
- Give a shift briefing.
- Lead small groups of fire fighters on an assignment.
- Construct a fireline.
- Reduce the threat of fire exposure.
- Secure an area of suspected fire origin.

- Serve as a lookout.
- Operate firing devices.
- Use a map and compass.
- Maintain and operate power tools, including chainsaws.
- Maintain and operate water delivery equipment, including pumps.
- Work as a member of a handcrew.
- Maintain and operate firing devices.

LISTEN UP!

Fire fighters have a responsibility to reduce errors and to learn and improve their performance at all times.

Social Media

Society has evolved over the years from relying solely on printed newspapers and the evening television news to digital platforms and social media that are updated in real time. Though **social media** can help connect fire fighters with the public and show the public a behind-the-scenes look at what wildland fire fighters do, it also poses several challenges to the wildland fire service.

A seemingly harmless photo with a group of fire fighters could be misinterpreted as disrespect for personal or real property, disrespect for victims or colleagues, or an embarrassment for a response agency, resulting in a stain on the reputation of the fire service. Wildfire response agencies need to understand the relationship between how social media can assist the industry in some situations and hinder it in others.

Regardless of rank, any wildland fire fighter presence on social media, on or off duty, must be professional. Most agencies have specific policies that outline social media use. These policies generally cover the following points:

- As a responder, you are always considered on duty. Anything captured on a personal device (such as a smartphone) is considered discoverable in a court of law. In other words, do not use a personal phone to obtain official work photos, videos, or audio.
- Make sure you check with the incident commander (IC), public information officer (PIO), or your supervisor before sharing any incident- or firefighting-related posts on *any* forms of social media.
- Do not take, share, or post any photos that contain imagery or content that could identify an address, a license plate, or an individual's personal or real property.
- During the initial stages of an emerging incident, do not post any information to any social media account, and do not share any information with friends. Leaking details of an emerging incident could disrupt the investigation process and cause irreparable harm or grief to victims or family members who view the post.
- When messaging is directed at the public, it is the responsibility of the IC and the PIO to make sure that all messages are accurate and consistent.
- In addition to the public ramifications of social media, government regulatory agencies are also watching. If the media you capture contain imagery that shows safety violations or unwarranted conduct, you could be held responsible for being involved as a result of your documenting an event that violates policies or procedures or local, state, or federal laws.

Other types of unprofessional conduct, such as hazardous attitudes and hygiene, are discussed in Chapter 3, *Safety on Wildland Fires*.

Wildland Firefighting as a Profession

The first fire fighters simply needed sufficient muscular strength and endurance to pass buckets or operate a hand pump. Today, fire fighters require formalized training, attention to safety, and good judgment.

Wildland fire fighters are versatile and usually cross-trained just as their structural-fire counterparts are. These cross-trainings can include training in areas such as medical, aviation, heavy equipment, leadership, and management. For example, a fire fighter with several years of experience can also be qualified as a squad boss or single resource boss (capable of managing a specific resource or a crew, equivalent to a fire department company officer). In some instances, fire fighters can also be trained and qualified in logistics, planning, and other leadership roles in the Incident Command System (ICS) hierarchy (see Chapter 2, *The Incident Command System and ICS Forms*).

There is not enough space to adequately cover all professional aspects of every position or role in wildland firefighting; just know that your attention to detail, your duty, your respect, and your integrity will propel you to be a professional.

The work you will do is laborious, and the hours will be long and will sometimes feel thankless. You will undoubtedly get dirty, and you will not be in a clean, crisp uniform all of the time. In some instances, you may be stuck in the woods, cut off from civilization for several days. Regardless, as a wildland fire fighter, you will always be a professional. In a way, this profession is similar to the quiet professionals of the military special operations community in that, most of the time, there are no TV cameras or crowds present.

LISTEN UP!

Make sure you know the expectations of your team and the leader's intent. Be in control of your actions and an example to other fire fighters and crews. Represent the wildland firefighting profession as a whole by always trying to do the right thing.

Qualifications and Employment

To become a wildland fire fighter, you must be at least 18 years old. Wildland fire fighter positions are generally advertised in the off-season (October to March, depending on the agency) for employment during the fire season (May to November, depending on the agency and part of the country you intend to work).

A physical fitness test is required by every federal agency before a wildland fire fighter qualification card can be issued. Different positions require different levels of fitness. The National Wildfire Coordinating Group sets the minimum training, experience, and physical fitness standards for specific wildland fire positions.

The Incident Qualification Card or the **red card** is a document that certifies that an individual is qualified to perform the tasks of a specific position on a wildland fire. A red card certifies your qualifications for 3 to 5 years (depending on the qualification), so long as you complete a required annual refresher course and a work capacity test each year. Some local agencies may issue letters of certification in place of a red card. Most red cards are now printed on white paper; however, a couple of decades ago, the cards were actually printed on red card stock.

The Incident Qualifications and Certification System (IQCS) tracks responder qualifications for land management agencies, the U.S. Air Force, and the U.S. Army. Once you have earned a red card, you will appear in this system. A second system, called the Resource Ordering and Status System or ROSS, tracks the availability of these individuals in the IQCS, as well as crews, equipment, and incident management teams. This system has the ability to facilitate resource ordering on state and federal fires. A red card that is not up-to-date could prevent you from being assigned to a fire or other incident.

Training

Depending on your agency, a considerable amount of time is put into training a wildland fire fighter, and like other professions in the firefighting industry, much of it will be spent on the fundamentals. In the wildfire environment, these fundamentals include a thorough understanding of fuel, topography, weather, human factors, tools and equipment, terminology, and safety. In some instances, training will include time spent in tents out in the woods to further the understanding of the potential for spiking out and being part of an extended attack assignment or campaign fire that will require you to be away from your home unit for two or more weeks, or several months. Training will also cover basic communications, diet, nutrition, physical fitness, first aid, and cardiopulmonary resuscitation (CPR). Initial training will be at least a week, but many agencies employ longer timelines to better prepare fire fighters.

Some crews have a trained emergency medical technician (EMT) as part of the crew. However, because most wildland firefighting is performed away from civilization, it is vital for all fire fighters to understand the basics of first aid in the event that they may need to stabilize someone before trained personnel can arrive. On larger incidents, much time is spent focusing on the safety and evacuation of fire fighters in the event that they get hurt. This is known as rapid extrication module support (REMS) or rapid intervention teams (RITs). This focus on REMS and RITs has created new positions within the ICS.

LISTEN UP!

Training, no matter how extensive, is not a substitute for experience. Training is, however, essential preparation to begin gaining experience in a safe manner.

Specializations

One of the remarkable components of wildland firefighting is that there is an incredible amount of diversity in the organization. Today's wildfire response force is constantly evolving to capture components that facilitate the integration of technology, procedures, and sound management to aid the suppression effort. While it still takes fire fighters with hand tools and other equipment to put out fires, there is an increasing need for supporting disciplines in the field and behind the scenes.

In recent years, there has been an incredible increase in the number of technical specialists involved in wildland firefighting. Many of these specialists end up focusing on areas that the typical fire fighter may never consider. A qualified fireline resource adviser, for example, observes the suppression effort from an ecological perspective, making sure that firefighting tactics are mindful of damage to land, sensitive species, and historical artifacts and areas.

Other areas of specialization include pilots of both manned and unmanned aircraft, buying teams, field observers, and weather forecasters. With the changing environment for data collection and map building, information technology (IT) is also an increasing presence in the wildland firefighting effort.

Station Life and Daily Routine

Routine is an integral component of developing your situational awareness for now and the future. It is important to note that there are many variations to what a daily routine could look like. Outlined next are a few common themes that those undertaking a full-time position on a fire crew will most likely experience.

Equipment Inspection

Upon arrival for work, one of the most important components to successful operations will be to inspect your equipment to make sure it is fire ready. If you are the driver/operator for your engine, it will be necessary to look at all of the tools and compartments and check the fire pump, emergency lights, radios (both handheld and mobile), and all operational aspects of the vehicles to be used. Do not be afraid to check all fluids, batteries, lights, tire pressures, treads on the tires, and fuel levels.

It does not matter whether you are assigned to a fire engine, tractor plow, or a crew that utilizes crew buggies; all aspects of each vehicle must be checked. Skipping a vehicle inspection could lead to equipment failure and may possibly result in you and your crew being taken out of service. More importantly, equipment failure on the fireline can put you and your crew in great danger.

Equipment and tool inspection is discussed in more detail in later chapters.

Information Gathering

Next, you will need to develop your situational awareness specific to the weather and resources in your agency and surrounding agencies or units. Weather is one of the key factors in safety and understanding fire dynamics (see Chapter 4, *Basic Wildland Fire Behavior*). It is not uncommon to view several weather forecasts. These forecasts could include local, regional, 48-hour trends and predictions, and weather for the next seven days. It is also important to know what your adjoining forces or area resources are doing and what their staffing capabilities are. This information will allow you to predict if or when an incident will require more resources.

Fitness

Wildland firefighting is a very physical demanding job and requires you to be in good physical condition. You are expected to carry up to 45 pounds as part of your line gear and work in extreme temperatures and high-stress environments while traveling on foot in varied terrain. As a result, wildland fire fighters need to be in good shape to be able to perform at the required performance level. Your position and assignment will dictate what level of fitness will be required and expected of you. Most agencies utilize at least the first hour of the workday for physical training. Physical training is paramount to your success as a wildland fire fighter by preparing you to meet the physical nature of the work. You must complete a physical fitness test in order to receive a red card.

The importance of physical fitness and fire fighters' health is discussed in later chapters.

LISTEN UP!

Fire fighter health and fitness are also important parts of your daily routine. Health, fitness, and safety are discussed in more detail in Chapter 3, *Safety on Wildland Fires*. Physical fitness testing is discussed in Chapter 9, *Human Resources*.

Daily Routines on Large Incidents

When involved in a large, multiple-day assignment, the daily routines are slightly different. In general, your clock for waking up and completing your daily assignments before heading to the line should be calibrated on when the briefing is delivered by the IC or the incident management team. Your routine will always vary based on your assignment as part of an engine crew, squad, or handcrew.

Most fire fighters do not participate in the morning briefing. In general, attendance at briefings is typically left to strike team leaders/task force leaders, division group supervisors, and other management positions. Once the supervisor receives the briefing and incident action plan, he or she will pass the information and tasks to his or her assigned resources.

While supervisors attend the briefing, crew members should be finishing morning equipment checks and logistical needs so they can be ready to move out on completion of the briefing. You will want to make sure that the apparatus or vehicle is properly checked over, tools are sharp and ready to be used, and lunches have been acquired from the lunch trailer or kitchen for all crew. In accordance with your agency's policy and the incident organization, any damaged equipment should be documented and exchanged at the fire cache before your fireline assignment.

LISTEN UP!

It is a good idea to request a pallet of water and sports drink for your cooler if your engine or crew truck has one as standard equipment. Depending on the size of the fire, you may be able to store a bag of ice to keep everything cool. If water pallets are not available, grab a couple of cubies (1-gallon [3.8-liter] potable water boxes) to refill your water bottles. The use of cubies is increasing as a result of camp crews' efforts to reduce costs and minimize trash generation.

Reference Materials

There are several critical references that the wildland fire fighter should own or have access to. It is important to familiarize yourself with these references to be able to quickly find information at a moment's notice. With the advent of smartphones and tablets, most of these reference documents can be stored in PDF format on your devices. With the overwhelming, diverse, and dynamic nature of wildland firefighting, having access to good reference documents can make a difference when having to calculate loads for aircraft, determine the probability of ignition, or manage risk, among many other important topics. Being able to retain everything learned is optimal but not practical with all of the requirements for the different positions. Referencing job aids, checklists, and other documents is what keeps fire fighters and managers safe in the wildland fire environment.

The following list of reference materials is not exhaustive. Some materials, such as the Red Book, are discussed in other chapters. Be sure to stay up to date with all of your references, and utilize the resources provided by your department or agency.

National Wildfire Coordinating Group

The **National Wildfire Coordinating Group (NWCG)** was established by the Department of Agriculture and the Department of the Interior in 1976 after a major fire season in California **FIGURE 1-5**. The NWCG comprises several wildfire management agencies and organizations. The group was created to coordinate programs of the participating agencies to standardize

FIGURE 1-5 National Wildfire Coordinating Group logo.
Courtesy of National Wildfire Coordinating Group.

trainings and policies, avoid duplicative efforts, and provide a means of constructively working together to establish efficient wildland fire response efforts.

Today, the NWCG member agencies include the Department of Agriculture (U.S. Forest Service), Department of the Interior (BLM), Department of Homeland Security (Fire Administration), Intertribal Timber Council, National Association of State Foresters, and International Association of Fire Chiefs. There are 12 working groups and 15 committees associated with the NWCG that are committed to achieving a national standard in wildland fire management as well as providing leadership for a "seamless response to wildland fire."

National Fire Protection Association

The **National Fire Protection Association (NFPA)** is a global organization, established in 1896, that is devoted to eliminating death, injury, property, and economic loss due to fire, electrical, and related hazards **FIGURE 1-6**. The NFPA develops and maintains nationally recognized minimum consensus standards on many areas of fire safety and specific standards on hazardous materials. These standards set the expected construction, performance, and operation of many aspects of fire service operations.

This book meets the job performance requirements outlined in Chapters 4 and 5 (Wildland Fire Fighter I and Wildland Fire Fighter II) of NFPA 1051, *Standard for Wildland Firefighting Personnel Professional Qualifications* (2020).

Incident Response Pocket Guide

The ***Incident Response Pocket Guide* (IRPG)** is a pocket-sized book that outlines standards and reference information for wildland fire incident response. It was first published for use in 1999 by the NWCG. This publication was intended to partially replace the *Fireline Handbook*, which has since been retired. More

FIGURE 1-6 National Fire Protection Association logo.
Logo of National Fire Protection Association.

LISTEN UP!

This book meets the job performance requirements and objectives outlined in the following standards:

- NFPA 1051, *Standard for Wildland Firefighting Personnel Professional Qualifications* (2020) (Chapters 4 and 5)
- NWCG S-190, *Introduction to Wildland Fire Behavior* (2020)
- NWCG S-130, *Firefighter Training* (2008)
- NWCG L-180, *Human Factors in the Wildland Fire Service* (2014)

importantly, it was intended to put a comprehensive collection of information at the fingertips of the fire fighter with the hope of enhancing decision-making, leadership, communication, leader's intent, human factors, and after-action review (AAR) information. Some of the specific topics include significant references, operational engagement components, specific hazards, all hazards response, aviation, and emergency medical information.

The IRPG is generally reviewed every 4 years. With each revision comes a new cover color: 2010 was orange, 2014 was green, and 2018 is purple **FIGURE 1-7**. The next revision will print in 2022. It is important to familiarize yourself with the content to be able to quickly find information in a moment's notice.

Wildland Fire Incident Management Field Guide

The ***Wildland Fire Incident Management Field Guide*** is a document published by the NWCG that encompasses critical and useful information that is relevant to entry-level fire fighters all the way up to the division supervisor level of responsibility. This document also serves all levels of ICS, from the Incident Commander Type 5 (ICT5) level to the Incident Commander Type 3 (ICT3 or extended attack IC) level.

FIGURE 1-7 *Incident Response Pocket Guide.*
Courtesy of National Wildfire Coordinating Group.

This document is a continuation of the *Fireline Handbook*, which was replaced in 2013. When the *Fireline Handbook* was conceptually created, it was meant to be a responder's guide for many facets of wildland fire response.

What was initially intended as a pocket reference guide grew to over 400 pages of information. As a result, the NWCG recognized that, despite the need to retire the *Fireline Handbook*, a similar document needed to exist. The majority of the pertinent information that needed to be observed on an ongoing basis during an incident was included in the IRPG. In 2014, the *Wildland Fire Incident Management Field Guide* was created. This document, which can be viewed in a PDF format, encompasses the history and importance of the *Fireline Handbook* as well as all other pertinent information that is required during the suppression of wildfires. The success of this newer document can be validated by the duplication of information in the *National Incident Management System (NIMS) Emergency Responder Field Operations Guide* (ERFOG), which is distributed by the Federal Emergency Management Agency (FEMA).

Position Task Books

A **Position Task Book (PTB)** is a document listing the performance requirements for a specific position in a format that allows for performance evaluation to determine whether an individual is qualified. PTBs are an efficient way to validate that students not only have retained the requisite knowledge but also have the ability to perform a specific task while on several assignments. All wildland fire fighters will experience and be part of the PTB certification system. Task books are assigned to new fire fighters or trainees for any new position to verify knowledge and proficiency. In most cases, PTBs are initiated by your agency. PTBs have been created for all wildland-specific positions by the NWCG. Some agencies have taken the base requirements of the PTBs and added other requirements based on guidelines and policies specific to their agency. A catalog of NWCG PBTs can be accessed at nwcg.gov/publications/position-taskbooks.

In almost every line-of-duty death, investigators dig deeply into the PTB to learn more about the fire fighter's background and training and experience requirements to develop a better picture of what went wrong. Therefore, it is crucial to be able to validate and prove that the candidate is capable and qualified. The integrity of the PTB system requires honesty from all participants. Failure to honestly evaluate a candidate can result in underqualified individuals rising to leadership roles for which they are unsuited, creating very real life safety issues for everyone around them.

PTBs are valid for 3 years. If a trainee takes longer than 3 years to complete consecutive assignments necessary for PTB certification, it would be impossible to truly verify competency at an evaluated level. Once PTBs are completed, recertification generally happens on a 2 or 3-year cycle, depending on the position.

Take the time to familiarize yourself with all the facets of PTB. PTBs and PTB processes will vary based on your agency. Knowing your agency's requirements will aid in a thorough PTB record keeping.

Skills Crosswalks for Structural Fire Fighters

Some agencies utilize NWCG skills crosswalks or gap courses to qualify personnel. **Skills crosswalks** are specific PTBs used in instances where structural fire fighters are looking to transition to a wildland fire fighter role. These fire fighters already have a basic knowledge of fire suppression and behavior and simply need to fill in the gaps to transition from structural to wildland firefighting. Crosswalks reduce the number of training hours needed to obtain qualification. Structural fire fighters, meeting NFPA 1001 and NFPA 1021 standards, can incorporate their existing knowledge and skills to obtain a qualification in one of the following wildland firefighting positions: Wildland Fire Fighter I, Wildland Fire Fighter II, single-resource engine boss, or strike team leader. Skills crosswalks require PTB sign-offs by a certifying official. In many cases, the fire fighter will also need to obtain a red card.

PTBs must be initiated before evaluations are entered. A PTB that has not been initiated is not a valid PTB.

General Information

General information and tips on PTBs are below:

- Most agency policies require a training officer or equivalent to initiate a PTB.
- Trainees must meet the prerequisites for the trainee position.
- PTBs are valid for 3 years from the date they were initiated. However, once the first evaluator enters an evaluator record, the 3-year clock starts all over again. This applies for only the first evaluation. All subsequent evaluations are required to be within the 3-year time frame.
- Evaluators must be fully qualified for the position they are evaluating.
- The number one spot on the evaluator's form should be filled out only if the trainee is ready for certification and only by a final evaluator.
- Trainee assignments should be representative of the skill or competency being evaluated. For example, an evaluation for a safety officer should be made when the lookout is actually performing in this capacity. The evaluation should not be made in a nonwildland situation, such as interior safety on a structure fire or burn to learn.
- Evaluation records must be complete. While paperwork is never an enjoyable component of training, accurately and completely filled out evaluation forms will lessen any confusion when the certifying official or committee is trying to verify that the trainee has met the requisite experience needed to satisfy the requirements of the trainee experience.
- PTBs should have a minimum of three trainee assignments/evaluation records. Some agencies require four or five with at least one of the assignments out of the local or home jurisdiction.
- Tasks that are identified in the PTB should have more than one signature by different evaluators.

This ensures that a trainee has been thoroughly evaluated by several people.

- The bulleted items in the PTB are only examples of tasks and are not a specific list of items that have to be evaluated. Look at the description of the event to validate performance of the skill and task to be evaluated.
- Certain tasks must be completed on actual wildfire incidents. Make sure you understand the codes.
  - W: Tasks labeled *W* must be performed on a wildfire incident. Simulated fires do not count as wildfires.
  - RX: Tasks labeled *RX* must be performed on a prescribed fire.
  - W/RX: Tasks labeled *W/RX* can be performed on either a prescribed fire or a wildfire.
  - R: Tasks labeled *R* mean "rare" events. These events include injuries, accidents, and aircraft or vehicle crashes. If a trainee is not able to experience one of these events, the evaluator can determine through an interview whether the trainee can actually perform the task. If the interview option is chosen, it is important to document this in the PTB.
- Many agencies utilize a task book committee to evaluate all positions at the ICT5/Squad Boss level and above. This process allows for a collective and comprehensive approach to a performance-based system. If the trainee is recommended for certification, the PTB is recommended to the certifying official for certification. Not all agencies utilize this system, but it is a growing trend.
- In addition to completing the evaluation forms, it is also important to receive a performance evaluation to mirror the PTB evaluation sheet. The ICS-225 *Incident Personnel Performance Rating* is often used for this purpose, but agency-specific forms may also be used. Make sure to keep the evaluation forms and the performance evaluation as part of your permanent record. On larger fires where there is a training specialist assigned to the incident, verification of these documents will be a prerequisite before being allowed to perform in the capacity of a trainee.
- Some PTBs cover more than one position. Make sure you fill out the appropriate final certification sheets when the trainee is ready and has completed all of the common requirements as well as the position-specific requirements. Examples include the NWCG Firefighter Type 1 (Squad Boss)/ICT5, and Single Resource Boss, which include Crew Boss, Engine Boss, Felling Boss, Firing Boss, and Heavy Equipment Boss.
- A good habit to get into when filling a trainee assignment is to make sure you identify yourself as a trainee during the incident, for example, Division Quebec from Crew 22 Alpha squad trainee. Identifying yourself as the trainee allows everyone on the incident to be able to tell the difference between the qualified individual and the trainee should the qualified personnel need to take over.

Fire Danger Pocket Cards

The **Fire Danger Pocket Cards**, sometimes simply referred to as Pocket Cards, outline potential fire dangers in a given area. They offer visual reference of current and past fire trends; historical worst and average conditions; and thresholds for temperature, humidity, wind speed, and fuel moistures associated with past problem fires **FIGURE 1-8**. Generally, Pocket Cards are printed and handed out to fire fighters. Pocket Card content is gathered by the **National Fire Danger Rating System (NFDRS)**. The NFDRS is a system that compares current and historic data related to weather, live and dead fuel moistures, and fuel types in order to identify the significance of burning conditions.

Pocket Card content can be correlated by an agency or regional group on the Fire and Aviation Management Information Technology (FAM-IT) website (famit.nwcg.gov) or other local land management–based system. Agencies can find cards for their specific regions on this website.

Pocket Cards are most effective when they are created using the NWCG standards. These standards are readily available on the FAM-IT website. In general, this product is completed over several months as a result of much analysis by experienced fire managers.

As originally conceived, Pocket Cards were created for fire fighters responding to an area, state, or region with which they were unfamiliar. Pocket Cards were part of the briefing process and were intended to be used as tools for out-of-area resources. Today, Pocket

LISTEN UP!

Regions with a distinctly bimodal fire season (spring/fall) may develop multiple Pocket Card versions for the same area to reflect seasonal conditions. In other regions, very different fuel types intermingled within a small geographic area are conveyed to fire crews using multiple Pocket Cards for different fuels. Sumter National Forest in South Carolina, for example, issues separate cards for co-occurring coastal and piedmont fuel types.

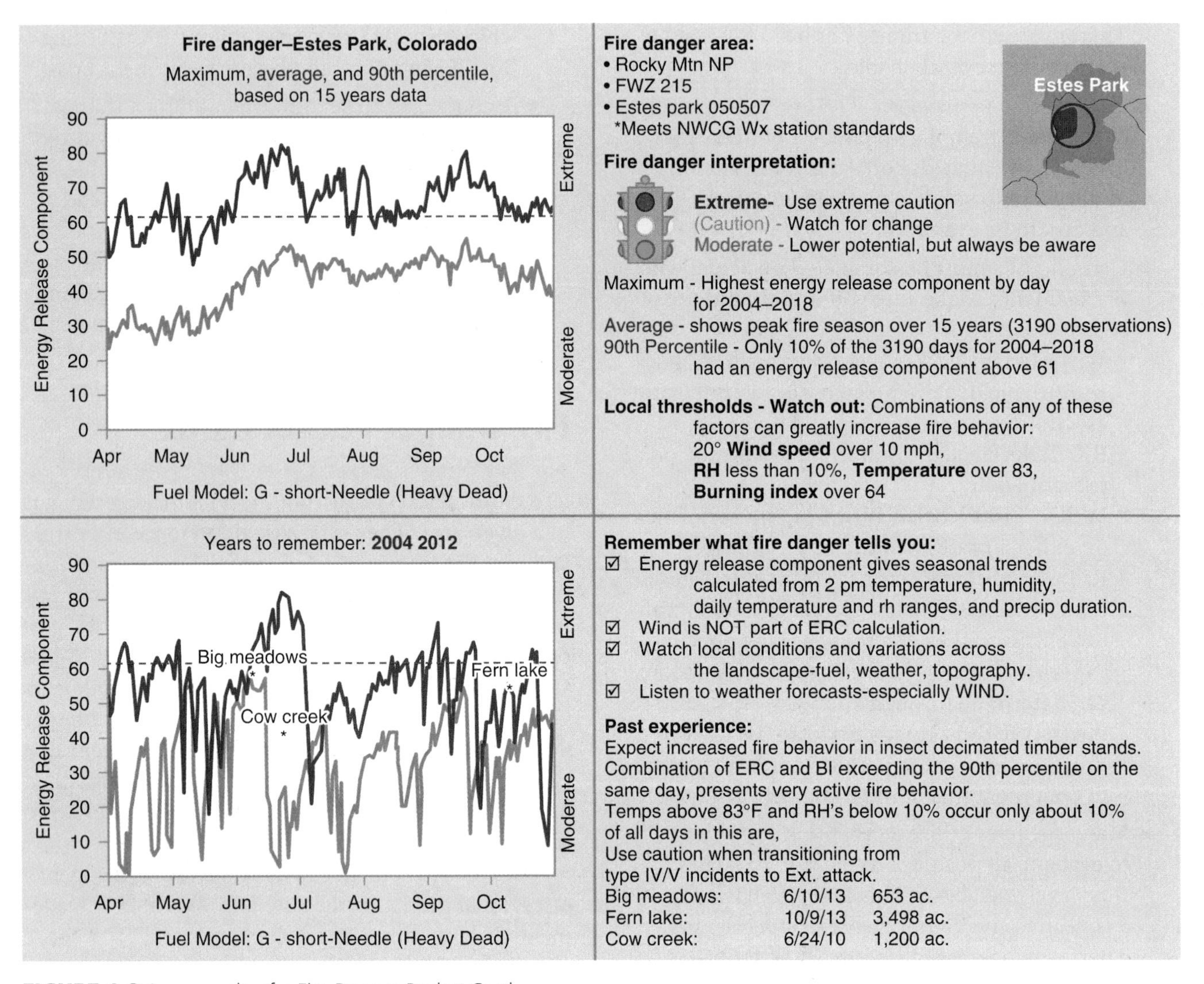

FIGURE 1-8 An example of a Fire Danger Pocket Card.

Design by NWCG Fire Danger Working Team, National Wildfire Coordinating Group. Retrieved from https://famit.nwcg.gov/sites/default/files/PC_Estes%20Park.jpg.

Cards can be utilized on a day-to-day basis to allow crews to understand current weather and fuel conditions in comparison to historical fire data.

Pocket Card Components

NWCG standard Fire Danger Pocket Cards include the following components:

- *Area.* Each Pocket Card is developed for a specific, defined geographic area. The area can encompass weather zones, dispatch areas, protection zones, fire climate zones, or predictive services. The area is usually clearly identified with a small map icon in the top right corner of the card.
- *Weather stations.* The area covered by the Pocket Card will determine which weather stations contribute data to its creation. The weather stations should be identified on the card in addition to the percentages applied to the multiple stations if not every station was given a value of 100 percent. In addition to complying with NWCG standards for weather data, the weather station is required to provide a data file encompassing at least 10 years of data.
- *Fuel model.* One of the challenges and limitations of the cards is that they are confined to only one fuel model at a time. Depending on the predominant fuel model of the area represented or historical fires given a specific fuel, one fuel will be determined. The fuel used for the Energy Release Component (ERC) chart and Pocket Card should be linked to the fire danger operating plan for the area represented. This fuel model is also used for configuring the daily response and staffing levels. Each region and area approaches this differently, so it will be important to know the details for the area you are working.

- *Seasonal trends*. Seasonal trends are identified in a graph on the Pocket Card. This graph depicts the length of the typical fire season and the worst all-time values of the components or index of the typical fire season for the area represented. It is not uncommon to see the Pocket Card cover 2 to 3 years, depending on the area. When looking at the fuel models and the NFDRS value for a given fire season, the Pocket Card will generally cover the ERC values spanning the 80th and 95th percentiles. This information coupled with historical fires will aid in developing your situational awareness.
- *Fire danger interpretation*. For the most part, the Pocket Card will depict the danger using specific colors. These colors span least danger to most danger with the most danger being represented in red. The color spectrum ranges from green (low danger) to yellow, brown, orange, and red (high danger). These colors are combined with a graphic to illustrate the potential danger for a given fire season.
- *Critical threshold values*. These values are generally expressed in written verbal terms that will include local pertinent information. The local information can include diurnal weather patterns, local resources, terrain, and specific fuel types. Other areas of consideration will include humidity, large fires, and specific fuel types.
- *Supplemental information*. This area of the card should be one of the key areas to consider when encountering a new fire. In this category, the component of years to remember or history of fire in the area help you to understand current conditions in the context of the covered area's most extreme events.
- *Currency*. The frequency at which Pocket Cards are updated is dependent on the local conditions or fuel. In other words, the Pocket Cards should be updated when necessary or more pertinent information is available. One component that is important to the Pocket Card system is to make sure the date is posted on the card to allow the user to identify how old or new the information is.

LISTEN UP!

Another important forecast tool can be found on the Internet on the National Interagency Fire Center (NIFC) predictive services website. The Geographic Area Seven-Day Significant Fire Potential Outlook is divided by geographic region. Based on the current conditions, a color-coded zone or area is associated with the calculated danger. The colors range from green (low danger) to yellow, brown, orange, and red (high danger) **FIGURE 1-9**.

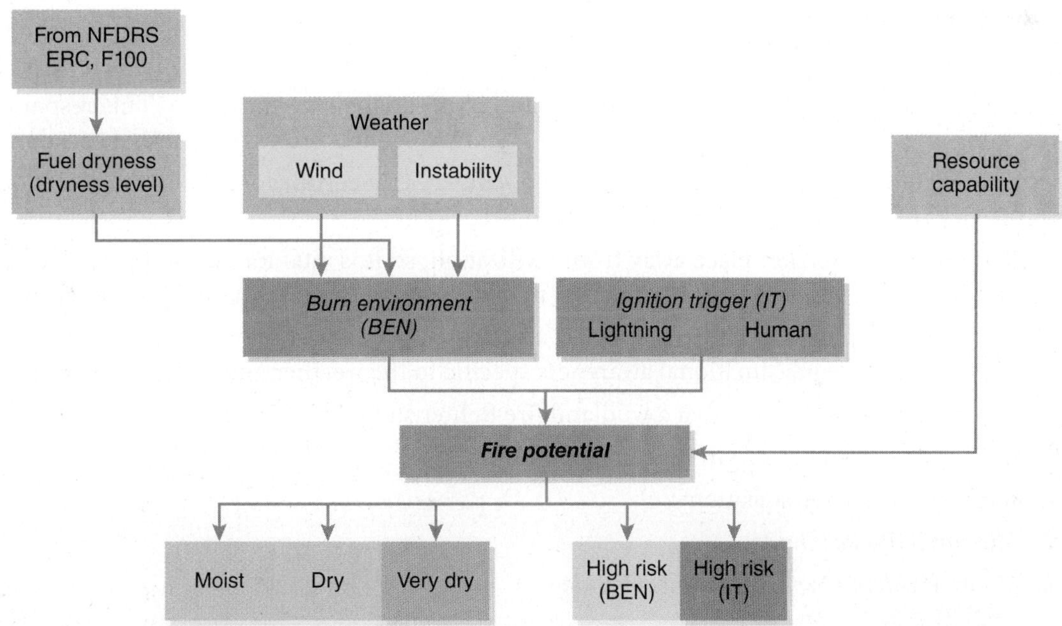

FIGURE 1-9 Fire potential model by color.

After-Action REVIEW

IN SUMMARY

- History has shown the contributions, benefits, and consequences of wildland fires.
- The establishment of the Civilian Conservation Corps (CCC) by President Franklin Delano Roosevelt, as a component of the New Deal, was integral in developing forest protection and what are today known as "hotshots."
- Education and training programs, investigative techniques, technological advances, and public awareness campaigns are all benefactors of our careful study of the history of wildland fires. Proper research and analysis can reduce or eliminate a repetition of the past.
- A greater percentage of fires are caused by humans than in previous history. This is in some part due to the migration of humans to the forest fringes and beyond.
- The Earth's changing climate is having a significant impact on wildland fire activity. Some examples of this include:
  - The frequency and ferocity of fires is increasing.
  - Fires are occurring in locations that have not historically seen fire activity.
  - Droughts are occurring for longer periods of time.
  - Temperatures are increasing and snowpacks are decreasing.
- Wildland firefighting is unique, yet it is also similar to structural firefighting. Training, daily routines, operational requirements, field activities, technology, equipment, and investigative efforts are all areas that one must consider when one considers the profession of wildland firefighting.
- A recent success strategy to combat fires has been employing the principles of high-reliability organizations (HROs) into the wildfire firefighting culture. These principles include adherence to and an understanding of:
  - Preoccupation with failure
  - Reluctance to oversimplify
  - Sensitivity to operations
  - Commitment to resilience
  - Deference to expertise
- Fire fighters have a responsibility to reduce errors and improve their performance. This responsibility includes defining and adhering to professional behavior and proper and professional use of social media.
- The Incident Qualification Card or the red card is a document that certifies that an individual is qualified to perform the tasks of a specific position on a wildland fire.
- Most wildland firefighting takes place away from civilization, so it is vital for all fire fighters to understand the basics of first aid in the event that they need to stabilize someone before trained personnel can arrive.
- Some of the most important components of being a successful wildland fire fighter are checking and maintaining equipment and having situational awareness specific to the weather and to available resources.
- There are several critical references that a wildland fire fighter should have access to:
  - National Wildfire Coordinating Group
  - National Fire Protection Association
  - *Incident Response Pocket Guide*
  - *Wildland Fire Incident Management Field Guide*
  - Position Task Books
  - Fire Danger Pocket Cards

KEY TERMS

Fire Danger Pocket Cards Outline potential fire dangers in a given area based on historical fire and weather data.

Incident Response Pocket Guide (IRPG) A pocket-sized book that outlines standards and reference information for wildland fire incident response.

National Fire Danger Rating System (NFDRS) Compares current and historic data related to weather, live and dead fuel moistures, and fuel types to rate the level of potential fire danger.

National Fire Protection Association (NFPA) Develops and maintains nationally recognized minimum consensus standards on many areas of fire safety and specific standards on hazardous materials.

National Wildfire Coordinating Group (NWCG) An organization made up of several wildfire management agencies that was designed to coordinate programs of the participating agencies to standardize trainings and policies, avoid duplicative efforts, and provide a means of constructively working together to establish efficient wildland fire response efforts.

Position Task Book (PTB) A document that lists the performance requirements for a position in a format that allows for performance evaluations to determine whether an individual is qualified.

Red card A document that certifies that an individual is qualified to perform the tasks of a specific position on a wildland fire; officially called the Incident Qualification Card.

Skills crosswalks Specific PTBs used in instances where structural fire fighters want to transition to a wildland fire fighter role to fill in training gaps to reduce the number of training hours needed to obtain qualification.

Social media Online websites on which people can publicly share personal information, pictures, news, and interests.

Wildland Fire Fighter I (NWCG Fire Fighter Type 2) The person at the first level of progression who has demonstrated the knowledge and skills necessary to function as a member of a wildland fire suppression crew under direct supervision.

Wildland Fire Fighter II (NWCG Fire Fighter Type 1) The person at the second level of progression who has demonstrated the depth of knowledge and skills necessary to function as a member of a wildland fire suppression crew under general supervision.

Wildland Fire Incident Management Field Guide A document that contains critical and useful information that is relevant to fire fighters at all levels of responsibility.

Wildland/urban interface The area where undeveloped land populated with vegetative fuels meets with manmade structures.

REFERENCES

Federal Emergency Management Agency (FEMA). "IS-100.C: Introduction to the Incident Command System, ICS 100." Modified June 25, 2018. https://training.fema.gov/is/courseoverview.aspx?code=IS-100.c.

Grant, Casey Cavanaugh. "History of NFPA." National Fire Protection Association (NFPA). Accessed December 13, 2019. https://www.nfpa.org/About-NFPA/NFPA-overview/History-of-NFPA.

Graves, Henry S. "A Policy of Forestry for the Nation." U.S. Department of Agriculture. December 1919. Accessed December 13, 2019. https://foresthistory.org/wp-content/uploads/2017/01/Graves_Policy.pdf.

National Fire Protection Association (NFPA). "NFPA 1051, Standard for Wildland Firefighting Personnel Professional Qualifications." 2020 ed. https://www.nfpa.org/codes-and-standards/all-codes-and-standards/list-of-codes-and-standards/detail?code=1051.

National Interagency Fire Center (NIFC). "High Reliability Organizing: What It Is, Why It Works, How to Lead It." March 2010. https://www.nifc.gov/training/HRO/HRO_2010trainingText.pdf.

National Interagency Fire Center (NIFC). "Interagency Standards for Fire and Fire Aviation Operations." NFES 2724. February 2019. https://www.nifc.gov/PUBLICATIONS/redbook/2019/RedBookAll.pdf.

National Wildfire Coordinating Group (NWCG). "L-180, Human Factors in the Wildland Fire Service, 2014." NFES 2985. Modified October 23, 2019. https://www.nwcg.gov/publications/training-courses/l-180.

National Wildfire Coordinating Group (NWCG). "NWCG Standards for Wildland Fire Position Qualifications." PMS 310-1. October 2019. https://www.nwcg.gov/sites/default/files/publications/pms310-1.pdf.

National Wildfire Coordinating Group (NWCG). "NWCG Standards for Course Delivery, PMS 901-1." Modified December 5, 2019. https://www.nwcg.gov/publications/pms901-1.

National Wildfire Coordinating Group (NWCG). "NWCG Standards for Interagency Incident Business Management." PMS

902. April 2018. https://www.nwcg.gov/sites/default/files/publications/pms902.pdf.

National Wildfire Coordinating Group (NWCG). "S-130, Firefighter Training, 2008." NFES 2731. Modified October 23, 2019. https://www.nwcg.gov/publications/training-courses/s-130.

National Wildfire Coordinating Group (NWCG). "S-190, Introduction to Wildland Fire Behavior, 2008." NFES 2902. Modified December 12, 2019. https://www.nwcg.gov/publications/training-courses/s-190.

National Wildfire Coordinating Group (NWCG). *Wildland Fire Incident Management Field Guide*. PMS 210. NFES 2943. January 2014. https://www.nwcg.gov/sites/default/files/publications/pms210.pdf.

Pyne, Stephen. "The Ecology of Fire." Nature Education Knowledge. 2010. Accessed December 13, 2019. https://www.nature.com/scitable/knowledge/library/the-ecology-of-fire-13259892.

Short, Karen C. "Spatial Wildfire Occurrence Data for the United States, 1992–2015." 4th ed. U.S. Department of Agriculture (USDA). 2017. https://www.fs.usda.gov/rds/archive/catalog/RDS-2013-0009.4.

U.S. Fire Administration (USFA) and Federal Emergency Management Agency (FEMA). "Wildland Training Skills Crosswalk for Structural Firefighters." 2019. https://www.usfa.fema.gov/training/other/#contents.

U.S. Fire Administration (USFA) and National Fire Academy. "Field Operations Guide." ICS 420-1. June 2016. https://www.usfa.fema.gov/downloads/pdf/publications/field_operations_guide.pdf.

U.S. Forest Service (USFS). "Smokey Bear Guidelines." March 2009. https://www.fs.usda.gov/Internet/FSE_DOCUMENTS/stelprdb5107991.pdf.

U.S. Forest Service (USFS). "Story of Smokey." https://www.smokeybear.com/en/smokeys-history?decade=1940.

Wildland Fire Fighter in Action

You have just completed your first day of your Wildland Fire Fighter I course. You have a quiz coming up, so you decide to join a small study group to review the material before the quiz. The discussion in your group generates a number of questions.

1. Which guide should be carried with you at all times?
   - **A.** IRPG
   - **B.** *Fireline Handbook*
   - **C.** IHOG
   - **D.** *Handy Dandy Firefighting Field Guide*
2. High-reliability organizations utilize how many principles?
   - **A.** 5
   - **B.** 4
   - **C.** 6
   - **D.** 7
3. Which U.S. fire had the largest loss of life?
   - **A.** The Peshtigo
   - **B.** The Great Fire
   - **C.** The Great Michigan Fire
   - **D.** The Thumb Fire
4. Fire Danger Pocket Cards are based on what system?
   - **A.** National Fire Danger Rating System
   - **B.** Wildland Fire Decision Support System
   - **C.** Wildfire Defense Systems
   - **D.** Wildland Detection System

Access Navigate for flashcards to test your key term knowledge.

CHAPTER 2

Wildland Fire Fighter I

The Incident Command System and ICS Forms

KNOWLEDGE OBJECTIVES

- Explain what an incident is. (p. 22)
- Explain what the Incident Command System (ICS) is. (p. 23)
- Describe the chain of command within the ICS. (pp. 23–24)
- Describe the general responsibilities of each section in the ICS:
    - Command section (pp. 24–27)
    - Operations section (pp. 27–30)
    - Planning section (pp. 30–31)
    - Logistics section (pp. 31–33)
    - Finance and administration section (p. 33)
- Describe the ICS forms used at an incident. (pp. 33–35)
- Identify the commonly used forms used in an incident action plan (IAP). (pp. 33–35)

SKILLS OBJECTIVES

This chapter contains no skills objectives for Wildland Fire Fighter I candidates.

Additional Standards

- **NWCG PMS 210 (NFES 2943)**, *Wildland Fire Incident Management Field Guide*
- **NWCG S-130 (NFES 2731)**, *Firefighter Training*

You Are the Wildland Fire Fighter

You have arrived at your first fire after completing your fire fighter training program and basic wildland fire training. You are part of a four-person engine company assigned to a rapidly expanding incident in South Dakota. You have spent the past two days driving to get there, and it is your first time in this state. You and your supervisor arrived just in time to attend the briefing. You obtain an IAP and listen intently as the incident management team fills everyone in on the objectives, weather, hazards, and accomplishments from the previous shift. As the briefing comes to a close and the division supervisors identify themselves and where they will be for the division breakouts, you realize that you and your engine are not assigned yet and are not on the IAP.

1. Which position(s) in the ICS organization would you speak with to find out where your engine will be assigned during this operational period?
2. Where would you find information on incident safety in the IAP?
3. Which ICS forms will have information on incident communications?
4. Which ICS forms are typically used in an IAP?

Access Navigate for more practice activities.

Introduction

An **incident** is a natural or human-caused occurrence that requires an emergency response to prevent or minimize loss of life or damage to property and natural resources. This chapter presents an overview of the chain of command within the Incident Command System (ICS) under the National Incident Management System (NIMS). It discusses how to best use the ICS and how the ICS builds from an initial attack fire to a large fire situation. Various incident-related ICS forms are described so the reader will be familiar with a document used on larger incidents, called an incident action plan (IAP).

Why the ICS Began

In the early 1970s, fire managers realized that a new approach for managing rapidly escalating wildfires was necessary. Larger and more complex fires exposed weaknesses in the organizational command structure. More homes were being built in the interface zone, and rapidly spreading, high-intensity fires often taxed incident commanders (ICs).

Without a better system, ICs were often faced with the following problems:

- Too many people were reporting to only one supervisor. As a result, it was realized that a supervisor could optimally manage up to five people, with seven people being the maximum number. This is known as **span of control**.
- Fire agencies were often working with different organizational structures, which created confusion when agencies had to work together.
- There was no common language between fire agencies, with each agency often using its own specific terminologies.
- Unclear lines of authority developed. Personnel were unclear as to whom they worked for or with.
- Incident objectives were often unspecified or unclear.
- Communications were a problem because radio systems were often incompatible.
- Information did not flow well within the organization, resulting in a lack of reliable situational awareness.
- There was no organizational framework coordinating planning in unified command situations.

To mitigate the adverse effects of these early command structures, a system needed to be developed. Finally, through a cooperative effort of local, state, and federal agencies, the ICS was established. This interagency development effort was called FIRESCOPE (Firefighting Resources of Southern California Organized for Potential Emergencies). In 1980, FIRESCOPE transitioned into a national program called the National Interagency Incident Management System (NIIMS).

Following the September 11, 2001 terrorist attacks, President George W. Bush issued the Homeland Security Presidential Directive-5 (HSPD-5) in February

2003. HSPD-5 directed the secretary of the Department of Homeland Security to establish a National Incident Management System (NIMS). In March 2004, the NIMS was released to the nation. One of the components of NIMS was the ICS. All federal agencies–including wildland fire agencies–were required to adopt the NIMS and its components, one of which is the ICS, as a condition of receiving any federal preparedness assistance (grant funding). Private sector organizations and non-governmental organizations were strongly encouraged to also adopt NIMS and its components.

The current **Incident Command System (ICS)** is an adaptable emergency management system that allows users to adopt an integrated organizational structure equal to the complexity and demands of a single or multiple incidents of any type or size. The ICS establishes a **chain of command**, or a series of management positions, in order of authority, through which decisions are made by building from the top down. The IC initially serves as the point at the top.

The ICS also provides for common terminology, a common organizational framework, and procedures that are understood by all those using it **FIGURE 2-1**. The system is simple to use and offers low operational costs. It is easier to understand if we first view all the major functions of the ICS and then expand our examination to its sublevels.

The Incident Command System

There are five major functional areas in the ICS. Smaller incidents may be managed by one individual, such as the fire captain of an engine company. That individual is responsible, as the IC, for all five major functional areas. As the incident expands, the IC may choose to staff other general staff functions or command staff positions. This provides for better incident management and accountability as the incident grows. One of the major advantages of the ICS is that it is adaptable to the size of an incident and only necessary ICS positions need to be filled. **FIGURE 2-2** shows the five major functional areas of the ICS:

- Command
- Operations
- Planning
- Logistics
- Finance and administration

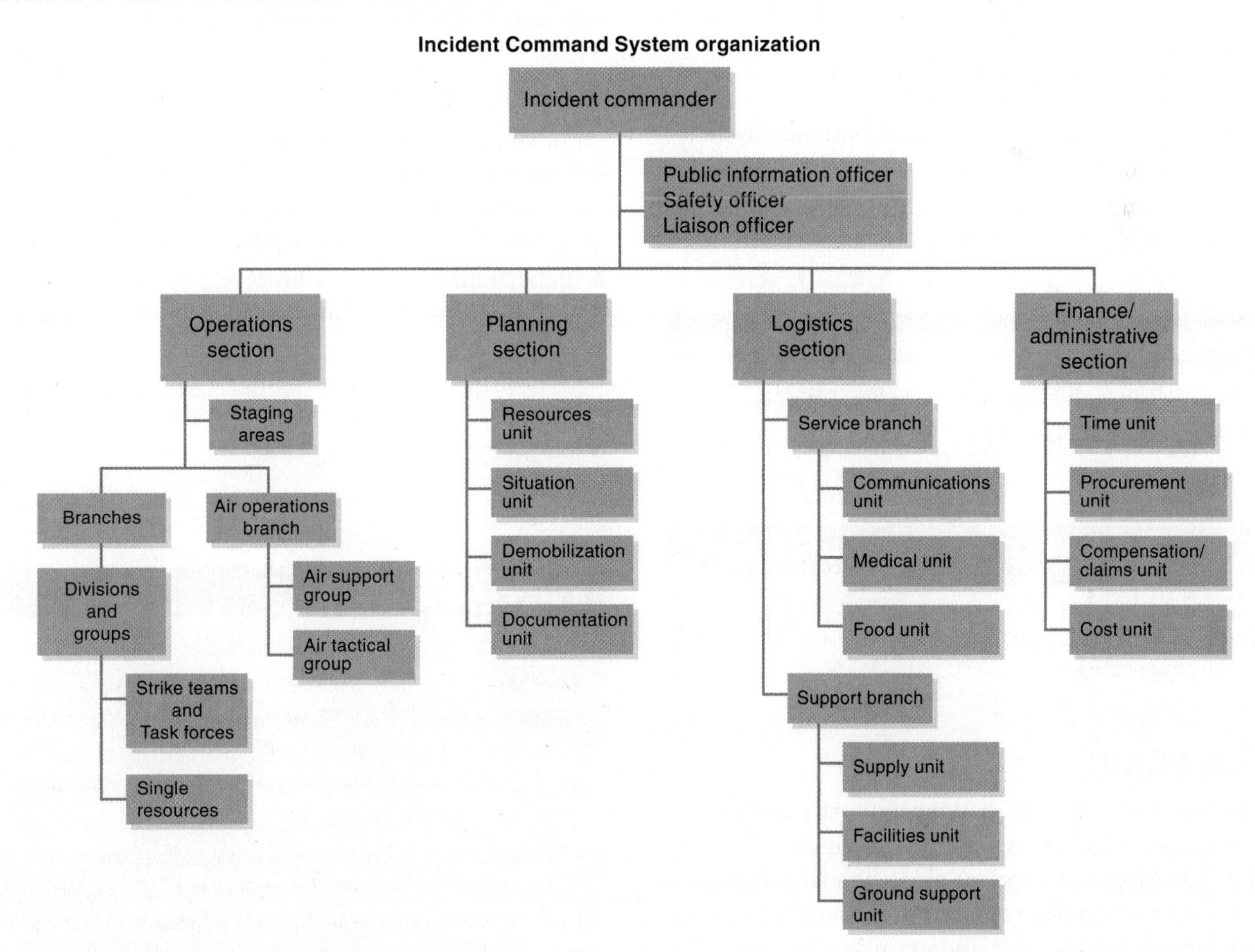

FIGURE 2-1 Major functions of the ICS broken down by section.

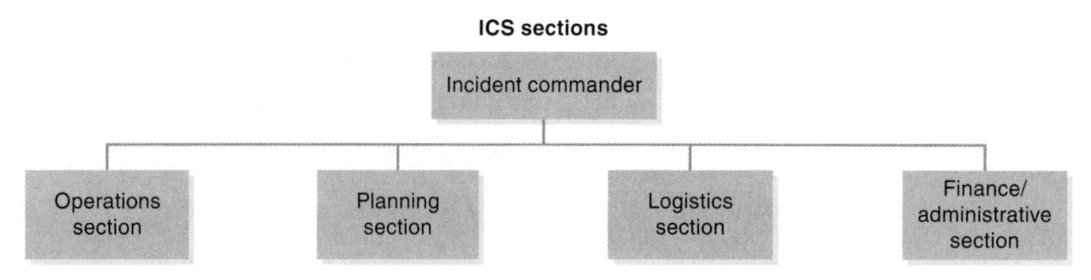

FIGURE 2-2 Five major components of the ICS.

Most agencies, through contract or agreements at the federal and state levels, are required to have all emergency response personnel complete the following Federal Emergency Management Agency (FEMA) courses:

- IS-100.C: Introduction to the Incident Command System
- IS-200.C: Basic Incident Command System for Initial Response
- IS-700.B: An Introduction to the National Incident Management System

It is highly recommended that you take the following FEMA courses if you plan to supervise other fire fighters and will have the responsibility for the ICS:

- ICS-300: Intermediate Incident Command System for Expanding Incidents
- ICS-400: Advanced Incident Command System for Complex Incidents

These courses are part of the national training curriculum for the ICS.

LISTEN UP!

Remember that the ICS is adaptable to the size of an incident. Only necessary ICS positions need to be filled.

LISTEN UP!

Statistically, 90 percent of U.S. wildland fires are extinguished by initial attack fire crews or engine companies.

Command

Command has decision-making authority and overall responsibility for management of the incident. Command develops the incident objectives and strategies, approves and directs resource orders, authorizes the development of the IAP, and approves the plan. It is important to note that on initial attack incidents, the incident action plan is usually an oral communication instead of a written IAP.

The IC on any incident, regardless of the size, has the ultimate responsibility for each of the five ICS functions. Once one of these functional areas requires independent management or the span of control exceeds the recommended number, the IC may assign an individual to manage certain functions within that area of responsibility. This is an example of the scalability of the ICS.

The command function is always staffed on every incident. On nearly every incident, command occurs in one of three ways: as a single command under one IC, as a unified command with two or more ICs, or, in rare circumstances, as an area command.

Single Command

On a single-command incident, the first-arriving officer usually assumes the role of IC and fills that position until there is a formal transfer of command. This transfer usually takes place upon the arrival of a higher-ranking officer; however, upon arrival, the higher-ranking officer may choose to let the original IC retain the post. If the higher-ranking officer does take command of the incident, he or she then often reassigns the original IC to a new position within the organization being set in place.

LISTEN UP!

Transfer of command is a formal process used to ensure vital information is passed on to the incoming IC. When there is a transfer of command, it is important to ensure that the incoming IC has a good understanding of the objectives that the previous IC has set, how the resources are deployed, any safety concerns, and what has been working and not working. Transfer of this information needs to occur before he or she assumes command of the incident. ICS Form 201, *Incident Briefing*, outlines this information and should be filled out by the outgoing IC to brief the incoming IC.

Unified Command

Unified command may be used on multiagency or multijurisdictional incidents or on planned events. It is a team effort in which all the stakeholders in an incident establish a common set of incident objectives **FIGURE 2-3**.

Unified command is usually established on multijurisdictional incidents. In unified command, the incident objectives are developed by the agency-designated officials working at the scene of the incident as a team. A single planning process takes place, and an **incident action plan (IAP)** is developed. IAPs are oral communications or written documents containing objectives that reflect the overall strategy and specific actions for managing an incident or planned event. IAPs may include several different attached forms that can include operational resources and assignments, directions, or other important information for managing the incident. These forms are discussed later in this chapter. How to form an IAP is discussed in Chapter 5, *Strategy and Tactics*.

Resource requirements that are needed at the incident by the operations section are determined at the tactics meeting. Incident tactical resources are managed by the operations section chief.

Area Command

Sometimes, area command is established when two or more large incidents occur in close proximity to each other. Other times, area command is established when a very large area has been affected by events, such as a hurricane, flood, or multiple fires. This type of command is utilized to assist in the appropriate allocation of resources and logistical support when these resources are in high demand. This also plays into demobilizing resources and reassigning them if resources need to be shifted. Area command will also work with all of the ICs to ensure that overall objectives are being met and establish priorities among incidents. Special training and involvement are needed for area command management to be utilized.

FIGURE 2-3 Unified command.
Courtesy of Jeff Pricher.

Incident Commander

The **Incident commander (IC)** is the person in charge of an emergency incident and has a wide variety of responsibilities; therefore, he or she must be fully qualified to do the job. Some of the major responsibilities include the following:

- Upon arrival at the scene, assessing the situation and/or obtaining an incident briefing from the current IC.
- Establishing the immediate priorities for the incident, with the safety of personnel being the first priority. The second priority is stabilization of the incident.
- Determining the incident objectives. This is a statement of intent that defines the goals. Incident objectives should be specific, measurable, achievable, and realistic as to what can be accomplished, but also flexible enough to allow strategic and tactical alternatives.
- Developing appropriate strategies and tactics to meet the incident objectives.
- Establishing the incident command post, from which the IC will work.
- Monitoring safety on the incident. On small incidents, the IC may do this; however, as the incident grows in size or complexity, this task is usually assigned to a safety officer.
- Establishing the incident organization. As the incident grows in size and taxes the IC's span of control, he or she sets up and monitors a management organization. The IC coordinates the activities of both the command and general staff.
- Ensuring appropriate meetings, such as the planning meeting and the operational period briefing, are scheduled as required.
- Approving requests for additional resources or the release of incident personnel or equipment.
- Authorizing and approving news media information. On larger incidents, the IC uses a public information officer.

Incident Command Typing

Incidents are typed from 5 (least complex) to 1 (most complex):

- Type 5. Generally, an initial attack that is short in duration and involves fewer than six people. The incident is usually limited to one

operational period (the period of time scheduled for execution of a given set of tactical actions, usually not exceeding 24 hours). These incidents have very limited or little complexity.

- Type 4. The IC at this level is generally responsible for all functions of the ICS (planning, logistics, operations, etc.). Because the incident is usually short in duration, command staff or general staff positions are only activated if needed. Usually, this type of incident is limited to one or two operational periods. In general, a written IAP is not needed, but an ICS 201 form may be used to document. Some people use incident organizers, such as tactical worksheets or command boards. Examples include a search and rescue incident, motor vehicle crash, or fire.
- Type 3. The IC at this level is looking at an extended event often lasting several operational periods. The complexity often requires some of the command staff and general staff positions to be filled, and likely includes divisions, groups, and task forces. Generally, these types of incidents require a written IAP. Examples include technical rescues, special events, and larger fires.
- Type 2. The IC at this level is overseeing the full complement of command and general staff. Many of the units under planning, logistics, and finance and administration sections are filled. Significant logistical support is needed, and the event is extended and complex. Several hundred people may be operating and large numbers of resources are used. In general, a base camp is established. Examples include major fires, natural or man-made disasters, multiday special events, and VIP visits.
- Type 1. The IC at this level is working at a national-capacity or very large multiagency incident. All command and general staff positions are filled and it is likely that all of the units under planning, logistics, and finance and administration sections are filled. The number of personnel in the operations section is generally between 500 and 1000 and, in some cases, exceeds 1000. On large fires in California, more than one fire camp could be established. Examples of incidents include large, complex fires and natural disasters.

As you can see, the IC has many responsibilities. To help, the IC may appoint a deputy IC to assume some duties to lessen the workload or when relief is needed. The deputy IC must have the same qualifications as the IC.

When an incident requires an IC to be present 24 hours a day, two operational periods will be needed. Having two operational periods reduces the workload for the command function by splitting the execution time in half. Instead of one 24-hour operational period, two 12-hour operational periods will exist with an overlap for briefing and debriefing of incoming and outgoing resources.

Initial Attack Incidents

An **initial attack (IA)** incident is one in which the fire or incident is controlled by the first dispatched resources without significant reinforcements **FIGURE 2-4**. The fire is controlled during the first 24 hours, when the fire spreads most rapidly (typically, from 10:00 AM to sundown). This is known as the **burning period**. Note that, in the IA fire organization, all resources are managed by the initial attack incident commander (IAIC). That same IC also fills all the roles of both the command staff and the general staff (discussed later).

An **extended attack (EA)** incident is one in which the fire exceeds the capabilities of the IA fire resources and requires additional resources **FIGURE 2-5**. EA fires can last several days to a week. Note that the IC is using a logistics section chief, who is a member of the general staff.

Command Staff

The **command staff** is shown in **FIGURE 2-6**. Persons who fill these positions are called officers. The command staff consists of the public information officer, the safety officer, and the liaison officer. These officers report directly to the IC. Command staff can have assistants.

The **public information officer (PIO)** is the point of contact for media and other organizations, enabling the IC to do his or her job without distraction from the numerous information requests that are received on a large incident or event.

The **safety officer** monitors and assesses the incident or event for unsafe or hazardous conditions. He or she also develops measures that ensure personnel safety on the incident. The safety officer corrects unsafe acts through the regular chain of command and has the

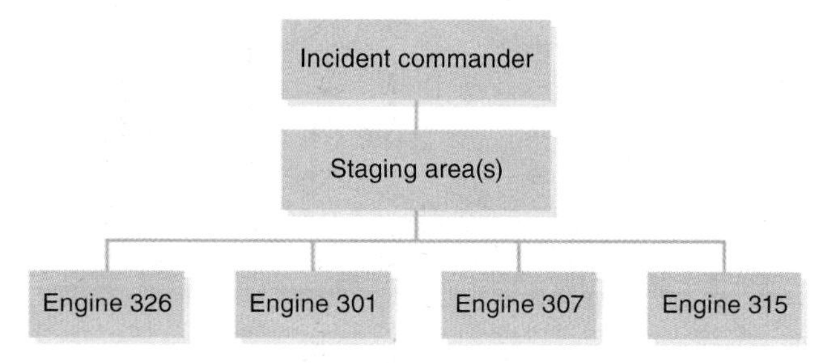

FIGURE 2-4 Initial attack incident organization.

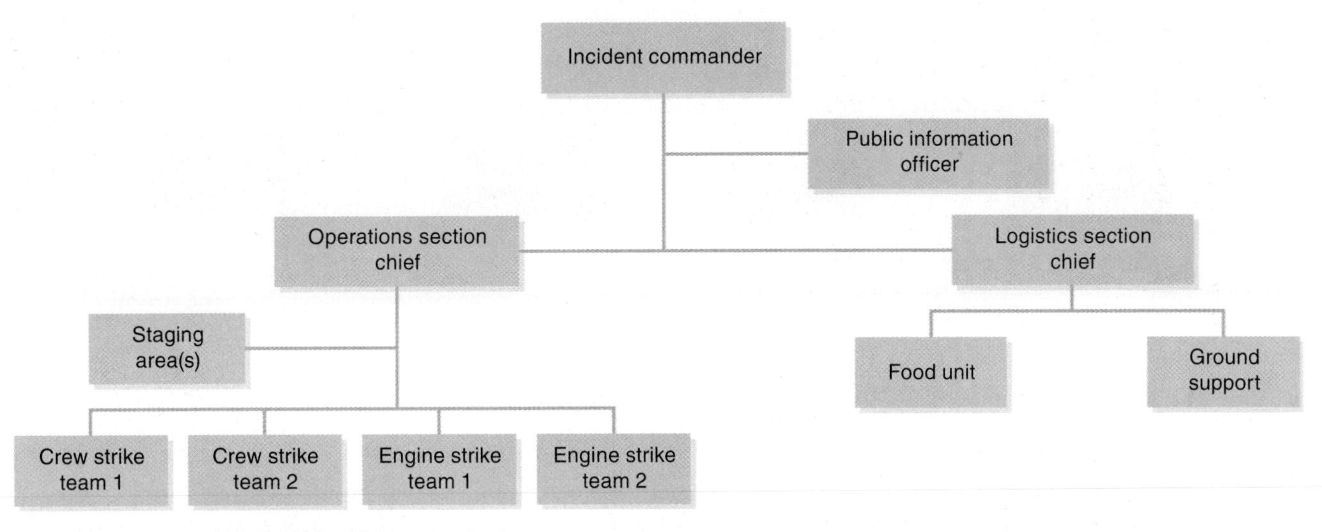

FIGURE 2-5 Extended attack incident organization.

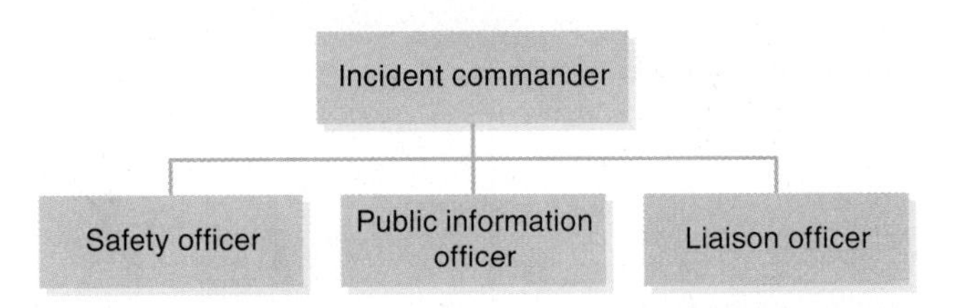

FIGURE 2-6 Incident command staff organization.

emergency authority to stop or prevent an unsafe act if immediate action is required. Large incidents or events may necessitate a safety officer with many assistants.

On large incidents or planned events where different agencies or jurisdictions are involved, a **liaison officer** is assigned. The liaison officer is its primary point of contact for assisting and cooperating agencies at an incident. Each agency usually assigns an agency representative (AREP) to represent its interests.

General Staff

Larger incidents or events that tax the IC's span of control require the use of people to fill positions in four major functional areas: operations, logistics, planning, and finance and administration. This group of personnel make up the **general staff** and report to the IC **FIGURE 2-7**.

The person in charge of each of these four sections is designated as a chief:

- Operations section chief
- Logistics section chief
- Planning section chief
- Finance and administration section chief

General staff may have deputies. Each deputy must be as qualified as the person for whom he or she works. There may be more than one deputy assigned to a position. A deputy can work in a relief capacity, be assigned a specific task, or work with the section chief assigned to one of the four general staff functions.

Operations Section

All strategic, tactical, and resource operations on an incident are managed and coordinated by the **operations section FIGURE 2-8**. The operations section chief's position is usually activated when the IC commits additional resources to the incident and the number of people reporting to him or her is exceeds the recommended span of control.

The operations section chief supervises and directs the suppression elements or tactical resources of the IAP. For each operational period on an incident or event, there is only one operations section chief. He or she can use deputies to assist. On large-scale incidents (Type 1 or Type 2), it is common to see two deputies filling the role of planning operations and field operations. These deputies assist in the coordination of the hundreds of people on the incident and assist the operations section chief with information gathering and managing span of control in the operations section. The planning operations section usually stays at the incident command post (the location at which primary functions are executed) and works with the general staff in managing and planning the resources. Field operations is the primary contact for the operations subsections: branches, divisions, and groups.

Branches

On a very large incident or an incident where there are numerous divisions and groups exceeding the

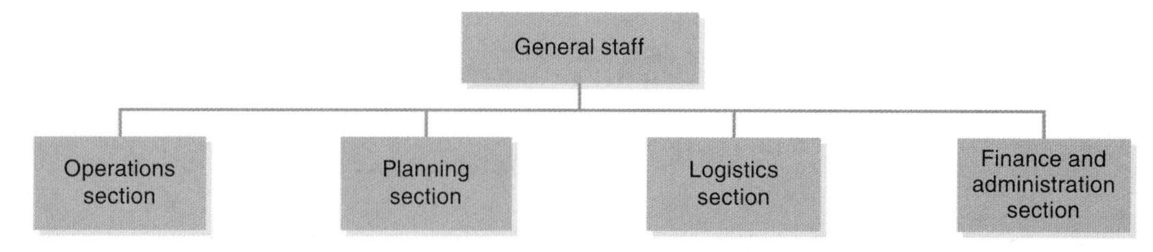

FIGURE 2-7 The general staff organization.

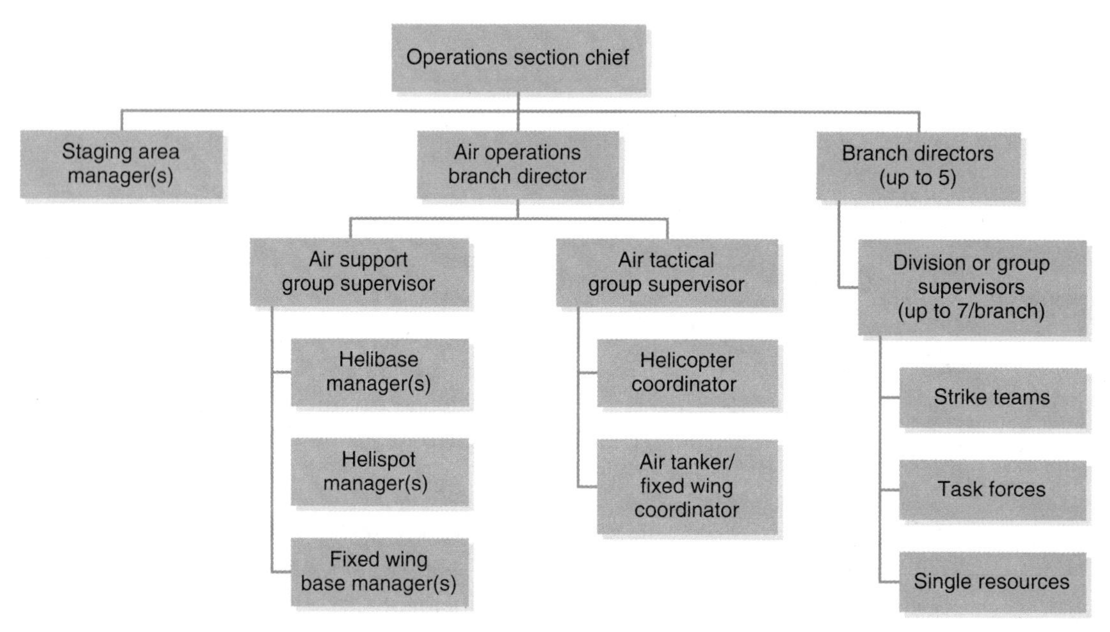

FIGURE 2-8 Sample organizational chart for operations section.

recommended span of control, branches may be established, and a branch director may be assigned. **Branches** help to manage and represent specific geographic or functional areas **FIGURE 2-9**. The branch director is responsible for the objectives identified in the IAP in his or her assigned branch and for the resources within that branch. The branch director is used when the operations section chief reaches the span-of-control limit. The operations branch director reports to the operations section chief.

When a branch is established to manage a geographic area, a roman numeral will be used (Branch I, for example). Geographic branches can be either delineated areas on the ground or set up by jurisdiction. If a branch is needed to manage a specific function, the designation will be based on the function (Search and Rescue Branch, for example). The most common example of a branch in the ICS is an air operations branch. The air operations branch director manages all aspects of the functions related to air operations.

Air Operations Branch. On larger incidents that have complex needs for aircraft use, a new branch level called the **air operations branch** is activated by the operations section chief. This position is ground based. The air operations branch director (AOBD) is responsible for preparation of the air operations portion of the IAP (ICS 220 form, Air Operations Summary). Upon approval of the plan, the AOBD is responsible for implementing the branch's tactical aspects and all logistical support operations.

Staging Areas. To temporarily locate available resources that are assigned to the incident, staging areas are established. **Staging areas** are branch locations set up at an incident where resources can be placed while awaiting a tactical assignment on a three-minute available basis. All resources assigned to staging are under the control of the operations section chief. Each staging area is run by a staging area manager (STAM) if more than one staging area is established.

Divisions

Divisions separate an incident into geographic areas of operation. They are established to regain and maintain span of control when available resources exceed the span of control of the operations chief. Divisions are managed by division supervisors. Their job is to implement the control objectives from the IAP for their division by using the assigned resources. The division supervisor (DIVS) reports directly to the operations section chief or an operations branch director when this position is activated.

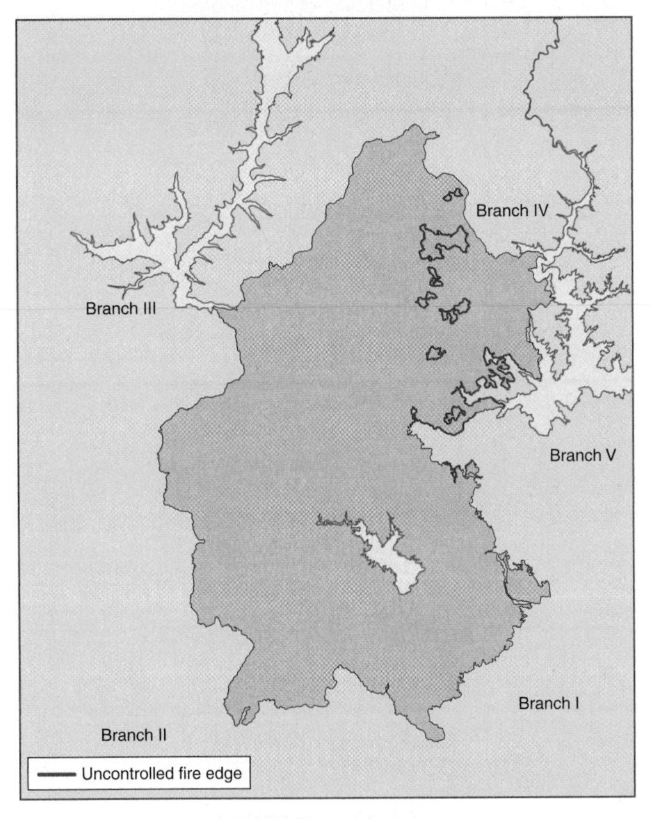

FIGURE 2-9 ICS branching.

Divisions allow the incident to be split up geographically. On wildland incidents, divisions are usually labeled by letters of the alphabet. Labeling starts with the letter *A* and extends clockwise around an incident (Division A, Division B, etc.) **FIGURE 2-10**. Incidents that involve buildings, such as high-rises, usually use the number of the floor as a division designator.

Groups

To divide an incident into functional areas of operation, **groups** are established. Groups are managed by a group supervisor (GRPS). Groups allow resources to perform a specific function. A structure protection group is one example. The needs of the incident dictate the kinds of groups that should be established.

Unlike divisions, group supervisors and resources are not bound to a specific geographic area; they can operate throughout the incident area. Group supervisors should coordinate with the appropriate division supervisor when operating in their division. Divisions and groups operate at the same organizational level in the ICS but do not supervise one another. Groups and divisions can be used together on the same incident. Again, on a large interface fire, divisions may be assigned to fight the wildland fire while the structure

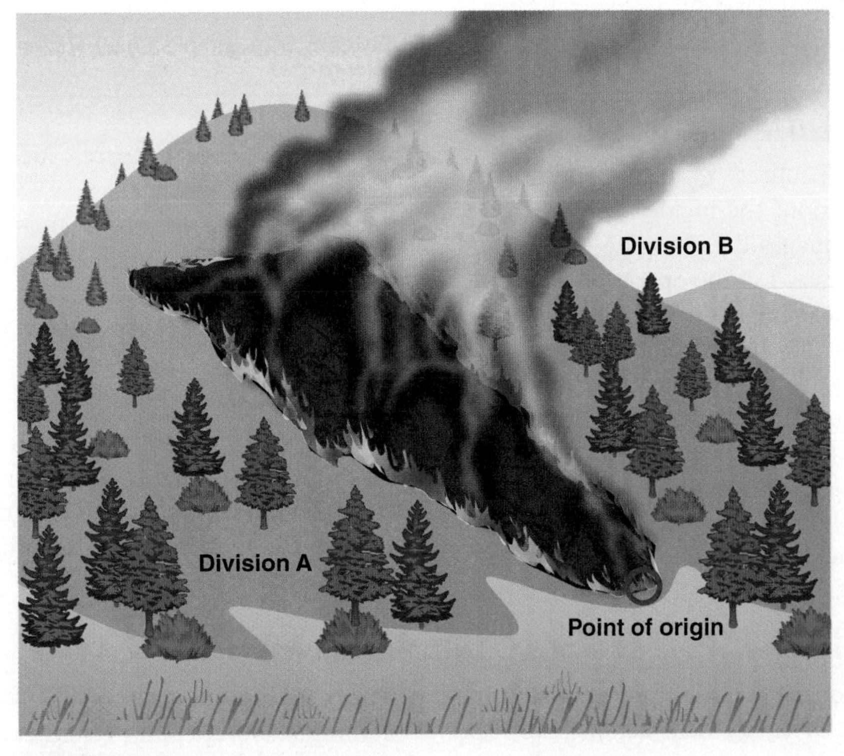

FIGURE 2-10 Division labeling example.

protection group protects a group of houses. In this case, the structure protection group supervisor works directly for the operations section chief or operations branch director if the incident is large enough.

On small incidents, where the operations section chief position has not been activated, the division or group supervisor reports directly to the IC.

Task Forces and Strike Teams

As an incident grows in size, additional fire equipment needs to be brought in. These additional resources can be organized as a task force or a strike team.

A **task force** is any combination of single resources assembled for a particular tactical need, with common communications and a task force leader (TFLD). A task force in wildland fires may include a dozer, an engine, and a handcrew, for example. It may be already established and sent to an incident but, generally, task forces are formed at an incident.

A **strike team** is a specified combination of the same kind and type of resource with common communications and a strike team leader. There are three primary types of strike teams in wildland firefighting operations: engines, handcrews and dozers. Other incidents may require other types of strike teams, such as ambulance strike teams.

Depending on the incident size and needs, these tactical resources can report to any of the following personnel:

- IC or operations section chief
- Division or group supervisor
- Operations branch director

Planning Section

The **planning section** collects, evaluates, and disseminates information about the incident **FIGURE 2-11**. This section also manages the planning process and develops the IAP for each of the operational periods. This major functional unit of the ICS also maintains information on the status of equipment and personnel resources, maintains all incident documentation, and is responsible for preparing the demobilization plan.

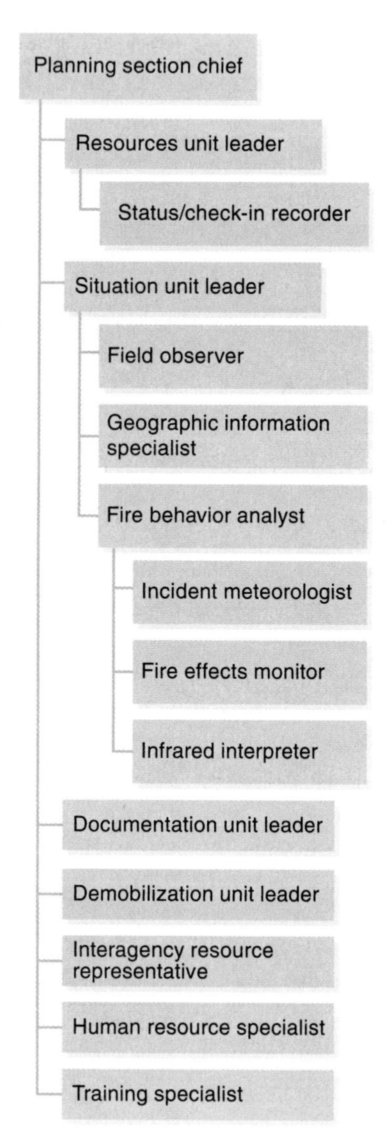

FIGURE 2-11 Sample organizational chart for the planning section.

The planning section chief is a member of the general staff; therefore, he or she can have deputies. The planning section chief reports directly to the IC. The planning section chief is responsible for the collection, evaluation, and distribution of all incident information (including resource status, resource effectiveness, and incident projections). The planning section chief is also responsible for the planning process that leads to the development of an incident action plan. The information gathered to help with these predictions comes from several sources, including technical specialists (discussed later).

The planning section can further organize into four unit-level positions: the situation unit, the resource unit, the documentation unit, and the demobilization unit. These four units are supported by technical specialists.

Situation Unit

The **situation unit** collects and processes information on the current incident status. Once this information is evaluated and analyzed, it is disseminated in the form of summaries, maps, and projections. It is run by the situation unit leader (SITL).

Resource Unit

The **resource unit** is responsible for checking in all incoming resources and personnel on the incident and maintaining the status of all assigned resources on the incident. Displays, charts, and lists reflect the current status and location of all incident resources.

A status check-in recorder, provided at each incident check-in location to account for incoming resources, works for the resource unit leader (RESL).

Documentation Unit

The **documentation unit** is important because it is responsible for maintaining complete and accurate incident files. All documents relevant to an incident are stored by this unit. Duplication services are also provided by this section. Incident files are stored and packed by this section for legal, historical, and analytical purposes. This unit is run by the documentation unit leader (DOCL).

Demobilization Unit

The **demobilization unit** is responsible for developing an efficient incident demobilization plan to ensure that incident personnel and equipment are released in an orderly, safe, and cost-effective method after they are no longer needed. This unit is run by the demobilization unit leader (DMOB).

Technical Specialists

Depending on the requirements of the planning section chief or incident needs, technical specialists may be assigned. **Technical specialists** (THSP) are personnel who have specialized skills. Technical specialists are typically managed by the planning section, but may be assigned to any section of the ICS that requires their knowledge and skills. For example, a structural engineer may be assigned to the operations section on an urban search and rescue operation during an earthquake.

There are many types of technical specialists. The most common include field observers, fire effects monitors, fire behavior analysts, incident meteorologists, and geographic information system specialists.

Field Observer and Fire Effects Monitor. The field observer (FOBS) and the fire effects monitor (FEMO) are technical specialists responsible for collecting site-specific information from personal observations and interviews at the incident and providing this information to the SITL, the DIVS, and other fireline resources as directed. They act as the eyes and ears for the section.

Fire Behavior Analyst. The fire behavior analyst (FBAN) is a technical specialist that develops projections on the fireline intensity levels and flame lengths, predicts how large the fire will grow, develops the rate of spread projection, and predicts where the fire will spread. These predictions are called a fire behavior forecast. The IMET, who provides site-specific weather forecasts, and field observers give the FBAN information needed to complete fire behavior forecasts.

Incident Meteorologist. The incident meteorologist (IMET) provides site-specific weather forecasts in real time. The IMET utilizes the weather that is documented on all of the divisions by fire fighters to assist in the local forecasting on a fire. When obtaining weather, make sure to include site-specific information and your GPS coordinates (latitude and longitude).

Geographic Information System Specialist. The geographic information system (GIS) specialist is a technical specialist responsible for providing timely and accurate digital information related to mapping, boundaries, ownership, and territories to the SITL. When any incident-specific data are shared (such as drop points, branch breaks, or sensitive resources), the GIS specialist adds them to one or more maps that are produced daily. With some of the digital collection devices now being deployed, GIS is becoming more and more important.

Other Technical Specialist Positions. Technical specialists will vary depending on the type of incident or expertise needed. Some other common technical specialist positions that you may see include the following personnel:

- Structural engineer
- Environmental impact specialist
- Flood control specialist

Logistics Section

The **logistics section** orders resources (facilities, services, and supplies) and develops the transportation, communications, and medical plans. The logistics section chief is a member of the general staff and can have a deputy position assigned to him or her. He or she directs units within the section to provide all services and supports needs related to the incident. The logistics section chief activates the six functional units, as necessary, on an incident. These six units include the communications unit, the medical unit, the food unit, the supply unit, the facilities unit, and the ground support unit. The logistics section chief may also activate two branches under the logistics section (the service branch and/or the support branch), if needed to provide better management and oversight of the logistics function.

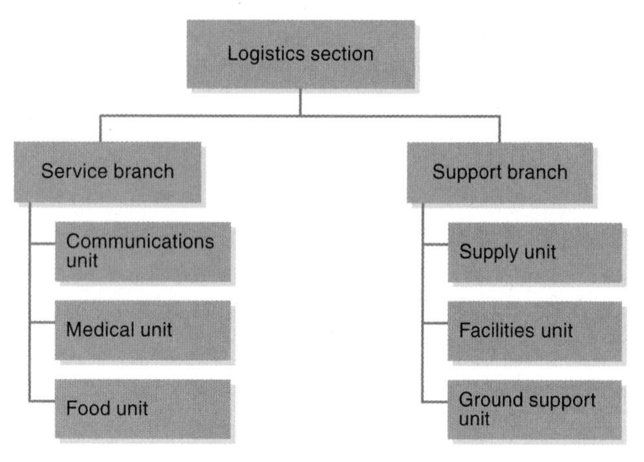

FIGURE 2-12 The logistics two-section branch structure.

It is important, as the incident grows in size, for the IC to recognize that a need may exist for logistical service and support to the incident. Once this has been determined, the IC should activate the logistics section chief position. It is important that the need for a separate logistics function be recognized. This will save both time and money on the incident.

The logistics section comprises two branch directors and six additional functional units. If necessary, the logistics section chief can establish a two-branch structure to help facilitate span of control **FIGURE 2-12**. This decision is determined by the size of the incident, the length of time the incident is expected to last, and the complexity of the support needs. On smaller incidents, only a portion of the six separate units are activated. On a major incident, it is likely that all six functional units will be established.

Service Branch Director

When activated, the service branch director (SVBD) position is under the supervision of the logistics section chief, who is responsible for all incident-related communications, as well as medical and food-related needs.

Communications Unit. The **communications unit** distributes and maintains all forms of communication equipment on the incident and is responsible for developing a plan, included in the IAP, for the most effective use of the communication equipment. The incident communication center, which can be housed in a building, trailer, or van, is also managed by this unit. The communications unit is managed by the communications unit leader (COML).

Medical Unit. The **medical unit** provides and maintains first aid and emergency medical treatment stations for incident personnel. Depending on the fire agency, these stations can be manned by emergency medical technicians, paramedics, or even doctors.

The medical unit leader (MEDL) is responsible for developing the incident medical plan. Included in the IAP, this document names the incident medical aid stations, addresses transportation issues, and gives information on hospitals.

Food Unit. Responsibility for determining and providing all food and hydration needs to the incident rests with the **food unit**. This responsibility can be handled in a variety of ways. On large agency fires, portable kitchens are set up and staffed by incident personnel. On type 1 or type 2 fires (and possibly on some type 3 fires), a national caterer may be brought in to feed all incident personnel. On smaller fires or for agencies that do not have portable kitchen capabilities, a catering service may be brought in. The food unit is managed by the food unit leader (FDUL).

Support Branch Director

When activated, the support branch director (SUBD) position is under the supervision of the logistics section chief, who is responsible for development and implementation of logistics plans in support of the IAP. This person supervises the supply unit, the ground support unit, and the facilities unit.

Supply Unit. The **supply unit** orders personnel, equipment, and supplies. In the ICS, all resource orders are placed through the logistics section's supply unit. If the supply unit is not activated, then the logistics section has the responsibility for all ordering requests. The unit also receives and stores all incident supplies, services all nonexpendable supplies and equipment, and maintains an inventory of supplies. The supply unit is managed by the supply unit leader (SPUL).

Facilities Unit. The **facilities unit** establishes and maintains all facilities required to support the incident, including sleeping facilities, security services, sanitation needs, lighting, trash removal and cleanup, and the setup of an incident command post trailer or other forms of shelter. The unit also provides incident base and camp managers. The facilities unit is managed by the facilities unit leader (FACL).

Ground Support Unit. The **ground support unit** is responsible for the maintenance and repairs of all vehicles assigned to the incident. This unit takes care of all the fueling needs of these vehicles and provides transportation services in support of incident operations. It does not provide for air services. The unit also implements the incident traffic plan. The ground support unit is managed by the ground support unit leader (GSUL).

Finance and Administration Section

Set up for incidents that require on-site financial management, the **finance and administration section** is under the supervision of the finance and administration section chief, who reports to the IC. This section analyzes, monitors, and manages all incident-related costs and establishes any necessary procurement contracts **FIGURE 2-13**.

The finance and administration section chief is a member of the general staff and is responsible for all financial and cost analysis aspects of the incident. The section may establish four additional units as necessary: the time unit, the procurement unit, the compensation and claims unit, and the cost unit **FIGURE 2-14**. On smaller incidents, the finance and administration section may not be activated. As the incident grows and a need is determined for the section, units are added.

Time Unit

The **time unit** is responsible for recording all personnel time and equipment time on an incident or event. The unit also manages the commissary operations. The time unit is managed by the time unit leader (TIME).

FIGURE 2-13 Finance and administration section setups can be in tents with signs, schools, or other buildings.
Courtesy of Jeff Pricher.

Procurement Unit

All financial contracts and the associated administrative paperwork related to vendors are administered by the **procurement unit**. The vendor contracts deal with incident equipment rentals and supply contracts. The procurement unit is managed by the procurement unit leader (PROC).

Compensation and Claims Unit

The **compensation and claims unit** is responsible for the management of all administrative matters pertaining to compensation for injury and claims related activities on an incident. The unit is responsible for ensuring that all forms required for a worker's compensation claim are filled out correctly. Second, the unit maintains records of injuries or illnesses incurred during the incident. The compensation and claims unit is managed by the compensation and claims unit leader (COMP).

Cost Unit

All cost data for the incident are collected by the **cost unit**. As incident costs are obtained and recorded, summaries are prepared. Cost estimates are shared with the planning section. The unit is also responsible for providing cost-saving recommendations to the finance and administration section chief. The cost unit is managed by the cost unit leader (COST).

ICS Forms

ICS forms are used to create the IAP to document actions taken at an incident and for incident management purposes.

TABLE 2-1 is a list of important ICS forms found in an IAP.

TABLE 2-2 includes other forms found in the IAP.

You should become familiar with the forms found in an IAP. Other routinely used forms include those listed in **TABLE 2-3**.

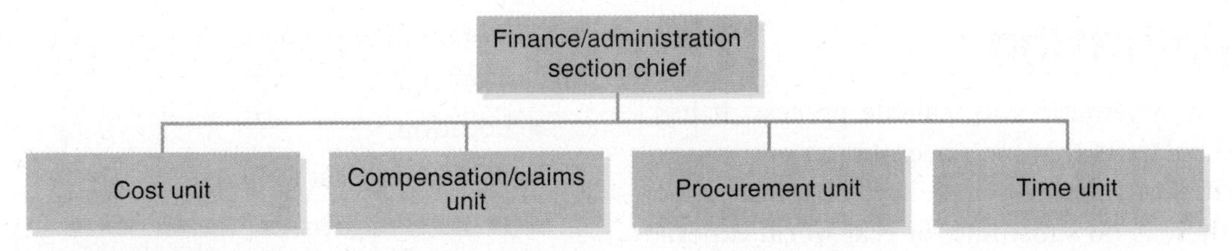

FIGURE 2-14 Example of finance and administration section organization.

TABLE 2-1 Important ICS Forms

ICS Form Number	Form Title	Description
202	Incident Objectives	▪ Used to communicate the incident objectives and guide tactical implementations ▪ Incident objectives developed by the IC or unified commanders ▪ Form prepared by the planning section (usually the resource unit leader)
203	Organizational Assignment List	▪ Identifies people in primary overhead positions on the management team ▪ Form prepared by the planning section (usually the resource unit leader)
204	Assignment List	▪ Identifies the work location, supervisor's name, and the resources assigned to the division ▪ Assigns tasks for the division (control operations) ▪ Defines the operational period ▪ Gives radio frequencies ▪ Gives any special instructions to the division ▪ Form prepared by the planning section (usually the resource unit leader)

205	Incident Radio Communications Plan	▪ Lists the radio frequencies assigned to the functional areas and divisions or groups working within the functional areas
		  ▪ Check the operational period because radio frequency changes do occur.
		▪ Lists the radio frequency allocations for the incident
		  ▪ Check the remarks on the form because they may contain special considerations for radio use.
		▪ Form prepared by the communications unit leader

206	Medical Plan	▪ Provides information on incident medical aid stations, transportation services, hospitals, and emergency medical procedures. ▪ Check the medical emergency procedures box because it differentiates the way medical emergencies shall be handled. ▪ Form prepared by the medical unit leader and reviewed by the safety officer
220	Air Operations Summary	▪ Summarizes aircraft missions planned for the operational period ▪ Defines key air operations section overhead personnel and the radio frequencies assigned to each ▪ Identifies aircraft assigned to the incident, as well as those available to it ▪ Form prepared by the air operations branch manager

ICS Application

The ICS is a systematic and scalable process. It has checks and balances that have been tested over many years of actual incident response. To truly grasp these concepts, there is no substitute for real-world experience. However, to begin to expand your knowledge, it is important to understand that the ICS is informed by the following:

- Location
- Potential threat to life
- Political influences
- Organizational and situational complexity

TABLE 2-2 Other ICS Forms

Form Title	Description
Safety Message/ Plan (ICS 208)	▪ Gives a narrative of general and specific hazards facing personnel assigned to the incident ▪ Remarks are not only restricted to fireline conditions; they address all incident safety concerns. ▪ Form prepared by the safety officer
Fire Weather Forecast	▪ Summarizes the weather influences for that operational period ▪ Provides fire weather forecasts for later operational periods ▪ Form prepared by the incident meteorologist
Fire Behavior Forecast	▪ Discusses past, present, and future fire behavior ▪ Identifies areas of potential concern ▪ Form prepared by the fire behavior analyst
Medical Incident Report	▪ Summarizes medical emergencies and evacuations ▪ Form prepared by the medical united leader as part of the IAP

- Boundaries of various jurisdictions
- Weather
- Values at risk

It is important to have a global perspective or, as some have coined, a 30,000-foot view of what is going on at an incident. When crews and fire fighters exhibit tunnel vision or a fixation on only one or two details, accidents happen and mistakes are made. Several ways in which preparation can assist with the progression of the ICS are discussed later.

Preparation

Several agencies begin the incident response process by establishing an incident command progression. The National Wildfire Coordinating Group (NWCG) has identified five levels of the incident command structure, with each level's associated responsibilities. Most agencies have adopted the NWCG knowledge skills and abilities that are needed for the performance-based positions.

Agencies that have adopted the NWCG process generally perform an annual refresher training, which includes a yearly delegation of authority. The delegation of authority outlines IC expectations at the various levels; decision-making processes; spending authority; and, for some jurisdictions, the Wildland Fire Decision Support System (WFDSS). As incidents transition from IAs to EAs or incidents that require large incident support, most land management agencies incorporate the WFDSS process into the refresher course and the IAP to assist in making specific ICS decisions.

TABLE 2-3 Other Routinely Used Forms

ICS Form Number	Form Title
201	Incident Briefing
207	Incident Organization Chart
209	Incident Status Summary
210	Status Change Summary
211	Check-In List
212	Incident Demobilization Vehicle Safety Inspection Form
213	General Message Form
214	Activity Log
215	Operational Planning Worksheet
215A	Incident Action Plan Safety Analysis
216	Radio Requirements Worksheet
217	Radio Frequency Assignment
218	Support Vehicle Inventory
219	Resource Status Card
221	Demobilization Check-Out
224	Crew Performance Rating
225	Incident Personnel Rating

Risk and complexity assessment (RCA) is the process that agencies utilize to transition from one type of ICS to another. The complexity is flexible in the flow path from increasing to decreasing an IC level. There are significant differences between local, state, and federal land management agencies in how they execute their decision processes with respect to ICS. Regardless of the specific process, there are some commonalities that exist, most of which revert back to span of control and RCA.

The NWCG book *Interagency Standards for Fire and Fire Aviation Operations* (shortened to Red Book) is a great resource to assist with understanding incident complexity, and it contains a form called the RCA. It discusses how to establish a decision process while transitioning from one level to another.

LISTEN UP!

It is vital that you understand the ICS process of the agency for which you are working.

Requesting Help

ICS is modular and has the flexibility to expand when needed, based on the directives of an agency or the experience of the IC. One of the most important yet least understood aspects of command and the ICS is the importance of requesting help. If you look at the headers Integrity and Communication Responsibilities in the *Incident Response Pocket Guide* (IRPG), there is one point that is common to the success of the ICS and the transition from an IA to an EA or large incident. The point is, know yourself and seek improvement when your skill level is challenged.

There have been several incidents when ICs have been relieved from blame as a result of requesting the next level of IC because their skill level or certification did not qualify them to be in charge. If an ICT5 (Incident Commander Type 5) requests an ICT4 or ICT3 as a result of complexity, the ICT5 cannot be held liable for abnormal requests or results of decision-making based on doing the best he or she can at the time with the existing resources. If an IC fails to make a request for the next level of IC and the incident complexity is beyond his or her control, he or she will be responsible for the resulting outcome.

Incident Development

Aside from local ICS statutes and ordinances, the typical progression of an incident is outlined in the following example case. This example is not the only process that is used; however, based on the typical progression as documented by the NWCG, this is what one could expect to experience.

Incident Development Example Case

Time: 1400, Day 1. Forestville Fire District is dispatched to a reported motor vehicle fire on the cross-country highway at mile post 44. Initial reports note that the vehicle is in close proximity to an urban interface area that is covered by state and federal assets. The time is 1400, there is a 10-minute ETE (estimated time en route), and engine 876 is the only resource assigned at this time, which is typical for this area.

As the engine is en route, a column is observed, with a mile left to respond to the scene. Engine 876 requests a second engine to assist.

Time: 1410, Day 1. Upon arrival, engine 876 establishes command and states that the fire has extended to grass and brush on a 30-degree slope. The fire is at 0.25 acre (0.1 hectare) and growing quickly as a result of wind and slope alignment. A tactical channel is established, and a simplex tactical frequency is assigned for resources engaged on the fire to utilize. Access will be from the cross-country highway, and the current hazards are the traveling public and power lines.

Because the fire is growing and the rate of spread is increasing by the minute, the IC (who is an ICT5) requests an upgrade to a second alarm and an ICT4 to respond to the incident due to structures in the path of the fire. A second alarm will bring a task force that will include four engines and a water tender. Because this area is a joint response for the state and federal land management agencies, they will also be sending equipment (engines, dozers, and aerial resources) and personnel to the fire.

Time: 1430, Day 1. Twenty minutes after arrival, the fire has grown to 15 acres (6.1 hectares); structures are threatened; and, including all the responding entities, there are over 11 pieces of equipment and a total of 33 people on the incident. At this point, the ICT5 has exceeded the span of control and needs to transfer to an ICT4 when one arrives. However, in the time between the realization of needing to go to the next level of IC and an ICT4 arriving, it is reasonable and acceptable for the ICT5 to prepare the ICT4 for success. In this instance, the ICT5 establishes two divisions: The left flank is labeled Division Alpha, and the right flank is labeled Division Zulu. This is consistent with the clockwise formation of divisions to manage span of control and allow for flexibility in this expanding incident. Using *Z* on the right flank allows for all of the other letters in the alphabet to be utilized as the fire continues to grow in size.

Time: 1505, Day 1. When the ICT4 arrives 45 minutes later, there is a period of time when the ICT5 and ICT4 brief and debrief before the ICT4 accepts and announces that the IC has transitioned. In this case, with the multiple jurisdictions and agencies responding, the ICT4 decides to establish a unified command with the state and federal response assets. This type of command will allow for easier ordering of aviation resources and handcrews to assist in the firefighting. At this point, the firefighting does not require a specific delegation of authority or the next level of IC. The ICT4 will need to review the RCA reference in the IRPG to determine whether a transition is needed. Generally, at the ICT4 level, a few divisions can be established, and air operations can be utilized without a dedicated air attack (assuming no retardant nor more than a single helicopter is needed). At this point, the fire is still in the first operational period, and the ICT4 needs to start looking at the long-term logistics, such as meals, equipment needs, supplies, fire behavior predictions, and whether the fire will be extending beyond the first 24 hours. If the fire can be controlled and contained in the 24-hour time frame and no additional resources are needed, the ICT4 structure will continue.

Time: 0200, Day 2. The ICT4 needs to make a decision regarding resource needs because the fire has not been caught. The increasing size of the fire and its logistical demands necessitate the ICT4 to order an ICT3. In general, all ICs need to be looking ahead 12, 24, and 36 hours and beyond for needs. Ordering for incidents that are continuing to go beyond the current and expected RCA requires more planning and logistical support.

Time: 0700, Day 2. At this point, it is clear that the fire is continuing to grow. The ICT4 has ordered additional resources, including food for the fire fighters, but it is clear that it will be necessary to expand ICS functions, such as operations, logistics, and planning, and add a PIO to the command staff. A request for either an ICT3 or Type 3 IC team has been placed.

Another challenge is that, in most areas of the country, mutual aid agreements cover only the first operational period. Beyond that time, significant costs occur, and the fire district would need to think about normal staffing for the fire district.

Time: 1100, Day 2. The ICT4 maintains command and begins the process of transitioning to the ICT3. The ICT4 is responsible for communicating with an agency administrator to express the needs of the incident and the projected status for the next 12 to 24 hours. The agency administrator (if that is how the local area is set up) will then begin the WFDSS process and establish a set of priorities. In some areas of the country, the land management agencies will begin the WFDSS process, and the structure agencies will work the state system for mobilization of structure resources for structure protection and additional fire crews to assist in the fire response needs. There are many names for this state process, but, in general, most states will have a declaration that will trigger a statewide coordinated response.

Based on the order for a team or an ICT3, as the ordered resources are responding to the location, the WFDSS and/or state declaration is being processed and decided on. Documents are drafted for the team or ICT3 to sign and assume responsibility through a delegation of authority. This delegation of authority will include spending limits, leader's intent, priorities, values at risk, and decision points for determining whether the incident needs to move to the next level of command.

Time: 1700, Day 2. At this point, the fire has grown to 478 acres (193.4 hectares). The ICT3 has assembled a limited command and general staff to assist in the span of control and to serve the needs of the incident. A fixed location is established for the incident command post for the IC and general staff to work out of. While the IC and operations make initial tactical decisions, a logistics chief and planning chief start the process of making a base camp and developing the IAP. The IC will then start the process of meeting with the local agency administrators and county officials and setting up meetings with the community. A PIO will be delegated to start delivering important messages to the public. If a team was not ordered, the IC will also most likely elect to establish a safety officer and implement safety oversight.

While all of this is occurring, the operations section chief will be determining all of the tactical needs to suppress the fire. An order list will be established, and resources will be ordered to the incident. Based on the new size of the incident, a couple more divisions will be created, and air operations will start to develop as more of a focal point for operations with the addition of an air attack.

The planning section chief will begin the process of creating the IAP for the next operational period and establish the meeting schedule for the planning. A more regimented schedule will be established.

The safety officer will begin the process of ensuring that medical support is available and, depending on the resources, terrain, and needs of the incident, possibly order a REM (rapid extraction module) team.

Time: 1300, Day 3. By this point, the Type 3 organization has been managing the incident for approximately 48 hours and had, at one point, almost contained the fire. Unfortunately, the fire has now grown to 650 acres (263 hectares) and is threatening more homes. After the IC reevaluates the RCA and what was contained in the WFDSS, a Type 2 team will need to be ordered. The Type 2 team will take 12 hours to mobilize and require a partial or full shift to shadow the Type 3 organization before taking over the fire. The ICT3 will need to plan on keeping the fire at least another 36 to 48 hours from the time of the order. During that time, a comprehensive IAP and accounting of all planned and expected actions will need to be documented for the transition.

All critical elements of the ICT3 are completed by the time the Type 2 team arrives. The fire is now at 1200 acres (485.6 hectares), and there is a total of 450 people on the fire and or ordered for the fire.

Time: 1200, Day 4. The Type 2 team has now taken over. All command and general staff positions have been filled. The general staff starts to build the organization with helispots, drop points, and the appropriate staff and equipment. The IC and the command and general staff focus on actionable objectives and provide clear leader's intent with task, purpose, and end state. The operations section is tasked with establishing the PACE (primary, alternate, contingency, and emergency) decision points. Depending on the size of the incident or incidents (sometimes, Type 2 teams are asked to manage several small fires), the appropriate number of divisions or branches (this will always be dependent on the depth of the team regarding qualifications) will be established based on span-of-control needs. Structure protection is established with triage and mapping. The safety group will establish a deliberate risk analysis (DRA) to determine whether safety mitigations are required for the incident. With more staff, there is more opportunity to meet the needs of the local organizations, the staff on the fire, and the public. This organization will continue until another RCA requires that a Type 1 team be required.

Other Outcomes

If the RCA and the WFDSS require the next level or the span of control is exceeded by the qualifications and staff of the Type 2 organization, the Type 1 team will come in. Currently, there are 16 Type 1 incident management teams available nationally. Several states have also developed Type 1 incident management teams in addition to the 16 national type 1 IMTs. Type 1 teams have the depth and experience in team members to handle the most complex and largest of incidents.

Depending on the complexity and scale of incidents, there are two more tiers of command. The first is the area command teams, of which there are three standing command teams in the country. They are called in to assist agency administrators or local entities when several incident management teams are managing incidents in close proximity. The second is the National Incident Management Organization (NIMO) teams, established by the U.S. Forest Service, of which there are four in the United States. NIMO teams are comprised of seven Type 1 command and general staff members. If needed, NIMO teams can expand to meet various complexity levels. Their primary responsibility is to manage complex wildland fires or long duration incidents, which can include fires that cross state or U.S. national boundaries.

As stated at the beginning of this section, the ICS is scalable and expandable. The hypothetical incident described is not the only way the ICS progresses. There are several instances when a fire will go from an ICT5 to an ICT2 or ICT1 in 24 to 36 hours. Many variables are factored into the levels of command. What is important to understand are the limits and available tools that can be used to determine what level is needed. Those tools are the IRPG; WFDSS; and the Red Book.

After-Action REVIEW

IN SUMMARY

- An incident is a natural or human-caused occurrence that requires emergency response to prevent or minimize loss of life or damage to property and natural resources.
- The Incident Command System (ICS) is an adaptable emergency management system.
  - It allows users to adopt an integrated organizational structure equal to the complexity and demands of single or multiple incidents of any type or size.
  - It is modular and establishes a series of management positions in order of authority (chain of command).
  - It is adaptable to the size of an incident, and only necessary ICS positions need to be filled.

- The ICS is built around five major functions:
  - Command
  - Operations
  - Planning
  - Logistics
  - Finance and administration
- Command has decision-making authority and overall responsibility for the management of the incident. Command develops the incident objectives and strategies, approves resource orders, and develops the incident action plan (IAP).
  - Single command
  - Unified command
  - Area command
  - Incident commander: The person in charge of an incident who has a wide variety of responsibilities; therefore, he or she must be fully qualified to do the job.
  - Levels of incident command:
    - Type 5. Generally, an initial attack that is short duration and involves fewer than six people. These incidents have very limited or little complexity.
    - Type 4. The IC at this level is responsible for all levels of the ICS, and the incident is usually short in duration. Examples include a search and rescue incident, motor vehicle crash, or fire.
    - Type 3. The IC at this level is looking at an extended event often lasting several operational periods. Examples include technical rescues, special events, and larger fires.
    - Type 2. The IC at this level is overseeing the full complement of command and general staff. Examples include major fires, multiday special events, and VIP visits.
    - Type 1. The IC at this level is working at a national-capacity or very large multiagency incident. Examples of incidents include large, complex fires and natural disasters.
  - Command staff: Consists of the public information officer (PIO), the safety officer, and the liaison officer. These officers report directly to the IC.
  - General staff: Larger incidents or events that tax the IC's span of control require the use of people to fill positions in four major areas: operations, logistics, planning, and finance and administration. This group of personnel report to the IC.
- Operations section: All strategy, tactical, and resource operations on an incident are coordinated by the operations section.
  - Branches: Help to manage and represent specific geographic or functional areas. Branches can include staging areas and air operations.
  - Divisions: Separate an incident into geographic areas of operation. They are established to regain and maintain span of control when available resources exceed the span of control of the operations chief. Divisions are managed by the division supervisor or group supervisor.
  - Groups: Divide an incident into functional areas of operation and allow resources to perform a specific function.
  - Task force: Any combination of single resources assembled for a particular tactical need, with common communications and a task force leader. It may be already established and sent to an incident or formed at an incident.
  - Strike team: A specified combination of the same kind and type of resource with common communications and a strike team leader. There are three primary types of strike teams: engines, handcrews and bulldozers.
- Planning section: Collects, evaluates, and disseminates information about the incident; manages the planning process; develops the IAP for each of the operational periods; maintains information on the status of equipment and personnel resources; maintains all incident documentation; and prepares the demobilization plan.
  - Resource unit: Responsible for checking in all incoming resources and personnel on the incident.

    - Situation unit: Collects and processes information on the current incident status.
    - Documentation unit: Provides duplication services and maintains, stores, and packs incident files for legal, historical, and analytical purposes.
    - Demobilization unit: Develops an efficient incident demobilization plan to ensure that incident personnel and equipment are released in an orderly, safe, and cost-effective method after they are no longer needed.
    - Technical specialists: Personnel who have specialized skills. May be assigned to the planning section, function within a unit, or form a separate unit within the planning section.
- Logistics section: Orders resources and develops the transportation, communications, and medical plans.
    - Service branch director: Responsible for all incident-related communications and medical- and food-related needs.
        - Communications unit: Distributes and maintains communication equipment on the incident; develops a plan, included in the IAP, for the most effective use of the communications equipment.
        - Medical unit: Provides and maintains first aid and emergency medical treatment stations for incident personnel.
        - Food unit: Determines needs for and provides all food and clean drinking water to the incident.
    - Support branch director: Develops and implements logistics plans in support of the IAP.
        - Supply unit: Orders personnel, equipment, and supplies.
        - Facilities unit: Establishes and maintains all facilities required to support the incident.
        - Ground support: Maintains and repairs of all vehicles assigned to the incident.
- Finance and administration section: Analyses, monitors, and manages all incident-related costs and establishes any necessary procurement contracts.
    - Time unit: Records all personnel time and equipment time on an incident or event; manages the commissary operations.
    - Procurement unit: Manages all financial contracts and the associated administrative paperwork related to vendors.
    - Compensation and claims unit: Ensures that all forms required for a worker's compensation claim are filled out correctly; maintains records of injuries or illnesses incurred during the incident.
    - Cost unit: Prepares incident cost summaries and estimates and shares them with the planning section.
- ICS forms are used to create the IAP to document actions taken at an incident and for incident management purposes.
- Incidents can develop in numerous ways. They require a global perspective.
- If an IC fails to request help, he or she will be responsible for the resulting outcome.
- Remember that incidents or events are not always static and your organizational structure needs may change. The beauty of the ICS is that it can be expanded to meet almost any need.

KEY TERMS

Air operations branch Ground-based branch assignment for management of air assets used for the incident.

Branches Organizational level having functional or geographical responsibility for major aspects of incident operations; can only be used in the operations section or the logistics section.

Burning period The part of each 24-hour period in which fires spread most rapidly, typically, from 10:00 AM to sundown.

Chain of command A series of management positions, in order of authority, through which decisions are made.

Command Has decision-making responsibility and overall responsibility for management of the incident.

Command staff Consists of the public information officer (PIO), the safety officer, and the liaison officer. These officers report directly to the IC.

Communications unit Distributes and maintains all forms of communications equipment on the incident and is responsible for developing an incident communications plan, included in the IAP, for the most effective use of the communications equipment.

Compensation and claims unit Responsible for ensuring that all forms required for a worker's compensation claim are filled out correctly and maintains records of injuries or illnesses incurred during the incident.

Cost unit Within the finance and administration section and responsible for tracking costs, analyzing cost data, making cost estimates, and recommending cost-saving measures.

Demobilization unit Responsible for developing an efficient incident demobilization plan to ensure that incident personnel and equipment are released in an orderly, safe, and cost-effective method after they are no longer needed.

Divisions Separate an incident into geographic areas of operation. They are established to regain and maintain span of control when available resources exceed the span of control of the operations chief. Divisions are managed by the division supervisor or group supervisor.

Documentation unit Provides incident duplication services, maintains, stores, and packs incident files for legal, historical, and analytical purposes.

Extended attack (EA) A fire that exceeds the capabilities of the initial attack fire resources and requires additional resources.

Facilities unit Within the logistics section and provides the layout, activation, and management of all incident facilities.

Finance and administration section Responsible for all incident costs and financial considerations; includes time, procurement, compensation and claims, and cost of units.

Food unit Within the logistics section and responsible for providing meals for incident personnel.

General staff Group of incident management personnel reporting to the IC, consisting of the operations section chief, planning section chief, logistics section chief, and financial and administration chief.

Ground support unit Within the logistics section and responsible for fueling, maintaining, and repairing vehicles and transporting personnel and supplies.

Groups Established to divide the incident into functional areas of operation.

Incident action plan (IAP) A document that contains objectives reflecting the overall incident strategy, as well as specific tactical actions and supporting information for the next operational period; the plan may be written or oral.

Incident commander (IC) The individual responsible for all incident activities, including the development of strategies and tactics and the ordering and release of resources.

Incident Command System (ICS) An adaptable emergency management system that allows users to adopt an integrated organizational structure equal to the complexity and demands of single or multiple incidents of any type or size.

Incident A natural or human-caused occurrence that requires emergency response to prevent or minimize loss of life or damage to property and natural resources.

Initial attack (IA) A fire that is controlled by the first dispatched resources, without significant reinforcements.

Liaison officer Member of the command staff responsible for coordinating with agency representatives from assisting and cooperating agencies.

Logistics section Orders resources (facilities, services, and supplies) and develops the transportation, communications, and medical plans.

Medical unit Provides and maintains first aid and emergency medical treatment stations for incident personnel.

Operational period Period of time scheduled for execution of a given set of tactical actions, usually not exceeding 24 hours.

Operations section Responsible for all tactical operations at the incident.

Planning section Collects, evaluates, and disseminates information about the incident. This section also manages the planning process and develops the IAP for each of the operational periods. Also, maintains information on the status of equipment and personnel resources, maintains all incident documentation, and prepares the demobilization plan.

Procurement unit Within the finance and administration section and responsible for financial matters involving vendor contracts.

Public information officer (PIO) A member of the command staff who acts as the central point of contact for the media.

Resource unit Responsible for checking in all incoming resources and personnel on the incident.

Safety officer Member of the command staff, reporting to the IC, who is responsible for monitoring and assessing hazardous and unsafe situations and developing measures for assessing personnel safety.

Situation unit Collects, processes, organizes and displays information on the current incident status.

Span of control Supervisory ratio of from three to seven individuals, with five individuals to one supervisor being established as optimal.

Staging areas Locations set up at an incident where resources can be placed while awaiting a tactical assignment, on a 3-minute available basis.

Strike team Specified combinations of the same kind and type of resources, with common communications and a leader.

Supply unit Within the support branch of the logistics section and responsible for ordering equipment and supplies required for incident operations.

Task force Any combination of single resources assembled for a particular tactical need, with common communications and a leader.

Technical specialists Personnel who have specialized skills. May be assigned to the planning section, function within a unit, or form a separate unit within the planning section. The most common specialists include field observers, fire effects monitors, FBANs, IMETs, and GIS specialists.

Time unit Within the finance and administration section and responsible for recording time for incident personnel and hired equipment.

REFERENCES

Department of Homeland Security. "Homeland Security Presidential Directive-5: Management of Domestic Incidents." Published February 28, 2003. https://www.dhs.gov/sites/default/files/publications/Homeland%20Security%20Presidential%20Directive%205.pdf.

Federal Emergency Management Agency (FEMA). "ICS Forms." Accessed October 3, 2019. https://training.fema.gov/icsresource/icsforms.aspx.

Federal Emergency Management Agency (FEMA). "ICS-420-1: Field Operations Guide." Published June 2016. https://www.usfa.fema.gov/downloads/pdf/publications/field_operations_guide.pdf.

Federal Emergency Management Agency (FEMA). "IS-100.C: Introduction to the Incident Command System, ICS 100." Published June 25, 2018. https://training.fema.gov/is/courseoverview.aspx?code=IS-100.c.

Federal Emergency Management Agency (FEMA). "IS-200.C: Basic Incident Command System for Initial Response." Published March 11, 2019. https://training.fema.gov/is/courseoverview.aspx?code=IS-200.c.

Federal Emergency Management Agency (FEMA). "IS-700.B: An Introduction to the National Incident Management System." Published June 25, 2018. https://training.fema.gov/is/courseoverview.aspx?code=IS-700.b.

Federal Emergency Management Agency (FEMA). "IT-300: Intermediate Incident Command System for Expanding Incidents." Accessed October 3, 2019. https://training.fema.gov/is/courseoverview.aspx?code=IT-300.

Federal Emergency Management Agency (FEMA). "IT-400: Advanced Incident Command System for Complex Incidents." Accessed October 3, 2019. https://training.fema.gov/is/courseoverview.aspx?code=IT-400.

Federal Emergency Management Agency (FEMA). "National Incident Management System." 3rd ed. October 2017. https://www.fema.gov/media-library-data/1508151197225-ced8c60378c3936adb92c1a3ee6f6564/FINAL_NIMS_2017.pdf.

Federal Emergency Management Agency (FEMA). "National Incident Management System (NIMS) Incident Command System (ICS) Forms Booklet." FEMA 502-2. September 2010. https://www.fema.gov/media-library-data/1425992150044-22337affef725b5f9d5fd8c7e9167ad8/ICS_Forms_508_12-7-10.pdf.

National Interagency Fire Center. "Interagency Standards for Fire and Fire Aviation Operations." NFES 2724. February 2019. https://www.nifc.gov/PUBLICATIONS/redbook/2019/RedBookAll.pdf.

National Wildfire Coordinating Group (NWCG). "NWCG Standards for Interagency Incident Business Management." PMS 902. Published April 2018. https://www.nwcg.gov/sites/default/files/publications/pms902.pdf.

National Wildfire Coordinating Group (NWCG). "S-130, Fire Fighter Training, 2008." Modified August 21, 2019. https://www.nwcg.gov/publications/training-courses/s-130.

National Wildfire Coordinating Group (NWCG). "S-200, Initial Attack Incident Commander, 2006." Modified August 16, 2019. https://www.nwcg.gov/publications/training-courses/s-200.

National Wildfire Coordinating Group (NWCG). "S-300, Extended Attack Incident Commander, 2008." Modified August 8, 2019. https://www.nwcg.gov/publications/training-courses/s-300.

National Wildfire Coordinating Group (NWCG). "Wildland Fire Incident Management Field Guide." PMS 210. NFES 002943. Published January 2014. https://www.nwcg.gov/sites/default/files/publications/pms210.pdf.

Wildland Fire Fighter in Action

You are a captain on a rapidly moving grass and brush fire. You are the initial IC and are receiving multiple resources at this incident because it borders on an area where there is federal jurisdiction, state jurisdiction, and fire district and city fire department response requirements. In a 10-minute time frame, you realize that the fire has spread over 10 acres (4 hectares) and there are at least 12 to 15 resources that are arriving to a staging area you designated before your arrival. The winds are blowing 25 to 30 mi/h (40.2 to 48.3 km/h), and the fire is moving toward a newer subdivision that has not yet been evacuated. After you have initially sized up the situation, the State Police and Department of Transportation approach you.

1. What type of incident is this?
- **A.** Type 1
- **B.** Type 2
- **C.** Type 3
- **D.** Type 4

2. What type of command would you establish?
- **A.** Single command
- **B.** Area command
- **C.** Unified command
- **D.** Joint information center

3. You decide to create several divisions. Which is the most appropriate designation for your divisions?
- **A.** Division 1
- **B.** Division IV
- **C.** Structural group division
- **D.** Division Alpha

4. Which one of these functions of the logistics section is correct?
- **A.** Provides incident maps
- **B.** Provides incident medical support
- **C.** Provides fire suppression
- **D.** Provides information to the public

Access Navigate for flashcards to test your key term knowledge.

CHAPTER

Wildland Fire Fighter I

Safety on Wildland Fires

KNOWLEDGE OBJECTIVES

After studying this chapter you will be able to:

- Describe how the 10 standard fire orders can be used to prevent future fire-fighter tragedies. (**NFPA 1051: 4.5.3**, pp. 46–48)
- Describe how the 18 watch outs can be used to determine whether an assignment is safe. (**NFPA 1051: 4.5.3**, pp. 48–49)
- Describe the parts of the LCES/LACES mnemonic. (**NFPA 1051: 4.1.2**, p. 50)
- Explain the importance of medical incident reports and incident within an incident procedures. (pp. 52–55)
- Inspect and maintain personal protective equipment. (**NFPA 1051: 4.1.1, 4.3.1, 4.3.2**, pp. 56–59)
- Maintain a personal gear kit according to the type and duration of the incident. (**NFPA 1051: 4.1.1, 4.3.1, 4.3.2, 4.3.4**, p. 59)
- Identify the contents of a personal gear kit. (**NFPA 1051: 4.3.1, 4.3.4**, p. 59)
- Understand the value of a fire shelter. (**NFPA 1051: 4.1.1**, pp. 59–66)
- Know what to do when trapped by fire in a vehicle or building. (pp. 66–68)
- Identify hazard trees and snags. (pp. 68–73)
- Explain the additional fire behavior watch outs. (pp. 74–76)
- Understand proper procedures for operating around aircraft. (pp. 76–77)
- Explain the benefits of maintaining a high level of physical fitness and health. (p. 79)
- Explain how proper nutrition and hydration can reduce fire-fighter fatigue. (pp. 79–80)
- Explain why it is important to recognize and manage mental stress. (p. 81)

SKILLS OBJECTIVES

After studying this chapter, you will be able to perform the following skills:

- Inspect and maintain personal protective equipment. (**NFPA 1051: 4.1.2, 4.3.1, 4.3.2**, pp. 56–59)
- Maintain a personal gear kit. (**NFPA 1051: 4.3.4**, p. 59)
- Deploy a fire shelter. (pp. 63–65)

Additional Standards

- **NFPA 1877**, *Standard on Selection, Care, and Maintenance of Wildland Fire Fighting Clothing and Equipment*
- **NFPA 1977**, *Standard on Protective Clothing and Equipment for Wildland Fire Fighting*
- **NWCG S-130 (NFES 2731)**, *Firefighter Training*
- **NWCG L-180 (NFES 2985)**, *Human Factors in the Wildland Fire Service*

You Are the Wildland Fire Fighter

It is mid-May—that time of the year when the preparation begins for the next fire season. You have completed your initial training and been assigned to an engine company in an area that is very prone to emerging incidents that go from very small to over a thousand acres in a very short amount of time. Your crew is serious about being in shape, but you did not prepare yourself over the winter months for the strenuous physical activities that will come your way. Your supervisor informs you about the expectations for the season, covers general safety topics and personal protective equipment (PPE) usage, and emphasizes that no one quits on his watch, no matter what. Your supervisor explains that the first week of your training will be all physical preparation. Essentially, the pack test is just a test, and you will be expected to do arduous hiking and lots of running.

On the third day of physical training, you start to develop a pain in your left calf. You do not want to be labeled a quitter, but you remember some of the safety training regarding a type of inflammation that can be very damaging or even fatal if you do not seek medical care. You are thinking about everything that has been thrown at you, and you are in pain.

1. What is the condition that is hinted at in the scenario?
2. Is this condition something that affects only people who are not in shape?
3. What type of testing is needed to determine the needed treatment?
4. How important is it that you notify your supervisor or other crew members?

Access Navigate for more practice activities.

Introduction

This chapter is one of the most important in this text. The concepts, practices, and information contained herein are the baseline fundamentals in the wildland fire environment. Your personal safety and that of your crew should be your prime consideration when operating at a wildland fire. Everyone is responsible for safety, and safe operation can occur only with a thorough and complete understanding of the concepts presented in this chapter and an understanding of fire behavior. Safety, as it relates to the wildland environment, is best described as being able to protect yourself from potential hazards and injury.

FIGURE 3-1 The 1949 Mann Gulch Fire.

Fire Orders and Watch Outs

10 Standard Fire Orders

The **10 standard fire orders** are a set of rules designed to reduce fire-fighter deaths and injuries and increase firefighting efficiency. These fire orders lead back to the "General Orders" of the military but were adapted in the late 1950s by the U.S. Department of Agriculture Forest Service Chief after a task force investigated 16 tragic wildland fires, most notably the Mann Gulch fire, which resulted in multiple fire-fighter fatalities **FIGURE 3-1**; the 10 standard fire orders were permanently established to prevent more fire-fighter tragedies from occurring during these types of fires.

The 10 standard fire orders remained unchanged until the 1980s, at which point they were changed to a mnemonic device spelling the word *fire orders*. Unfortunately, after years of research and further fire-fighter fatalities, in 2009 it was determined that the fire orders needed to be returned to their original order; they were then broken down into three easy-to-remember categories:

- Fire behavior
- Fireline safety
- Organizational control

If the first nine orders are in place, then you would be able to complete the 10th fire order. They are as follows:

Fire Behavior

1. Stay informed on fire weather and forecasts. One of the biggest influences on wildland fires is the weather. Wind brings in a fresh supply of oxygen to the fire, contributes to the rate of spread, and gives the fire direction. The weather must be tracked to have successful outcomes on a wildland fire.
2. Know what your fire is doing at all times. A fire fighter needs to know what the fire is doing, where it is going, and what kind of fireline intensities it is generating before a sound plan can be developed.
3. Base all actions on current and expected fire behavior. An understanding of wildland fire behavior is needed to meet your tactical objectives and work safely on the fireline. From this foundation, current fire behavior factors can be analyzed, and objectives can be developed based on behavior expectations. Also, consider unexpected circumstances and develop backup plans.

Fireline Safety

4. Identify safety zones and escape routes and make them known. One of the first things to do when getting to the fire is to establish escape routes and safety zones (discussed later in this chapter). Also, remember that, as the fire progresses, both the escape route and safety zone may change. Escape routes and safety zones need to be easily accessible to all, including your slowest fire fighter. Keep in mind the condition of the crew when planning escape routes. Are the crew members fresh, or have they been on the line for a long time? Evaluate all these factors when planning escape routes and safety zones.
5. Post lookouts when there is possible danger. The role of lookout should be filled by experienced and competent crew members. They need to be alert and always focused on watching the fire and the weather. They need to see when a potential problem is developing and sound the alarm early, thereby enabling your crew to start toward the safety zone early or to change fireline tactics.
6. Be alert, keep calm, think clearly, and act decisively. A wildland fire is a dynamic event, making it imperative to stay alert to changes in fireline conditions. As a supervisor, stay calm, or the crew will pick up on your excitability and react to it. Think clearly, take fatigue factors into account, and rest when necessary. Also, limit the number of people supervised and the number of radio channels monitored. If overload occurs, then a fire fighter will not function appropriately. Once you make a decision that is based on firm facts, act decisively.

Organizational Control

7. Maintain prompt communication with your crew members, supervisor, and adjoining forces. Keep your team informed and let them inform you. If fire fighters do not communicate with each other, then the system starts to break down. Tactical objectives are not met, and fire crews get injured. A good example of a communication breakdown is when one crew is involved in a firing operation that an adjoining fire crew knows nothing about. The first firing operation will severely affect that adjoining force.
8. Give clear instructions and ensure they are understood. If fire crews or fire fighters do not understand the task at hand, then they cannot contribute effectively to plan of attack and they risk the safety of themselves and their crew members. Once instructions are given, make sure they are understood. When laying out a tactical objective for a fire crew, check with the members periodically to see where they are, the progress they are making, and whether they have any support needs.
9. Maintain control of your forces at all times. Make sure your instructions are clear, concise, and understood. Know the locations of the crew members assigned to you, make sure your crew members know your location, and stay in communication with your crew. Be aware of what the fire is doing at all times.

If numbers 1–9 are in place, then number 10 is intuitive:

10. Fight fire aggressively, having provided for safety first. Safety is always the first consideration on a wildland fire, taking precedence over

all other actions. Fight fire aggressively; however, always analyze fire behavior factors before taking hasty action. Fireline intensities may be too great for a direct attack.

18 Watch Outs

The **18 watch outs** were developed not too long after the 10 standard fire orders. They were created as an expansion of the 10 standard fire orders to identify specific areas of concern in the wildland environment. It is believed that, if a fire fighter is following all the orders and the situations that shout "watch out," the risk reduction as a result of these principles will be significant.

Another way of looking at this is that you should not engage in firefighting efforts if the fire orders are not in place. However, if you encounter more than five situations that shout watch out that cannot be mitigated, you should reconsider engaging in your assignment.

The 18 situations that shout watch out are warning signs that a potentially dangerous situation is developing. They are as follows:

1. The fire has not been scouted and sized up. In this case, tactics are developed based on no information, and failure is bound to result. You have *no* idea about the fire's size or intensity levels. Should you do a direct or an indirect attack? What resources are required? Where should the safety zones be? How can any of these questions be addressed without scouting the fire?
2. You are in country that you have not seen in daylight. At night you will not be able to see all the factors (such as topography or fuel types) necessary to develop a sound plan. You may not be able to see dangerous terrain features where people or equipment can become trapped, such as drainage ditches.
3. Safety zones and escape routes are not identified. Sometimes, the fire behavior factors change, making it necessary to use an escape route and safety zone. Most often, injury or death results if the plan did not include escape routes or safety zones.
4. You are unfamiliar with local weather and other factors that may influence fire behavior. Weather is one of the most significant influences on a fire. If a local area has unique weather patterns, then be aware of them. When working a fire in an unfamiliar area, ask the locals about their weather patterns. Also ask fire crew members who are coming off the line about the fireline conditions and weather that they experienced.
5. You are uninformed on strategy, tactics, and hazards. Do not freelance on a fire. Understand the overall objectives and plan of action before starting work on the fireline. The action taken can affect all crews working in proximity. A good example of this would be a poorly coordinated firing operation that jeopardizes adjacent crews.
6. The instruction and assignments you were given are not clear. How can anyone carry out a plan he or she does not understand? If incident objectives are not understood, ask for clarification, and only then implement the plan.
7. You have no communication link with crew or supervisor. Successful communications are the key to having a successful outcome on a wildland fire. Important tactical changes cannot be made if communication with the crew or supervisor does not take place. The fire may have changed direction or a fixed-wing aircraft may be preparing to drop on the location. Dangerous situations may be developing, and the fire fighter must be aware of them.
8. You are constructing a fireline without a safe anchor point. A fireline is built from a safe anchor point. If it is not, the fire may cross open spots in the fireline and catch a fire fighter off guard. If an anchor point is not established, the chances for bad outcomes increase.
9. You are building a fireline downhill, and there is fire below you. The key here is that there is active fire below you. This situation is dangerous and should only be attempted if there is no other tactical alternative. If you must construct line downhill, follow the Down Hill Checklist found in the *Incident Response Pocket Guide* (IRPG).
10. You are attempting a frontal assault on the fire. The most intense part of a fire is the head, which is also where the flame lengths are the greatest. It is better to start from an anchor point, work up the flank, and then suppress the head. Exceptions do exist, and, in these cases, remember to be constantly aware of what the fire is doing—and have several experienced lookouts. In the case of structure protection, do not ever place yourself between a high-intensity fire and the structure being protected.
11. There is unburned fuel between you and the fire. Overall, one of the safest places to be is on the fireline, with one foot in the burned area and one foot in the unburned area. With

low-intensity fuel types, you can step into the burned area and be safe should the fireline condition change to an increase in fire activity.

If you are in the green, unburned area with the fire advancing toward that location, then you run the risk of the fire spotting around you. These new spot fires may cause a fire that entraps you and your crew.

12. You cannot see the main fire and are not in contact with anyone who can. How can you make a tactical decision or plan escape routes or safety zones? You have no idea of the size or intensity of the fire. It is better to move to a safe location and wait until this information is available.
13. You are on a hillside where rolling material can ignite the fuel below you. If this potential exists, post lookouts and constantly monitor the fire's activity.
14. The weather is getting hotter and drier. Fine fuel moistures will be getting lower. Fine fuels are the primary carriers of the fire. Expect to see a change in your fireline conditions as it gets hotter and drier.
15. The wind is increasing or changing direction. Generally, wind gives the fire direction and brings in a fresh supply of oxygen. An increase in wind speed increases the rate of fire spread. Evaluate the fire and make the necessary tactical changes.
16. You are getting frequent spot fires across the line. This situation is another indication of a change in fire behavior. Evaluate the fireline factors present, and make appropriate changes in your tactics. It may be necessary to abandon the original plan and move to the secondary control line.
17. Terrain and fuels make escape to your safety zone difficult. This is a real safety problem. Either construct new escape routes and safety zones or move out of the area.
18. You feel like taking a nap near the fireline. This is a potential problem for you and your crew. If you are in the unburned fuels, then the potential exists to get burned should the fire spot, rekindle, or jump the control line. Additionally, heavy equipment, such as dozers or tractor plows, may be working, and the potential exists for an injury from either of these pieces of equipment. When you feel this way, consider fatigue factors. It may be time to rest yourself and your crew.

The 18 watch outs can be found on the back cover of any edition of the IRPG.

Safety Zones and Safety Systems

Safety Zones and Escape Routes

A **safety zone** is an area cleared of flammable materials and used as refuge when a fire has been determined to be unsafe. You should not need a fire shelter in a safety zone. There are three safety zone categories:

- The burn or the black. An area that has already burned.
- Natural. A green field or a rocky area, for example.
- Constructed. An area created by humans, such as a road, or by fire fighters with burnout methods, dozers, or other equipment.

The distance between safety zones and fires should be at least four times the maximum flame height. If wind or slope is present, the distance should be even greater than this. Distance is measured from the center of the safety zone to the nearest fuels. The size of a safety zone can vary, but, with no wind or slope affecting a fire, it usually follows the measurements listed in the IRPG and in **TABLE 3-1**. It is important to note that these IRPG guidelines assume flat terrain and no wind. They are based on still heat (radiant heat) only and do not account for heat that may be carried by wind or terrain, such as steep slopes. If wind or slope is present, the distance separation will need to be significantly increased.

LISTEN UP!

One acre is approximately the size of one football field.

TABLE 3-1 Safety Zone Construction Requirements

Flame Height	Separation Distance	Safety Zone Size
10 ft (3.0 m)	40 ft (12.2 m)	1/10 acre (0.04 ha)
20 ft (6.1 m)	80 ft (24.4 m)	1/2 acre (0.2 ha)
50 ft (15.2 m)	200 ft (61.0 m)	3 acres (1.2 ha)
100 ft (30.5 m)	400 ft (121.9 m)	12 acres (4.9 ha)
200 ft (61.0 m)	800 ft (243.8 m)	46 acres (18.6 ha)

Incident Response Pocket Guide 2018. URL: https://www.nwcg.gov/sites/default/files/publications/pms461.pdf.

An **escape route** is a preplanned route that fire fighters take to move to a safety zone. **Escape time** is the time it takes for crew members to make it to a safety zone. Loose soil, difficult terrain, uphill routes, downed trees, heavy smoke, rivers and creeks, and dangerous wildlife are travel barriers that could increase escape time.

A **deployment site** is a last-resort location where a fire shelter must be deployed. It is used when access to escape routes and safety zones have been compromised.

LISTEN UP!

Most entrapments do not occur while fire fighters are in a safety zone, but, instead, while they are using the escape route to get to a safety zone.

LCES and LACES

In 1991, Paul Gleason, the superintendent of the Zigzag Interagency Hotshot Crew (United States Forest Service, Oregon) developed a mnemonic device to help combine and simplify all the safety-related components that fire fighters are expected to remember. The acronym was **LCES**, standing for lookouts, communication, escape routes, and safety zones. LCES is a safety system used by fire fighters to routinely assess their current situation for hazards. It was adopted by all agencies in short period of time due to its simplicity. This system was developed in conjunction with another tragic loss of life in the wildfire environment, the Dude Fire in Tonto National Forest, Arizona. The theory was that, if you did not have the LCES tactic in place, you should not engage on any fire because the components of LCES are the core safety principles that are derived from the 10 standard fire orders and the 18 watch outs, previously discussed. Each fire fighter must be briefed on these components before engaging in suppression activities.

The components of LCES follow:

- Lookouts need to be in a position where both the objective hazard and the fire fighters can be seen. Lookouts must be trained to observe the wildland fire environment and recognize and anticipate changes in fire and weather behavior. When the objective hazard becomes a danger, lookouts relay the information to their crew members so they can reposition to the safety zone or a safer area. How to serve as a lookout and communicate condition changes is discussed in Chapter 9, *Human Resources*.
- Communication must be prompt, concise, and clear. All personnel should be on the same or assigned radio frequencies. Effective communication ensures that everyone is aware of approaching hazards.
- Escape routes are the paths fire fighters take from a dangerous location to a safety zone. Because the effectiveness of an escape route is dependent on the behavior of the fire, escape routes are probably the most elusive component of LCES. Fire fighters should make sure that more than one escape route is always available and the escape route is known to all.
- Safety zones are planned locations of refuge to which fire fighters can retreat where they will not need to use a fire shelter. Fireline intensity, safety zone topography, and the fire fighter's location relative to the fire determine the safety zone's size and effectiveness.

LACES is an adaptation of the LCES mnemonic that is used in various parts of the country. The main elements of LCES, described previously, are still current and relevant, but an *A* has been added for an additional safety-related focus. The *A* component may have three different meanings, depending on the area in which it is being used: anchor points, attitude, or awareness. It is important to note this difference when interacting with fire fighters from Canada or other areas of the United States that use this terminology. Depending on which area references are used, establishing an anchor point can be paramount to preventing entrapment. In some situations, a poor attitude has created dangerous situations for fire fighters in all areas of the firefighting effort. A negative attitude can lead to carelessness, complacency, taking shortcuts, or even serve as a distraction from the task at hand. Having a specific reference to attitude could prevent serious injuries or fatalities. Awareness is a key component to firefighting. Without a good awareness of what is going on around you, the chances for bad things to happen increase significantly.

Remember to plan for the components of LCES/LACES before firefighting in the wildland environment. Everyone must be informed of the escape routes and safety zones, and everyone should adhere to the operational plans.

LCES/LACES are the operational components of the 10 standard fire orders. LCES/LACES are the actions you must do to help control and minimize hazards.

Human Factors and Risk Management

Fire fighters have a responsibility to learn, reduce errors, and improve their performance. Understanding human factors in the wildland fire environment will increase fire-fighter awareness and prevent accidents or errors from occurring. Almost all accidents or fatalities can be traced back to a human error.

Understanding the importance of situational awareness, communication, attitude, and decision-making can best sum up human factors.

DID YOU KNOW?

The Wildland Fire Lessons Learned Center (LLC) website at www.wildfirelessons.net, established in 2002 by the National Wildfire Coordinating Group (NWCG) and partner agencies, hosts forums for fire fighters to share, discuss, and debrief lessons learned in the wildland environment. The LLC also hosts a podcast, a YouTube channel, a newsletter, and more. Use this website as a resource for learning and improving your performance.

A testament to learning lessons from human factors is embedded in the LLC logo. The 14 small stars are an homage to the 14 fire fighters killed in the 1994 South Canyon Fire in Colorado. The 14 stars surround a larger center star, representing the wildland fire community. The logo depicts the South Canyon 14 watching over the wildland fire community.

Situational Awareness

It is critical that you always maintain good situational awareness on a wildland fire. It can help keep you out of harm's way. **Situational awareness**, or situation awareness, both abbreviated SA, is the foundation for all decision-making. It is the process of making sure you understand what is going on around you. SA involves gathering information through observation and communication. The SA process needs to be honed over time and continually updated. SA is specific to the individual. In most cases, you do this every day; you just don't know the process.

On a wildland fire, good SA will help you to stay aware of the fire's current situation at all times by allowing you to think clearly and globally. However, there are several factors that can negatively affect the SA process, and they are described next.

LISTEN UP!

Situational awareness is the process of making sure you understand what is actually going on around you (the reality) *not* your perception of what is happening. It is critical to maintain SA at all times. Dangerous conditions can develop rapidly.

Communication

Communication plays a role in all human factors in the wildland fire service. It is the key to good SA and success. Without effective communication, orders will get lost, tasks will not be timely, and fire fighters will not know what is going on around them. Fire fighters have a responsibility to communicate effectively. Effective communication can be identified by five communication responsibilities as outlined by the NWCG:

1. Brief others as needed. Briefing others is paramount when conditions change or new equipment and people are working in your area. The act of conducting a briefing is a skill that should be practiced regularly. A briefing checklist can be found on the inside back cover of the IRPG.
2. Debrief your actions. At the end of the shift and or the incident, it is important to identify lessons learned and things that went right as well as things that did not go well that can be improved. This step is essential in preventing future mistakes. A reference for after-action reviews can be found in the IRPG.
3. Communicate hazards to others. Everyone is responsible for everyone else's safety, to a point. If you identify something that is hazardous or dangerous, it is important to let others know. This can be accomplished in a written format, over the radio, face-to-face, or with signs or flagging **FIGURE 3-2**.
4. Acknowledge messages. This communication is sometimes referred to as *echoing*. When someone passes a message on to you, you need to repeat it back to the sender. If you incorrectly repeat something back, the sender will then have the opportunity to repeat the correct information to you.
5. Ask if you don't know. In the firefighting world, there are many factors that affect what we do. In some instances, we do not want to let people down, or people have hazardous

FIGURE 3-2 "Do not cross" hazard flagging.
Courtesy of Jeff Pricher.

attitudes from a know-it-all type of personality. Trying to accomplish a task that you are unfamiliar or uncomfortable with will, in most cases, negatively affect the outcome of the assignment. It is better to let your supervisor or the person asking you to do something know that you don't know exactly what to do so that he or she can assist you or find someone else to assist you.

Additional communication techniques, such as radio communications, are discussed throughout this book.

LISTEN UP!

Speak up if you have questions or need help! Trying to accomplish a task that you are unfamiliar or uncomfortable with can negatively affect the outcome of the assignment.

LISTEN UP!

The PLOWS method is an alternative to the after-action review. PLOWS stands for plan, leadership, obstacles, weaknesses, and strengths. PLOWS allows fire fighters to reflect on actions from a safety standpoint.

Hazardous Attitudes

Attitude is everything. A positive attitude can be the difference between success and failure in difficult situations. It can mean survival when you are fatigued, stressed, or in mental overload.

A **hazardous attitude** is a human characteristic that can affect your situational awareness and be detrimental to your safety, your crew's safety, and the success of the assignment. Poor morale, poor stress management, lack of confidence, fear, complacency, overconfidence, invulnerability, sexism, racism, and bias are just a few examples of hazardous attitudes.

If you recognize a hazardous attitude from yourself or your crew members, do not act impulsively. Instead, stop and use previous experience and technical skills as a base to draw from in the decision-making process. Remember to use the acronym STOP:

- S—Slow down
- T—Think
- O—Observe
- P—Predict

Understanding and identifying hazardous attitudes will help you avoid accidents and fatalities in the field. Understand that human beings are fallible, but learn from your mistakes. Always work at improving technical skill levels, evaluating and learning from experiences, and keeping egos in check.

Risk Management and Decision-Making

Fire fighters have a responsibility to minimize risk. **Risk management** is the process of identifying, observing, and acting on potential risks to hinder or eliminate those risks. This process is based on a job aid found in the IRPG. The job aid allows you to identify specific areas that may pose a problem during the operation and provide an opportunity to effectively manage the risk factors in wildland firefighting. This process is called recognition-primed decision-making (RPD or RPDM).

Every fire fighter needs to understand the risk management process to effectively communicate potentially hazardous situations.

There are five steps to the risk management process:

1. Situational awareness (gathering information)
2. Hazard assessment (estimate fire behavior, identify tactical hazards)
3. Hazard control
4. Decision point
5. Evaluation

The IRPG offers a decision tree that you can follow to help you determine how much risk to take or not take. Other references that can be used for this process include the 10 standard fire orders and the 18 watch outs, LCES/LACES, and establishing a solid and safe anchor point before committing too much to your fire attack. Decision-making and the risk-management process are discussed in more detail in Chapter 9, *Human Resources*.

Incident Reports

On July 25, 2008, Andy Palmer was killed during a tree-felling operation on the Eagle Fire in California. This tragic loss of life resulted in the establishment of two new sets of requirements for wildland firefighting: the 9-line, now referred to as the medical incident report (MIR), and the incident within an incident (IWI) procedure.

Medical Incident Reports

The 9-line report, now known as the **medical incident report**, or MIR, is a written document (or sometimes a verbal report) that outlines details of an accident, such as when an injury occurs on a scene **FIGURE 3-3**.

MEDICAL PLAN (ICS 206 WF)

1. Incident/Project Name	2. Operational Period
Taylor Creek and Klondike East	Date/Time 08/28/2019 to 08/28/2019 ALL SHIFTS

3. Ambulance Services

Name	Complete Address	Phone & EMS Frequency	Advanced Life Support (ALS) Yes	No
AMR	Josephine County – Taylor Creek	911 555-222-2833	X	
AMR	Grants Pass/Cave Junction - Klondike	911 555-222-5081	X	

1. Air Ambulance Services

Name	Phone	Type of Aircraft & Capability
Mercy Flights – Taylor Creek, Klondike	911 or 555-222-1019 (dispatch)	Bell 407 GX Night Vision Capable
Cal/Or Life Flight -Agness Spike	555-222-6661 (Curry County)	Bell 407 Night Vision Capable
Wenatchee Short Haul	555-222-2505 Rogue Disp.	B3 ASTAR
Oregon Air National Guard Call in order listed.	555-222-6236 Pilot 555-222-5331 Co-Pilot 555-222-2800 ORNG JOC	Oregon National Guard Medivac with night vision

2. Hospitals

Name Complete Address	GPS Datum – WGS 84 DD° MM.MMM' N - Lat DD° MM.MMM' W - Long		Travel Time from ICP Air	Gnd	Phone	Helipad Yes	No	Level of Care Facility
Asante Rogue Regional Medical Center 2825 Barnett Rd, Medford	Lat:	42° 19.04904 N	20	1 hr	555-222-7132	☒	☐	Level 2 Trauma
	Long:	122° 49.8696 W						
Three Rivers Hospital 500 S.W. Ramsey Ave Grants Pass	Lat:	42° 25.27086 N	N/A	35 min	555-222-7069	☐	☒	Level 4 Trauma (minor medical)
	Long:	123° 20.55882 W						
Providence Hospital 1111 Crater Lake Ave, Medford	Lat:	42° 20.37792 N	20	1 hr	555-222-6440	☒	☐	Level 3 Trauma
	Long:	122° 51.732 W						
Legacy Emanuel Hospital 2801 N Gantenbein, Portland	Lat:	45° 32.6502 N	90	3 hrs 45 min	555-222-4121	☒	☐	Level 1 Trauma Burn Center
	Long:	122° 40.1811 W						
Sacred Heart Medical 3333 Riverbend Dr Springfield, OR	Lat:	44° 04.871 N	70	2 hrs 45 min	555-222-6929 Or 555-222-5111	☒	☐	Level 2 Trauma
	Long:	123° 01.591 W						

3. Division | Branch | Group	Area Location Capability			
Selmac ICP	EMS Responders & Capability:	Eamonn Ryan MEDL, Diana Braun MEDL (T), John Andrews, AEMT		
Div A/V	EMS Responders & Capability:	Medic 6 Ambulance		
	Equipment Available on Scene:	ALS equipped 4x4 Ambulance, ALS gear		
	Medical Emergency Channel:	Command		
	ETA for Ambulance to Scene:			
	Air:	10 min		
	Ground:	Travel time from position of staging determined by Division		
	Approved Heli spot:	H-32	H-40	H-49
	Lat:	42° 27.767' N	42° 34.076' N	42° 35.476' N
	Long:	123° 36.025' W	123° 44.581' W	123° 38.885' W

FIGURE 3-3 MIR form.

MEDICAL PLAN (ICS 206 WF)

Controlled Unclassified Information//

Medical Incident Report

FOR A NON -EMERGENCY INCIDENT, WORK THROUGH CHAIN OF COMMAND TO REPORT AND TRANSPORT INJURED PERSONNEL AS NECESSARY.

FOR A MEDICAL EMERGENCY: IDENTIFY ON SCENE INCIDENT COMMANDER BY NAME AND POSITION AND ANNOUNCE "MEDICAL EMERGENCY" TO INITIATE RESPONSE FROM IMT COMMUNICATIONS/DISPATCH.

Use the following items to communicate situation to communications/dispatch.

1. **CONTACT COMMUNICATIONS / DISPATCH** **(Verify correct frequency prior to starting report)**
   Ex: "Communications, Div. Alpha. Stand-by for Emergency Traffic."
2. **INCIDENT STATUS:** *Provide incident summary (including number of patients) and command structure.*
   Ex: "Communications, I have a Red priority patient, unconscious, struck by a falling tree. Requesting air ambulance to Forest Road 1 at (Lat./Long.) This will be the Trout Meadow Medical, IC is TFLD Jones. EMT Smith is providing medical care."

Severity of Emergency / Transport Priority	☐ **RED / PRIORITY 1 Life or limb threatening injury or illness. Evacuation need is IMMEDIATE** *Ex: Unconscious, difficulty breathing, bleeding severely, 2° – 3° burns more than 4 palm sizes, heat stroke, disoriented.* ☐ **YELLOW / PRIORITY 2 Serious Injury or illness. Evacuation may be DELAYED if necessary.** *Ex: Significant trauma, unable to walk, 2° – 3° burns not more than 1-3 palm sizes.* ☐ **GREEN / PRIORITY 3 Minor Injury or illness. Non-Emergency transport** *Ex: Sprains, strains, minor heat-related illness.*	
Nature of Injury or Illness & Mechanism of Injury		*Brief Summary of Injury or Illness (Ex: Unconscious, Struck by Falling Tree)*
Transport Request		*Air Ambulance / Short Haul/Hoist Ground Ambulance / Other*
Patient Location		*Descriptive Location & Lat. / Long. (WGS84)*
Incident Name		*Geographic Name + "Medical" (Ex: Trout Meadow Medical)*
On-Scene Incident Commander		*Name of on-scene IC of Incident within an Incident (Ex: TFLD Jones)*
Patient Care		*Name of Care Provider (Ex: EMT Smith)*

3. **INITIAL PATIENT ASSESSMENT:** *Complete this section for each patient as applicable (start with the most severe patient)*

Patient Assessment: See IRPG page 106

Treatment:

4. **TRANSPORT PLAN:**

Evacuation Location (*if different*): (*Descriptive Location (drop point, intersection, etc.) or Lat. / Long.*) Patient's ETA to Evacuation Location:

Helispot / Extraction Site Size and Hazards:

5. **ADDITIONAL RESOURCES / EQUIPMENT NEEDS:**

Example: Paramedic/EMT, Crews, Immobilization Devices, AED, Oxygen,Trauma Bag, IV/Fluid(s), Splints, Rope rescue, Wheeled litter, HAZMAT, Extrication

6. **COMMUNICATIONS: Identify State Air/Ground EMS Frequencies and Hospital Contacts as applicable**

Function	Channel Name/Number	Receive (RX)	Tone/NAC *	Transmit (TX)	Tone/NAC *
COMMAND					
AIR-TO-GRND					
TACTICAL					

7. **CONTINGENCY:** ***Considerations:*** ***If primary options fail, what actions can be implemented in conjunction with primary evacuation method? Be thinking ahead.***

8. **ADDITIONAL INFORMATION:** *Updates/Changes, etc.*

REMEMBER: Confirm ETA's of resources ordered. Act according to your level of training. Be Alert. Keep Calm. Think Clearly. Act Decisively.

FIGURE 3-3 MIR form. (*Continued*).

The MIR acts as a communication tool by providing the incident commander and responding personnel with a baseline of information to determine the needed medical treatment, rescue resources, and specialized equipment on a fire scene. This protocol also helps incident personnel triage the injury, which in turn helps the evacuation process.

The MIR is broken down into eight bullet points. Each one of the bullet points directs you to collect specific information that will help you determine and obtain the medical resources most suited for your incident. MIRs require record of contact with a dispatcher or the incident base, a summary of the incident, a status of the injured person, the name of the person in charge at the incident, a patient assessment, a transportation plan, any additional resources or equipment needs, communication channels or frequencies you plan to use, a contingency in case you need to change your original plan, and an update for the incident commander or incident base.

The medical incident report can be found in the emergency medical care section of the IRPG (pink section). On large incidents or incidents that span more than one operational period, the MIR will be found in the incident action plan (IAP). In most IAPs, the MIR will be the next to last page so that it is easy to find. For areas that use a preprinted incident organizer, the MIR will also be found in that document.

Incident Within an Incident Procedure

An **incident within an incident (IWI)** is an accident or emergency that occurs on the scene of a fire (hence, the name, an incident occurring within a larger fire incident). When an accident or injury occurs on a fire, it is important to understand the protocol, or plan of action, to follow. It is human nature to want to stop what you are doing and race off to see who is hurt or injured. It is imperative that everyone understands the IWI procedure in order to stay on task and avoid working without direction.

The nearest operations branch director (OPBD) or division group supervisor (DIVS) assigned for that operational period shall be initially assigned to the emergency situation as the on scene incident commander to mitigate the emergency. The closest medical provider will be the primary point of contact for all patient-related activities. Using the MIR format and the preestablished procedures set by your agency, appropriate resource ordering, crew coordination, communications, and notifications will be made through a coordinated effort. It is important to know what this procedure will entail if you are joining a larger incident or what your home agency utilizes for this approach.

It is crucial for all fire fighters, contractors, and incident personnel to be familiar with this process. Knowing how to call in an accident or injury could save your life or someone else's.

Rapid Extraction Module Support Units

Over the past several fire seasons, medical support on large fire incidents has increased with more trained personnel and specially trained **rapid extraction module support (REMS) units**. These units are still evolving as you read this text, and the standardization for the equipment and deployment will change over time. As more REMS units are created, the protection of fire crews will increase dramatically.

A typical REMS unit will be equipped with a pickup that is hauling a utility vehicle on a trailer **FIGURE 3-4**. The minimum crew size is four. The crew's skills and equipment will include basic- and paramedic-level care and low- and high-angle rescue capabilities. Crew members will be in top physical shape and required to complete the arduous pack test. (Physical fitness testing for wildland fire fighters

LISTEN UP!

During the 2018 fire season, a REMS unit was successful in removing a patient involved in a chainsaw accident on a ridge in very difficult terrain. The accident occurred at 4800 ft (1463.0 m), access points were at 4200 and 5200 ft (1280.2 and 1585.0 m), and the ambulance was located at 3800 ft (1158.2 m). From the time of the accident to the time the patient was in the back of the ambulance, the REMS unit members completed their assignment in 45 minutes.

FIGURE 3-4 REMS units often utilize utility vehicles.
Courtesy of Jeff Pricher.

FIGURE 3-5 REMS units are deployable to any challenging circumstances.
Courtesy of Jeff Pricher.

FIGURE 3-6 Full PPE for a wildland fire fighter.
Courtesy of National Interagency Fire Center.

is discussed in more detail in Chapter 9, *Human Resources*). These specialized units are intended to be deployable on a fire where challenging circumstances present themselves in rescuing and transporting fire fighters **FIGURE 3-5**.

Personal Protective Equipment and Clothing

Personal protective equipment (PPE) and clothing are the fire fighter's first line of defense against a wildland fire. PPE must be worn at all times on a fireline **FIGURE 3-6**. It enables a person to survive under conditions that might otherwise result in death or serious injury. Understanding the design, application, and limitations of PPE is critical. The more you know about the protection your PPE can provide, the better you will be able to judge conditions that exceed its limitations.

PPE should meet the requirements of NFPA 1977, *Standard on Protective Clothing and Equipment for Wildland Fire Fighting*. Avoid the use of structural PPE for wildland firefighting. Wearing structural PPE for long periods of time will result in the retention of body heat and will greatly increase the chances of dehydration, heat exhaustion, and heatstroke.

Fire fighters are responsible for maintaining their PPE. PPE must be cleaned, maintained, and inspected after each assignment to ensure that it will continue to provide the intended degree of protection when it is needed. It should be replaced if holes or tears exist.

Helmet

The helmet or hard hat protects your head from falling objects and falls. There are several helmet styles,

FIGURE 3-7 Single-brimmed, cap-style helmet (top) and full-brimmed helmet.
Courtesy of Jeff Pricher.

but you should not wear structural helmets because they are heavier and can be fatiguing after extended use **FIGURE 3-7**. In high winds, the beavertail brim of the structural helmet can act as a sail and have a greater chance of blowing off a fire fighter's head. It is important to clean your helmet and its straps with soap and water. If significant scrapes or dents are present, it may be time to retire your helmet.

Eye Protection

Eye protection can be in the form of clear American National Standards Institute (ANSI)-approved

A

B

FIGURE 3-8 Eye protection. **A.** From left: clear glasses, mesh glasses, sunglasses. **B.** Goggles.
Courtesy of Jeff Pricher.

lenses or, in many cases, ANSI-approved sunglasses **FIGURE 3-8**. Protecting your eyes is just as important as protecting your head. Eye injuries during wildland firefighting are common. Factors contributing to these injuries include dust blowing in the wind; bark and other debris falling from trees or other objects; dirt and rocks flying back at you during mop-up; and other high-risk activities, such as chainsaw work. There are several styles of eye protection, so make sure you use the style that meets your authority having jurisdiction (AHJ) equipment standards.

Clean with water and replace when broken, scratched, or ripped.

Flame-Resistant Clothing

Shrouds are pieces of fire-resistant fabric worn around the face, ears, and neck. Some of them attach to helmets **FIGURE 3-9**. Shrouds help to keep heat, small debris, and embers off of these areas.

The flame-resistant shirt, coat, and pants issued to you protect your torso and legs **FIGURE 3-10**. All items must be free of holes and tears and fit in a way that is not too tight or loose. You need a little space for your upper body to move around in. Also, the small air pockets provide some insulation from radiant heat when you are engaged in hot-line construction.

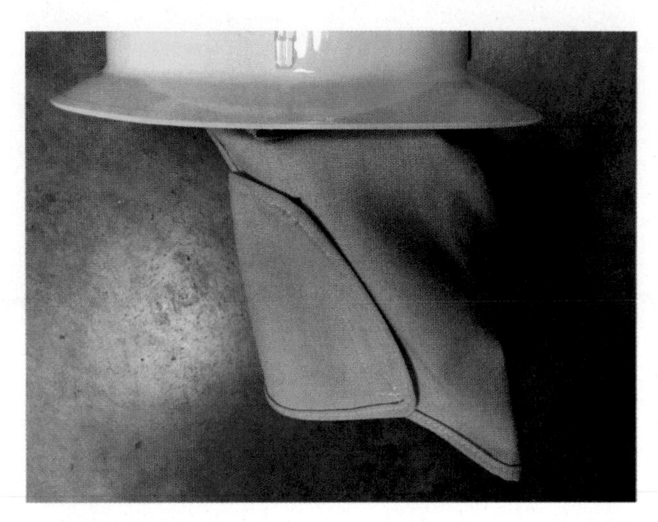

FIGURE 3-9 Shroud.
Courtesy of Jeff Pricher.

Clean your flame-resistant clothing according to the washing instructions on the tag. Replace at signs of rips, tears, or wear.

Boots and Socks

Boots should be formfitting, not too tight nor too loose, and the uppers should be free of defects. The most common boot type is the leather boot **FIGURE 3-11**. Leather boots should have a height of no less than 8 in. (20.3 cm) and should not have a steel toe or midsole. Boots with metal in them can absorb heat and burn your feet.

Boots should be regularly cleaned and, if they are leather, oiled. There are several products that can be applied to condition the leather. Check with your boot manufacturer to make sure you choose the appropriate product.

Even though socks are not typically thought of as PPE, they are an important component of taking care of your feet. Change your socks as often as feasible. A good wool or wool-blend sock will be one of the best investments you can make. While wool socks may be more expensive than socks made from other materials, they will last longer and have better moisture-wicking properties. Some fire fighters wear multiple pairs of socks, such as a liner sock and an outer sock or a running sock and an outer sock. Employ a strategy that works for you.

Gloves

Gloves should be leather and made specifically for fighting fires. There are different types of leather

A

B

C

FIGURE 3-10 A. Flame-resistant coat. **B.** Flame-resistant shirt. **C.** Flame-resistant pants.
Courtesy of Jeff Pricher.

FIGURE 3-11 Boots.
Courtesy of Jeff Pricher.

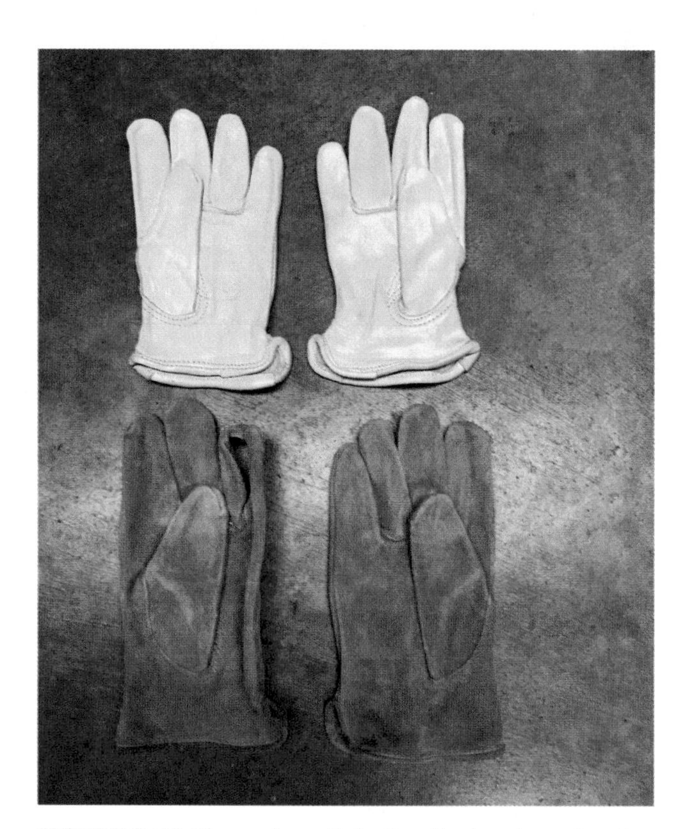

FIGURE 3-12 Gloves. Smooth leather (top) and rough outer leather (bottom).
Courtesy of Jeff Pricher.

gloves to accommodate for fire fighter preference **FIGURE 3-12**. Gloves should be worn *at all times.* Gloves help prevent blisters, make it easier to handle hot debris during mop-up and cold trailing, protect from cuts when maintaining hand tools or sharpening chainsaw chains, and prevent burns.

Gloves should be cleaned according to the manufacturer's instructions whenever they are dirty. They should be replaced upon tearing or wear.

FIGURE 3-13 Hearing protection can include foam ear plugs.
Courtesy of Jeff Pricher.

FIGURE 3-14 Web gear/fire pack.
Courtesy of Jeff Pricher.

Hearing Protection

Hearing protection should be worn around power equipment, such as helicopters and chainsaws **FIGURE 3-13**. However, do not use foam earplugs in high temperatures close to a fire. The high temperatures could cause them to melt to your body. Replace foam earplugs according to the manufacturer's instructions. Mend earmuff pads as needed, and replace when worn.

Web Gear/Fire Pack

Not all agencies issue web gear/fire packs as part of the PPE ensemble, but, for those that do, it is crucial that all parts of the web gear/fire pack be fire ready. A web gear/fire pack is a belt and harness used to carry the fire shelter and canteens **FIGURE 3-14**. Fire shelters are discussed later.

Personal Gear Kit

A wildland fire fighter should be able to be self-sufficient for 24 to 48 hours if needed. This is where your personal gear kit comes into play. This kit should include the following items:

- Lighter or matches
- Pocket knife
- Compass
- First aid kit
- Food and water
  - Meal ready-to-eat (MRE), or other easy meals
  - Trail mix or jerky
  - Baby food or applesauce
- Headlamp
- Flashlight

Your personal gear kit can also contain other items, such as spare clothing, boot laces, eyeglasses, or prescriptions **FIGURE 3-15**.

Fire Shelter

The fire shelter is a very important piece of fire-safety equipment for the fire fighter. A **fire shelter** is a reflective cloth tent that offers protection from fire, heat, smoke, and ember showers in an entrapment situation. It reflects 95 percent of a fire's radiant heat and absorbs 100 percent of the convective heat from flames and hot gases. It also traps cooler, breathable air **FIGURE 3-16**. It is mandatory for all fireline and support personnel to carry a fire shelter. Never enter a fire area without a fire shelter

Remember that your highest priority is to avoid entrapment in the first place. Carrying a fire shelter should never be considered an alternative to safe firefighting, but, in an emergency, a fire shelter can save your life.

LISTEN UP!

Carry your fire shelter where you can reach it, never inside your pack.

Storage

A fire shelter comes in a case that should be attached onto the web gear while in the fire area **FIGURE 3-17**. It should be worn on the web gear in a way that allows it to be easily removed from the case. It should never be left on the engine.

A

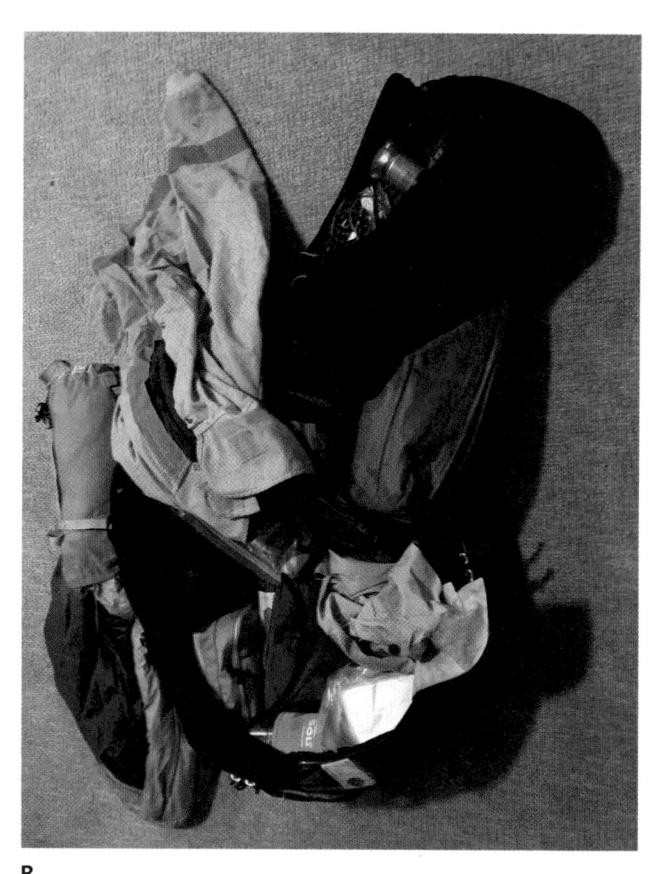

B

FIGURE 3-15 Personal gear kit. **A.** Kit. **B.** Contents.
Courtesy of Jeff Pricher.

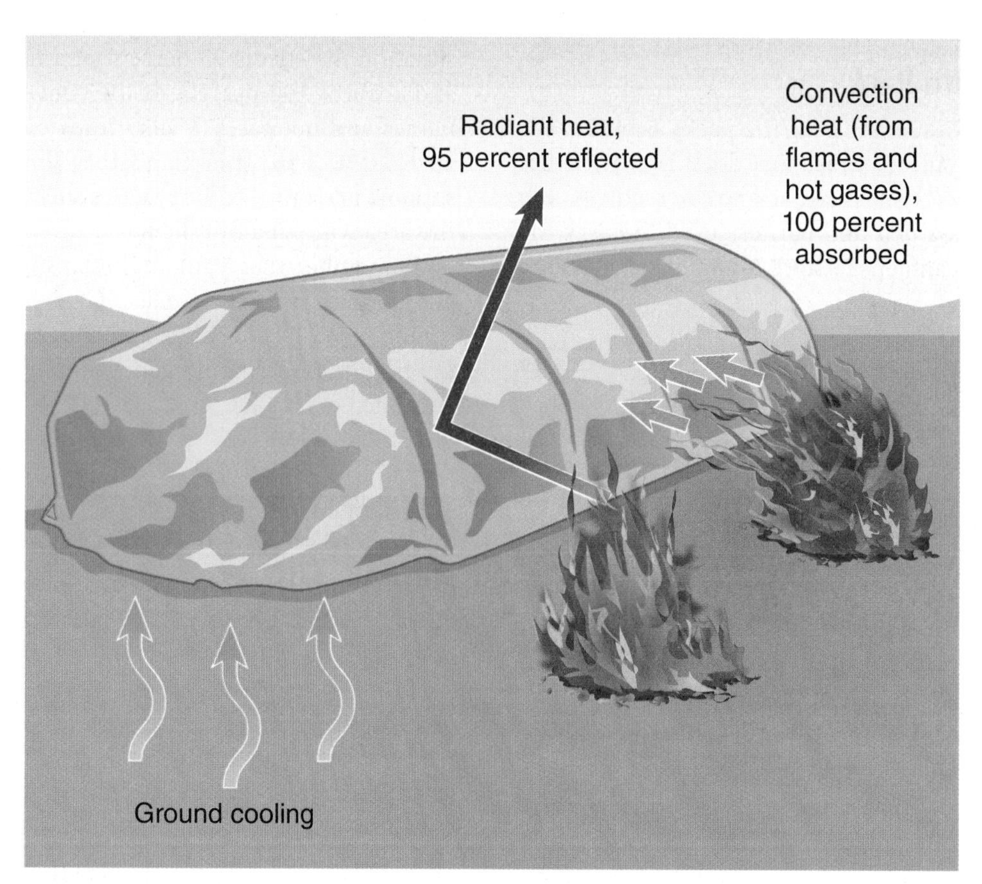

FIGURE 3-16 A fire shelter reflects 95 percent of a fire's radiant heat and absorbs 100% of the convective heat (from flames and hot gases).

FIGURE 3-17 Note that all of these fire fighters are wearing web gear and a fire shelter. The web gear is the dark colored strapping that can be seen on the fire fighters, and the fire shelter can be seen at the bottom of the pack.
Courtesy of Joe Lowe.

FIGURE 3-18 Fire shelters should be stored in a plastic liner and then in a carrying case.
Courtesy of Jeff Pricher.

The fire shelter case should have a Velcro closure and quick-pull strap **FIGURE 3-18**. The shelter is in a plastic enclosure inside the case. It has a red pull tab that works like a zipper. Pull the red ring on the plastic bag down to the bottom and up the other side. When the shelter has been removed from the plastic, it is ready for use and can be deployed.

A

B

FIGURE 3-19 A. New-generation (left) and old generation fire shelters in their cases. **B.** New generation (left) and old generation fire shelters deployed.
Courtesy of Jeff Pricher.

New-Generation Fire Shelters

After January 1, 2010, all agencies, cooperators, and contracted resources were required to carry the M-2002 new-generation fire shelter on wildfires **FIGURE 3-19**. The main differences between the new and old fire shelter are the shelter shape, the materials used, and the construction features. All these features offer better protection for the fire fighter.

The new-generation fire shelter protects in a similar way to the old shelters—by reflecting radiant heat and trapping breathable air. Like the old shelter, the new shelter has two layers. The outer layer is aluminum foil bonded to woven silica cloth. The foil reflects radiant heat, and the silica material slows the passage of heat to the inside of the shelter. An inner layer of aluminum foil laminated to fiberglass prevents heat from reradiating to the person inside the shelter. When these layers are sewn together, the air gap between them offers further insulation benefits. The outer layer of foil reflects about 95 percent of the radiant heat that reaches the fire shelter. Because only 5 percent is absorbed into the shelter materials, the temperature of the material rises slowly. Unlike radiant heat, convective heat (from flames and hot gases) is easily absorbed by the fire shelter, causing the temperature of the material to rapidly rise. When the material reaches about 500°F (260°C), the glue that bonds the layers begins to break down. The layers can separate, which allows the foil to be torn by turbulent winds. Without the foil, the shelter loses much of its ability to reflect radiant heat. The silica material will slow heat transfer but offers significantly less protection without the foil. That is why it is extremely important to clear an area free of fuels and away from heavier fuels when it is necessary to deploy a fire shelter.

The new-generation fire shelter protects its occupant with a more heat-resistant material than the old-style fire shelter and with a shape that is more aerodynamic and rounded to better reflect radiant heat. Destructive testing conducted at the Missoula Technology Development Center showed that, even though the new-generation fire shelter has less airspace inside, it performs much better than the old-style shelter **FIGURE 3-20**.

First and foremost, remember that all firefighting tactics must be selected to ensure fire-fighter safety at all times. Everyone must know the escape routes to safety zones, and the routes must be continually reevaluated. The fire shelter is your last resort; it is not a guarantee of safety. If the fire shelter must be deployed, it is extremely important to deploy in an area where flames will not contact the shelter.

FIGURE 3-20 Grab handle performance was updated in the new-generation fire shelters. Old-generation (top) and new-generation fire shelter grab handles are identified with pink flags. Notice that the new shelter handles are tucked closely to the sides of the shelter. This modification was made to prevent straps from impeding the user from entering the shelter.

Courtesy of Jeff Pricher.

LISTEN UP!

All fire shelters come with instructions from the manufacturer. Take the time every fire season to read these instructions.

Deployment

A fire shelter should be deployed in 25 seconds or less in an area that is clear of fuels by scraping the area down to mineral soil. Clear an area of at least 4 × 8 ft (1.2 × 2.4 m) or larger. The shelter cannot tolerate sustained heat from high-intensity, heavy fuels; this is extremely important to know because, the glue that bonds the layers of the shelter will start to break down.

There are two ways to deploy the fire shelter: static deployment, which is done in place, and dynamic deployment, which is done on the run. Dynamic fire-shelter deployment was developed as a result of the 1994 South Canyon Fire in Colorado, where 14 people died trying to escape. Thereafter, many fire agencies developed fire-shelter techniques to be performed on the run.

When the fire is close at hand and fire fighters are caught between the safety zone and the flaming front, follow the steps in **SKILL DRILL 3-1** for the dynamic deployment technique.

LISTEN UP!

Always take your gloves, helmet, radio, and water with you into a fire shelter. Discard flammable items and hand tools.

Now, let's look at a static deployment technique. This is the technique most fire fighters are familiar with. It is done in place, not on the run **SKILL DRILL 3-2**. Note that the fire shelter pictured here is a practice fire shelter, as indicated by its green color and orange case. Fire shelters with orange cases are intended for training purposes only. Fire shelters intended for real fire purposes come in a blue case.

If the fire is close and a decision must be made on how to deploy, remember that it is better to be on the ground before the flaming front arrives than standing up and trying to get into the fire shelter.

SKILL DRILL 3-1
Dynamic Fire-Shelter Deployment Technique

1. Wear gloves and a hard hat, and cover your face and neck if possible. Discard flammable items and any hand tools. Start moving in the direction of the safety zone.
2. Unsnap the web gear, and remove the shelter from its carrying case.
3. While picking up speed and starting to run, slip out of the web gear, and let it fall to the ground.
4. Once you have done this, hold the shelter with both hands, and remove it from its plastic case. The shelter is now ready to deploy.
5. There will be no time to clear an area, so look for a place to deploy that is clear of vegetation or has low-intensity fuels, such as a short-grass model. When a spot is located, hit the ground faceup, and deploy the fire shelter while lying on the ground. The optimal survival zone is within 1 ft (0.3 m) of the ground. Use arms and legs to enter the shelter.
6. Once inside the shelter, make sure the shelter has airspace away from the body, and roll over on your stomach. Your face should be covered and airway protected. Remember, always use the wind to your best advantage. Place feet toward the fire if the wind permits.

SKILL DRILL 3-2
Static Fire-Shelter Deployment Technique

1. Pick the largest available clearing. Wear gloves and a hard hat, and cover your face and neck if possible. Discard flammable items and any hand tools. Scrape away flammable litter if you have time.

(continued)

SKILL DRILL 3-2 Continued
Static Fire-Shelter Deployment Technique

2 Pull the red ring to tear off the plastic bag covering the shelter.

3 Grasp the handle labeled "Left Hand" in your left hand and the handle labeled "Right Hand" in your right hand.

4 Shake the shelter until it is unfolded. If it is windy, lie on the ground to unfold the shelter.

SKILL DRILL 3-2 Continued
Static Fire-Shelter Deployment Technique

5 Slip your feet through the hold-down straps.

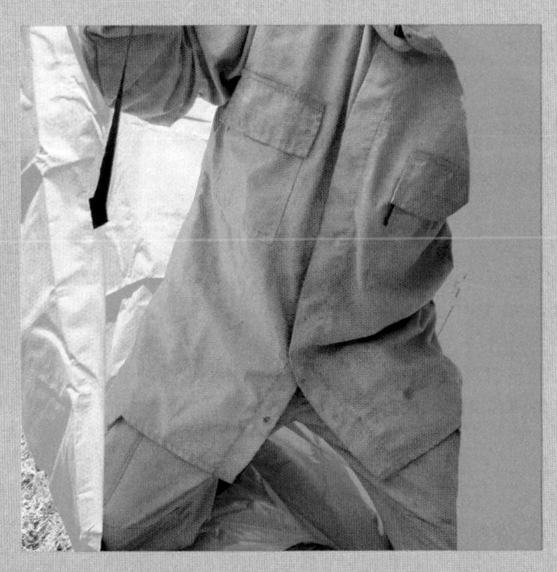

6 Slip your arms through the hold-down straps.

7 Lie down in the shelter with your feet toward the oncoming fire. Push the sides out for more protection, and keep your mouth near the ground.

Courtesy of Jeff Pricher.

Shelter Maintenance and Safety

Here are some safety and maintenance points to remember when handling or deploying the fire shelter:

- Pack away from sharp objects.
- Avoid rough handling the shelter.
- Do not sit on top of the fire shelter or use it as a pillow.
- Never deploy the fire shelter under power lines.
- Do not deploy the fire shelter under a snag or widow-maker.
- Deploy in an area that is as free of fuels as possible. Avoid areas where the shelter could sustain direct flame contact.
- Watch terrain features; avoid deploying in saddles or chimneys.
- Always protect the airway.
- Deploy in a group if possible. This action results in mutual shielding and grants supervisors better control of their crews.
- Never exit the fire shelter until the flame front has passed.
- Do not come out of the shelter until you get an all clear from the supervisor.
- If you are alone, wait for the shelter to cool.
- If time permits, clear as large an area as possible.
- Discard all flammable items, such as fusees, fuel bottles, and chainsaws.
- Always check the fire-shelter seal once in the shelter and on the ground. Blowing embers should not enter the shelter.
- Keep the shelter away from the body by maintaining an airspace inside the shelter.
- Communicate often with other crew members while in the shelter.
- Limit movement while inside the shelter.

Shelter Inspection

Inspecting a fire shelter should occur once a week during the fire season if you are part of a handcrew or engine crew. These inspections are necessary to account for the stress that is put on the fire shelter as it is jostled inside the plastic liner.

A fire shelter inspection is primarily visual and should take only a few minutes. To inspect your fire shelter, remove it from its protective case. Starting with the red plastic rings, make sure that the pull handles are intact and have not started to tear away. Make sure that the tear channels (the areas that connect the pull handles and allow the shelter to be opened) are intact. The shelter should look as pristine as it was the day it was issued to you. The shelter should also show no signs of being crushed or squished (a sign that the white plastic liner was not used). If the shelter is not symmetrical, there is a chance it had pressure applied to it. If you observe this, bring it to your supervisor for a potential exchange or removal from service.

There should be no tears, rips, or holes in the clear plastic liner. If you observe any significant discoloration of the clear plastic liner or silver flakes inside the clear plastic liner, bring this to the attention of your supervisor for a potential exchange or removal from service.

Never discard the fire shelter after it has been used until it has been replaced with a new one. It may be needed again while leaving the fire area. Once in camp or at the station, get a new one immediately.

A shelter greatly improves survival chances, so never hesitate to deploy it should it become necessary. Waiting could mean the difference between life and death.

LISTEN UP!

A fire shelter is the most important piece of fire-safety equipment, but it is not always fail-proof. Avoid situations where a fire shelter is needed, but be prepared to deploy and use one when necessary.

Vehicle and Building Entrapments

It may be necessary to use a vehicle or building for an area of safe refuge. In all cases, evaluate whether it is better to be in a vehicle, a building, or a fire shelter as a barrier. Sometimes, this is not a simple decision.

Vehicle Entrapment

There may come a time when you become entrapped in a vehicle and using that vehicle as a temporary refuge area may be your only option. Should this occur, follow the guidelines from the LLC:

- Park the vehicle in an area devoid of vegetation.
- Burn out around the vehicle if there is time.
- If you do not have time to clear the area of fuels or burn out, park in the light fuels. Heavy fuels will produce more heat and will last longer.
- Park behind a natural barrier or structure. Consider that a structure could become involved, which could severely affect nearby exposures/vehicles.
- Do not park on the downhill side of a road. If you do, you will be exposed to the fire moving upslope. Park away from the toe of the slope.

- Watch your overhead. Do not park under power lines or overhanging vegetation.
- Stay out of saddles and draws, which will channel heat upslope.
- Position the vehicle in a direction that provides the area occupied by crew personnel with the maximum protection from an approaching flame front.
- Set the parking brake, leave the motor running at a high RPM, and keep the vehicle lights on.
- Roll up the windows and do not lock the doors because someone else might need to get in.
- Cover windows with fire shelters, with the reflective material placed against the window. If your engine has breathing apparatuses, take them in the cab with you and use them.
- If for some reason the fire shelter is not in the cab, then get to the floor of the vehicle and cover up with a brush jacket or turnout coat if available.
- Do not panic. In light fuels, the flaming front will pass in about 30 seconds. However, in heavier fuels, the flame and intense heat will pass in 2 to 5 minutes.
- You must protect your airway; remain as low in the vehicle as possible, and use a dry bandanna to cover your nose and mouth.
- Expect the following conditions if you are trapped inside the vehicle:
  - Temperatures may reach over 200°F (93.3°C).
  - Smoke and sparks may enter the vehicle.
  - Plastic parts may start to melt and give off toxic gases.
  - Windows may start to crack.
  - Exposed skin may receive radiant-heat burns.
- If the vehicle catches fire or the windows blow out and you must exit the vehicle before the fire has passed, then:
  - Each crew member should cover himself or herself with a fire shelter.
  - Exit the vehicle from the side away from the greatest heat.
  - Stay together, and get as low to the ground as possible, moving away from the vehicle.
  - Deploy your shelter in a safe area **FIGURE 3-21**.

Building Entrapment

In some instances, it may be necessary to use a house as a temporary refuge area (TRA). Generally, any area that can temporarily protect a fire fighter from flames and heat can be considered a TRA.

FIGURE 3-21 Fire fighter opening the fire shelter and preparing to deploy.
Courtesy of Joe Lowe.

First, evaluate whether the house is built well and is not too close to large concentrations of heavy fuels. If the answer is yes to both questions, then it is safe to use as a temporary barrier against the fire.

If you plan to use the house as a TRA, all doors and windows should be unlocked and closed. Immediately turn off any LPG tanks. This should be done well in advance of the main fire. If time permits before the flaming front arrives, clear away any concentrations of heavy fuels around the outside perimeter of the house.

Just before the flaming front arrives, enter the house with a fire shelter and web gear on. If your engine carries breathing apparatus, ensure that you take yours with you inside the house. Also take a shovel because it may be needed later as a suppression tool in the event that the apparatus is burned when the flame front arrives.

Once inside, check again to see that all doors and windows are closed. Immediately move the furniture away from the windows. Remove any flammable curtains. Then, find a place near a closed doorway, on the side away from the fire, and sit down. Remember to stay low and be prepared to exit the house if necessary.

Should it become necessary to exit the house, wrap up in the fire shelter and open the door slowly. Once outside, stay low and find a deployment area. Once in this area, immediately fall to the ground and enter the fire shelter.

Remember that a well-built house will afford protection before heavy fire involvement occurs. The flaming front usually will have passed before the house burns to the ground.

Hazard Trees

It is important to understand basic tree anatomy when identifying hazard trees **FIGURE 3-22**. A **hazard tree** is a tree with a structural defect. These defects can cause the entire tree or part of the tree to fall. Identifying hazard trees is a skill that is very important to your safety and survival in the fire environment. Hazard trees and falling operations have contributed to several fatalities. In recent years, failure to identify danger or hazard trees has killed professionals, such as smokejumpers, fallers, and crew members. The 2012 Steep Corner Fire in Idaho is one example where a fire fighter was killed by a falling tree.

FIGURE 3-22 Tree anatomy.

Identification

Hazard trees are created by several conditions, including fire damage; weather and lightning damage; insect damage; fungus, root rot, bole or trunk rot; and hang-up situations where a tree is lodged into another.

One of the most common hazard-tree characteristics is the snag. A **snag** is a dead or dying tree that is still standing **FIGURE 3-23**. Sometimes, healthy trees or snags can have a widow-maker. A widow-maker, also called a hang-up or fool killer, is a broken tree branch that is caught in the tree or other trees, waiting to fall **FIGURE 3-24**.

If you are working in an area with numerous burning trees, expect snags. Long after the passage of the flaming front, trees will continue to smolder and may finally burn through and fall, especially if a fire has burned through in previous years and left deeply scarred trees with little wood left to support them. In addition, the root systems may burn out and leave stump holes, which can also cause the tree to be unstable. If aircraft are making drops in the area or dozers are working nearby, then watch for additional instability of trees with structural defects.

Additional types of hazard-tree characteristics to look out for include the following **FIGURE 3-25**:

- Browning or decaying leaves, needles, or branches
- Broken tops and dead limbs overhead or an accumulation of downed limbs
- Broken or loose bark
- Exposed sapwood
- Large cracks, gashes, or splits on the trunk
- Decay, rot, or erosion on the trunk or roots
- Conks or cankers on the trunk
- Leaning, uprooted or partially uprooted trees
- Insect or wildlife damage
- Collapsed snag
- High-risk tree species (those that are known for rot and shallow root systems) in the area

FIGURE 3-23 Snag.

- Timbered areas that have been burning for an extended period
- Snag on fire
- Stump hole
- Weather forecast with high winds or wind impacting the tree canopy

Each one of these conditions can create an environment that is ripe for a tree to fail. Understanding how to identify these trees can save your life.

FIGURE 3-24 A widow-maker is a branch waiting to fall.

Failure Zone

Most tree-failure situations happen without warning and leave little time for escape. In general, if a hazard tree has been identified, it is important to keep your distance and stay out of the potential failure zone.

The **failure zone** is any area that can be reached by any part of the failing tree. To get out of the potential failure zone, keep your distance at least one and a half times the height of the tree because, when a tree fails, it can strike other trees and cause them to fall or toss other materials. On a steep slope, rocks, debris, and the actual percentage of the slope can multiply the size of the potential failure zone. When a hazard tree has been identified, it is important to mark it and communicate the hazard to everyone.

Flagging and Felling

Always identify hazard trees in the work area by flagging them with orange and black danger-tree flagging (skull and cross bones) tape **FIGURE 3-26**. This should

FIGURE 3-25 Other hazard-tree characteristics. **A.** Forked top. **B.** Dead top. **C.** Loose bark. **D.** Exposed sapwood.

Courtesy of Pacific Northwest Hazard Tree Working Group.

E

F

G

FIGURE 3-25 (*Continued*) **E.** Crack or split in the trunk. **F.** Canker on the trunk. **G.** Partially uprooted roots.

Courtesy of Pacific Northwest Hazard Tree Working Group.

H

I

J

K

L

FIGURE 3-25 *(Continued)* **H.** Exposed roots. **I.** Wildlife damage. **J.** Collapsed snag. **K.** Cottonwood branches. **L.** Dwarf mistletoe infection.

Courtesy of Pacific Northwest Hazard Tree Working Group.

FIGURE 3-25 (*Continued*) **M.** Dwarf mistletoe infection up close. **N.** Snag on fire. **O.** Stump hole.
M: © Hanjo Hellmann/Shutterstock; N: Courtesy of Pacific Northwest Hazard Tree Working Group; O: Courtesy of Joe Lowe.

FIGURE 3-26 A. Danger-tree flagging tape. **B.** Hazard tree marked with flagging tape.
Courtesy of Jeff Pricher.

FIGURE 3-27 A hotshot firefighter cuts down a tree.
Courtesy of National Interagency Fire Center.

FIGURE 3-28 Chimney.
Courtesy of Joe Lowe.

be done before mop-up crews enter the burned area and start their operations.

Give hazard trees a large safety radius, and stay clear of them until they are brought down by a saw team. It is also important to post lookouts and account for and communicate with all fire crews working within the hazard area before felling a snag. A qualified faller should bring down trees that are assessed as complex operations **FIGURE 3-27**. Specially trained fallers are considered FAL2 (Intermediate) or FAL1 (Advanced). Additional information on safe chainsaw maintenance and operation is discussed in Chapter 14, *Handcrew Operations*.

LISTEN UP!

Here is a list of hazard-tree watch outs to help maintain safety when working in a forested area:

- Increased wind in an area with numerous burning trees
- Leaning dead trees
- An area that is experiencing air drops from fixed-wing aircraft or helicopters
- Dozers working in the area
- Mop-up operations that cut root systems.
- Using strong hose streams that wash away supporting root systems.

Fire Behavior Watch Outs

Understanding wildland fire behavior is the foundation upon which everything else is built. Without such a foundation, poor tactical decisions can be made that will put others at risk. Certain terrain features cause fires to spread at a greater rate. Terrain and weather features (Chapter 4, *Basic Wildland Fire Behavior*) will be discussed here from a safety perspective. They are cause for concern because they enhance fire spread.

FIGURE 3-29 A saddle.
Courtesy of CAL Fire.

Chimney

A chimney is a steep drainage on the side of a mountain **FIGURE 3-28**. This terrain feature channels all the convective energy of the fire and is responsible for rapid rates of fire spread. Recognizing this terrain feature and avoiding it could prevent you from descending or ascending in the area where the convective column would travel. Getting caught in the channel where the convective energy travels would likely cause severe burns or worse.

Saddle

A saddle is a low topography point between two high points in a mountain range **FIGURE 3-29**. Saddles are points of least resistance for winds and convective

FIGURE 3-30 Narrow canyons are cause for concern. The canyon walls are so close together that spotting can easily occur on the opposite canyon wall.

FIGURE 3-31 Box canyon.

Courtesy of Joe Lowe.

energy. A fire pushes through this low point at a rapid rate during uphill fire runs. The wind speed generally increases as the fire passes through the constricted area. This increase in wind speed could potentially create a dangerous situation for the fire fighter if he or she were in that area.

Narrow Canyon

A narrow canyon is a terrain feature that requires great care **FIGURE 3-30**. The problem with a narrow canyon is that the two canyon walls are so close together that spotting can easily occur on the opposite canyon wall, trapping the fire fighter between two bodies of fire. Area ignition can also easily occur in a narrow canyon. Narrow canyons may incorporate some of the dangers of a chimney and a saddle. Depending on the time of day, wind will travel up the canyon or down the canyon. In some situations, fire can spread from one side of the canyon to the other as a result of spotting or radiant heat. This poses an inherent danger to the fire fighter by potentially cutting off an escape route.

Box Canyon

Box, or dead-end, canyons, are areas where intense updrafts and extreme fires occur **FIGURE 3-31**. The steep canyon walls help force the convective column aloft. In addition, with this terrain feature, there is usually only one way in and one way out. Always have plenty of lookouts and maintain heightened awareness when working in a box canyon.

Midslope Road

A midslope road is a road that appears in the middle of a slope or between two slopes. When working on a midslope road, always remember to watch the fuels on the unburned side. The problem with midslope roads is that there will be active fire above or below the location, and there is a danger of spotting occurring in the unburned fuel bed. If the fire spots in the unburned fuel, there is a potential for trapping fire fighters between two bodies of fire. If spotting occurs in the unburned fuels, plan on leaving the area.

Eddy Effect

As wind is blowing across a mountaintop, a rolling current effect, called the eddy effect, can take place on the lee side (side away from the wind) of the mountain. The danger is that spotting can occur below on the lee side of the ridge opposite the main fire. If working near the top of the ridge on the lee side, the fire fighter may become trapped by the advancing flame front, which is driven by the effect of slope, wind, and fuels. If personnel are available, it is good to place a lookout on both sides of the ridge.

Thunderhead

A thunderhead is a violent local storm produced by dense, vertical clouds called cumulonimbus clouds **FIGURE 3-32**. The danger to fire fighters is the downdrafts and erratic winds that occur when the mature thunder cell passes over the fire. These winds will drive the fire in many directions. Thunderstorms and thunderheads may also be accompanied by lightning. If lightning is occurring, follow these rules:

- Stay out of dry creek beds.
- Do not use radios or telephones.
- Put down all tools and remove caulk boots.
- Sit or lie down if in open country.
- Avoid grouping together.

FIGURE 3-32 Thunderheads forming over a mountain range.

- Do not handle flammable materials in open containers.
- Use the vehicle as a shelter whenever possible.
- Turn off machinery and electric motors.
- Where there is no shelter, avoid high objects, such as lone trees. If only isolated trees are nearby, the best protection is to crouch in the open at a distance of twice the height of the tree. Keep away from wire fences, telephone lines, and electrically conductive elevated objects.
- Avoid ridgetops, hilltops, wide-open spaces, ledges, rock outcroppings, and exposed shelters.
- Advise crew that, if they feel an electrical charge—their hair stands on end or their skin tingles—lightning may be about to strike. Drop to the ground immediately.
- Adjustments to tactics and heightened awareness become necessary as thunderheads occur.

Weather Front

A weather front is a boundary between two air masses of differing properties. As a front approaches, expect the wind to increase. Winds are strongest during the passage of the front; as it passes, the winds start to change direction. Winds blow counterclockwise in a low-pressure area and clockwise in a higher-pressure area. For this reason, it is easy to see why a wind shift takes place with the passage of a front. Remember to adjust tactical plans in accordance with the wind shift. Weather fronts are discussed in more detail in Chapter 4, *Basic Wildland Fire Behavior*.

Safety Watch Outs

The safety watch outs in this section deal with accountability of incident personnel and working around equipment and hazards that are found on firelines. Safety watch outs are necessary for the safety of personnel and operations at a wildland fire.

Accountability of Incident Personnel

Accounting for incident personnel is a major safety function at a wildland fire. **Personnel accountability** is the ability to account for the location and well-being of your crew. The locations of all known incident resources must be tracked from the time of check-in at the fire to the release of resources. Without a system that tracks personnel, safety issues start to appear. Crew members may become lost if they are isolated on the fireline. This could be a problem when they need to be fed or rested. Without food and rest, fatigue sets in, and safety problems start to occur. Not knowing the whereabouts of incident resources can also set up entrapment or burn-over situations where a fire can overtake personnel and equipment.

Accountability begins with proper check-in. Resources are sent either directly to the incident (immediate-need resources) or used at another time (planned-need resources). Check-in takes place at the following locations:

- Staging area
- Division/group supervisor (immediate-need resources that are part of a strike team)
- Base or camp
- Helibase
- Incident command post resource unit

Once resources have been properly checked in, they are tracked with a tag system such as e-ISuite, with T-cards, or with a combination of these things. These systems are used to account for all personnel assigned to the fire until they are released. While personnel are on a fireline, supervisors are responsible for keeping track of their subordinates at all times.

Aircraft Safety

Aircraft, such as helicopters, single-engine air tankers, and helitankers, are usually deployed to drop retardant on any large wildland fire. Retardant is a substance that coats and reduces the flammability of fuels. It is heavy and can lift and throw rocks, dirt, and loose fuels on landing. Fire fighters working with aircraft in

LISTEN UP!

Personal accountability is also important. Own up to your mistakes and learn from them for the betterment of yourself and your crew.

close proximity could potentially be in the drop zone. Though you should avoid operating in a drop zone at all costs, it is important to know what to do if you find that you are in a drop zone.

If this potential exists, then consider moving out of the area if time permits. If time does not permit and a drop is about to occur, then evaluate the situation in relation to overhead objects. Stay out from under large, older trees because the treetop or limbs may break due to the force of the retardant drop.

During a retardant drop, find a place to lie down behind a large stationary object, such as a large rock. Then, lie down on your stomach in the direction of the oncoming drop. Leave helmet and goggles on with your face to the ground. Put your arms out, with any hand tools away from your body, and feet apart for additional stability. Prepare for the impact of the retardant or water **FIGURE 3-33**.

After the drop has taken place, get up slowly and make sure all crew members are unhurt. Retardant is slippery. So move out of the area slowly, and remember to clean any hand tools that may have been soiled.

Aircraft operations are discussed in Chapter 8, *Ground and Air Equipment*.

LISTEN UP!

The goal is to never be anywhere near an aircraft while it is dropping water or retardant.

FIGURE 3-33 Aircraft dropping retardant.
Courtesy of National Interagency Fire Center.

Dozer Safety

Dozers are used on any size of wildland fire because they are extremely effective at removing flammable vegetation in the fire's path **FIGURE 3-34**. They are to be respected because these large pieces of equipment are capable of dislodging huge rocks, which can roll downhill and injure fire crews. In addition, the operator may have poor visibility due to dust, smoke, and working with a helmet on.

Here is a safety checklist to use when working around a dozer:

- If working below a dozer, then watch for large rolling rocks that are displaced. It is best not to work below a dozer if given a choice.
- When working around a dozer, keep a 200-ft (61.0-m) safety radius around it. A dozer can change direction very quickly.
- Do not get right in front of or behind a dozer. The operator has poor visibility due to the blade and the tractor's protective cage.
- Never sleep or sit down close to or under a dozer.
- Never get on or off a moving dozer.
- Always remain in full view of the operator.
- During night operations, always have a good headlamp or reflective vest on.
- Use predetermined hand signals for direction and safety. The operator cannot hear verbal orders.
- Do not ride on a dozer. The operator is the only one who has a proper seat.
- Never approach the operator unless he or she knows your proximity and you were asked to do so.

FIGURE 3-34 Dozers.
Courtesy of South Dakota Wildland Fire.

- When requested to do so by the operator, approach from a 45-degree angle only. The operator has a better view from this angle.
- If climbing up to talk with the operator, then make sure the blade is down, and use the handholds.

Dozer operations are discussed in more detail in Chapter 8, *Ground and Air Equipment*.

Liquified Petroleum Gas Tanks

In many rural areas in the United States, the primary source of cooking and heating fuel is liquified petroleum gas (LPG). LPG is usually stored in pressurized gas tanks located on the property next to the house. Note their location before the fire arrives, and clear flammable fuels around them. Turn the tank's connection to the house off.

Keep LPG tanks cool and avoid direct flame impingement. If direct flame impingement on the LPG tank is evident, move out of the area. It will take fire water flow, which is probably not available, to properly cool the tank, including small barbecue LPG tanks.

Narrow Bridges

Rural areas usually have narrow bridges over streams and creeks **FIGURE 3-35**. Some are able to support the weight of a fire apparatus and others are not; as such, they can become a trap for a fire engine. Some agencies have such bridges denoted on preplans of the area. Use this information if it is available. If not, evaluate the bridges carefully if there is any question as to their weight limitations. Get out and look before crossing.

Power Lines

During the heat of a wildland fire, power lines may burn through and fall. Watch overhead at all times **FIGURE 3-36**. Do not park under or deploy a fire shelter under power lines. They present a dangerous electrical hazard.

If the wires are down upon arrival at a wildland fire, assume that they are live, and cordon off the area. Notify fire department dispatch of their location.

If power lines are downed or drooping, do not spray water directly at them or at the base of the power-line pole (unless absolutely necessary). If a power company representative is on scene, follow all of his or her directions. If a representative is not on scene, request one to respond immediately.

Hazardous Materials

While providing structure protection during a wildland fire, always check the garage or outbuildings for hazardous or flammable materials. Also, check the brush area near an outbuilding for old tires, polyvinyl chloride (PVC) piping, or other hazardous materials.

If these items are found, try to keep the fire away from them. If this is not possible, then either evacuate the area or stay upwind of any smoke. Consult a hazardous material guide and follow the recommendations found there. If there are any doubts, then move back a safe distance and deny entry to anyone.

FIGURE 3-35 Narrow bridge.
Courtesy of Jeff Pricher.

FIGURE 3-36 Watch overhead for power lines.
Courtesy of Jeff Pricher.

Fire Fighter Health

Safety also involves taking care of yourself before, during, and after a wildland fire. It is important that you stay physically fit and hydrated, eat a balanced diet, and understand the symptoms associated with heat stress, carbon monoxide poisoning, and fatigue. Any of these conditions can debilitate a fire fighter working on a fireline.

Physical Fitness

Being in good physical condition is an important factor in survival at a wildland fire. Establish a good exercise program before and during the fire session. This program should last at least an hour each day and include weight training, cardiovascular workouts, and stretching. Check with a personal or departmental physician before entering into any new workout routine

At some handcrew training programs, new recruits hike 3 mi (4.8 km) in full gear with a hand tool each day before class. They also do push-ups as a crew to add upper-body strength. Exercise programs that simulate actual fireline work conditions are best.

Physical fitness must be a career-long activity. Firefighting is a stressful job that demands that you maintain a good fitness level throughout your career.

Evaluating your crew members' readiness and physical fitness tests is discussed in Chapter 9, *Human Resources*.

LISTEN UP!

Maintain your physical fitness! Your life and the lives of your crew depend on it!

Rhabdomyolysis

Though it is important to push yourself and maintain your physical fitness, it is also important to listen to your body and stop when it tells you to, especially in the field. Due to extreme, prolonged overexertion, fire fighters are at risk for rhabdomyolysis. Rhabdomyolysis, also known as "rhabdo," is a condition where muscle tissue begins to break down and release damaging proteins in the blood. If left untreated, rhabdomyolysis can be fatal or cause permanent muscle and kidney damage.

If you are experiencing pain or muscle aches that are out of proportion to the amount of exercise you are doing or you have urine that is brown or a very dark yellow, do not ignore these symptoms. Get tested for rhabdomyolysis immediately. Treatment includes rest and rapid intravenous fluids.

To prevent this condition, when you are just starting a new workout regimen, try to limit the amount of caffeine, over-the-counter medications, and antibiotics that you consume, which can worsen the condition.

Nutrition

Diet is another important aspect of personal health. Eating well and staying hydrated can reduce fire-fighter fatigue. A healthy menu includes fruits, vegetables, healthy fats, whole grains, and lean protein. Pay attention to portion sizes. Unfortunately, most people eat larger portions than their bodies need. Replace high-calorie desserts and alcoholic beverages with healthier choices, such as fruit.

Hygiene

Hygiene is another important part of personal health. When entering all food lines, it is imperative that you utilize the hand-washing stations that are provided. With the close proximity of all the fire fighters in the small pop-up village or city, failure to regularly wash hands can lead to the transmission of disease. The best way to avoid this is to wash your hands.

Hydration and Heat Stress

Wildland firefighting is done in conditions that subject the fire fighters to heat-stress injury. Make sure to start drinking fluids before your work assignment on the fireline. Once on the fireline, consume at least 1 quart (0.9 L) of water each hour. Sports drinks, also called carbohydrate/electrolyte replacement drinks, can be used as a water replacement. However, only 25 to 50 percent of your total water intake should be from a sports drink.

When fire fighters are mobilized from a temperate climate area to an area experiencing high temperatures, they will not be able to work as many hours as they are used to working. Fire fighters need to monitor their condition and make sure their work assignments do not compromise their health and safety.

Remember to set a responsible pace, take frequent breaks, and drink fluids, even if you are not thirsty. It is also extremely important to know the symptoms of heat stress and to monitor yourself and your colleagues.

Heat-stress disorders are divided into four categories: heat cramps, heat exhaustion, dehydration exhaustion, and heatstroke.

- Heat cramps are painful muscle cramps that may be caused by lack of fitness or failure to replace salt lost in perspiration. The victim should be treated by resting in the shade and drinking a carbohydrate/electrolyte replacement drink, lightly salted water, lemonade, or tomato juice.

- Heat exhaustion is caused by a failure to hydrate properly. The symptoms include general weakness, an unstable gait or extreme fatigue, wet clammy skin, headache, nausea, and possible collapse. The victim needs to be moved to the shade and given carbohydrate/electrolyte replacement drinks. The victim should also lie down with the feet elevated slightly.
- Dehydration exhaustion is another heat-stress disorder caused by a failure to replace water in the system. The symptoms are weight loss and excessive fatigue. Treatment includes increasing fluid intake and resting until body weight is restored.
- Heatstroke is a true medical emergency that is caused by a total collapse of the body's temperature-regulating mechanism. Call for emergency medical assistance at once because heatstroke is a life-threatening emergency. Immediately start cooling the victim by either immersing in cold water or soaking his or her clothing with cool water and fanning him or her.

The victim presents with the following signs and symptoms:

- Hot, often dry, skin
- A high body temperature (106°F [41.1°C] or higher)
- Mental confusion
- Delirium
- Loss of consciousness
- Convulsions

Fatigue and Sleep Deprivation

Fatigue affects fire fighters by reducing their alertness and situational awareness. It causes errors in judgment and makes people physiologically vulnerable. Fatigue of fire crews happens in many ways.

First, incident personnel often work too many consecutive hours, over too many consecutive days. Often, those same fire crews are then moved to yet another fire. These factors need to be mitigated. Guidance can be found in the NWCG PMS 902, *Standards for Interagency Incident Business Management.*

Fire fighters often pay little attention to their nutritional needs and do not hydrate properly. Both of these factors can cause fatigue because the energy requirements of incident personnel are high. Make sure you eat right and hydrate properly.

Lack of rest is a major factor that contributes to fatigue. Incident personnel must receive adequate rest. Incident base camps are usually busy places with a lot of noise, including from portable generators, which usually provide base-camp lighting. For security reasons, the areas are well lit, which can make sleep difficult. Night-shift personnel are particularly vulnerable to the effects of chronic sleep deprivation because they must sleep during the day.

Chronic sleep deprivation affects a fire fighter's mood and reduces cognitive ability and motor skills. Get quality sleep when off shift. In incident base camps, find a dark, quiet area to set up a tent or sleeping bag. Get as much sleep as possible.

Transporting fire fighters over long distances also contributes to fatigue. Fire fighters on large incidents are often flown across the country and then placed immediately on the fireline. Similarly, incident personnel are often requested at night or very early in the morning. Sometimes, personnel are on the road all night and, upon arrival at the incident, immediately assigned to the fireline. Rest guidelines must be adhered to. Follow your agency policy.

When checking in, be honest about a need for rest. If exhausted, do not be afraid to speak up. Let your supervisor know. Remember, this is a serious safety problem that needs to be addressed.

Carbon Monoxide Poisoning

Carbon monoxide is produced from the combustion of forest fuels or the burning of gas and oil in an internal combustion engine. In a wildland fire, carbon monoxide can coexist with smoke. It becomes a problem when fire fighters are exposed to heavy concentrations of carbon monoxide for an extended period of time. This invisible, odorless, and colorless gas is absorbed by the body. The carbon monoxide molecules attach to the hemoglobin molecules in the blood and displace oxygen needed by the body. Carbon monoxide is absorbed by the body at a rapid rate for the first hour of exposure, after which the rate drops slightly for the next 4 to 8 hours.

The symptoms of moderate carbon monoxide levels in the blood are a possible headache, nausea, and increased fatigue and drowsiness. The behavioral symptoms include increased impairment of alertness, unclear vision, problems with physical coordination, and errors in judgment of time. Very high concentrations of carbon monoxide in the bloodstream can cause convulsions and respiratory difficulty.

If you or one of your crew members is experiencing these symptoms, leave the work site and go to a carbon monoxide–free area to rest. It may also be necessary to administer oxygen to the person exhibiting these symptoms. It will take about 8 hours in an uncontaminated environment to purge the carbon monoxide from the body.

Mental Stress

Stress is the body's reaction to mental or emotional strain that results from adverse or very demanding circumstances. We respond to these demands in three stages:

- Alarm reaction. In this stage, the body recognizes the stressor and prepares for a fight-or-flight response. Stress causes physical, mental, and behavioral reactions.
- Resistance. In this stage, the body repairs any damage and may adapt to some stressors, such as heat, hard work, or worry. During our lifetime, we cycle through these first two stages on a daily basis.
- Exhaustion. When a stressor continues for a long duration, it will cause the body to remain in a constant state of readiness. Eventually you will be unable to keep up with the adrenaline and awareness demand, which leads to exhaustion.

Some stress is good. Performance and awareness improve with a moderate amount of stress; however, performance and awareness both decrease rapidly with high amounts of stress.

Stress can be caused by many factors. Many factors are within your control, but most are not. The first step toward developing controls to manage stress is to know the causes of stress and to know what your triggers are.

You must learn how to deal with your specific stress-reaction tendencies. As a fire fighter, it is certain that you will encounter stress. Know that it will occur. Identify situations where you know the stress levels will be high so you can be prepared when encountering those situations.

Hazardous attitudes and their effect on situation awareness are discussed earlier in this chapter.

Do a stress-recognition exercise with your team members. Ask all team members to define what they look like under high stress. The goal is to have individuals define how high-stress situations affect them and what signs show on them. When the exercise is completed, everyone will know what everyone else looks like when they are under high stress, which enables team members to offer each other assistance once that point has been reached.

LISTEN UP!

Fire fighters have a responsibility to recognize and minimize stress. It is important to look out for yourself and your team.

After-Action REVIEW

IN SUMMARY

- The 10 standard fire orders are designed to reduce fire fighter injuries and fatalities, and increase firefighting efficiency.
  1. Stay informed on fire weather and forecasts.
  2. Know what the fire is doing at all times.
  3. Base all actions on current and expected fire behavior.
  4. Identify safety zones and escape routes and make them known.
  5. Post lookouts when there is possible danger.
  6. Be alert, keep calm, think clearly, and act decisively.
  7. Maintain communication with crew members, supervisors, and adjoining forces.
  8. Give clear instructions and ensure they are understood.
  9. Maintain control of your forces at all times.
  10. Fight fires aggressively, providing for safety first.
- The 18 watch outs expand on the 10 standard fire orders and are warning signs that a potentially dangerous situation is developing.
  1. The fire has not been scouted and sized up.
  2. You are in a country you have not seen in daylight.

3. Safety zones and escape routes are not identified.
4. You are unfamiliar with local weather and other factors that may influence fire behavior.
5. You are uninformed on strategy, tactics, and hazards.
6. The instructions and assignments you were given are not clear.
7. You have no communication link with your crew or supervisor.
8. You are constructing a fireline without a safe anchor point.
9. You are building a fireline downhill, and there is fire below you.
10. You are attempting a frontal assault on the fire.
11. There is unburned fuel between you and the fire.
12. You cannot see the main fire and are not in contact with anyone who can.
13. You are on a hillside where rolling material can ignite the fuel below you.
14. The weather is getting hotter and drier.
15. The wind is increasing or changing direction.
16. You are getting frequent spot fires across the line.
17. Terrain and fuels make escape to your safety zone difficult
18. Taking a nap near the fireline.

- There are three categories of safety zones: the burn or the black, natural, and constructed
- LCES/LACES is a safety system created to simplify all of the safety-related components fire fighters need to remember.
  - Lookouts; Anchor points, Attitude, Awareness; Communication; Escape routes; Safety zones
- Understanding the importance of situational awareness, communication, attitude, and risk management can best sum up human factors.
  - Situational awareness is the process of making sure you understand what is going on around you
  - Effective communication is the key to situational awareness and assignment success
  - A positive attitude can be the difference between success and failure in difficult situations.
  - Risk management is the process of identifying, observing, and acting on potential risks.
- Medical incident reports outline details of an accident that occurs on a scene.
- REMS units are specially trained units that deploy to a fire scene where challenging circumstances present themselves in rescuing and transporting fire fighters.
- You must be familiar with the design, application, maintenance, and limitations of your PPE.
- Personal gear kits enable wildland fire fighters to be self-sufficient for 24-48 hours, and typically include: lighter or matches, pocket knife, compass, first aid kit, food and water, headlamp, and flashlight.
- Fire shelters are reflective cloth tents that protect from fire, heat, smoke, and ember showers.
  - Fire shelters can be deployed in place (static deployment) or on the run (dynamic deployment)
  - Be sure to follow proper safety and maintenance protocols for handling and deploying fire shelters to avoid damage to the equipment.
- It may be necessary to use a vehicle or building for an area of safe refuge. In all cases, evaluate whether it is better to be in a vehicle, a building, or a fire shelter as a barrier.
- Hazard trees are trees with structural defects that can cause part or all of the tree to fall.
- Certain terrain features cause fires to spread at a greater rate.
  - A chimney is a steep drainage on the side of a mountain; it channels the convective energy of the fire upwards.
  - A saddle is a low topography point between two high points in a mountain range.
  - A narrow canyon features canyon walls so close together that spotting can easily occur on the opposite canyon wall.

- Box canyons have steep walls where intense updrafts and extreme fires can occur.
- A midslope road is a road that appears in the middle of a slope or between slopes.

- Certain weather features can be hazardous to fire fighter safety and cause fires to spread.
  - The eddy effect is a rolling current effect that occurs as wind blows across a mountaintop. Spotting can occur below on the lee side of the ridge, opposite the main fire.
  - A thunderhead is a violent local storm produced by cumulonimbus clouds. Downdrafts and erratic winds occur when the thunder cell is over the fire.
  - A weather front is a boundary between two air masses of differing properties.
- Accounting for incident personnel is a major safety function at a wildland fire. The locations of all known incident resources must be tracked from the time of check-in at the fire to demobilization.
- Fire fighters working with aircraft in proximity could potentially be in the drop zone, and risk being injured or killed by falling retardant.
- Dozers are capable of dislodging huge rocks, which can roll downhill and injure crew members. Follow safety procedures when working with dozers.
- When dealing in areas with LPG tanks, note the location of the tanks before the fire arrives, clear flammable fuels around the tanks, and disconnect the tank's connection to the residence.
- Evaluate narrow bridges carefully. Some may be able to support the weight of a fire apparatus, and others may not. Check with local agencies to see if bridges have been denoted on preplans of the area.
- During the heat of a wildland fire, power lines may burn through and fall. If the wires are down upon arrival at a wildland fire, assume that they are live, and cordon off the area.
- While providing structure protection during a wildland fire, always check the garage or outbuildings for hazardous or flammable materials.
- It is important that you stay physically fit and hydrated, eat a balanced diet, and understand the symptoms associated with heat stress, carbon monoxide poisoning, and fatigue. Any of these conditions can debilitate a fire fighter working on a fireline.
- Being in good physical condition is an important factor in survival at a wildland fire.
  - It is also important to listen to your body and stop when it tells you to, especially in the field. Due to extreme, prolonged overexertion, fire fighters are at risk for rhabdomyolysis.
  - Eating well and staying hydrated is an important aspect of personal health and can reduce fire-fighter fatigue.
- With the close proximity of all the fire fighters in the small pop-up village or city, failure to regularly wash hands can lead to disease transmission.
- Wildland firefighting is done in conditions that subject the fire fighters to heat-stress injury. Make sure to start drinking fluids before your work assignment on the fireline.
- When checking in with your supervisor, be honest about a need for rest. If exhausted, do not be afraid to speak up.
- In a wildland fire, carbon monoxide can coexist with smoke. Be aware of the symptoms of carbon monoxide poisoning.
- Performance and awareness improve with a moderate amount of stress; however, performance and awareness both decrease rapidly with high amounts of stress.

KEY TERMS

10 standard fire orders A set of rules designed to reduce fire-fighter deaths and injuries, and increase firefighting efficiency.

18 watch outs An expansion of the 10 standard fire orders that identifies specific areas of concern in the wildland environment.

Deployment site A last-resort location where a fire shelter must be deployed; used when access to escape routes and safety zones has been compromised.

Escape route A preplanned route that fire fighters take to move to a safety zone.

Escape time The time it takes for crew members to make it to a safety zone.

Failure zone Any area that can be reached by any part of a falling tree.

Fire shelter A reflective cloth tent that offers protection from fire, heat, smoke, and ember showers in an entrapment situation.

Hazardous attitude A negative human characteristic that can affect situational awareness, crew safety, and assignment success.

Hazard tree A tree with a structural defect that can cause the entire tree or part of the tree to fall.

Incident within an incident (IWI) An accident or emergency that occurs on the scene of a fire.

Medical incident report (MIR) A written document or verbal report that outlines the details of an accident.

Personal protective equipment (PPE) Clothing and equipment designed to protect fire fighters in conditions that might otherwise result in death or serious injury.

Personnel accountability The ability to account for the location and well-being of one's crew.

Rapid extraction module support (REMS) unit A specially trained unit that specializes in basic- and paramedic-level medical care and high-angle rescue skills.

Risk management The process of identifying, observing, and acting on potential risks to hinder or eliminate them.

Safety zone An area cleared of flammable materials that is used as refuge when a fire has been determined to be unsafe.

Situational awareness The understanding of what is going on around one's self through observation and communication.

Snag A dead or dying tree that is still standing.

REFERENCES

National Fire Protection Association (NFPA). "Standard on Protective Clothing and Equipment for Wildland Fire Fighting." 2016. NFPA 1977. https://www.nfpa.org/codes-and-standards/all-codes-and-standards/list-of-codes-and-standards/detail?code=1977.

National Fire Protection Association (NFPA). "Standard on Selection, Care, and Maintenance of Wildland Fire Fighting Clothing and Equipment." NFPA 1877. https://www.nfpa.org/codes-and-standards/all-codes-and-standards/list-of-codes-and-standards/detail?code=1877&tab=nextedition.

National Wildfire Coordinating Group (NWCG). "Hazard Tree Identification: 6 Minutes for Safety." Modified November 2019. https://www.nwcg.gov/committee/6mfs/hazard-tree-identification.

National Wildfire Coordinating Group (NWCG). "L-180, Human Factors in the Wildland Fire Service, 2014." (NFES 2985). Modified October 23, 2019. https://www.nwcg.gov/publications/training-courses/l-180.

National Wildfire Coordinating Group (NWCG). "The New Generation Fire Shelter." PMS 411 (NFES 2710). March 2003. https://www.fs.fed.us/t-d/pubs/pdfpubs/pdf03512803/pdf03512803dpi300.pdf.

National Wildfire Coordinating Group (NWCG). "NWCG Report on Wildland Firefighter Fatalities in the United States: 2007–2016." PMS 841. December 2017. https://www.nwcg.gov/sites/default/files/publications/pms841.pdf.

National Wildfire Coordinating Group (NWCG). "S-130, Firefighter Training, 2008." (NFES 2731). Modified October 23, 2019. https://www.nwcg.gov/publications/training-courses/s-130.

National Wildfire Coordinating Group (NWCG). "S-404, Safety Officer, 2013." Modified October 23, 2019. https://www.nwcg.gov/publications/training-courses/s-404.

National Wildfire Coordinating Group (NWCG). "Wildland Fire Leadership Development Program." Modified December 11, 2019. https://www.nwcg.gov/wfldp.

Thorburn, W. R., and M. E. Alexander. "LACES versus LCES: Adopting an "A" for "Anchor Points" to Improve Wildland Firefighter Safety." 2001. https://bigbendvalleyvfd.org/bbvvfd-docs/LACES-versus-LCES.pdf.

United States Department of Agriculture (USDA) Forest Service. *Field Guide for Danger Tree Identification and Response.* 2008. https://osha.oregon.gov/OSHAPubs/reserve-trees.pdf.

United States Department of Agriculture (USDA) Forest Service. *Field Guide for Hazard-Tree Identification and Mitigation on Developed Sites in Oregon and Washington Forests.* 2014. https://www.fs.usda.gov/Internet/FSE_DOCUMENTS/stelprd3799993.pdf.

United States Department of Agriculture (USDA) Forest Service. "Hazard Trees." Accessed December 30, 2019. https://www.fs.usda.gov/visit/know-before-you-go/hazard-trees.

United States Fire Administration (USFA), Federal Emergency Management System (FEMA). "Command and Control of Wildland Urban Interface Fire Operations for the Structural Chief Officer." F0612. May 2019. https://apps.usfa.fema.gov/nfacourses/catalog/details/98.

Wildland Fire Lessons Learned Center (LLC) Risk Management Committee. "Safety Gram 2016." Accessed December 30, 2019. https://www.nwcg.gov/sites/default/files/safety-gram/rmc-sg-2016.pdf.

Wildland Fire Fighter in Action

It is July 22. You are part of a two-person crew assigned to clear a helispot high up on a ridge in the Salmon-Challis National Forest. You and your supervisor start your cutting operation at 1000. You are cutting large-diameter trees and heavy brush at the top of a drainage that could be identified as a chimney. There is no fire activity where you are, based on the current weather. But knowing what you know about this forest, weather can change quickly in the afternoon, and the fire history in this area has recorded several large fires that made long runs in a short amount of time. You are out in front, and several other handcrews are working their way up the ridge to your location. The initial assessment was that it was going to take only a couple of hours to complete your assignment and then you were to be picked up by a helicopter in the helispot you were creating to go back to fire camp.

Between 1000 and 1400, there was minimal fire behavior, which included a couple of spot fires. The 60 other people working the line were starting to experience increased fire behavior. The fire grew so intense that the crews had to pull back. At 1423, the incident commander contacted the forest supervisor and stated that the fire was about to make some pretty big runs as a result of the increased fire activity. Between 1423 and 1440, an active flaming front developed at the bottom of the drainage where you and your crewmember were cutting a helispot. At 1505, you start to notice significant smoke coming up the drainage below you. The smoke was not there before. What should you do next?

1. Which of the 10 standard fire orders are not in place?
   A. 1, 2, 3, 4, and 5
   B. 5, 6, 7, 8, and 10
   C. 1, 3, 5, 7, and 9
   D. 7 and 9
2. Which of the 18 watch outs are being violated?
   A. 3, 4, 5, 7, 8, 11, 12, 14, 15, and 17
   B. 1, 3, 5, 7, 9, 12, 14, 16, and 18
   C. 2, 4, 6, 8, 9, 12, 13, 17, and 18
   D. 1 and 5
3. Do you have LCES/LACES in place?
   A. Yes, but everything is dynamic and mitigated.
   B. No, there are several components that need to be reevaluated.
   C. Yes, all seems well except the ship is late.
   D. No, but it does not matter because the job is almost finished.
4. Which human factors might be affecting this situation?
   A. Complacency
   B. Invulnerability
   C. Macho attitude
   D. All of the above
5. How would you apply the risk management process in this situation, and would it change your efforts to complete your assignment?

Access Navigate for flashcards to test your key term knowledge.

CHAPTER 4

Wildland Fire Fighter I

Basic Wildland Fire Behavior

KNOWLEDGE OBJECTIVES

After studying this chapter, you will be able to:

- Describe the elements of the fire triangle. (pp. 89–90)
- Describe the methods of heat transfer. (pp. 90–91)
- Describe the types of wildland fuels:
    - Ground fuels (**NFPA 1051: 4.1.1, 4.5.6**, p. 90)
    - Surface fuels (**NFPA 1051: 4.1.1, 4.5.6**, p. 90)
    - Aerial fuels (**NFPA 1051: 4.1.1, 4.5.6**, p. 91)
- Describe how different fuel characteristics determine a fire's rate of spread:
    - Horizontal continuity (**NFPA 1051: 4.1.1, 4.5.6**, pp. 91–92)
    - Vertical fuel arrangement (**NFPA 1051: 4.1.1, 4.5.6**, p. 92)
    - Compactness (**NFPA 1051: 4.1.1, 4.5.6**, p 92-93)
    - Live-to-dead fuel ratio (**NFPA 1051: 4.1.1, 4.5.6**, pp. 92–93)
    - Fuel moisture (**NFPA 1051: 4.1.1, 4.5.6**, pp. 93–95)
    - Fuel temperature (**NFPA 1051: 4.1.1, 4.5.6**, pp. 95–96)
- Identify wildland fuel classification models. (**NFPA:1051: 4.1.1, 4.5.6**, pp. 97–98)
- Explain how topography can influence a wildland fire. (**NFPA 1051: 4.1.1, 4.5.6**, pp. 99–102)
- Explain how weather factors can influence a wildland fire. (**NFPA 1051: 4.1.1, 4.5.6**, pp. 102–115)
- Describe extreme fire behavior and fire behavior watch outs. (**NFPA 1051: 4.1.1, 4.5.6**, pp. 115–116)
- Identify extreme fire behavior watch outs. (**NFPA 1051: 4.1.1, 4.5.6**, pp. 115–116)

SKILLS OBJECTIVES

This chapter contains no skills objectives for Wildland Fire Fighter I candidates.

Additional Standards

- **NWCG PMS 210 (NFES 2943)**, *Wildland Fire Incident Management Field Guide*
- **NWCG S-190 (NFES 2902)**, *Introduction to Wildland Fire Behavior*
- **NWCG PMS 437**, *Fire Behavior Field Reference Guide*

You Are the Wildland Fire Fighter

It is your first day on the type 3 engine. You have just moved from one of the engines in the city because you wanted to be closer to the interface zone to learn more about urban interface firefighting. You go about your daily routine of checking out your apparatus and all the associated equipment to become familiar with the location of everything and the new types of equipment you will have to master. Before lunch, dispatch notifies your station that a Red Flag Warning has been issued for your response area.

At 1624, you and your crew are dispatched to a reported brush fire 2 mi (3.2 km) up the hill from the station in the forested part of your response area. As you leave the station, you notice a column of smoke that towers up to 7000 ft (2133.6 m) and shears off at the top, creating an anvil-looking shape. Surrounding the column is topography that presents with patchy fuels ranging from 1-hour all the way up to 1000-hour fuels. Gusty winds are coming from the east in your general direction. Your engine boss starts making assignments, talks about lookouts, escape routes, and safety zones. As you are actively listening to your supervisor, you all notice a fire whirl roll off the top of a ridge. Your supervisor asks you where you think the fire is going to move next.

1. How important will topography be in influencing this fire?
2. How important is fire weather and fire behavior to you now?
3. What steps will you need to follow to determine the probability of ignition?
4. You have heard the term *red flag* before. What does it mean to you now?
5. Do you think having a lookout in this situation is important?

Access Navigate for more practice activities.

Introduction

This chapter discusses the elements of the fire triangle, how heat transfers, and how wildfire spreads. It also discusses extreme fire behavior watch-out situations and how the alignment of fuels, weather, and topography can dramatically change fire behavior **FIGURE 4-1**.

The Fire Triangle

To safely and effectively extinguish wildland fires, fire fighters need to understand the various factors that cause fire ignition and affect the growth and spread of wildland fires. The **fire triangle** provides a model for explaining the behavior of wildland fires **FIGURE 4-2**.

The fire triangle consists of three elements: fuel, oxygen, and heat. First, a combustible fuel must be present. Second, oxygen must be available in sufficient quantities to feed the fire. Third, a source of ignition (heat) must be present. If we graphically place these three components together, the result is the fire triangle. Fire will burn until one or more of these elements are removed.

Fuel

The first side of the fire triangle is fuel. **Fuels** are materials that store energy. The primary fuel for wildland and ground-cover fires is the vegetation (grasslands, brush, and trees) in the area. The amount of fuel in an area may range from sparse grass to heavy underbrush and large trees. Some fuels ignite readily and burn rapidly when dry, whereas others are more difficult to ignite and burn more slowly. Wildland fuels are discussed in more detail later in this chapter.

Oxygen

The second side of the fire triangle is oxygen. In wildland fires, which are usually free burning, oxygen is rarely an important variable in the ignition or spread of the fire. Unlike in structure fires, oxygen is available in unlimited quantities in these open-air fires. Nevertheless, air movement around wildland fires will influence the speed at which the fire moves. Wind blowing on a wildland fire brings more oxygen to the fire, accelerates the process of combustion, and influences the direction in which the fire travels.

Heat

The third side of the fire triangle is heat. Sufficient heat must be applied to fuel in the presence of adequate oxygen to produce a fire. Wildland fires may be ignited by natural, accidental, or intentional causes.

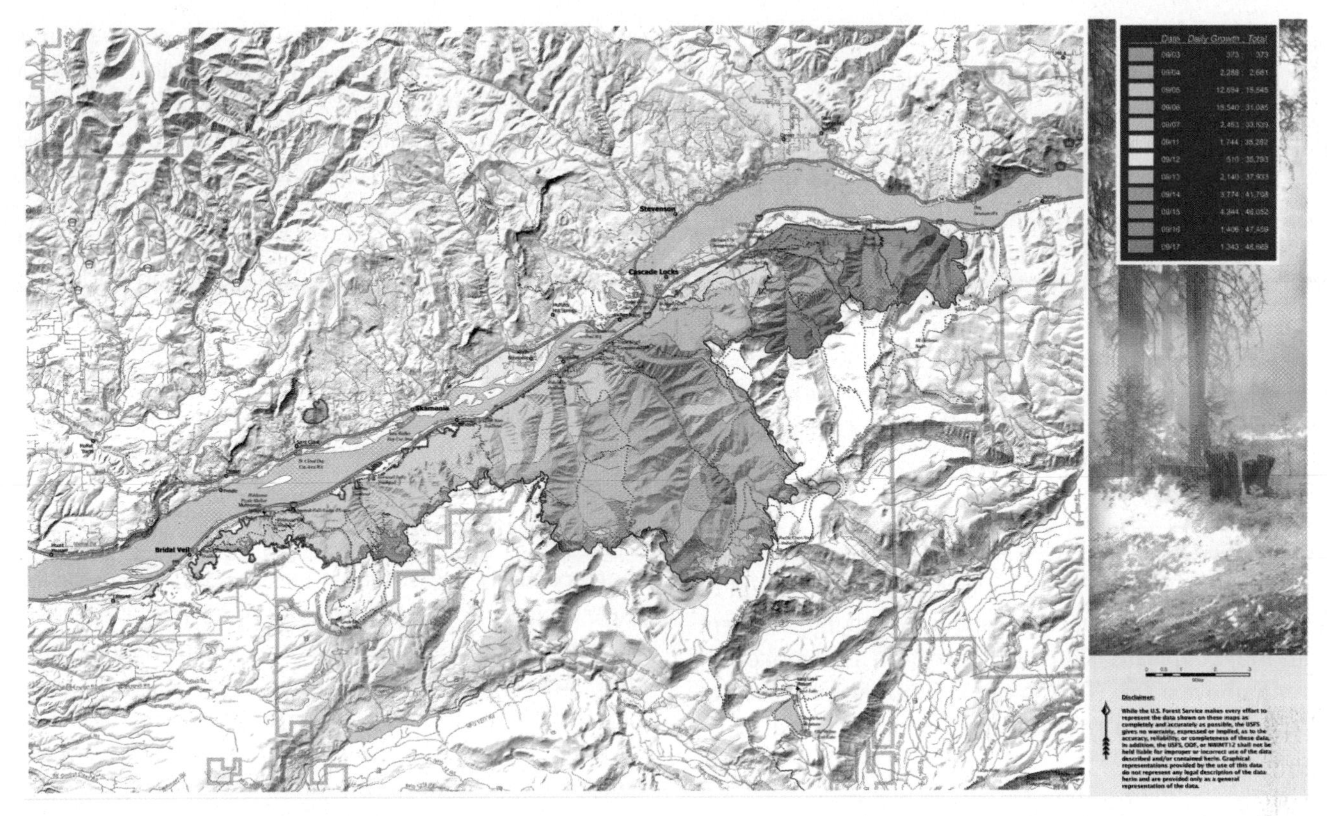

FIGURE 4-1 This map depicts the progression of fire in the 2017 Eagle Creek wildfire in Oregon and Washington.
Courtesy of Shawn Bushnell NW IMT 12 GISS, USDA.

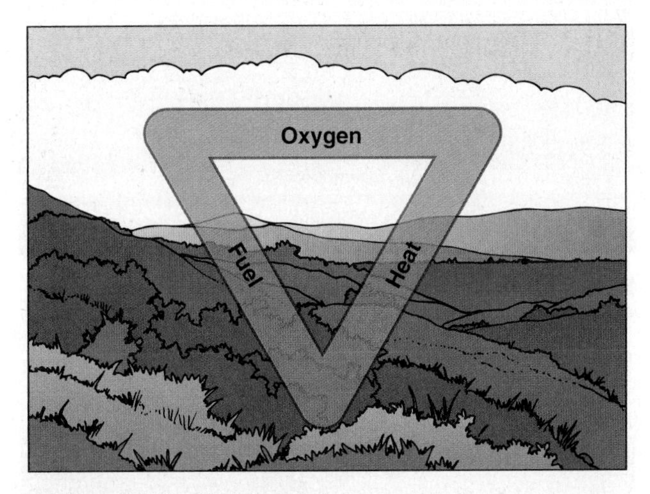

FIGURE 4-2 The fire triangle.

Lightning is the source of almost all naturally caused wildland fires. Weather factors are discussed later in this chapter.

Sometimes, a variety of accidental causes may result in wildland fires. These causes may include discarded smoking materials, improperly extinguished campfires, and downed electrical wires. Sometimes, intentional acts of arson are a cause of wildland fires.

LISTEN UP!

Break the fire triangle by removing one of the sides. Remove fuel by clearing space. Remove oxygen by restricting the oxygen supply, for example, with foam or dirt. Remove heat by applying, for example, water or dirt.

Fire Spread and Heat Transfer

Fire follows the laws of physics. To predict how and where a fire may spread and the effects of your chosen tactics, you need to understand the basics of fire behavior and the influence of fire spread and heat transfer. **Heat transfer** is the process by which heat is imparted from one body to another. There are three primary mechanisms by which heat is transferred: conduction, convection, and radiation.

Conduction

Conduction is the process of transferring heat to and through one solid to another. This transfer occurs when two solids are in contact with each other and one has a higher temperature than the other. Think of conduction as a metal spoon in a hot drink. Conduction occurs when you touch the spoon handle and feel its warmth. Heat transfers from the liquid, through the spoon, and to your skin upon direct contact.

Metals are better conductors than wood, meaning that heat will not travel through wood easily. The conduction process is less of a factor in fire spread on wildland fires compared to convection and radiation.

Convection

Convection is the transfer of heat by the movement of a gas or liquid. Convection happens when lighter warm air moves upward. Think of convection as a smoke column above a fire. The smoke column is composed of hot gases. As the smoke moves, it can dry and ignite other fuels.

Radiation

Radiation is the transfer of heat in a straight line through emission of energy. The sun radiates energy to Earth, for example. You can feel radiant heat as you stand in the sunlight or near a campfire. Radiant heat can dry surrounding fuels and sometimes ignite them.

Wildland Fuels

Fuels are materials that store energy. Think of the vast amount of heat that is released during a large fire. The energy released in the form of heat and light had been stored in the fuel before it was burned. The release of the energy in a gallon of gasoline, for example, can move a car many miles down the road. A wildland fuel can be any organic material, either living or dead, that can ignite and burn. **Organic material** is any natural matter, such as leaves, rotting logs, or deep duff (artificial material includes manmade items, such as houses, or fences).

Fuels can be loosely grouped into three levels: ground, surface, and aerial.

Ground Fuels

Subsurface fuels, or **ground fuels**, are combustible materials lying on or beneath the ground or subsurface litter. Ground fuels include root systems, deep duff, rotting buried logs, and other woody materials in various states of decomposition **FIGURE 4-3**. Fires in ground fuels are slow moving and do not play a major role in fire spread. They become important in line construction and mop-up operations because they can carry heat to more flammable fuel sources. Root systems have carried fire across control lines, and **smoldering** fires (fire burning without flame and barely spreading) below ground have been known to hold fire through the winter. In the Great Lakes states, the Pacific Coast, the Southeast, and Alaska, organic soils, especially during drought periods, play an important part in fire spread. Consider this especially when constructing a fireline and mopping up fires.

Surface Fuels

Surface fuels are combustible materials lying on or near the surface of the ground and a major cause of wildland fire spread. Surface fuels include ground litter, such as leaves, needles, bark, grasses, tree cones, and brush up to 6 ft (1.8 m) in height **FIGURE 4-4**. These fuels act as kindling for heavier fuels. Most fires start and spread in surface fuels.

FIGURE 4-3 Ground fuels.

FIGURE 4-4 Surface fuels.

Aerial Fuels

Fuels that are greater than 6 ft (1.8 m) in height are considered **aerial fuels** **FIGURE 4-5**. Trees, branches, foliage, and tall brush fall into this category. The main concern with these fuels is the crown, or canopy, closure, that is, how close the fuels are to each other. Timber stands that have open canopies usually will support the spread of surface fires with individual torching of trees. **Torching**, or candling, occurs when tree foliage burns from the bottom to the top. Torching is intermittent and may be an indicator that fire behavior is increasing.

Closed canopy stands greater than 6 ft (1.8 m) in height that occur in tall brush or trees present the greatest opportunity for running **crown fires**, or fires that run from treetop to treetop or shrub top to shrub top **FIGURE 4-6**. Crown fires burn independently from the surface fire. They are generally the fastest spreading of all wildland fires and burn with the greatest intensity.

FIGURE 4-5 Aerial fuels.
© Andrew Orlemann/Shutterstock.

FIGURE 4-6 Crown fire.
© David McNew/Getty Images News/Getty Images.

Crown fires are not confined just to trees; they occur in the crowns of other aerial fuels as well. Examples of brush models where crown fires could occur include oak brush, as found in the southern Rocky Mountains; an old-age chaparral bed found in Southern California; or palmetto and gallberry, as found in the coastal plains of the Southeast. Assess your region and identify fuels that have this potential. Do not be caught by surprise.

LISTEN UP!

Old-age chaparral is chaparral that has not burned for a long period of time. In general, older stands of chaparral can carry fire more efficiently than younger stands.

LISTEN UP!

Of special concern are aerial fuels that have been heated and dried by ground fires. Under certain conditions, such as high winds, these pretreated aerial fuels can burn explosively.

Fuel Characteristics

The speed at which a fire moves away from the site of origin is called the **rate of spread**. Rate of spread is usually expressed in chains per hour. Fuel characteristics determine a fire's rate of spread and how intensely the fire burns.

Horizontal Continuity

Fuel continuity refers to the degree of continuous distribution of fuel particles in a fuel bed. **Horizontal continuity** refers to the way that wildland fuels are distributed at various levels. Continuity of the fuel bed determines where the fire will spread and whether it will travel along the surface fuels, the aerial fuels, or both.

Horizontal distribution of fuel refers to patchy or continuous fuel beds. Patchy fuel beds are areas of widely distributed fuels **FIGURE 4-7**. A strong wind or steep slope is required for a fire to travel through these fuel beds. Continuous or uniform fuel beds provide ample fuels for burning and the opportunity for the fire to spread rapidly, especially when large concentrations of continuous fire fuels are present **FIGURE 4-8**. Horizontal continuity also applies to aerial fuels, such as treetops.

FIGURE 4-7 Patchy fuels.
Courtesy of Joe Lowe.

FIGURE 4-8 Continuous fuels.
Courtesy of National Interagency Fire Center.

Vertical Fuel Arrangement

Vertical fuel arrangement refers to the different heights of the fuels that are present at a fire. Vertical fuel arrangement can allow a fire to reach different fuels, such as when fires moving through surface fuels reach any available aerial fuels. The term **ladder fuels** defines this stepping effect **FIGURE 4-9**. Ladder fuels allow a fire to climb from ground fuels to surface fuels to aerial fuels. Fires usually start in light fuels, such as grasses, and

LISTEN UP!

Standing agricultural vegetation, such as corn, wheat, or sugar cane, has its own burning characteristics. Be familiar with local fuels and how they burn. This knowledge will help you make better decisions on the initial attack of the fire.

FIGURE 4-9 Ladder fuels.
Courtesy of Joe Lowe.

then spread up into the brush. Once the brush starts burning, the potential exists for the fire to move into aerial fuels.

Compactness

The spacing between fuel particles can be defined as their **compactness**, or density. The more loosely compacted the fuel bed, the more easily oxygen can move in and around the fuels **FIGURE 4-10**. Usually, the rate of spread is greater as a result of an increased oxygen supply. Tightly compacted fuels usually have slower rates of spread because oxygen cannot move freely around the fuels **FIGURE 4-11**.

Live-to-Dead Fuel Ratio

The amount of dead fuels present in a fuel bed is important when evaluating fire potential **FIGURE 4-12**. This concept is called the **live-to-dead fuel ratio**. The dead fuels increase as the fuel bed matures. Overmature fuel complexes, such as frost kill, prolonged drought, insect damage, previous fire damage, disease,

FIGURE 4-10 Loosely compacted fuel.
Courtesy of Joe Lowe.

FIGURE 4-11 Tightly compacted fuel.
Courtesy of Joe Lowe.

FIGURE 4-12 Live-to-dead fuel ratio.
© Keldridge/Shutterstock.

FIGURE 4-13 Cheatgrass is an example of herbaceous vegetation.
Courtesy of Joe Lowe.

wind, and snow increase the amount of dead fuels present in a fuel bed. Dead fuels move the fire and assist it in heating the live fuels to their ignition point. The more dead fuels that are present, the greater the chance is of the living fuels igniting.

Fuel Moisture

Fuel moisture is an important factor in how easily a fuel ignites, burns, and spreads. In general, when fuel moisture is high, a fire will burn poorly. If the fuel moisture is low, fires start easily and then spread and burn rapidly.

There are two types of fuel moisture: live fuel moisture and dead fuel moisture. **Live fuel moisture** is the ratio of the amount of water to the amount of dry plant material in living plants. Live fuels have a much higher moisture content than dead fuels and are more affected by seasonal changes than daily changes in weather. Live fuel moisture is highest when foliage is fresh and the plant is in its growing cycle. It is usually at its lowest point in autumn or late summer.

Living fuels can be categorized into either woody or herbaceous plants. Woody plants produce wood and usually have stems that are covered in bark. Examples include trees and shrubs. Herbaceous plants are soft and do not develop woody stems or tissue. Examples include flowers and grasses **FIGURE 4-13**. Before a fire can grow in herbaceous fuels, at least one-third of the plant must be dead. These fuels usually cure in early summer (although some never cure) and are responsible for much of the fire spread in other fuel types. Perennial vegetation regrows each year from its base. Woody vegetation reacts to seasonal changes, and, normally, live fuel moisture decreases as the growing season comes to an end **FIGURE 4-14**. Of most concern are the small twigs, leaves, and needles that would be consumed by the **flaming front zone** of the fire, where continuous flaming combustion is primarily taking place.

LISTEN UP!

Every fire season is different. It is important to understand and differentiate between woody and herbaceous plant growing cycles.

Dead fuel moisture is the moisture found in dead plants, such as perennial grasses. Fine dead fuels are affected by daily weather, as opposed to seasonal changes. They react rapidly to precipitation or changes in the relative humidity and temperature. If it rains, they reach their saturation point quickly. When humidity levels are high, these fuels absorb moisture. When humidity levels are low, these fuels release moisture. These properties make fine dead fuels a good fire conductor.

The *timelag principle* demonstrates how long it takes the fine dead fuel to reach equilibrium moisture content (i.e., how long it takes grass to absorb rainwater or atmospheric moisture to capacity). This principle uses stock fuel diameters as a reference **TABLE 4-1**.

FIGURE 4-14 Manzanita is an example of woody vegetation.
Courtesy of Joe Lowe.

When a fuel type approaches 60 to 70 percent of its equilibrium moisture content with the surrounding air, the equilibrium change is said to be complete. Fire fighters are most concerned with fine dead fuels that are less than 0.25 in. (0.6 cm) in diameter. These 1-hour fuels are most susceptible to changes in relative humidity and are the prime carriers of a wildland fire.

The size and shape of wildland fuels are also an important consideration. Smaller fuels have a greater surface-to-volume ratio. Smaller fuels ignite easier because less heat is required to drive off the moisture in the fuels and raise them to their ignition temperature.

Think of lighting a fireplace or campfire. Small fuels are needed to ignite the larger pieces of wood. Larger fuels with a smaller surface-to-volume ratio are harder to ignite and do not sustain combustion as readily. Fuel size and surface area examples are shown in **FIGURE 4-15**.

Many wildland fuels contain chemicals and minerals that can enhance or retard combustion. Volatile substances, such as oil, resin, wax, and pitch, are

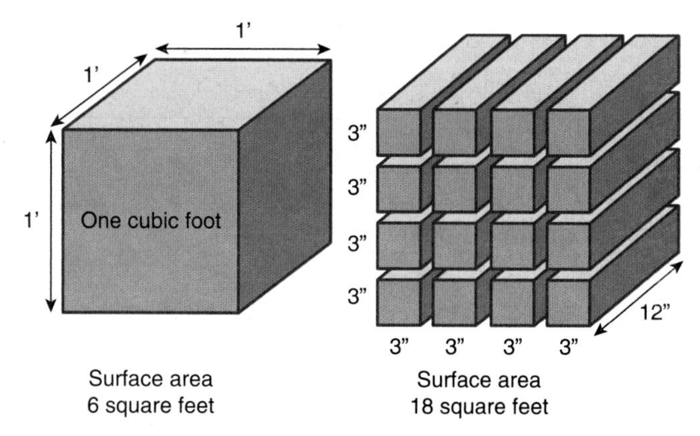

FIGURE 4-15 Surface-to-volume ratio. All fuel pictured here, as blocks, has a volume of 1 ft^3 (0.03 m^3). However, because of their shape and size, the surface area varies greatly.

TABLE 4-1 Timelag of Fine Dead Fuel Types

Fine Dead Fuel Type	Fuel Diameter	Time It Takes to Reach Equilibrium Moisture Content (Timelag)	Examples
1-hour fuels	Less than 0.25 in. (0.6 cm)	1 hour	Grass, leaves, needles
10-hour fuels	0.25 to 1 in. (0.6 to 2.5 cm)	10 hours	Small branches, twigs, brush
100-hour fuels	1 to 3 in. (2.5 to 8 cm)	100 hours	Branches, small trees, brush
1000-hour fuels	3 to 8 in. (8 to 20 cm)	1000 hours	Trees, brush piles

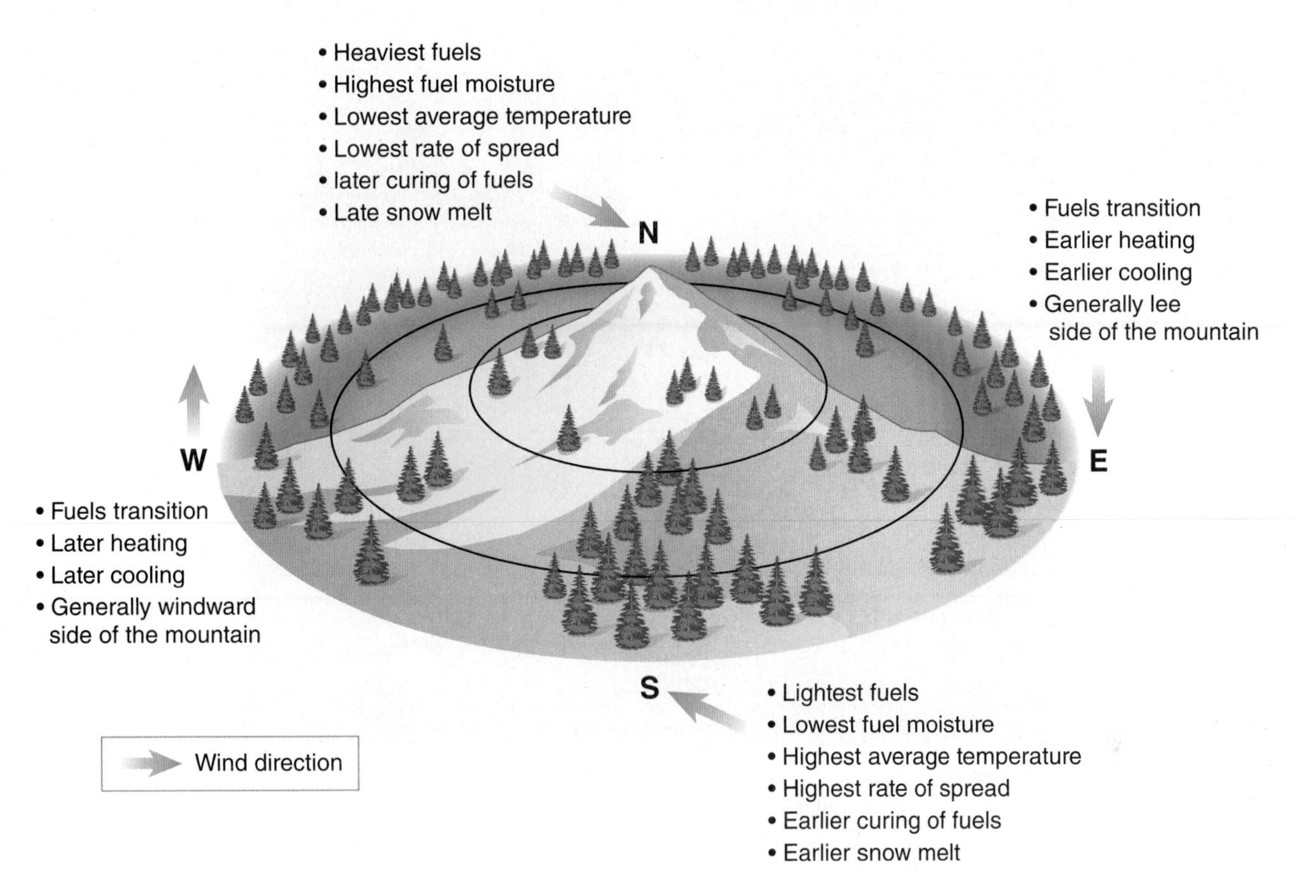

FIGURE 4-16 Fuel types, fuel temperature, and moisture related to aspect.

present in many wildland fuels throughout the world. They have been responsible for many high-intensity, rapidly spreading fires. Conversely, fuel with high mineral content retards the spread and intensity of the fire.

Fuel Temperature

Fuel temperature also plays a big part in fuel moisture levels. Each day, the sun heats Earth's surface and the fuels close to it. As the ground surface and heated fuels become warmer, they begin to warm the air close to the ground. This warming of the air causes a decrease in humidity and wind and an increase in air temperature.

The first slopes to be heated by the sun's energy are those facing east. The direction of the slope's face in relation to the cardinal compass points is called its **aspect**. Fuel types also vary by aspect **FIGURE 4-16**. Each aspect receives its peak solar influence at a certain time of the day. Eastern aspects (east-facing slopes) peak at around 9:00 AM, and western aspects (west-facing slopes) peak around 3:00 PM. Both northern and southern aspects (north-facing and south-facing slopes) peak at around noon; however, northern aspects have the highest fuel moisture and the lowest daytime temperatures. Peak burning times may vary due to seasonal changes. You can expect the fine dead fuels in shaded areas to have higher moisture contents, as much as 8 percent, than those found in direct sunlight **FIGURE 4-17**.

When present, smoke, haze, and clouds interfere with the radiant energy from the sun, thus cooling the fuels on the ground's surface. Changes in elevation also affect fuel moisture. The higher one goes on a mountainside, the cooler it gets. Cooler temperature means higher relative humidity and higher fuel moisture. Under normal atmospheric conditions, each 1000-ft (305-m) change in elevation represents a 3.5-degree temperature change **FIGURE 4-18**.

Slope also has an effect on fuel moisture; surfaces perpendicular to the sun absorb more of the solar energy. South- and west-facing slopes receive considerably more of the incoming solar radiation than those facing north or east; thus, they have higher temperatures and lower fuel moisture. The angle of the sun's radiation changes throughout the day and with the seasons.

Wind can also affect fuel moisture levels. During calm periods, the air that is saturated with water vapor significantly increases the moisture of fine fuels. Fuels reach equilibrium moisture content with the air at a faster rate when it is windy. Stronger winds during the day may prevent surface temperatures from rising and

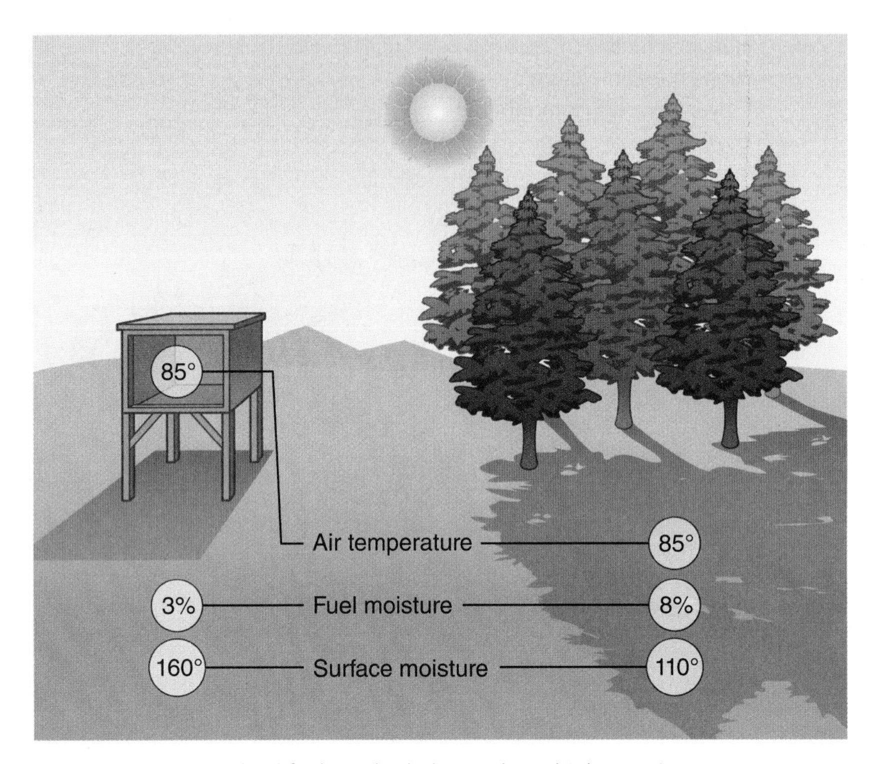

FIGURE 4-17 Fine dead fuels in shaded areas have higher moisture contents than those in direct sunlight.

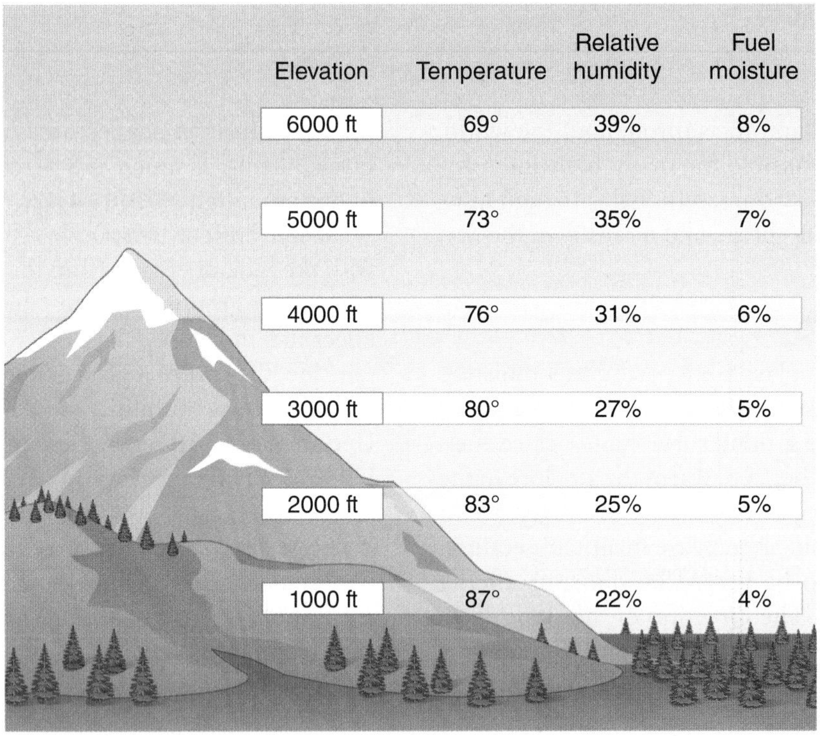

Elevation	Temperature	Relative humidity	Fuel moisture
6000 ft	69°	39%	8%
5000 ft	73°	35%	7%
4000 ft	76°	31%	6%
3000 ft	80°	27%	5%
2000 ft	83°	25%	5%
1000 ft	87°	22%	4%

FIGURE 4-18 Elevation affects temperature and fuel moisture. This is a daytime example.

thus actually cool the fuels. This cooler air raises the relative humidity.

Night winds, frequently seen in canyons and valleys as colder air drops, can cause a turbulent mixing of air, which may prevent surface temperatures from reaching the dew point. A wind that causes fuels to dry rapidly is called a foehn wind. Foehn winds are discussed later in this chapter.

LISTEN UP!

Grasses are a primary carrier of wildland fires.

Fuel Classification Models

For the purposes of this text, we will be basing fuel discussions on the original 13 fuel models described by the National Wildfire Coordinating Group. The original 13 models were designed to estimate and calculate wildfire behavior under peak fire conditions. Fire fighters are usually battling wildfires under peak fire conditions.

In June 2005, an additional 40 stand-alone standard fuel models were added. These models were developed to facilitate analysis of fire use and fuel modification treatments. As fire science and fire modeling software evolve to best represent the different fuel types, fuel models have been added to the software to increase the success rate of predictions. Adding these fuel models to the modeling software in turn assists fire fighters by increasing their situational awareness and allowing very accurate fire behavior predictions, which are used for critical and tactical decision-making priorities.

Grass Models

Grasses are a primary carrier of wildland fires **FIGURE 4-19**. The depth of grasses varies between 1 and 2.5 ft (0.3 to 0.8 m). Fires in grasses burn rapidly, and the fuel is usually entirely consumed. Wind greatly affects the spread of fire in grass models. These fine fuels also respond quickly to changes in relative humidity. Intensity levels are usually low compared to other fuel models. Fires in grass fuel model 3 are the most intense and have higher rates of spread than fires other fuel models. Grasses are found in most areas; however, they are more prevalent in desert and range areas. Fire behavior fuel models 1, 2, and 3 are grass models, as defined in Appendix B: *Fire Behavior* of the NWCG Fireline Handbook.

Shrub Models

Fuel bed depths from 2 to 6 ft (0.61 to 1.8 m) are found with shrub models **FIGURE 4-20**. A good example of this type of fuel is the mixed chaparral fuels found in the Southwest. Most of these fuels are less than 1 in. (15.2 cm) in diameter, with small leaves. Some of the fuels in this grouping contain volatile compounds that allow the fuel to burn in higher fuel moisture. A good example of this is model 7 fuels, such as the palmetto and gallberry, which are found in Florida. Fire intensity levels are higher than those found in grass models. Shrub fuel models are 4, 5, 6, and 7, as defined by the fire behavior fuel model descriptions found in Appendix B: *Fire Behavior* of the NWCG Fireline Handbook.

Timber Litter Models

Timber litter models are composed largely of leaves; mixed litter; and, occasionally, twigs and large branch wood **FIGURE 4-21**. The fuel bed compactness can range from loose to tight. Fires generally run through the surface and ground fuel **FIGURE 4-22**; however, when the fire finds a large concentration of dead fuel, crowning out and torching are possible. Timber litter models are 8, 9, and 10, as defined by the fire behavior fuel model descriptions in Appendix B: *Fire Behavior* of the NWCG Fireline Handbook. Flare-ups can be expected for fires in fuel model 8 if the fire finds a heavy centration of fuel. Fires in fuel models 9 and 10 can contribute to possible torching out of trees, crown fire activity, and spotting.

FIGURE 4-19 Firing operation in a grass model.

Courtesy of Joe Lowe.

FIGURE 4-20 Fire in a shrub model.

Courtesy of California Department of Forestry and Fire Protection.

FIGURE 4-21 A timber litter fuel model.
Courtesy of Joe Lowe.

FIGURE 4-22 A surface fire burning in a stand of timber.
Courtesy of Joe Lowe.

Logging Slash Models

Logging slash is the debris left after a logging operation **FIGURE 4-23**. Slash can also be produced by thinning and pruning operations. Fires in these fuel models tend to have moderate to higher rates of spread. They also tend to produce moderate- to higher-intensity fires. Fires burning in areas where many trees have been blown down, such as along the East Coast after the passage of a hurricane, will burn like models 11, 12, and 13, which represent the logging slash models, as defined by the fire behavior fuel model descriptions found in Appendix B: *Fire Behavior* of the NWCG Fireline Handbook.

Each fuel model burns at a different rate of spread and produces varying fireline intensities. **TABLE 4-2** summarizes these fuel models. Evaluate your area, and identify the different fuel types. Predict what type of fire intensities and rates of spread you would expect to see. This evaluation will better prepare you to make appropriate tactical decisions in a wildland fire.

FIGURE 4-23 Logging slash.
Courtesy of National Interagency Fire Center.

TABLE 4-2 Fuel Classification Models

Model Type	Model Number	Description
Grass models	Model 1	Short grass
	Model 2	Timber
	Model 3	Tall grass, agricultural vegetation
Shrub models	Model 4	Chaparral
	Model 5	Brush
	Model 6	Dormant brush, hardwood slash
	Model 7	Southern rough
Timber models	Model 8	Closed timber litter
	Model 9	Hardwood litter
	Model 10	Timber litter and understory
Logging slash models	Model 11	Light logging slash
	Model 12	Medium logging slash
	Model 13	Heavy logging slash

Modified from *NWGC Fireline Handbook*, Appendix B: Fire Behavior, April 2006, https://www.nwcg.gov/sites/default/files/products/appendixB.pdf.

DID YOU KNOW?

In recent years due to warmer winters, our pine forests have come under attack from a mountain pine beetle epidemic. In 2011, it was estimated that the Rocky Mountain region of the U.S. Forest Service had 250,000 acres of wildland/urban interface threatened by the mountain pine beetle. The beetles tunnel into tree trunks, where they mate and lay eggs under the bark. This eventually kills the tree. The death of large areas of pine trees poses new hazards for fire fighter safety, including heavy fuel loads, unpredictable fire behavior, abundant snags, and microclimate changes.

Much research is being done on fires in these forests since the epidemic. Fire hazard and fire behavior are still a matter of debate. Some questions will require more study, but here are some management implications to be aware of:

- Standard fire behavior model predictions are highly unreliable in mountain pine beetle forests since the epidemic.
- Expect more intense fire behavior and more rapid transitions from surface to crown fires.
- Expect greater than normal spotting distances in red needle–stage forests.
- Enforce larger safety zones when conducting fire operations in red needle–stage forests.
- Fire operations on the ground may not be feasible due to longer flame lengths and greater heat release.

Topography

Topography represents the physical features of the land in a given region; it can be divided into four categories: slope, aspect, elevation, and terrain features.

Slope

Fires on steep slopes spread more rapidly because the flames are closer to the fuels, thus preheating (drying out the fuels so fire can spread easier) any upslope fuels **FIGURE 4-24**. These upslope fuels are then more susceptible to spotting from aerial firebrands. **Spotting** occurs when small burning embers and sparks are carried by the wind and land outside the fire perimeter, which starts new fires beyond the zone of direct ignition. As the fire burns upslope, the products of combustion move in an upslope direction, and oxygen rushes in at the fire's base to feed the growing fire. As the slope factor increases, the rate of spread and flame length also increase. Fires on steep slopes burn in wedge-shaped patterns. If wind is present, fire may burn across the hill instead of in the wedge-shaped pattern. This is because the effect of wind is greater than the effect of slope.

On steep slopes, you also need to watch for burning, rolling material because it may roll across the control line and start a fire below it, potentially trapping you between two fires.

Where the fire started on a slope is also an important factor when considering the rate of spread. Fires that start close to the base of a slope have more available fuel to burn and preheat upslope fuels. Fires on a slope are also influenced by the normal upslope daytime winds. Fires starting near the top of a slope generally do not have as much available fuel to burn and therefore do not grow to be large in size.

The steeper the slope, the greater all three heat transfer methods combine to influence fire spread.

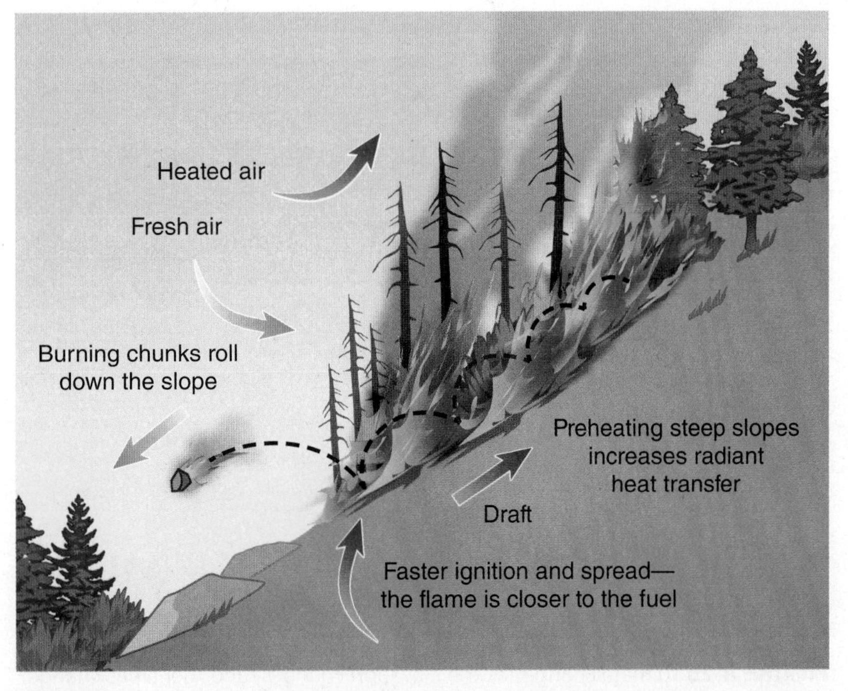

FIGURE 4-24 Preheating upslope fuels.

LISTEN UP!

A firebrand is any source of heat that becomes airborne through wind, convection currents, or gravity and causes additional fires as it comes into contact with unburned fuels.

Aspect

Aspect (the direction a slope faces) is an important factor to consider when evaluating fire behavior. Each aspect heats at different times of the day. This heating raises the fuel temperature, reduces the humidity (see the previous discussion under the section titled Fuel Temperature), and starts the upslope winds. In addition, each of these aspects has a different type of fuel. A good example of this type of fuel variation occurs on a southern aspect, which generally has lighter fuel types. Fires on these slopes usually become large fires quickly. Northern aspects, however, have the heaviest fuels, highest fuel moisture, and the lowest average temperature of all the aspects.

Elevation

Elevation is measured using mean sea level. **Mean sea level** is the average height of the sea for all stages of the tide over a 19-year period. Elevation above mean sea level influences fuel types and their ability to burn. There are also climate changes, especially at higher elevations, such as in mountain ranges **FIGURE 4-25**. On the lower slopes, grass models are the predominate fuel types. As the elevation increases, shrub models start developing. Farther up the slope, timber models develop. Above the tree line, found at very high elevations, there is an absence of fuels, if any vegetation at all as a result of the extreme weather factors found there. Besides a difference in fuel types, there is also a decline in temperature as elevation increases. The accepted figure for temperature change is a 3.5-degree decrease per each 1000-ft (305-m) increase in elevation. It is important to remember that, as the temperature decreases, there is a corresponding change in the relative humidity. The cooler the temperature, the higher the relative humidity. Higher relative humidity in fine dead fuels makes these fuels harder to ignite.

Terrain

Certain terrain features cause fires to spread at a greater rate. A **chimney** is a terrain feature that has a channeling effect on the convective energy of the fire **FIGURE 4-26**. In a chimney, extreme rates of fire spread and spotting are likely to occur. **Saddles** are low topography points (also called depressions or passes) between two high topography points in a mountain range **FIGURE 4-27**. The wind speed generally increases as the wind passes through these constricted areas and then spreads out on the lee side. A circular-like flow of air or water drawing its energy from a flow of much larger scale, called an **eddy**, generally occurs on the lee side. **Narrow canyons** are steep canyon walls that are close together. Narrow canyons are a problem for fire crews because a fire on one canyon wall can easily

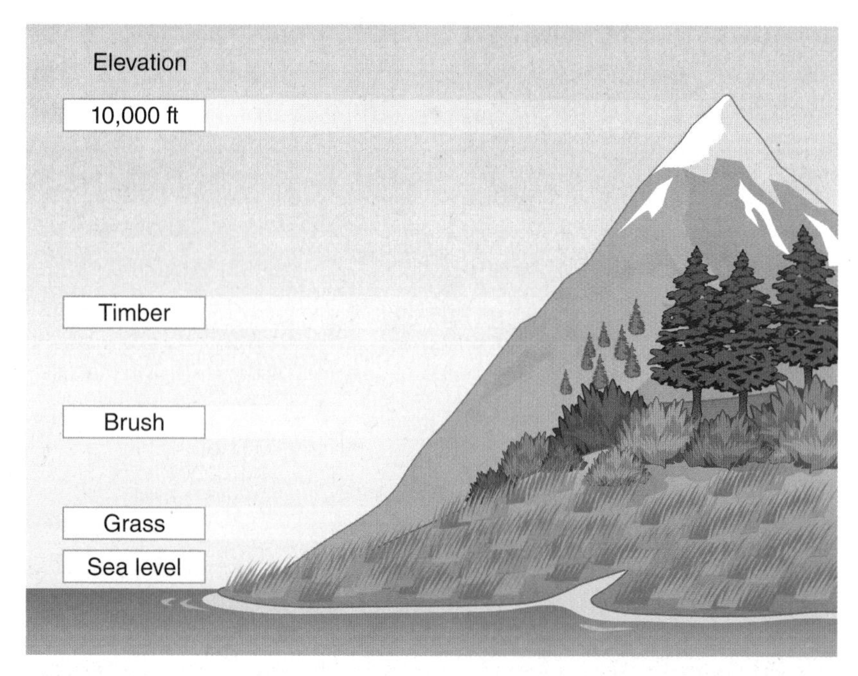

FIGURE 4-25 Climate changes as the elevation changes, and fuel types change as a result. Note that there is an absence of fuels above the tree line.

A

B

FIGURE 4-26 A. A chimney running to the top of a mountain. **B.** Chimney effect.

A: Courtesy of Joe Lowe.

FIGURE 4-27 Saddle.

Courtesy of California Department of Forestry and Fire Protection.

FIGURE 4-28 Narrow canyon.

© Serg Zastavkin/Shutterstock.

A

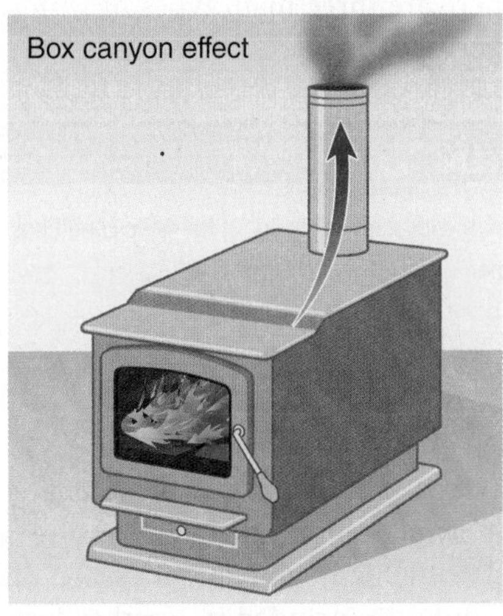

B

FIGURE 4-29 A. Box canyon. **B.** Box canyon effect.

A: © Marko Ignjatovic/Alamy Stock Photo.

jump to the opposite canyon wall as a result of radiant heat or spotting **FIGURE 4-28**. This proximity can also cause area ignition. **Area ignition**, also called simultaneous ignition, occurs when several individual fires throughout an area cause the main body of fire to produce a hot, fast-spreading fire condition. **Box canyons**, also known as dead-end canyons, are canyons with three steep walls, which create very strong upslope winds **FIGURE 4-29**. These upslope winds create rapid fire-spread conditions. As the convective column rises, air is drawn up from the canyon bottom. Box canyons are areas where extreme fire can occur.

Safety hazards concerning these terrain features are discussed in greater detail in Chapter 3, *Safety on Wildland Fires.*

LISTEN UP!

As the temperature decreases, there is a corresponding change in the relative humidity.

Weather

Weather is the state of the atmosphere over the Earth's surface. Weather factors are of the greatest concern to a fireline because they are the most difficult to predict and have the greatest effect on the course of a wildland fire. Many different weather factors can influence a wildland fire. These factors are discussed next.

Wind

Wind drives the fire, gives it direction, causes spotting, preheats fuels ahead of the fire, and brings a fresh supply of oxygen to a fire. Winds are responsible for a rapid rate of spread on wildland fires. Wind direction is defined as the direction from which the wind is coming; an east wind blows from the east, for example. There are three main types of winds: general, local, and critical.

LISTEN UP!

Always ask the local residents about the wind conditions that occur in their area. This information will help you to develop better technical plans and enable you to operate safely on the fireline.

General Winds

General winds, also known as gradient winds, are large-scale winds produced by pressure gradients associated with high- and low-pressure differences. General winds are included in weather forecasts and result from temperature variations. Areas of high and low pressure will always try to equalize each other. In North America, these pressure cells typically move from west to east. Air in a high-pressure cell moves outward and in a clockwise direction. In low-pressure cells, the air moves toward the center in a counterclockwise direction **FIGURE 4-30**. These winds are common higher in the atmosphere, where there is little friction from terrain. The pressure gradients flow

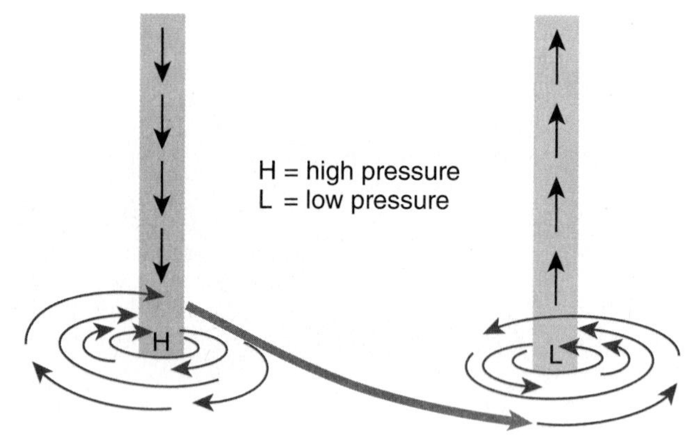

FIGURE 4-30 A high-pressure cell moves wind in a clockwise direction, whereas a low-pressure cell moves wind in a counterclockwise direction.

in a curved pattern and are influenced by the Coriolis effect and centrifugal force. In the northern hemisphere, the flow will be in a clockwise direction and counterclockwise in the southern hemisphere. Based on the curved nature of the pressure gradients, wind shifts are gradual or nonexistent.

Local Winds

Smaller-scale winds that occur daily are called **local winds**. Local winds are the result of local temperature differences and are most influenced by terrain factors. A good example of local winds would be the upslope winds that form as the result of the heating of Earth's surface. Upslope winds are covered in the following section on slope winds.

Local winds play an important role in fire behavior and must be accounted for when making tactical decisions on firelines. These winds occur daily. Local winds include three types: sea and land breezes, slope winds, and valley winds.

Sea and Land Breezes. Sea breezes develop daily along coastal areas as the result of the differential heating that occurs locally. Each day, the land mass warms, and the ocean remains cooler than the land mass. The cooler air from the ocean moves onshore to replace the heated, lighter air rising off the warmer land mass **FIGURE 4-31**. The result is wind. This sea breeze (wind) begins between midmorning and early afternoon. These times vary due to seasonal changes and location. Typical wind speeds are approximately 10 to 20 mi/h (16.1 to 32.2 km/h). They can be stronger in different regions of the country. For example, winds along the California, Oregon, and Washington coasts can attain speeds of 20 to 30 mi/h (32.2 to 48.3 km/h). Similar winds also occur on the Eastern shoreline.

Land breezes are the result of the land mass cooling more quickly than the water surfaces at night

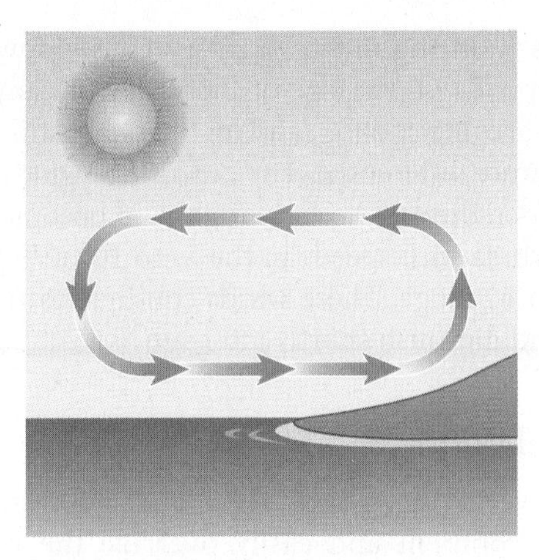

FIGURE 4-31 Sea breeze development.

FIGURE 4-33 Upslope winds.

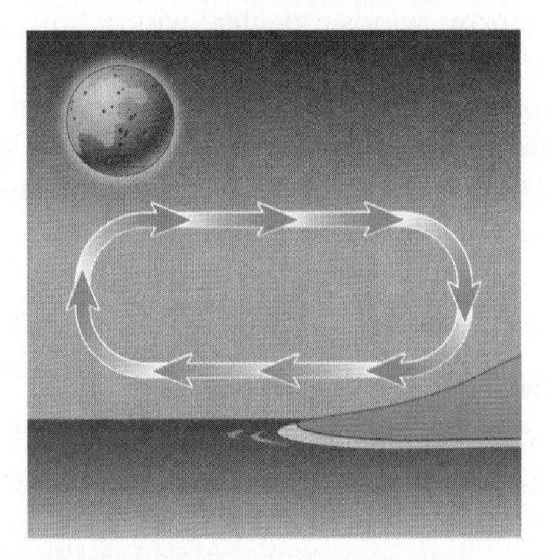

FIGURE 4-32 Land breeze development.

FIGURE 4-34 Downslope winds.

FIGURE 4-32. This difference in air pressure causes the air to flow from the land to the water. These land breezes develop approximately 2 to 3 hours after sunset. The nighttime land breeze is not as strong as the daytime sea breeze. Wind speeds are typically between 3 and 10 mi/h (4.8 and 16.1 km/h).

Understanding sea and land breezes assists with the predicted fire behavior and knowing that, at two specific times of the day, the fire will more likely than not experience a change in behavior.

Slope Winds. Upslope winds occur during the day as a result of the heating of the land surfaces **FIGURE 4-33**. The air in the valleys becomes warmer than the air on the mountaintops. This differential causes the movement of air upslope. Upslope winds start on eastern aspects as a result of the sun's exposure on the land mass. Maximum upslope winds occur about midafternoon. Upslope winds are stronger than downslope winds and range between 3 and 8 mi/h (4.8 and 12.9 km/h).

Downslope winds occur at night and are weak, superficial winds that blow downslope **FIGURE 4-34**. These winds are usually no stronger than 2 to 5 mi/h (3.2 to 8.0 km/h). Gravity is the principal force behind these winds. Downslope winds continue during the night until the slope warms again and the upslope flow reoccurs. Calm periods do exist in the evening and early morning hours when these winds are changing direction.

Slope winds affect fire by increasing the rate of spread. A contributing factor to the rate of spread is the percentage of slope. In general, the steeper the slope, with the right wind and slope alignment, the greater the rate of fire spread. One condition to watch out for is a strong downslope wind. Should the downslope wind become strong enough, it can in some cases overcome the effect of slope, bringing a reversal in the direction of the fire. Effect of slope has been attributed to large growth on smaller fires and has led to several near misses. This is another example of why it will be important for you to constantly monitor the fire weather components of the 10 standard fire orders (numbers 1, 2, 3, and 5) and evaluate what the fire is doing at all times.

Valley Winds. **Valley winds**, which are daily (diurnal) winds that flow up a valley during the day and down a valley at night, are the result of local **pressure gradients** caused by differences in temperature between the air in the valley and the air at the same elevation over an adjacent plain or in a larger valley **FIGURE 4-35**. Upslope winds start within minutes after the sun strikes a mountain slope. The upvalley winds do not start until the whole mass of air within the valley becomes warmed. That is why upvalley winds do not start until late morning or early evening. Common upvalley wind speeds range from 10 to 15 mi/h (16.1 to 24.1 km/h).

The transition to downvalley winds starts gradually. First to appear are light 2- to 5-mi/h (3.2- to 8.0-km/h) downslope winds in the canyon bottoms. This takes place early in the night, shortly after sundown. The time depends on the size of the valley or canyon, on factors favoring cooling, and on the establishment of a temperature differential. The downslope winds gradually deepen during the early night and become downvalley winds with speeds in the 5- to 10-mi/h (8.0- to 16.1-km/h) range. These winds continue through the night and diminish shortly after sunrise.

A

B

FIGURE 4-35 Diurnal valley winds. **A.** Day. **B.** Night.

Critical Winds

Critical winds are winds that totally dominate the fire environment and easily override the upslope/downslope and upvalley/downvalley winds. These winds cause great concern on firelines. Foehn winds, thunderstorm winds, glacier winds, and frontal winds are examples of critical winds. This chapter focuses on Foehn winds and thunderstorm winds, which are discussed next in the Critical Fire Weather section. Critical winds are discussed in greater detail in an S-290 (Intermediate Wildland Fire Behavior) course.

Critical Fire Weather

Critical fire weather is weather that can quickly increase fire danger and cause extreme fire behavior. Weather fronts such as cold fronts, thunderstorms, and foehn winds are considered critical fire weather.

Weather Fronts

A weather **front** is a boundary between two air masses of differing properties. Most increased wind activity is in this boundary area. Weather fronts can be caused by unusually cold or warm temperatures or unusually dry or moist air masses. Weather fronts can be classified as critical fire weather.

Cold and Warm Fronts. The passage of a cold air mass, also called a cold front, causes the most problems for fire crews. Cold fronts can cause strong, erratic winds that can change the direction and intensity of a fire. Pressure gradients are tight in cold fronts, and strong upper winds are more easily mixed down to the surface as unstable air, which results in an increase in wind speed and a radical change of direction as the front approaches and strong gusty winds as the front passes.

Warm air masses or warm fronts are weaker systems than cold fronts and are not as great a concern to fire fighters. Though not as dramatic as cold fronts, warm fronts can still cause winds that can change the direction and intensity of a fire.

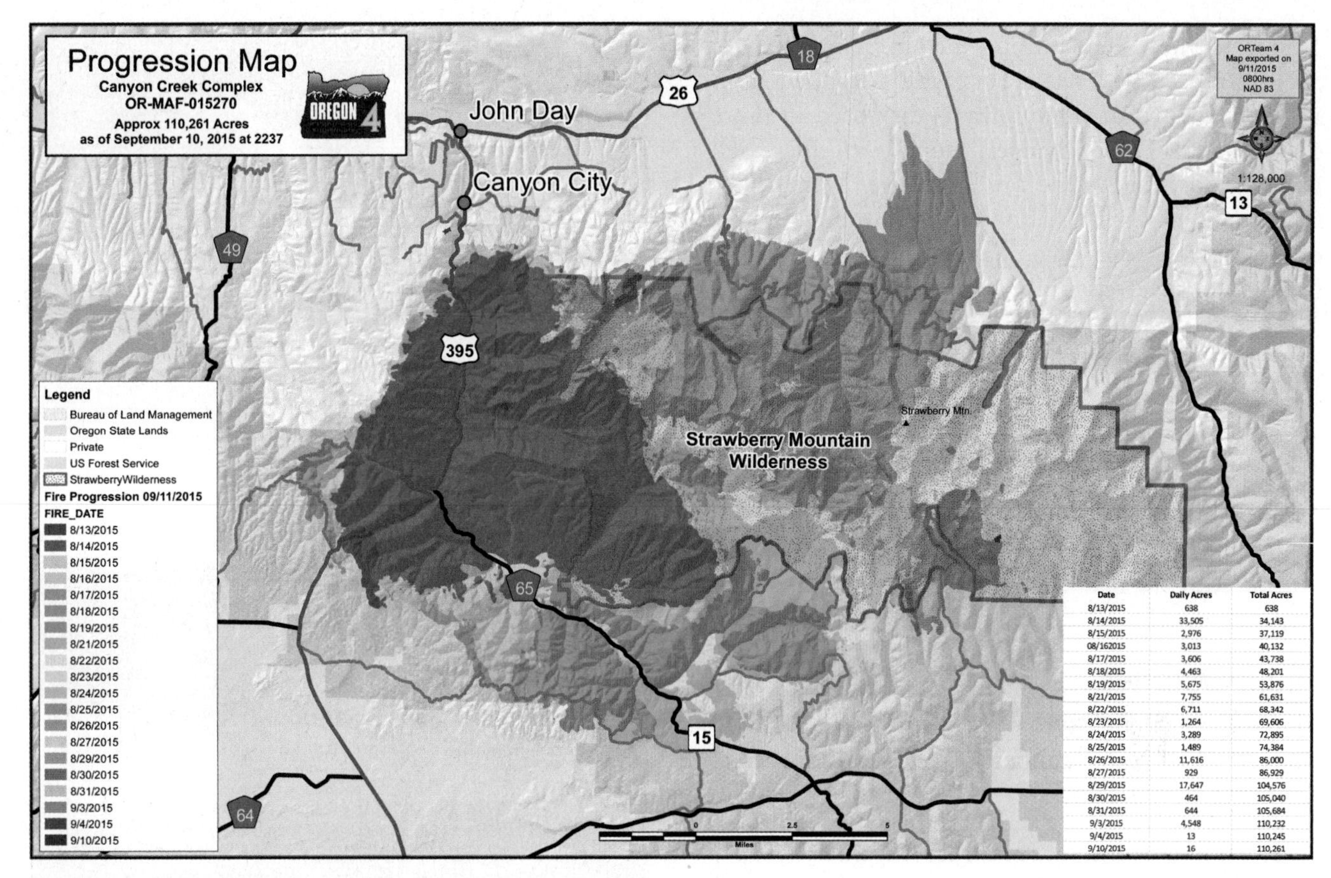

Date	Daily Acres	Total Acres
8/13/2015	638	638
8/14/2015	33,505	34,143
8/15/2015	2,976	37,119
08/162015	3,013	40,132
8/17/2015	3,606	43,738
8/18/2015	4,463	48,201
8/19/2015	5,675	53,876
8/21/2015	7,755	61,631
8/22/2015	6,711	68,342
8/23/2015	1,264	69,606
8/24/2015	3,289	72,895
8/25/2015	1,489	74,384
8/26/2015	11,616	86,000
8/27/2015	929	86,929
8/29/2015	17,647	104,576
8/30/2015	464	105,040
8/31/2015	644	105,684
9/3/2015	4,548	110,232
9/4/2015	13	110,245
9/10/2015	16	110,261

FIGURE 4-36 Map showing the progression of the Canyon Creek Complex Fire in 2015. As a result of a frontal passage, the fire grew significantly.

Courtesy of United States Forest Service.

LISTEN UP!

Expect the winds to increase and change direction during a frontal passage **FIGURE 4-36**. A downburst produces downdraft winds that can push the fire in many directions and cause it to jump over already constructed control lines. Heighten your awareness, think safety, and anticipate needed changes in tactics.

Dry Line Front. A **dry line front**, also called a dew point line, is a boundary separating moist and dry air masses **FIGURE 4-37**. The difference between a cold front and a dry line front is that the temperature on either side of a dry line does not vary much, as opposed to a cold front, where the temperatures can vary a great deal. Near the ground's surface, warm, dry air is more dense than warm, moist air of a lesser temperature, and thus the warm, dry air wedges under the humid air like a cold front. At higher elevations, the warm, moist air is less dense than the cooler, drier air, and the boundary slope reverses. Dry lines typically form in the Midwest and the Great Plains states and almost always move slowly from west to east, usually in the spring and summer months. A dry line will move eastward during the afternoon, retreat at night, and often re-form again the next day. As they develop, dry lines often produce heavy thunderstorms, especially just to the east of the dry line.

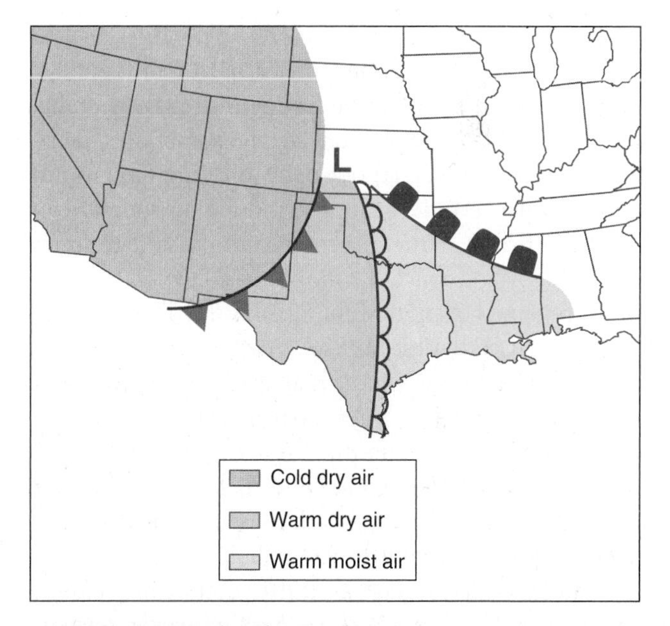

FIGURE 4-37 Dry line front.

The 2011 Coal Canyon Fire in Hot Springs, South Dakota, is a good example of how dry line fronts can quickly cause extreme fire behavior and endanger fire fighters. A dry line front occurred on the fire and, in

1 hour, the dew points dropped from 52°F (11°C) to 39°F (4°C) and the relative humidity dropped from 27% to 17% (relative humidity is discussed in more detail later). These changes, along with increased wind speeds, caused the fire to change dramatically. One fire fighter was killed on the fire, two fire fighters experienced serious burn injuries, and two additional fire fighters received minor burn injuries.

LISTEN UP!

Dry line fronts can cause extreme or abnormal fire behavior, such as a fire pushing downslope in the middle of the day. This can cause a fire to unexpectedly grow and become unmanageable. It is important to carefully monitor weather indicators and react appropriately.

Thunderstorms

Thunderstorms are a type of critical fire weather. Two characteristics make thunderstorms a great concern to fire fighters on fire lines:

1. Thunderstorms are often accompanied by cloud-to-ground lightning, which often starts wildland fires.
2. Thunderstorms are often accompanied by strong downbursts.

Downbursts are commonly referred to as downdrafts; however, these terms, while having a similar meaning, are different. **Downbursts** are strong, sometimes, damaging winds that spread out upon meeting the ground and produce erratic gusting winds. Downbursts can push the fire in many directions and cause it to jump over already constructed control lines. Downbursts can produce winds over 50 mi/h (80.5 km/h). The term *microburst* is used to describe the size of the downburst.

Downdrafts are winds that are part of the normal flow pattern of a thunderstorm. Downdrafts and updrafts are what make up the structure of a thunderhead. While they contribute to the turbulent air common with a thunderhead, they are not classified as the more dangerous downbursts.

Thunderstorms start as fluffy, cotton-like clouds called cumulus clouds **FIGURE 4-38**; however, few cumulus clouds develop into thunderstorms. If there is instability of the air mass only in the lower atmosphere while there is stability aloft, then the convectional activity essential for a thunderstorm is not present. However, if the air is conditionally unstable through a deep layer, well beyond the freezing level, then dense, vertical clouds called cumulonimbus clouds can develop **FIGURE 4-39**.

FIGURE 4-38 Cumulus clouds.

FIGURE 4-39 Cumulonimbus clouds.

Lifting. The triggering mechanism necessary to release that instability is usually one of four forms of lifting: thermal, orographic, frontal, or convergence **FIGURE 4-40**. **Thermal lifting** is the result of strong heating of the air near the ground. The heated, moisture-laden air rises high enough in the atmosphere to form cumulus clouds. **Orographic lifting** occurs in mountainous areas where the heated moist air is forced up as a result of the presence of a slope and topographic features. It is common for thermal lifting and orographic lifting to work together in mountainous areas. If you have ever flown into the Denver airport in the summertime, the bumpy ride you experience on the final approach is the combination of thermal and orographic lifting.

Frontal lifting occurs when a moving, cooler air mass pushes its way under a warmer air mass and forces the air up a slope. **Convergence** occurs when horizontal air currents merge together. Lifting by convergence occurs when there is more horizontal air movement into an area than movement of air out. When this happens, the air is forced upward. It always occurs around a low-pressure system. Convergence is

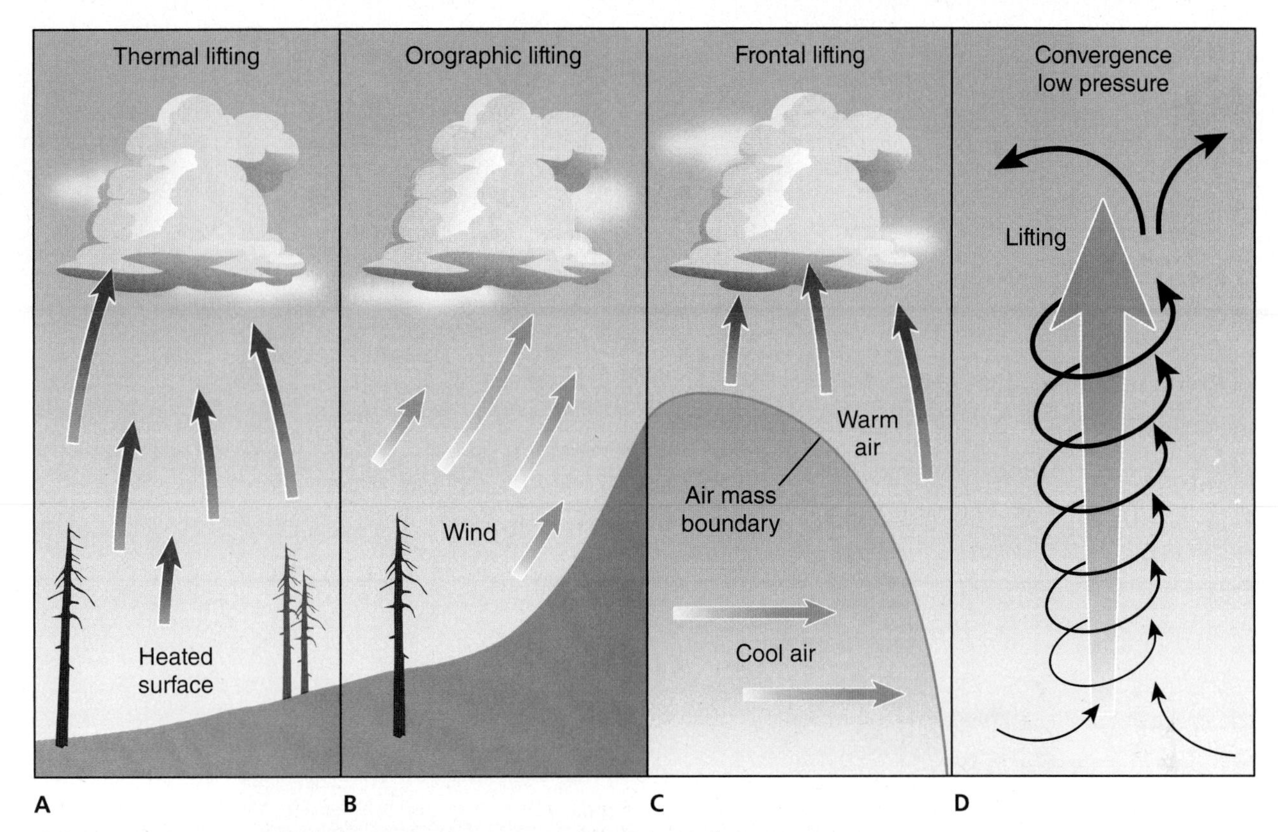

FIGURE 4-40 Types of lifting. **A.** Thermal. **B.** Orographic. **C.** Frontal. **D.** Convergence.

present in the aforementioned lifting systems but can also occur independently.

Thunderstorm Stages. Typically, thunderstorms have three stages of development and decay: cumulus, mature, and dissipating.

The cumulus stage starts with a rising column of moist air. That column then develops into a cumulus cloud that expands vertically into a cauliflower-like appearance. During this stage, the cloud has strong updrafts (air drawn inward and upward) **FIGURE 4-41**. If such a cloud passes over a wildland fire, the convective energy from the fire may mingle with the updraft of the cloud, and they may reinforce each other. This in turn may strengthen the inflow at the ground's surface, causing an increase in surface winds and fire activity.

The mature stage of a thunderstorm causes the most concern on firelines because of the strong downdrafts that are present in the cumulonimbus clouds **FIGURE 4-42**. Downbursts are the result of the moist air condensing into rain droplets or ice particles, depending on the development. As these raindrops and ice particles grow to such an extent that they can no longer be supported by the updraft, they fall and can create the downburst winds. There is a downdraft in part of the cloud and an updraft in the rest of the cloud. When these two opposing forces create enough friction, lightning occurs. The downburst winds are strongest near the bottom of the cloud. Downbursts that reach the ground result in cooler, gusty surface winds that can increase the fire's activity and cause it to change direction or jump control lines.

Sometimes, we see streaks of rain falling out of the cloud but evaporating before it reaches the ground. This phenomenon is called **virga** and can be found in some thunderstorms or rainclouds. These dry thunderstorms are a problem because rain does not wet the ground when lightning strikes, often causing lightning-started wildland fires.

LISTEN UP!

Downbursts that reach the ground result in cooler, gusty surface winds that can increase the fire's activity or cause it to change direction or jump control lines.

Thunderstorms generally travel in the direction of the winds aloft. You can note their direction of travel by the direction that their anvil-shaped tops are pointing (see Figure 4-45); however, to observe this you must be a considerable distance from the actual thunderstorm activity.

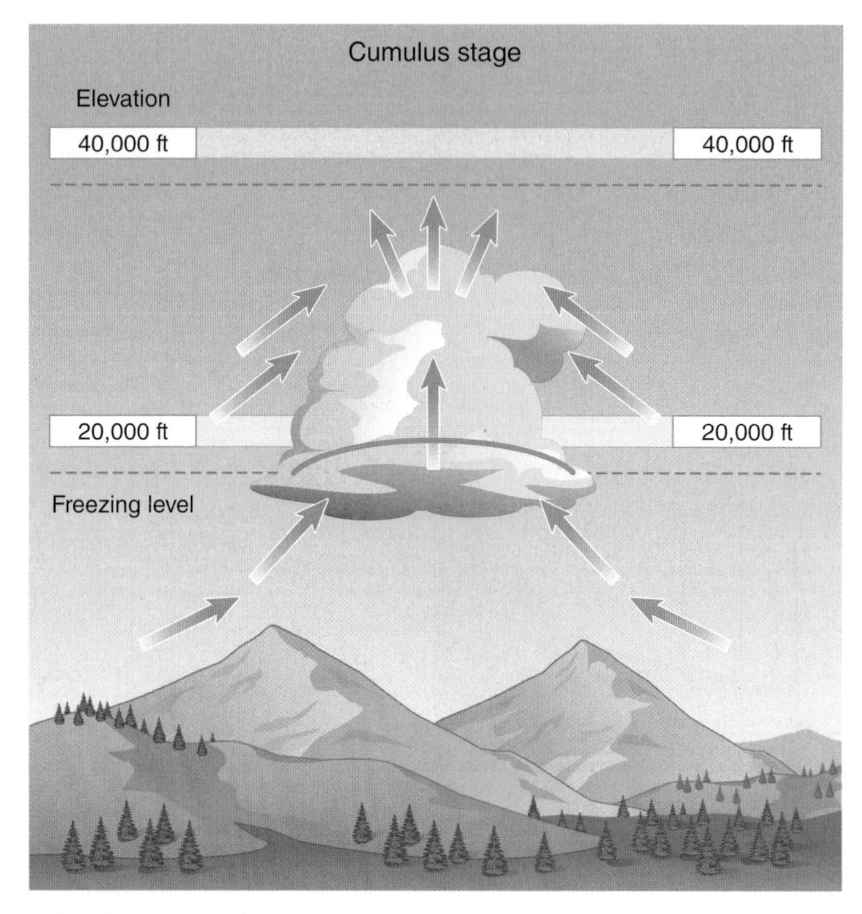

FIGURE 4-41 Cumulus stage.

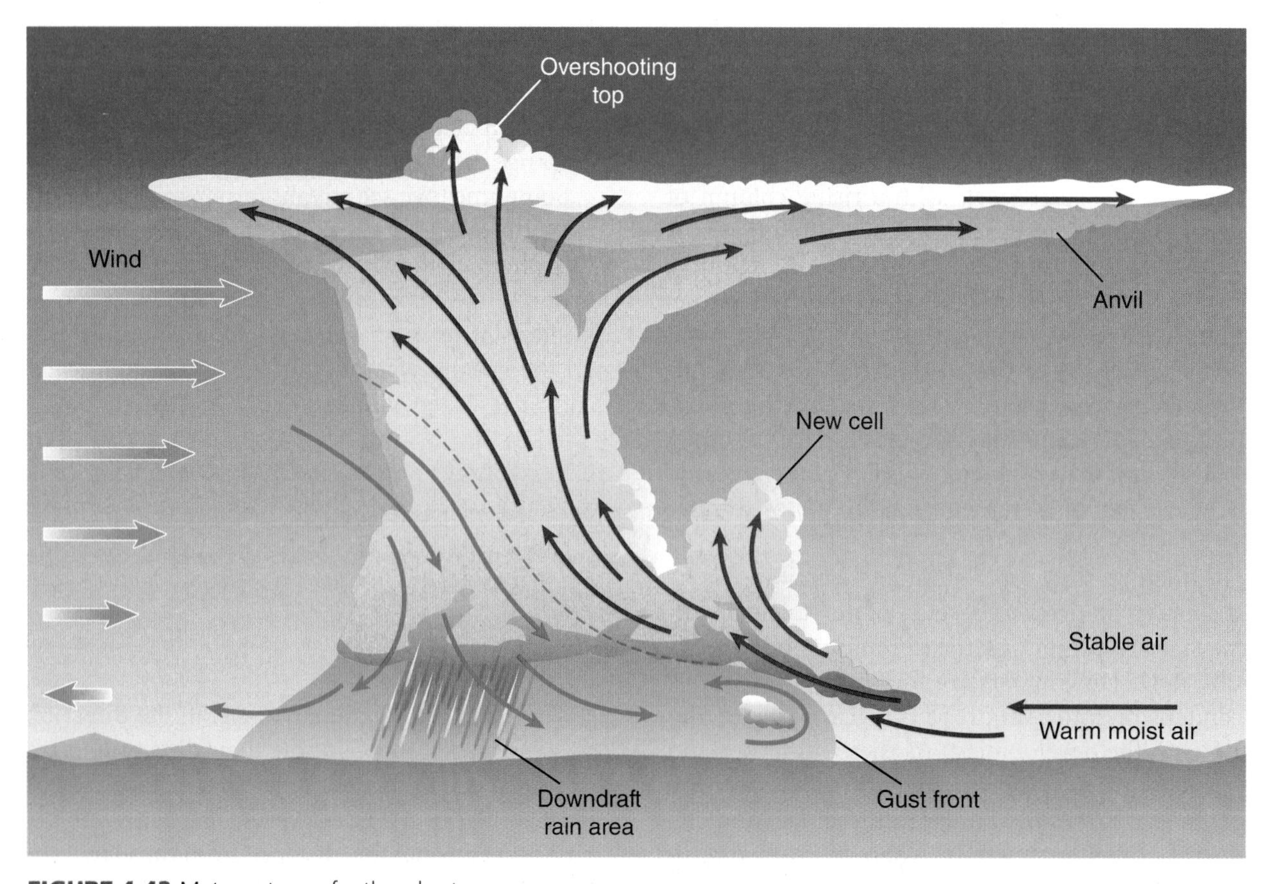

FIGURE 4-42 Mature stage of a thunderstorm.

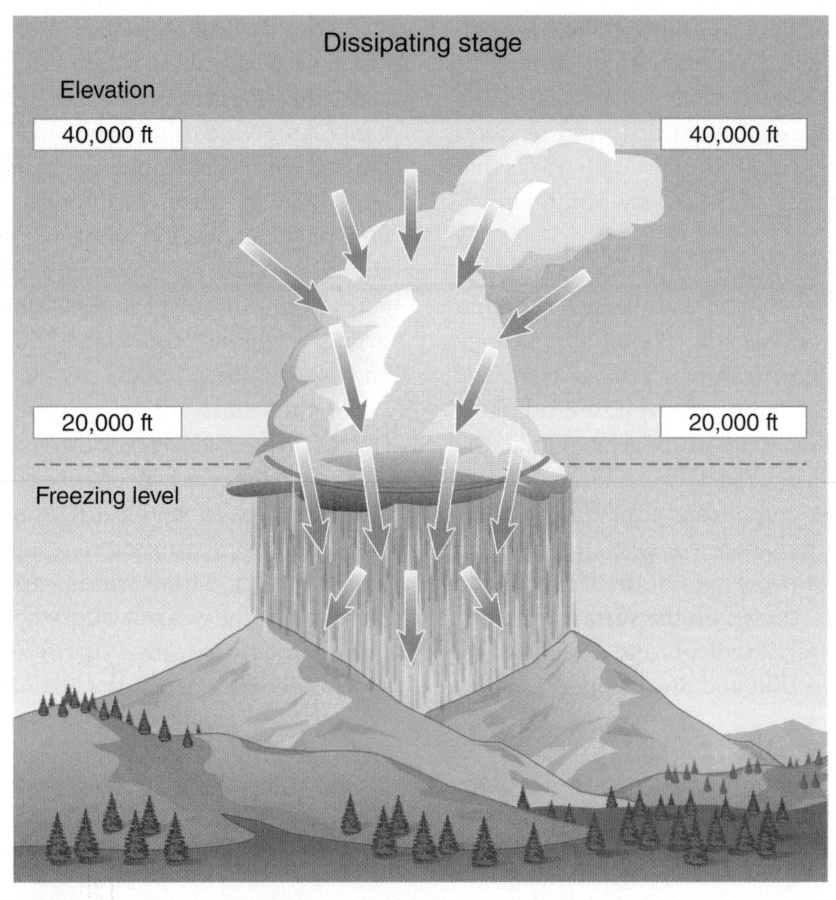

FIGURE 4-43 Dissipating stage of a thunderstorm.

Thunderstorms end in the dissipating stage **FIGURE 4-43**. At that time, the downdrafts continue to develop and spread both vertically and horizontally. Once this happens, the entire thunderstorm becomes an area of downdrafts. The storm cell now enters the dissipating stage. In this stage, the updrafts—which are the source of moisture and energy—end, and the cell's growth and activity is cut off. The downdraft then gradually weakens, the rain ceases, and the clouds start to dissipate.

Thunderstorm safety is discussed in Chapter 3, *Safety on Wildland Fires.*

Foehn Winds

Foehn winds, or gravity winds, are considered critical winds and a type of critical fire weather. Foehn winds are known by several names specific to the area where they occur. They are called Chinook winds in the Rocky Mountains, Santa Ana winds in Southern California, north winds in Northern California, Wasatch winds on the west side of the Rockies, and east winds in most parts of Washington and Oregon. Foehn winds also occur in other parts of the country, such as on the eastern slopes of the Appalachian Mountains.

Foehn winds are downflowing winds that are usually warmer and drier as a result of compression as the air descends the leeward side of a mountain range. These winds occur when heavy, stable air pushes across a mountain range and descends rapidly on the lee side (the side of the mountain range away from the wind). This rapid descent compresses the air and makes it hotter and dryer, in turn, reducing the fuel moisture. If these winds combine with the nightly local downslope winds, they can become quite strong. These winds usually occur from September to April; they do not occur frequently during the summer months.

There are two types of foehn winds. The first type is the result of moist Pacific air being forced upward and over a mountain range. We find this wind occurring in the Sierra–Cascade range and the Rocky Mountain range. As the air is forced upward on the windward side, it reaches its maximum condensation level. When it lifts farther, clouds are produced, along with precipitation. Upon passing over the mountain range, precipitation is lost, and the air then starts to descend and warm. As it descends the leeward side of the mountain range, it is possible for these pressure gradients and air movements to change in temperature up to 25 degrees in a matter of minutes (5.5 degrees per

1000 feet [305 m] of fall). As this wind pushes through the mountain passes or other constricted topographical features, it gains speed. When it arrives in the lowlands, it is a strong, gusty, and dry wind. A good example of this type of wind is the Chinook winds that occur on the eastern side of the Rocky Mountains in the fall or winter.

The second type of foehn wind occurs when a cold, dry, stagnated high-pressure air mass is centered over the Great Basin in Nevada. The mountain barrier separates this high-pressure air mass from the low-pressure center located on the opposite side. It is important to remember that, with foehn winds, the mountains block the flow of surface air. When this occurs, the airflow must come from aloft. This airflow from the high-pressure mass or winds aloft moves toward the area of lower pressure. On the leeward side of the mountain range, the air on the surface is forced out of the way by pressure gradients and is replaced by heavy air flowing from aloft and downslope.

These downslope winds are also influenced by a downward or sinking motion of air in the atmosphere called **subsidence**. Winds like these are typified by the Santa Ana winds that occur in California during the fall and winter and cause many of the rapidly spreading large wildland fires common in Southern California. One of the greatest examples of subsidence to date is the 2018 California Camp Fire, which burned 120,000 acres in a 48-hour period **FIGURE 4-44**. When it was over, the fire had consumed 158,000 acres, destroyed 18,804 structures, and 87 people died.

Foehn winds are known for their sustained and strong surface winds (winds blowing near Earth's surface). Surface wind speeds are commonly in the 40 to 60 mi/h (64.4 to 96.6 km/h) range. In a severe Santa Ana condition, winds can reach as high as 90 mi/h (144.8 km/h). These winds can last for days before they gradually weaken and stop. In some instances, these winds can cause significant damage and loss, such as experienced in the 2017 and 2018 fire seasons

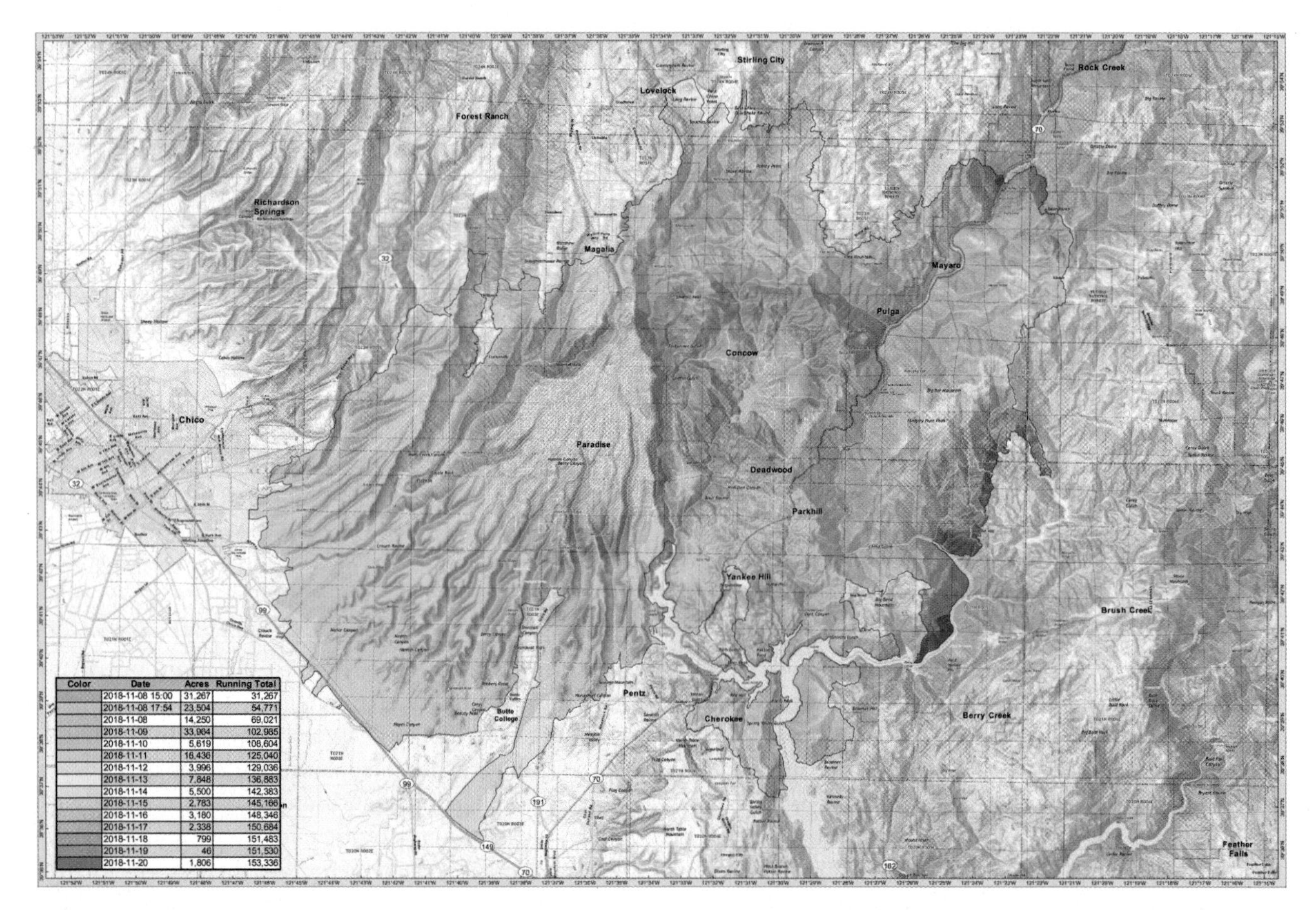

Color	Date	Acres	Running Total
	2018-11-08 15:00	31,267	31,267
	2018-11-08 17:54	23,504	54,771
	2018-11-08	14,250	69,021
	2018-11-09	33,964	102,985
	2018-11-10	5,619	108,604
	2018-11-11	16,436	125,040
	2018-11-12	3,996	129,036
	2018-11-13	7,848	136,883
	2018-11-14	5,500	142,383
	2018-11-15	2,783	145,166
	2018-11-16	3,180	148,346
	2018-11-17	2,338	150,684
	2018-11-18	799	151,483
	2018-11-19	46	151,530
	2018-11-20	1,806	153,336

FIGURE 4-44 This map shows the progression of the deadly Camp Fire that destroyed the Paradise, California, community in November 2018. Of significance is the rapid pace at which the fire burned. Several factors contributed to this, but one of the takeaways is the critical fire behavior that occurred in such a short time.

with the Atlas, Nuns, Tubbs, Camp, and Woolsey fires. These fires have disrupted communities and challenged the way we will be fighting fires in the future. The five fires destroyed close to 30,000 homes and caused over 115 deaths.

LISTEN UP!

Surface wind speeds are measured using an anemometer **FIGURE 4-45**.

Air Mass Stability

The stability of the air mass, which has to do with the resistance of air to vertical motion, can greatly affect a wildland fire. **Atmospheric stability** may either enhance or suppress vertical motion, which in turn affects a fire.

If there is an unstable air mass, the fire will develop a well-defined smoke column that rises vertically to great heights **FIGURE 4-46**. This strong vertical movement of converted energy creates a great need for oxygen to be updrafted into the fire's base. Wildland fires burn hotter and with more intensity when the air mass is unstable because a stable air mass suppresses the convective energy of the fire (like putting a lid on a pan). After a limited rise, a smoke column will drift apart and appear as if it has been capped off **FIGURE 4-47**. Fireline intensities are not as great because of the air's resistance to vertical lifting. Visual

FIGURE 4-45 Measuring wind speeds with an anemometer.
Courtesy of Jeff Pricher.

FIGURE 4-46 Unstable air mass indicator—a well-defined smoke column.

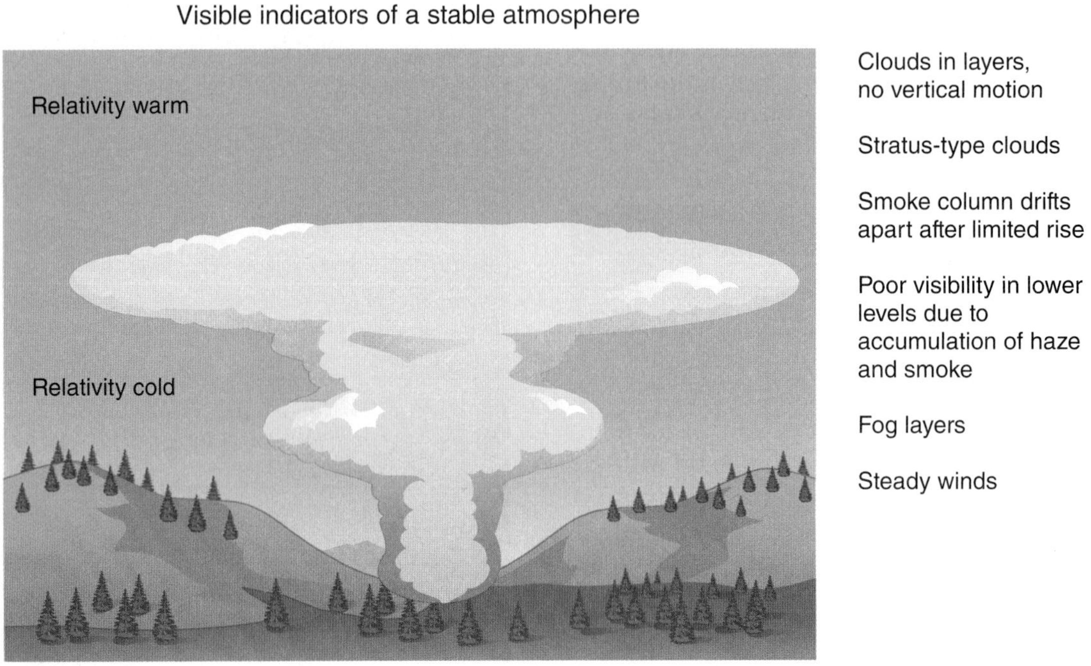

Visible indicators of a stable atmosphere

Relativity warm

Relativity cold

Clouds in layers, no vertical motion

Stratus-type clouds

Smoke column drifts apart after limited rise

Poor visibility in lower levels due to accumulation of haze and smoke

Fog layers

Steady winds

FIGURE 4-47 Stable air mass indicators. Note the smoke column drifts apart after limited rise.

TABLE 4-3 Stable and Unstable Air Mass Indicators

Stable	Unstable
Clouds in layers	Clouds grow vertically, and smoke rises to great heights
Stratus clouds	Cumulus or cumulonimbus clouds
Smoke layer or plume that drifts apart after a limited rise	Towering smoke plume, dust devils, or fire whirls
Poor visibility due to haze, smoke, or fog	Good visibility
Steady winds	Gusty or erratic winds

signs that are present on the fireline indicate the stability of the air mass. **TABLE 4-3** displays characteristics of stable and unstable air masses. Use these indicators to give you additional clues as to the fireline intensities to expect.

Inversion Layers

An **inversion** layer acts like a lid or cap over a fire. The smoke column rises and then flattens out and spreads horizontally. The fire will not burn as rapidly if a stable air mass is present.

When the inversion layer breaks or dissipates, expect to see a change in the convective column and an increase in fire activity. As the inversion dissipates or breaks, unstable conditions develop.

There are three types of inversion layers that can affect a fire: night inversion, marine inversion, and subsidence inversion.

Night Inversion

Night inversion, or radiation inversion, is the most common type of inversion and is found predominantly in mountainous terrain and inland valleys. During night inversion, air is cooled as it comes in contact with Earth's surface. At night, Earth loses its heat through radiation, and the air in contact with the ground cools and becomes dense. This cold, dense air readily flows downslope and gathers in air pockets and valleys. Night inversions deepen as the night progresses, which creates a condition of cool, heavier air on the valley floor, below warm air. In mountain areas, topography plays a decisive role in both the formation and intensity of night inversions.

Conditions usually start to reverse after sunrise. As the sun rises, surface heating takes place and begins to warm the cold air. Expansion of the air takes place, and the inversion top may rise slightly. Heating starts to destroy the inversion along the slopes, and upslope winds start to develop. As this heating and mixing of the air continues, the inversion layer continues to dissipate until final dissipation is complete. When this occurs, unstable atmospheric conditions develop, and fire activity increases.

Marine Inversion

Marine inversions are common around large bodies of water. In this type of inversion, the cool, moist air from the ocean spreads over the low-lying land. The cool air mass is capped off by a layer of much warmer, drier, and relatively unstable air. The layer of cool air can vary in depth from a few hundred feet to several thousand feet. A deep marine layer can spread over coastal mountains into the inland valleys of the central basin. Marine inversions are strongest at night and dissipate as the day goes on. If the cold air is shallow, fog will form; however, if the cold air is deep, stratus clouds (low clouds with horizontal layering) are likely to form. One of the greatest challenges to the firefighting efforts affecting the fire is when a marine layer lifts. The change can occur dramatically, oftentimes, increasing the fire behavior, which makes predictability very challenging.

Subsidence Inversion

Subsidence inversions are caused by warm high pressure in the upper atmosphere. The sinking air in the higher-pressure area warms and dries as it descends. This subsiding air may reach the ground's surface with only a slight change in moisture. The air is also warmed and dried as it sinks. Mountaintops are affected first, even though coastal slopes may be under the influence of a massive humid layer. Subsiding air is also responsible for foehn winds, which contribute to some of the most significant weather to be found anywhere.

Thermal Belt

A **thermal belt** is any area in a mountainous region that typically experiences the least changes in temperature on a daily basis **FIGURE 4-48**. Here, we find the highest daily minimum temperatures and the lowest nighttime fuel moisture.

Thermal belts are found in mountainous areas below the main ridges, typically, at the middle third of a slope where an inversion layer comes in contact with the mountain slope. The location of a thermal belt also varies from night to night in these areas. At night, air comes in contact with the upper slopes and cools. This dense, cool air flows downhill into the low-lying mountain valleys. An inversion layer then develops above the pool of cool air. This area contains relatively warmer air.

At night, wildland fires remain quite active in the thermal belt. Identify and plan appropriate tactics for these areas.

Relative Humidity

Relative humidity (sometimes abbreviated as RH) is the ratio of the amount of moisture present in the air compared to the maximum amount of moisture that the air can hold at a given temperature and atmospheric pressure **FIGURE 4-49**. Warm air has a higher capacity for moisture content than cool air does.

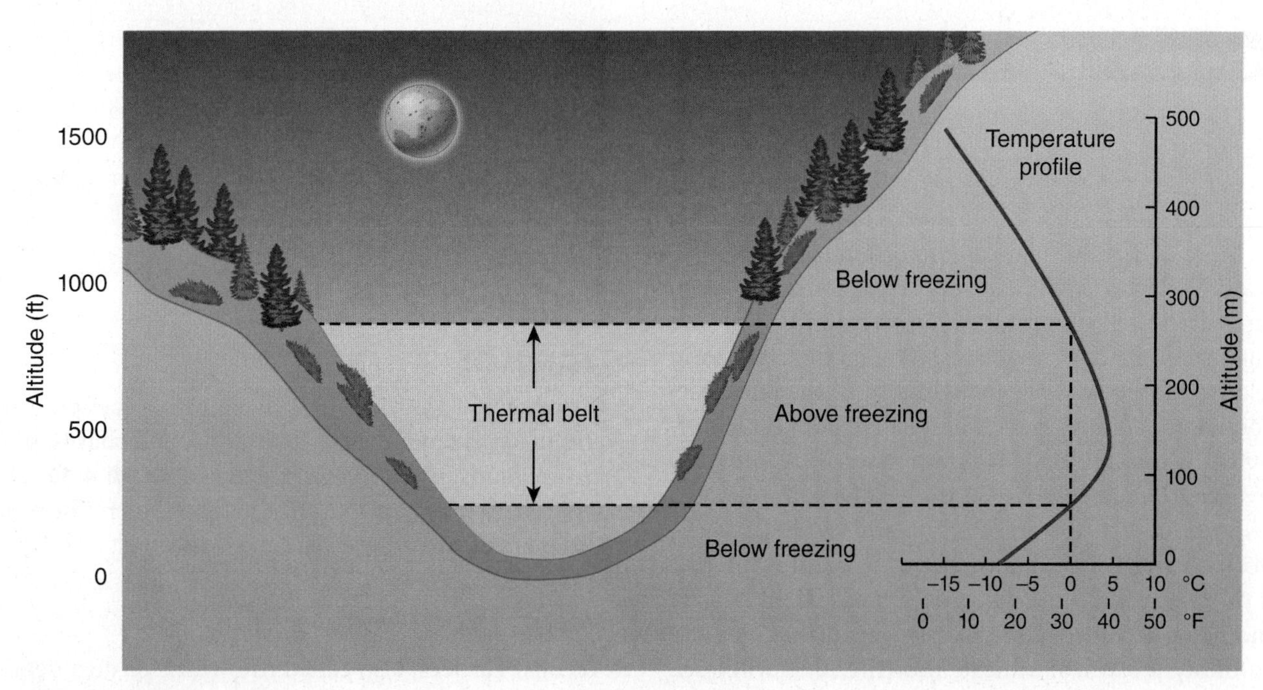

FIGURE 4-48 Thermal belt.

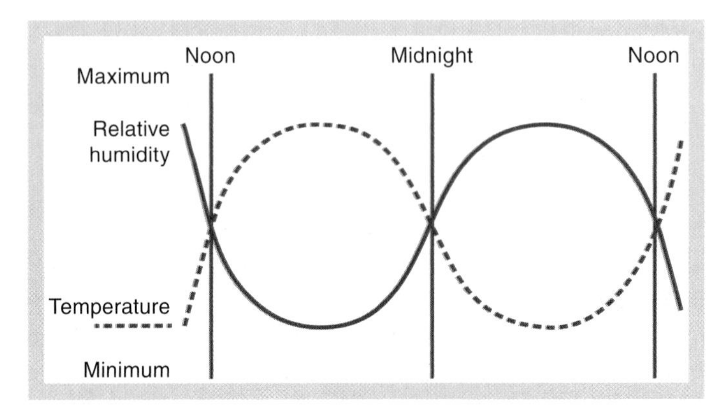

FIGURE 4-49 Temperature and relative humidity chart.

Relative humidity is expressed as a percentage. If the air contains the maximum amount of moisture possible, the relative humidity is said to be 100 percent. In dry climates, the relative humidity may be 20 percent or less, which means the moisture contained in the air is only 20 percent of the maximum amount that the air is capable of holding.

Relative humidity is a major factor in wildland fire behavior. When relative humidity is low, fire behavior increases because fine fuels dry out, making them more susceptible to ignition. Conversely, when relative humidity is high, fire behavior decreases because moisture from the air is absorbed by the fuels, making them less susceptible to ignition. Changes in relative humidity affect fine fuels more dramatically than they affect heavy fuels. If the relative humidity drops and stays at 10 percent, for example, eventually all fuels will equalize at that level, but fine fuels will reach that level more quickly than heavy fuels.

LISTEN UP!

Temperature and relative humidity have an inverse relationship; when one goes up, the other goes down.

Time of day also plays an important role in relative humidity levels. In the early-morning hours, temperatures are lower when the relative humidity levels are higher. As the day progresses, temperatures rise, and relative humidity levels fall. Usually, in the late afternoon, temperatures peak, and relative humidity levels are at the minimum. It is important to check humidity levels periodically. A rise in the relative humidity can help to slow the spread of a wildland fire and aid the work of fire crews who are trying to control the fire. A drop in relative humidity can increase fire intensity and hamper control efforts. Instruments that measure humidity levels will indicate how fire fuels will burn and carry the fire.

In some parts of the United States, the season of the year can also play an important role in relative humidity levels. These changes contribute to variations in the moisture content of vegetative fuels, a factor that partly explains why many regions in the United States demonstrate seasonal patterns of wildland fires. The seasons with the most wildland fires are often the seasons with the lowest relative humidity.

DID YOU KNOW?

Temperature and moisture have several kinds of relationships:

- Dry bulb temperature is the temperature of the air measured in shade 4 to 8 ft (1.2 to 2.4 m) above the ground. It is the temperature measured with a thermometer and referenced in our day-to-day life as Fahrenheit or Celsius.
- Wet bulb temperature is the lowest temperature air can be cooled by evaporating water. It is measured with a wet bulb thermometer. Wet bulb temperature is a good indicator of atmospheric pressure but is not a direct measurement.
- Dew point temperature is the temperature to which air must be cooled to reach saturation. It is determined with an electronic weather meter or a sling psychrometer and psychrometric table references (refer to NWCG PMS 437, *Fire Behavior Field Reference Guide*, for these tables). Dew point temperature is one of the most reliable methods for measuring atmospheric moisture.

Vortices

A **vortex** is a whirling mass of air with a low-pressure area in its center that tends to draw fire and objects, such as firebrands, into the vortex action **FIGURE 4-50**. The vortex action becomes strongest at the center. There are two types of vortices: vertical and horizontal.

Vertical Vortices

Vertical vortices, also called fire whirls or dust devils, are spinning columns of air rising within a vortex and

FIGURE 4-50 A large vortex.

carrying aloft smoke, debris, and flames. Vertical vortices can be triggered by the following factors:

- Thermally driven vortices are triggered by intense heating of the land mass or intense burning of portions of the wildland fire.
- Wake-type vortices are triggered by physical obstructions to wind movement, such as ridgetops, convective columns, and trees. They occur on the lee side of the physical obstruction.
- Convective column vortices are produced by unequal convective activity in parts of the convective column.

Vertical vortices aid in fire spread through heat and mass transfer as the vortex (fire whirl) moves along the surface, lifting firebrands into the air. The fire whirl action concentrates the localized wind. Large fire whirls have the intensity of a small tornado.

Horizontal Vortices

Horizontal vortices are a phenomenon associated with extreme burning conditions and are further divided into two types. One occurs on the surface along the flanks of wildland fires, and the other occurs in the convective column, where it can affect air operations on the fire. Horizontal vortices tend to form more readily over flat or gentle terrain and at low to moderate wind speeds.

Horizontal vortices are difficult to predict and can therefore make a safe flank of the fire a dangerous place.

Extreme Fire Behavior Watch Outs

If several of these indicators exist simultaneously, then a high-intensity fire is likely to occur. Always observe the indicators; they will help you predict a change in the fire's behavior early on.

Fuels

- Unusually low fuel moisture (in the fine dead fuels and live fuels)
- Large amounts of fine dead fuels on slopes, especially if the fuels are continuous
- Dried crown foliage as the result of a previous surface fire
- Large concentrations of snags
- A large portion of the fuel bed containing dead fuels (frost kill, bug kill, or drought conditions)
- Many ladder fuels present
- Fuel bed containing fuels with highly flammable oils

Topography

- Slope aspects with increasing fuel temperatures
- Steep slopes
- Narrow canyons
- Box canyons
- Chimneys
- Saddles
- Thermal belts, which can burn with intensity at night
- Tops of mountains—fires burning on slopes, under the influence of local winds, may come under the influence of gradient winds at the ridgetop

Weather

- Low relative humidity
- Thunderstorm activity in the area
- Approaching cold fronts
- Foehn winds
- Inversion layers
- Battling winds or sudden calm, which may indicate a wind shift
- Lenticular clouds, indicating strong winds aloft—these winds may become surface winds on the lee side of the mountain range
- High winds aloft, as indicated by high, fast-moving clouds that are blowing in the opposite direction of the surface winds
- High temperatures and low humidity in the morning

Unstable Air Mass Indicators

- Good visibility
- Well-defined smoke column
- Fire whirls or dust devils
- Frequent spot fires
- Cumulus clouds
- Gusty winds

Red Flag Warnings

Red Flag Warnings and Fire Weather Watches are a significant type of forecast warning issued by the National Weather Service (NWS) for particularly dangerous weather conditions. These warnings are issued to inform area fire fighters and fire agencies (land management, municipal, and rural fire agencies) that conditions are lined up for rapid fire growth and fire spread. The NWS issues these warnings when specific weather parameters (e.g., winds and relative humidity), seasonal dryness (past rainfall, Keetch–Byram Drought Index, and Haines Index [index for dry, unstable air]), and fuel moisture harmonize with some of the following conditions:

- Lightning after an extended dry period
- Significant dry frontal passage
- Strong winds
- Very low relative humidity
- Dry thunderstorms

LISTEN UP!

Even small fires can rapidly spin out of control during Red Flag Warnings, so extra vigilance must be used. Do not expect normal fire behavior during periods when these warnings are issued.

Fire Suppression Techniques

Although there are no physical fire suppression measures that can be derived from this chapter, there are skills that need to be learned to aid in the fire suppression decision process. An example of one of these skills is recording weather. Weather is one of the fundamental components of establishing your situational awareness. On most larger fires, it is very common for a crew member to be assigned to record the weather every hour and broadcast the information over the radio to all of the fire fighters in the area closest to the weather recording. This allows all fire fighters and supervisors to maintain awareness on the current and expected fire behavior. In addition, the incident meteorologists utilize the weather recordings to provide location-specific forecasts. Weather can be obtained via a belt weather kit or an electronic weather device.

Recording weather is a skill that every fire fighter needs to have. Having this skill is important when serving in the capacity of a lookout, assisting the IC in developing a weather history to obtain a spot weather forecast, and determining the probability of ignition (PIG) for the area where you are fighting the fire. Understanding the trends for the location you are working can aid in suppression or establishing valuable and time-critical decision points for when crews need to leave. Lookouts serve an essential role in maintaining LCES/LACES (lookouts, awareness communication, escape routes, and safety zones), the 10 standard fire orders, and the 18 watch outs.

The most common tools used for recording weather include a belt weather kit (or a digital weather-collecting instrument) and the IRPG. The IRPG has tables in the reference list for determining humidity and dew point, as well as tables for calculating the PIG. The PIG is very useful in determining the rate of spread of the fire and trying to predict the expected fire behavior.

Chapter 5, *Strategy and Tactics*, discusses wildland fire suppression tactics and strategy in more detail.

DID YOU KNOW?

Developing news: Since September 2019, at least 27 million acres of Australia have burned in one of the continent's worst wildfire seasons to date. At least 27 people have been killed, including at least three volunteer fire fighters, and many more have gone missing; an estimated 2,500 homes have been destroyed or damaged; and, according to an estimate by the University of Sydney, more than 1.25 billion animals have lost their lives due to the fires (including an estimated 25,000 koalas). The wildfires in Australia have burned more than 14 times the area that burned in the 2018 California Camp Fire (the worst wildfire season in California history) **FIGURE 4-51**.

FIGURE 4-51 Australian fire fighters battling the continent's worst wildfire season to date.

Courtesy of the National Interagency Fire Center.

After-Action REVIEW

IN SUMMARY

- Fire behavior is the result of three factors: fuel, topography, and weather. Fire spread is usually caused by a combination of all three. The fire triangle represents the three components required for combustion in wildland fires: fuel, oxygen, and heat.
- Fuels are materials that store energy.
  - A wildland fuel can be any organic material, either living or dead, that can ignite and burn.
  - Fuels can be loosely grouped into three levels: ground, surface, and aerial.
- Fuel characteristics determine how fast a fire travels (rate of spread) and how intensely it burns.
  - Continuity: The degree of continuous distribution of fuel particles in a fuel bed.
    - Horizontal continuity refers to the way that wildland fuels are distributed at various levels.
    - Continuity of the fuel bed determines where the fire will spread and whether it will travel along the surface fuels, the aerial fuels, or both.
  - Vertical fuel arrangement: The different heights of the fuels that are present at a fire.
  - Compactness: The spacing between fuel particles, or density.
  - Live-to-dead ratio: The amount of dead fuel present in a fuel bed.
  - Fuel moisture:
    - Live fuel moisture is the moisture found in living plants (either woody or herbaceous).
    - Dead fuel moisture is the moisture found in dead plants, such as perennial grasses.
  - Fuel temperature
- There are 13 fire behavior fuel models, grouped into four broad classes:
  - Grass models (1, 2, and 3) are a primary carriers of a wildland fire.
  - Shrub models (4, 5, 6, and 7) usually burn with greater intensity than grass models. Contain both dead and live fuel moisture.
  - Timber litter models (8, 9, and 10) are composed largely of leaves, mixed litter, occasional twigs, and large branch wood. This litter is naturally occurring, as opposed to the result of a logging operation.
  - Logging slash models (11, 12, and 13) are the result of logging operations. Fires in these fuel models tend to have moderate to high rates of spread. They also tend to produce moderate- to high-intensity fires.
- Topography represents the physical features of the land. Slope, aspect, elevation, and certain terrain features aid in fire spread.
- Weather factors are the primary elements that cause changes in a wildland fire.
  - Wind brings a fresh supply of oxygen to the fire, gives the fire direction, and causes increased spread rates. There are three types of winds:
    - General winds (gradient winds)
    - Local winds
    - Critical winds
  - Weather fronts bring with them increased wind spreads and changes in wind direction.
  - Thunderstorms are often accompanied by cloud-to-ground lightning and strong downbursts.
  - Stability of an air mass has to do with the resistance of air to vertical motion. Atmospheric stability may either encourage or suppress vertical motion, which in turn affects fire behavior.
  - An inversion layer acts like a lid or cap over a fire. The smoke column rises and then flattens out and spreads horizontally.
    - There are three types of inversion layers that can affect a fire: night inversion, marine inversion, and subsidence inversion.

- A thermal belt is any area in a mountainous region that typically experiences the least changes in temperature on a daily basis. Here, we find the highest daily minimum temperatures and the lowest nighttime fuel moisture.
- Relative humidity is the ratio of the amount of moisture present in the air compared to the maximum amount of moisture that the air can hold at a given temperature and atmospheric pressure.
- A vortex is a whirling mass of air with a low-pressure area in its center that tends to draw fire and objects, such as firebrands, into the vortex action. There are two types of vortices: vertical and horizontal.

- To thoroughly grasp fireline safety concepts, you need to have a good knowledge of wildland fire behavior. You must understand wildland fire behavior before responding to wildland fires because tactical decisions are based on an understanding of these precepts.
- Recording weather is a skill that every fire fighter needs to have.

KEY TERMS

Aerial fuels Fuels that are greater than 6 ft (1.8 m) in height. Include trees, branches, foliage, and tall brush.

Area ignition Occurs when several individual fires throughout an area cause the main body of fire to produce a hot, fast-spreading fire condition. Also called simultaneous ignition.

Aspect The direction of a slope face in relation to a cardinal compass point.

Atmospheric stability The degree to which vertical motion in the atmosphere is enhanced or suppressed.

Box canyons Canyons with three steep walls that create very strong upslope winds. Also known as dead-end canyons.

Chimney Terrain feature that has a channeling effect on the convective energy of fire.

Compactness Refers to the spacing between fuel particles, sometimes, called density.

Convergence Occurs when horizontal air currents merge together.

Critical fire weather Weather that can quickly increase fire danger and cause extreme fire behavior, such as weather fronts and foehn winds.

Critical winds Winds that totally dominate the fire environment and easily override the upslope/downslope and upvalley/downvalley winds; include foehn winds, thunderstorm winds, glacier winds, and frontal winds.

Crown fires Fires that advance from the tops of trees or shrubs, more or less dependent on a surface fire.

Dead fuel moisture The amount of water contained in dead plant tissue.

Downbursts Strong and, sometimes, damaging winds that spread out upon meeting the ground and produce erratic gusting winds.

Downdrafts Winds that are part of the normal flow pattern of a thunderstorm.

Dry line front A boundary separating moist and dry air masses. Also called a dew point line.

Eddy A circular-like flow of air or water drawing energy from a flow of much larger scale.

Fire triangle Model for explaining the behavior of wildland fires. Consists of three elements: fuel, oxygen, and heat.

Flaming front zone Where continuous flaming combustion is taking place.

Foehn winds Warm, dry, and strong general winds that flow down into the valleys when stable, high-pressure air is forced across and then down the lee slopes of a mountain range. Also known as Chinook, east, north, Santa Ana, and Wasatch winds.

Front A boundary between two air masses of different properties.

Frontal lifting Occurs when a moving, cooler air mass pushes its way under a warmer air mass and forces the air up a slope.

Fuel continuity The degree of continuous distribution of fuel particles in a fuel bed.

Fuel moisture Refers to the amount of moisture present within a fuel source. It is an important factor in how easily a fuel ignites, burns, and spreads.

Fuels Materials that store energy and therefore are combustible.

General winds Large-scale winds produced by pressure gradients associated with high- and low-pressure differences. Also known as gradient winds.

Ground fuels Combustible material lying on or beneath the ground or subsurface litter. Ground fuels include root systems, deep duff, rotting buried logs, and other woody materials in various states of decomposition.

Heat transfer The process by which heat is imparted from one body to another, through conduction, convection, and radiation.

Horizontal continuity Refers to the way that wildland fuels are distributed at various levels.

Inversion A layer in the atmosphere that acts like a lid or cap over a fire.

Ladder fuels Refers to the vertical arrangement of fuels; allow a fire to climb from ground fuels to surface fuels to aerial fuels.

Live fuel moisture Ratio of the amount of water to the amount of dry plant material in living plants.

Live-to-dead fuel ratio The amount of dead fuel present in a fuel bed.

Local winds Result of local temperature differences and are most influenced by terrain factors.

Marine inversions Result of cool, moist air from the ocean spreading over low-lying land. Common around large bodies of water.

Mean sea level Average height of the sea for all stages of the tide over a 19-year period.

Narrow canyons Steep canyon walls that are close together, possibly creating problems for fire crews if fire from one canyon wall jumps to the opposite canyon wall.

Night inversion Results from air cooling as it comes in contact with Earth's surface. It is the most common type of inversion, found predominately in mountainous terrain and inland valleys. Also called a radiation inversion.

Organic material Any natural matter, such as leaves, rotting logs, or duff.

Orographic lifting Occurs in mountainous areas where the heated moist air is forced up as a result of the presence of a slope and topographic features.

Pressure gradients The differences in atmospheric pressure between two points on a weather map.

Rate of spread Speed at which a fire is moving away from the site of origin.

Relative humidity The ratio of the amount of moisture present in the air compared to the maximum amount of moisture that the air can hold at a given temperature and atmospheric pressure.

Saddles Low topography points between two high topography points.

Smoldering Burning without flame and barely spreading.

Spotting Occurs when small burning embers and sparks are carried by the wind that land outside the fire perimeter, which starts new fires beyond the zone of direct ignition.

Subsidence Downward or sinking motion of air in the atmosphere.

Subsidence inversions Inversions caused by subsiding air, often resulting in very limited atmospheric mixing conditions.

Surface fuels Combustible material lying on or near the surface of the ground and a major cause of wildland fire spread. Surface fuels include ground litter, such as leaves, needles, bark, grasses, tree cones, and brush up to 6 ft (1.8 m) in height.

Thermal belt Any area in a mountainous region that typically experiences the least changes in temperature on a daily basis.

Thermal lifting The result of strong heating of the air near the ground. The heated, moisture-laden air rises high enough in the atmosphere to form cumulus c louds.

Torching Occurs when tree foliage burns from bottom to top. Torching is intermittent and may be an indicator that fire behavior is increasing. Sometimes, referred to as candling.

Valley winds Daily (diurnal) winds that flow up a valley during the day and down a valley at night.

Vertical fuel arrangement Refers to the different heights of the fuels that are present at a fire.

Virga Precipitation that falls from a cloud but evaporates before reaching the ground.

Vortex A whirling mass of air with a low-pressure area in its center that tends to draw fire and objects, such as firebrands, into the circling action.

REFERENCES

Campbell, Doug. *The Campbell Prediction System: A Wildland Fire Prediction System*. 4th ed. Ojai, CA, 2016.

Colorado State Forest Service. "Mountain Pine Beetle." Accessed October 4, 2019. https://csfs.colostate.edu/forest-management/common-forest-insects-diseases/mountain-pine-beetle/.

Kaye, Byron, and Colin Packman. "Australia evacuates parts of its capital as bushfire conditions return." *Reuters*. January 2020. https://www.reuters.com/article/us-australia-bushfires/australia-evacuates-parts-of-its-capital-as-bushfire-conditions-return-idUSKBN1ZK2T5?il=0.

Landscape Fire and Resource Management Planning Tools (LANDFIRE). "40 Scott and Burgan Fire Behavior Fuel Models." Accessed October 4, 2019. https://www.landfire.gov/fbfm40.php.

National Fire Protection Association (NFPA). "NFPA 1144, Standard for Reducing Structure Ignition Hazards from Wildland Fire." 2018 ed. https://www.nfpa.org/codes-and-standards/all-codes-and-standards/list-of-codes-and-standards/detail?code=1144.

National Wildfire Coordinating Group (NWCG). "S-190: Introduction to Wildland Fire Behavior, 2008." Modified October 2, 2019. https://www.nwcg.gov/publications/training-courses/s-190.

National Wildfire Coordinating Group (NWCG). "S-190: Introduction to Wildland Fire Behavior, 2019." Modified October 2, 2019. https://www.nwcg.gov/publications/training-courses/s-190/test/overview.

National Wildfire Coordinating Group (NWCG). "S-290: Intermediate Wildland Fire Behavior, 2010." Modified August 16, 2019. https://www.nwcg.gov/publications/training-courses/s-290.

National Wildfire Coordinating Group (NWCG). "S-390: Introduction to Wildland Fire Behavior Calculations, 2007." Modified August 21, 2019. https://www.nwcg.gov/publications/training-courses/s-390.

National Wildfire Coordinating Group (NWCG). "S-490: Advanced Fire Behavior Calculations, 2010." Modified September 4, 2019. https://www.nwcg.gov/publications/training-courses/s-490.

National Wildfire Coordinating Group (NWCG). *Fire Behavior Field Reference Guide*, PMS 437. Modified April 7, 2019. https://www.nwcg.gov/publications/pms437.

National Wildfire Coordinating Group (NWCG). "*NWCG Fireline Handbook* Appendix B: Fire Behavior." PMS 410-2. NFES 2165. April 2006. https://www.nwcg.gov/sites/default/files/products/appendixB.pdf.

National Wildfire Coordinating Group (NWCG). *Wildland Fire Incident Management Field Guide*. PMS 449-1. NFES 2943. January 2014. https://www.nwcg.gov/sites/default/files/publications/pms210.pdf.

Scott, Joe H., and Robert E. Burgan. "Standard Fire Behavior Fuel Models: A Comprehensive Set for Use with Rothermel's Surface Fire Spread Model." June 2005. https://www.fs.fed.us/rm/pubs/rmrs_gtr153.pdf.

Wildland Fire Fighter in Action

It is June 11 around 1550. You are part of a five-person engine crew that has been assigned to support a burnout operation along a road that is being led by a hotshot crew. The fire you are on started near a campground up in the mountains between 3000 and 4000 ft (914.4 and 1219.2 m) and has been burning for three days. This fire started during a foehn wind event. Over the past 3 days, the fire intensity has increased and calmed down. Earlier in the morning, the wind was calm, but the weather usually changes during the afternoon due to the drainages and steep terrain of 40 to 60 percent slopes. At and near the burn operation, there are multiple aspects and intersecting drainages. The forecast was for the weather to become hotter and drier, but it did not meet the conditions for a Red Flag Warning. The fuel type is cured knee-high grass that has existed in an extended drought. Adjacent fuels are oak, pine, and brush, which have a patchy continuity. The overnight humidity recovery was poor at or less than 10 percent.

Within 30 minutes from the time of ignition, the wind intensifies at the base of the column and the edge of the road. You are not able to see the entire fire but suddenly notice spot fires across the road, which is what you are trying to suppress. At this time, you notice the winds picking up to about 20 to 30 mi/h (32.2 to 48.3 km/h) and more spot fires are beginning to occur.

1. With your being in a supporting role in the burn operation, what is the most important component to have in place?

A. Lookouts, communications, escape routes, and safety zones

B. Wearing all of your personal protective equipment

C. Water in the tank of your fire engine

D. A radio to communicate with

2. What type of weather event can you expect to have happen at this time of the day?

A. Extreme fire behavior, including rotational vortices, rapid spot fires, and rapid fire growth

B. Minimal fire activity because the fire has been quiet the past day or two

C. Cold front moving in

D. Wasatch winds followed by decent humidity recovery

3. What would you do if you heard over the radio earlier in the day that a crew member from another crew observed fire whirls?

A. Expect that there is a potential for erratic fire behavior.

B. Do nothing because your supervisor hasn't said anything.

C. You are nowhere near the reported fire activity. Essentially you are safe.

D. Base your decisions on current and expected behavior of the fire, and talk this over with your supervisor.

4. If you were in this situation, what would you say to your supervisor?

A. Nothing

B. Casually mention this because it wasn't near you

C. Keep track of the weather trends throughout the day and check to see whether you can get a spot weather forecast.

D. Let's see if we can get another engine over here to assist us.

Access Navigate for flashcards to test your key term knowledge.

CHAPTER 5

Wildland Fire Fighter I

Strategy and Tactics

KNOWLEDGE OBJECTIVES

After studying this chapter, you will be able to:

- Explain the differences between objectives, strategy, and tactics. (p. 124)
- Explain the differences between strategy and tactics. (p. 124)
- Explain the activities required to confine and extinguish a wildland fire. (**NFPA 1051: 4.5.1, 4.5.4**, pp. 124–125)
- Describe the size-up process. (**NFPA 1051: 4.5.2, 4.5.3**, pp. 124–127)
- Describe how to form an incident action plan. (pp. 127–129)
- Describe the parts of a wildland fire. (p. 129)
- Explain the importance of an anchor point. (p. 129)
- Describe common types of attacks on wildland fires. (**NFPA 1051: 4.1.1**, pp. 131–132)
- Describe the difference between a fire control line and a fireline. (p. 133)
- Describe how to construct different types of fireline. (**NFPA 1051: 4.5.4**, pp. 133–135)
- Describe safety considerations for fireline construction. (**NFPA 1051: 4.5.5**, pp. 133–138)
- Describe hazards to an existing control line. (p. 135)
- Describe how to secure a fireline. (**NFPA 1051: 4.5.4, 4.5.5**, pp. 135–136)
- Describe the importance of mop-up. (**NFPA 1051: 4.5.7**, pp. 136–137)
- Describe the importance of patrolling a fire area. (**NFPA 1051: 4.5.8**, pp. 137–138)
- Explain the importance of a decision point. (pp. 137–138)
- Explain the importance of Take 5@2. (p. 138)

SKILLS OBJECTIVES

After studying this chapter, you will be able to perform the following skills:

- Construct a one-lick hand line. (pp. 134–135)
- Construct a leapfrog hand line. (p. 135)
- Perform mop-up operations. (**NFPA 1051: 4.5.7**, pp. 136–137)
- Perform patrolling operations.(**NFPA 1051: 4.5.8**, pp. 137)

Additional Standards

- **NWCG S-190**, *Introduction to Wildland Fire Behavior*
- **NWCG S-130 (NFES 2731)**, *Firefighter Training*

You Are the Wildland Fire Fighter

You are a new engine boss who is responding as mutual aid to a neighboring jurisdiction. You oversee a crew of six and operate out of a Type 3 engine with a chase truck. You are responding as part of a preplanned automatic dispatch for a structure fire that is starting to extend into the forest. The area you are responding to is part of a newer subdivision tucked up on a hillside with narrow roads that have very steep slopes. This area has been identified as a target hazard due to the numerous structures that abut a very sensitive forest that is managed by one of the Native American tribes. Over the past couple of years, the fires that have occurred in this area have experienced rapid fire growth in the first operational period. As you leave your station, you see several miles off in the distance a column of smoke that seems to be growing rapidly that starts to lean over at 9000 ft (2.7 km). Weather conditions are as follows: It is 97°F (36.1°C), 17 percent humidity, and the winds are steady at 15 mi/h (24.1 km/h) with an occasional gust of 23 mi/h (37 km/h).

1. What factors would you consider on the way to your fire assignment?
2. What factors would you evaluate upon arrival at the fire?
3. On what factors would you base your incident action plan?
4. What is the difference between a strategy and a tactic?

Access Navigate for more practice activities.

Introduction

Incident objectives, strategies, and tactics are three fundamental pieces of a successful incident. **Objectives** state what is to be accomplished. **Strategy** can be defined as a general plan, usually including the leader's intent, to meet a set of predetermined objectives. **Tactics** describe the specific steps that need to be taken to support and achieve the strategy. Tactics are the operational aspects of fire suppression. The incident commander (IC) or operations section chief is responsible for establishing the tactical direction on the incident **FIGURE 5-1**.

This chapter discusses all the skills, methods, and activities for a fire fighter to safely confine and extinguish a wildland fire.

FIGURE 5-1 A battalion chief takes command of an initial fire attack.
Courtesy of Joe Lowe.

Size-Up

A **size-up** is the process of gathering and evaluating information from a fire to determine a course of action and prepare for response. It involves developing a mental image of a fire's past, present, and future behavior and conditions. This information can be derived from what you have or another person has observed or from a dispatcher. A size-up is an ongoing process because fire is dynamic and needs to be constantly reevaluated. The first component of a good size-up is knowing and understanding the facts. The inside front cover of the *Incident Response Pocket Guide* (IRPG) has a size-up checklist that can be used. **TABLE 5-1** outlines this checklist.

For emerging incidents, the size-up checklist is a crucial component of how you will begin to develop your situational awareness (SA) for the incident you are responding to. SA is what will allow you to constantly evaluate how you are engaged on an incident. It will also allow you to determine decision points and other components of LCES/LACES, the 10 standard fire orders, and the 18 watch outs.

LISTEN UP!

The size-up is an ongoing process, not a one-time action. A fire should be constantly reevaluated.

Dispatch

The size-up should start on the way into the fire station before the shift assignment begins. On the way to

TABLE 5-1 Size-Up Report

The size-up report is a checklist to help with establishing and reporting the following information:

- Incident type (wildland fire, vehicle accident, hazmat spill, search and rescue, etc.)
- Location/jurisdiction
- Incident size
- Incident status
- Incident command and fire name
- Weather conditions
- Radio frequencies
- Best access routes
- Assets/values at risk
- Special hazards or concerns
- Additional resource needs

This report is intended to assist in establishing key information regarding incident conditions when first arriving on scene. All agencies will have specific information requirements that may involve additional reports.
Taken from *Incident Response Pocket Guide,* 2018, https://www.nwcg.gov/sites/default/files/publications/pms461.pdf.

work, observe the weather factors. Are they different from the day before?

When the alarm goes off, look at the area where the fire is burning on a map, and compare it with a mental picture of that area formed from previous knowledge of that location. The fire officer will try to recall the terrain features of the area, the fuels that are present, the daily wind patterns, and predicted fire dangers in the area.

En Route

En route to the fire, the fire officer looks again at the weather factors present. Which way is the wind blowing? What is the wind speed? Are the winds gusty or steady? Are there thunderheads or other cloud types forming that indicate unstable conditions? Does the humidity seem to match the forecast for the day?

Other factors that should be included in an en-route size-up include the safest access or route to the fire, whether the property is privately or publicly owned, and any vehicles or people leaving the fire scene. It is also a good idea to assign tactical channels and communications while en route.

While coming closer to the fire, the fire officer looks at and evaluates the smoke column. The fire officer checks the smoke column for direction, shape, size, height, and color. If the smoke column is leaning, it is a good indication that the fire is wind driven. The wind gives the fire direction and increases its rate of spread. Wind also increases the potential for spotting activity.

Smoke columns can help to determine weather factors and a fire's intensity. If the column has an anvil top to it, high winds are present aloft. A large, well-developed column can indicate a plume-dominated fire, which indicates that the thermal energy being released upward is overpowering the local winds. With a plume-dominated fire, its rate of spread and direction will be unpredictable, and there will be strong indrafts and downbursts. This fire will be dangerous. A well-defined smoke column that rises to great heights indicates that the air mass is unstable, which contributes to intense fireline conditions.

The smoke color is usually a reliable indication of the type of fuel in which the fire is burning. Light-colored smoke indicates that the fire is burning in light flashy fuels, such as grass, and dark-colored smoke indicates that the fire is burning in heavier fuels, such as brush or timber.

Once the company officer is in a position to observe the fire, it should be evaluated. Begin by looking at the flame lengths. Flame lengths indicate the correct method of attacking the fire, either directly or indirectly, and also help to define the intensity of the fire.

Next, look at the rate of spread of the fire and the direction in which it is burning. Will it be moving into a different type of fuel soon? Are there any natural or manmade fire barriers? Will there be a change in topography? How large a fire is it? Is it spotting?

Arrival

Upon arrival at the fire scene, the fire officer is still gathering facts and then starts forecasting, which deals with probabilities. Once the initial size-up has been completed, the fire officer reports the conditions via radio, which gives incoming fire units information on the size of the fire, fuel type, rate of spread, direction of fire spread, exposures involved, and the mode of operation that the first-arriving unit is in. Additional resource requests are also placed by radio. Any resources that might be needed should be ordered. If you think you need something, order it. A typical report on conditions would sound like this: "Orange County Engine 318 on scene of a vegetation fire in light fuels with a rapid rate of spread. Structures are threatened, and the fire is approximately a quarter acre and moving easterly toward a group of homes on Overhill Drive. Engine 318 will be Peters Canyon IC. I would like to request one additional Type III strike team, two Type II handcrews, two Type 3 dozers, and two Type 2 helicopters."

After giving a report on conditions and ordering any additional resources, the company officer needs to make sure that LCES/LACES is in place and then announce that over the radio. If LCES/LACES cannot be established, fire fighters should not engage the fire.

Once the escape route and safety zone have been identified and communicated to the crew, it is time for the fire officer to evaluate factors pertinent to the initial attack fire operations. At this time, the fire officer either takes command of the fire or works with the fire crew to extinguish the fire. If the fire officer is working with the crew, command is usually taken by the next arriving officer. The decision to take command of the fire is based on size-up factors. If the first-in fire officer feels that his or her involvement would allow for extinguishment of the fire, then he or she should remain with the crew and suppress the fire. Should the fire require multiple alarms (extended attack fire), then the fire officer usually becomes the IC.

The critical factors that must be looked at to complete the fire size-up are based on the observed fire behavior, the fuels present, the weather, the topography of the area, and the time of the day. By observing the flame lengths, you can judge the kind of fireline intensity levels to expect.

LISTEN UP!

The critical factors that must be looked at to complete the fire size-up are based on the observed fire behavior, the fuels present, the weather, the topography of the area, and the time of day.

Fuel Factors

Fuels provide energy for the fire. A size-up must evaluate fuels for the following factors:

- What type of fuel is the fire burning in, and what type will it be moving into?
- What is the continuity of the fuel bed?
- Is the fuel bed an old-age class fuel bed with a large dead component (live-to-dead fuel moisture ratio)?
- What ladder fuels are present?
- Can the fire step up into the crowns of tall brush or trees?
- What is the horizontal continuity of those crown fuels?
- Are there snags or broken tree limbs in the area?
- What is the fuel moisture of the 1-hour fine dead fuels?
- Is there a reburn potential, especially if in a previous year there was an understory burn only?
- Is it a fuel that contains flammable oils, such as chaparral, palmetto, gallberry, pine, or eucalyptus? These fuels produce greater fireline intensities.

Topography Factors

Topography factors must also be evaluated during a size-up because they can influence the rate of spread and direction of the fire. These topography factors include:

- Steepness of the slope.
- Origin of the fire on the slope (bottom of the hill, midslope, or ridgetop).
- Aspect. Is it a hot or cold slope? What time of day is it?
- Terrain features that could intensify or increase the rate of spread (chimneys, saddles, and box canyons).
- Location. Is the fire in a narrow canyon? Can it easily spot to the other side?
- Elevation of the fire.
- Slope and elevation in relation to interface areas or structures.

Evaluating Resources

Next, the IC must evaluate the resources available to mitigate the hazard. Are they the right kind or type? Are there enough resources present? How long will it take to get additional ones and how much will they cost?

All fire fighters, supervisors, and higher-ranked personnel need to be constantly evaluating crew readiness. While the concept seems relatively simple, there are several complexities that need to be evaluated both internally and externally. For example, is a local, state, or federal entity responsible for a fire? Each entity can have its own standards regarding the following:

- Work-to-rest ratio standards (the number of hours that can be worked before personnel are required to take a break)
- Crew cohesion (the overall readiness of the crew and ability to accept the assignment based on fitness factors)
- Time for the crew to assemble (assuming several entities make up a crew))
- Who is capable of operating agency vehicles

Most of the information mentioned here is determined before the fire season; however, between incidents or when a crew or resource is released from one incident to head to another, all of these components need to be considered.

With respect to human factors, if fire fighters do not have adequate supplies and rest before acceptance of an incident, there is a greater chance for an accident to occur. Following are questions that need to be answered before acceptance of an assignment:

- Is the crew prepared for response?
- Has the crew just finished another fireline assignment without a period of rest?
- Has the crew traveled a great distance to the fire?
- When was the last time the crew was fed?
- If crew members were working a 48-hour shift, did they have adequate rest during the first 24 hours? If not, what can be done to mitigate this?
- If there are several agencies that make up a strike team or task force or are part of a regional handcrew where will they meet to form a packaged resource before responding to the incident?

One resource that should never be overlooked is water. The responding fire resources need to evaluate the water supplies, taking into account how far they are from the fire, what the sources are, and how much water there is. It is important to look beyond a 24-hour period to make sure that adequate water and food are available.

Finally, the IC looks at the actions that are already being taken. Can the operation that has started be supported, or does it need to be scrapped to develop a new plan?

Weather Factors

Weather factors have a pronounced effect on a fire. Because they are the most changeable, they must be evaluated constantly. Evaluate these weather factors during a size-up:

- What is the wind speed and direction? Are the winds gusty or steady?
- Can any thunderstorm activity be observed?
- Are there any indicators of a frontal passage?
- Are there indications of instability of the air mass? Is the visibility clear? Is the smoke column well defined and rising to great heights?
- Are there fire whirls or dust devils?
- What is the temperature?
- What is the humidity level?
- Are there indications of an inversion layer?
- What are the normal wind patterns for the area?
- Did the winds suddenly become calm

Forecasting and Weather Observations

One of the most important decisions to be made is evaluating life hazards. Will the fire be moving toward the structures soon? Have the structures been evacuated? If not, should evacuation take place? The decision to evacuate residents or shelter them in place is based on a size-up of the current observed and expected future fire behavior. The fire officer also must determine whether the evacuation route is safe from the fire while the evacuation is in progress.

Part of the forecasting that is done involves the availability of critical support items, such as aircraft, bulldozers, handcrews or additional engine companies. The IC also needs to know whether there are restrictions on suppression tactics, such as the use of bulldozers in wilderness areas.

Once the evaluating and forecasting have been done, the IC can start developing a plan.

The fire officer also needs to collect current weather observations and understand, based on the previous information gathered, where the fire is going. What will its intensity levels be? Where will it be in 30 minutes or 1 hour (depending on the size before arrival)?

Weather observations can be accomplished with the use of the standard belt weather kit or other electronic (calibrated) multipurpose weather instrument. Knowing the temperature and current elevation allows you to determine humidity. This information aids in determining the probability of ignition, as well as provides a base of information to submit to your dispatch agency or the fire behavior analysist on a larger fire for submission to the National Weather Service for spot weather forecasting.

A **spot weather forecast** is issued upon request to fit the time, topography, and weather of a specific incident. Generally, these forecasts require precise global positioning system (GPS) locations or a legal description of where you are (township, range, and section). (See Chapter 11, *Orienteering and Global Positioning Systems*, for more detailed information on GPS devices.) These forecasts are more detailed, timely, and specific than zone forecasts. Lastly, all of the weather information gathered can be used in conjunction with the tables found in the IRPG to calculate the probability of ignition, or PIG.

Forming an Incident Action Plan

Developing an incident action plan (IAP) is the next tactical step. IAPs are oral reports or written documents containing general objectives that reflect the overall strategy and specific actions for managing an incident or planned event (see Chapter 2, *The Incident Command System and ICS Forms*). They can include several different attached forms. An IAP is based on the following information:

- Facts
- Available tools
- Completed forecasting

At the start of an incident, the situation may be chaotic and complete information may be difficult to obtain. The IC's initial plan may be developed very quickly and with incomplete information. During this initial stage of the incident, the IC develops a simple plan with the current information and communicates that plan through concise oral briefings. The IC will usually start using ICS Form 201, *Incident Briefing*, to keep track of resources allocated to the incident, where they are assigned, and basic information on the incident situation. This form will be part of the incident's permanent record. The ICS Form 201 will later be used to transfer command of the incident. As the incident grows in size or complexity, more of the command and general staff positions are filled and a more formal planning process takes place to develop a written IAP.

Objectives

The IC must decide where to attack the fire. The first consideration and the highest priority in developing an IAP is always life safety. Stabilization of the incident should be the second priority. The third priority is conservation of the homes and property in the fire's path.

From these incident priorities, the IC sets incident objectives that outline what is to be accomplished. Objectives need to be SMART: specific, measurable, achievable and action oriented, realistic, and timely.

Strategy

Next, the IC decides on a strategy. How can the incident objectives be attained? Time can be used as a control tool and tactics can be time tagged. If the strategy is not working after a given time period has elapsed, then it may be time to switch to another tactic.

Objectives form the strategy. Objectives and strategy need to be communicated appropriately (leader's intent). One way to do this is to use the task, purpose, and end-state method for idea sharing:

- Task. This is what needs to be done or accomplished.
- Purpose. Why the assignment needs to be done and how the task fits into the purpose. Without a purpose, motivation may be lost, and expectations may not be met.
- End state. What the final product needs to look like when the assignment is successfully completed. A misunderstanding of this step could mean the difference between saving or losing a structure.

These three components are the building blocks to developing an effective strategy. If fire fighters do not understand these components, the objectives will not be met, and the assignment will not be successful.

Tactics

Once the objectives and strategy have been established, the IC must develop tactical priorities listed in their order of importance. Tactical priorities are tasks that must be performed to meet incident objectives. The situation and progress being made must constantly be evaluated, and personnel must be accounted for and their needs supported.

Risk Management

While forming an IAP, risk management needs to be constantly evaluated. Risk management is generally a responsibility for the IC or the resource boss; however, the five components that make up this process are important for every fire fighter to understand. The components are as follows:

- Identify hazards
- Assess hazards
- Develop controls and make risk decisions
- Implement controls
- Supervise and evaluate

The risk management process can be found on the first green page of the IRPG. As well as forming an IAP, all fire fighters have a responsibility to identify and participate in the first three components listed previously. There will be many situations in which you will see or hear something that a supervisor may not be able to. In this case, if you see or hear something, say something to your supervisor. For example, if you were to identify a hazard tree that several people have walked by, say something! In addition to communicating the information to your supervisor, make sure you flag off the area (using killer tree tape or some other identifiable surveyor flagging) so that no one else will take an unneeded risk **FIGURE 5-2**.

FIGURE 5-2 Flag off hazards with tape as you see them.
Courtesy of Joe Lowe.

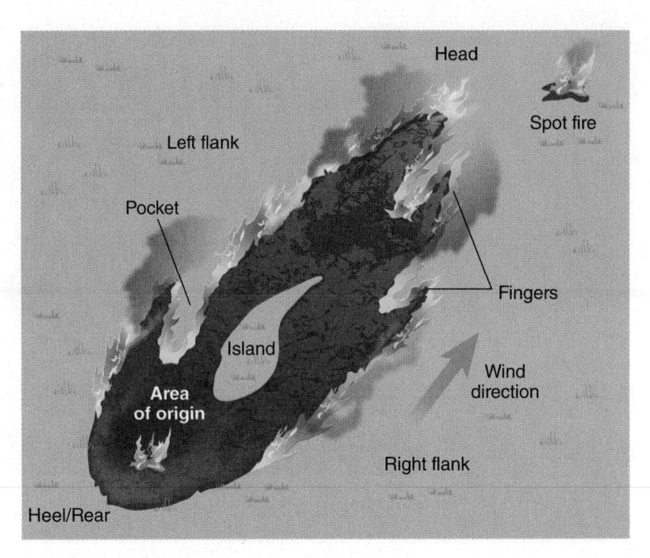

FIGURE 5-3 The parts of a wildland fire.

Parts of the Wildland Fire

Before giving an explanation on where to attack the fire, the parts of the fire must be identified **FIGURE 5-3**. Knowing the parts of the fire is necessary to understanding the remainder of this chapter and to being successful as a wildland fire fighter.

The location where a wildland fire begins is called the **area of origin**. As the fire grows and moves into new fuel, the most rapidly spreading area and intense part of the fire is called the **head** of the fire. As the fire gets bigger, the area close to the area of origin may be referred to as the **heel**, or rear, of the fire. The heel of the fire usually spreads directly into the wind or downslope. The **flank** of the fire is the edge between the head and heel of the fire that runs parallel to the direction of the fire spread.

As the fire grows, a change in weather, topography, or fuel may cause it to move in such a way that it projects out into a long, narrow extension called a **finger**. A finger can produce a secondary direction of travel for the fire. The unburned area between a finger and the main body of the fire is called a **pocket**. A pocket is a dangerous place for fire fighters because this area of unburned fuel is surrounded on three sides by fire. An area of land that is left untouched by the fire but is surrounded by burned land is called an **island**. A **spot fire** is a new fire that starts outside the perimeter of the main fire. Spot fires usually begin when flaming vegetation, in the form of an ember or spark, is picked up by convection currents generated by the fire and dropped some distance away from the original fire.

Attacking the Fire

After size-up, the plan must be communicated to the crew. This plan should include task, purpose, end state, LCES/LACES, and other items noted on the inside back cover of the IRPG. An anchor point must also be established from which to start the fire attack. An **anchor point** is an area close to where the fire started and from which vegetation has already been burned. Anchor points are physical barriers to fire, such as a roadway surface, that provide an area of safety for fire fighters. Anchor points help prevent a situation in which the fire might change direction, circle back on fire fighters, and trap them in a point of danger.

LISTEN UP!

Remember that fire will burn when all three elements of the fire triangle are present. The fire triangle can be broken by removing one or more of the following elements:

- Oxygen. Suffocate the fire.
- Heat. Cool the fire.
- Fuel. Remove fuel to prevent combustion.

Determining an Attack Point

The IC must then decide where to attack the fire. In the case of an interface fire, the IC needs to evaluate whether to try to contain the fire first or to protect property. The fire attack may require resources to attack the fire at the flanks, heel, or head of the fire, or a combination of all three. This decision is determined by the fireline intensity level and rate of spread of the fire. A good way to judge a fire's intensity is to look at the flame lengths. **Flame length** is a measurement from the average flame tip to the middle of the flaming zone at the base of the fire **FIGURE 5-4**. It is measured on a slant when the flames are tilted due to effects of wind and slope. An evaluation of the flame length

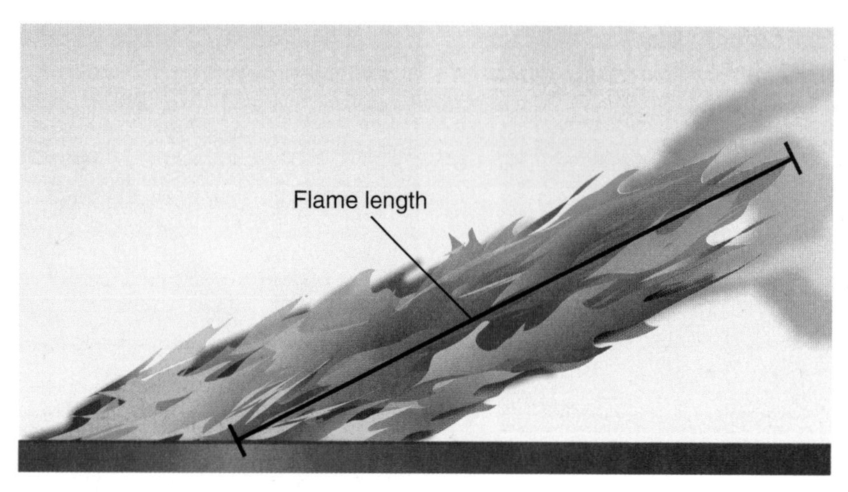

FIGURE 5-4 Flame length.

TABLE 5-2 Fire Suppression Limitations Based on Flame Length

Flame Length	Intensity	Method of Attack
0–4 ft (0–1.2 m)	100 BTUs/ft/sec	Direct attack hand tools
4–8 ft (1.2–2.4 m)	500 BTUs/ft/sec	Hoselays, retardant, bulldozers
8–11 ft (2.4–3.4 m)	1000 BTUs/ft/sec	Direct control difficult, control efforts at head may be ineffective
More than 11 ft (3.4 m)	1000 BTUs/ft/sec	Indirect attack, control efforts at head ineffective, spotting, major runs

Abbreviation: BTU, British thermal unit
Data from: USDA. Forest Service's Table of Fire Suppression Limitations Based on Flame Length.

helps the IC determine which part of the fire to attack in both initial-attack and larger fires.

An effective tool that the IC can use to help him or her select an attack method is the USDA Forest Service's Table of Fire Suppression Limitations Based on Flame Length, shown here as **TABLE 5-2**.

The top priority is to ensure the safety of both fire fighters and citizens. All fire fighters should make sure that they know the escape route and escape plan before attacking the fire. The second priority is to minimize the property losses and environmental damage from the fire as much as possible.

LISTEN UP!

Minimum impact suppression tactics (MIST) allow for effective fire suppression with the least environmental, cultural, and social impacts. MIST should always be taken into consideration when responding to a fire. MIST is discussed in more detail in Chapter 6, *Wildland/Urban Interface Considerations*.

Backup Plans

The IC always has an alternate, or backup, plan for fire control when the fire cannot be contained promptly, which includes having a secondary control line in mind. In some cases, having only a backup control line is not enough. Recent fires prove that alternate, or backup, plans frequently have to be backed up by yet another plan, or contingency plan. If the primary, alternate, and contingency plans all fail, emergency actions need to be established. A good mnemonic tool to help you remember this process is the word PACE: *primary*, *alternate*, *contingency*, and *emergency*.

The Fire Behavior Hauling Chart found in the IRPG is a good reference for understanding the intensity of

LISTEN UP!

Regardless of the specific tactics identified by the IC, all crew members need to be thinking about the readiness and accessibility of escape routes.

the fire and how it will affect your decision-making. This chart is a simple and quick reference for understanding what to expect based on the fire activity and flame lengths.

Types of Attacks

Wildland fires can advance and change directions quickly. Given their unpredictability, they present a very hazardous environment for fire fighters. For these reasons, it is important for the IC to match the type of firefighting attack to the conditions present at the fire. There are three main types of attack: direct, indirect, or parallel. The type of attack is determined by the types and amounts of resources available as well as the fire size and behavior. If ample resources are readily available, then offensive mode can be used to accomplish the strategic goals of the incident. Fire behavior, intensity levels, and the kind of resources available determine whether the fire is attacked directly or an indirect method is used.

Fire attacks should always start from an anchor point, that is, an area close to where the fire started and from which vegetation has already been burned. This strategy helps prevent a situation in which the fire might change direction, circle back on fire fighters, and trap them in a point of danger.

Direct Attack

A **direct attack** on a wildland fire is any treatment applied directly to burning fuel such as wetting, smothering, or chemically quenching the fire or by physically separating the burning from unburned fuel. A direct attack is mounted by containing and extinguishing the fire at its burning edge. A direct attack can be made with a handcrew and hand tools, a structural engine company, a wildland engine company, a plow, or an aircraft. Fire fighters might smother the fire with dirt, use hose lines to apply water or class A foam to cool the fire, or remove the fuel. A direct attack is typically used on a small wildland fire before it has grown to a large size.

Any direct attack needs to be coordinated through the IC. Before a direct attack is started, the IC will assess the resources available and determine the best way to attack the fire.

A direct attack is dangerous because fire fighters must work in smoke and heat close to the fire. However, this type of attack has the advantage of accomplishing quick containment of a fire. By extinguishing the fire while it is small, fire fighters reduce the risk posed by the fire. Most wildland and ground-cover fires are kept small because the initial crews used an aggressive direct attack on the fire.

Two major types of direct attacks are used on wildland and ground-cover fires: the anchor, flank, and pinch attack and the flanking attack.

Pincer Attack. The **pincer attack** or the **anchor, flank and pinch attack** (also known as the pincer action) is a method of direct attack that requires two or more fire attack units. The pincer attack can be accomplished with engines, heavy equipment, or air equipment. This is the safer and more reliable method of direct attack. The key to this method is the establishment of two solid anchor points. One attack unit establishes an anchor point on the left side of the fire, near the point of origin, and mounts a direct attack along the left flank of the fire, working toward the head of the fire. A second attack unit establishes an anchor point on the right side of the fire, near the point of origin, and mounts a direct attack along the right flank of the fire, also working toward the head of the fire. As the units advance, the fire gets "pinched" between them, which reduces its growth. By starting at a safe anchor point, the risk to fire fighters is minimized. A successful attack using this method requires the availability of sufficient personnel to be able to mount a two-pronged attack **FIGURE 5-5**.

Flanking Attack. The **flanking attack** (sometimes called the flank attack) is a method of direct attack that is used for moderately intense fires moving at a moderate rate of spread. The flanking attack generally requires only one unit of fire fighters. Fire attack starts at an anchor point and attacks along one flank of the fire. Such an attack is also used when sufficient

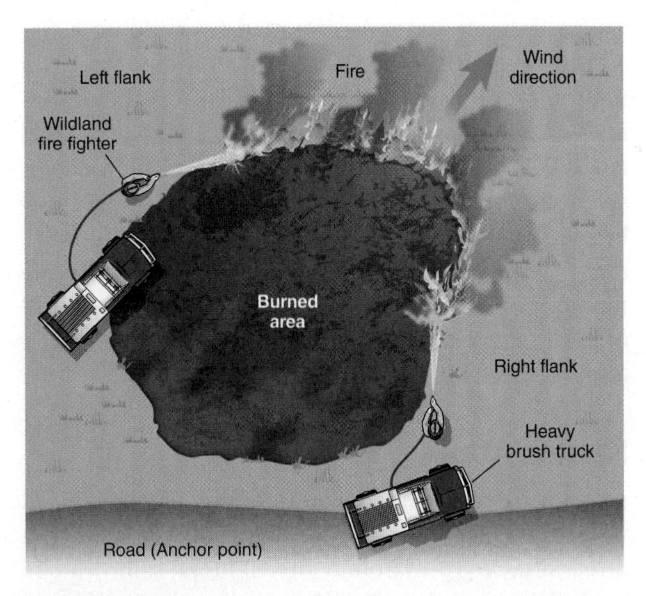

FIGURE 5-5 The pincer attack involves engagement on both flanks of a wildland fire.

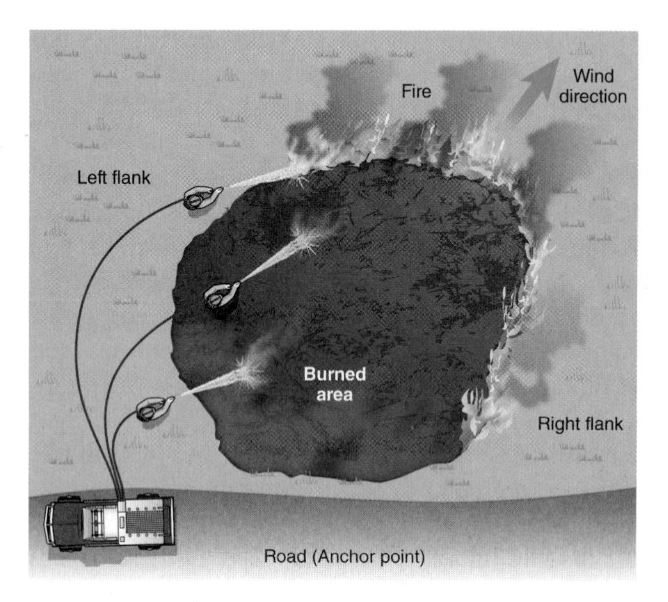

FIGURE 5-6 A flanking attack is made on the left flank or right flank of a wildland fire.

FIGURE 5-7 An indirect attack is made along natural fuel breaks, at favorable breaks in the topography, or at a considerable distance from the fire.

resources are not available to mount two simultaneous attacks or when the risks posed by the fire are high on one flank of the fire and lower on the other flank of the fire. The decision regarding which flank of the fire to attack is based on the determination of which side of the fire poses the greater risk **FIGURE 5-6**. After progress has been made on the flank, additional resources can be placed at the other flanks.

Indirect Attack

An **indirect attack** is most often used for large wildland and ground-cover fires that are fast-spreading or of high-intensity. It is also appropriate when the topography or weather conditions pose a high risk for fire fighters' safety.

An indirect attack is mounted by building a fire control line along a predetermined route, based on natural fuel breaks or favorable breaks in the topography, and then burning out the intervening fuel **FIGURE 5-7**. This strategy is similar to using a defensive attack on a structure fire. The fire control line may be constructed fairly close to the fire or several miles away. On smaller incidents, you may be closer to the flame front. An indirect attack requires only one unit of fire fighters and can be mounted using either hand tools or mechanized equipment.

Parallel Attack

As with the indirect attack, a **parallel attack** is used when the fire edge is too hot to approach through a direct attack. To maintain fire fighters' safety, a fire control line is constructed parallel to the fire's edge, with a distance of about 5 to 50 ft (1.5 to 15 m) between the control line and the fire's edge—far enough from the fire that workers and equipment can continue to work effectively if the intensity of the fire grows **FIGURE 5-8**. In some cases, the fire control line can be shortened by being constructed across the fire's unburned fingers. The unburned fuel usually burns out as the fire moves alongside the control line, but it can also burn out unassisted with the main fire as long as the weather and terrain do not present a threat to the fire control line.

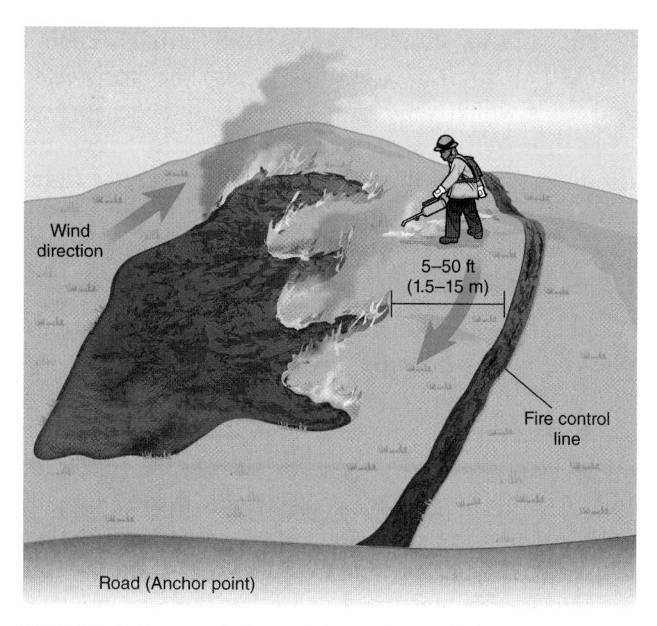

FIGURE 5-8 A parallel attack is made parallel to the fire's edge.

Fireline Construction

A **control line**, or fire control line, is an inclusive term for all constructed or natural barriers used to control a fire. A **fireline**, also called a constructed fireline, is a barrier purposefully constructed by fire fighters to control a fire. One of the most basic and fundamental skills for the fire fighter to master is fireline construction. Fireline construction is the process of making a box around a fire to put it out and keep the fire from spreading. This is accomplished using several methods. Generally, the fire fighter and the fire officer choose the method that is most appropriate for the fuel type, terrain, and weather. The goal is to remove the fuel from the fire triangle. This prevents the fire from spreading. In most cases, the removal of fuel will require removing all vegetation down to the dirt or mineral soil. Remember to establish an anchor point before you start your fireline.

The types of constructed fireline and natural control lines are discussed next.

Constructed Fireline

A constructed fireline is a fireline intentionally created by fire fighters or handcrews. Some constructed firelines include hand lines, saw lines, mechanized lines, retardant lines, and explosives.

Scratch Line

A **scratch line** is an unfinished, temporary control line that was quickly constructed or scratched into fuels as an emergency measure to check the spread of fire. A scratch line is used as an expedient control line until a wider, more permanent control line can be constructed. Oftentimes, scratch lines are later incorporated into the main control line.

LISTEN UP!

Hot spotting is a method used to reduce or stop fire spread and intensity at points with particularly rapid rates of spread (usually fingers). Hot spotting involves small crews that work in front of the fire to slow its pace. These crews create scratch lines around the hot spots. Hot spotting is dangerous because fire fighters are working without an anchor point.

Hand Line

A **hand line** is a fireline constructed with hand tools. This line is most effective when employed by a team or crew of fire fighters working in tandem. A variety of tools are used to construct the line, including shovels, Rineharts, Pulaskis, rakes, fire swatters, hazel hoes, McLeods, and other specialized tools. Spacing of the tools and variety is very important in order to be the most effective. More often than not, a handcrew follows a saw team in constructing a fireline. Hand lines, saw teams, and proper methods of carrying, passing, and using hand tools are discussed in more detail in Chapter 14, *Handcrew Operations.*

Cold Trailing

Cold trailing, also called a feel-out line, is a technique where fire fighters feel ashes and fuels with their bare hands to make sure a fire is out **FIGURE 5-9**. If any live spots are found (determined by the feeling of heat), fire fighters dig it out and extinguish it.

Saw Line

A **saw line** is created by a team or groups of teams used to clear away trees and brush so that the remaining fuel can be removed with hand tools. This is usually accomplished by creating a fuel break and buffer area in the fuel. Saw teams also clear any ladder fuel adjacent to the line to prevent vertical movement of the fire. In Alaska, saw lines are very effective in helping to control the spread of fire when the fire is burning in a combination of tundra and spruce or tundra and brush.

Saw teams will be in front of the crew putting in the hand line. If saw teams are working ahead of you, make sure you are two and a half tree lengths away to allow for the required safety spacing. A good saw team can be very effective and significantly enhance the forward progress of the fireline construction.

The density of the canopy and adjacent fuel may pose a challenge for saw teams. If the fuel is very thick, progress may be slowed by the need to create pockets to store the cut fuel. This will take time if multiple saw teams are not available.

FIGURE 5-9 Cold trailing.
Courtesy of Rocky Mountain National Park.

Machine Line

A **machine line**, or mechanized line, is completed by using a variety of heavy equipment. The area of the country, topography, and fuel type will dictate what type of heavy or mechanized equipment can be used. Some of the more common equipment includes dozers of varying sizes, tractor plows, feller bunchers, graders, and skidgens.

Dozers are used to make quick work over long distances. They can scrape down to mineral soil with ease, remove brush and trees, and create roads for wheeled fire equipment. A line constructed by dozers is called a dozer line.

Tractor plows have a successful history in the Southeast with the Southern rough fuel, for example. Feller bunchers are becoming more common in timber country to create lines in dense forests where using hand fallers would take too long. Depending on the terrain and fuel type, graders can be used to slow down fast-moving grass fires. A skidgen acts as a combination dozer and fire engine. Chapter 8, *Ground and Air Equipment*, goes into further detail on these types of vehicles.

Wet Line

A **wet line** is a line of water or a mixture of water and foam or retardant that is sprayed along the ground as a control line. Wet lines are used in light fuels. For more information, see Chapter 12, *Water Supplies and Operations*.

Retardant Line

A **retardant line** is made up of a chemical solution that retards, or slows, the forward progress of the fire **FIGURE 5-10**. In most instances, fixed-wing aircraft distribute retardant **FIGURE 5-11**. However, rotor-wing aircraft can also distribute it. Retardant is only as effective as the coverage level for a specific fuel type. It is important that the two variables be in sync. If they are not, the retardant will not be very effective. Retardants are discussed in more detail in Chapter 13, *Class A Foam and Fire-Blocking Gels*.

FIGURE 5-10 Retardant line.
Courtesy of Joe Lowe.

FIGURE 5-11 Fixed-wing aircraft usually distribute retardant.
Courtesy of Joe Lowe.

Detonation Line

A **detonation line**, also called a detonation cord line or blast line, is a fireline created by explosives. Detonation lines can be very effective in difficult terrain where fuels are light to moderate. In some situations where a fire is burning on a cliff line that is too steep to use conventional hand tools or retardant, a special team can be brought in to use explosives to blast a line through the vegetation.

Natural Control Line

A natural control line is generally established by using natural barriers, such as a river, scree fields, cliffs, rockslides, or other fire resistive or fireproof features. In some cases, previously constructed barriers, such as roads, canals, or train tracks, can be considered natural control lines.

Fireline Crew Construction

Fireline crews are diverse teams of wildland fire fighters. Crews are typically made up of 18 to 20 fire fighters. Crews that are assembled as needed can vary greatly in size. Regardless of size, it is important that crews work together. There are two types of coordinated crew methods for fireline construction: the one-lick method and the leapfrog method.

One-Lick Method

The **one-lick method** is a progressive system used to build fireline without crew members changing positions in the line. Each crew member does one to several strokes (or licks) of work and then moves forward a specific distance to make room for the person behind him or her. Distances are determined by the number of crew members, types of tools, and number of licks needed per area of fireline to complete the work for that tool.

Leapfrog Method

The **leapfrog method**, also called the bump-up method, is a system of building fireline in which each crew member is assigned a task on a specific section of line until the task is completed. When the task is completed, the crew member loudly states "Bump up," and everyone ahead of the caller moves ahead one or more positions, leaving unfinished fireline for those coming up behind them. Crew members then start working where the previous members left off.

Securing the Fireline

There are many threats to an existing fireline. Once a control line has been established, you need to make sure the fire does not leave the control line or fireline you have established around the fire. Common threats to existing control lines include the following:

- Spotting. Wind can carry embers across the control line.
- Rolling debris. Fires on steep slopes require trenches to catch embers and debris before it rolls across the control line.
- Creeping. Root systems may carry fires across the control line.
- Radiant heat. Fire inside the control line may grow hot enough to ignite fuels outside the control line.

There are many ways to secure the fireline, including blacklines, hoselays, mop-up operations, patrolling, and combination methods. The technique used to secure the fireline will always be based on available manpower and equipment. It is very common for crews to retrace the steps of initial-attack fire fighters in attempts to contain a fire and establish a newly secured fireline. All decisions that are made to secure the fireline need to take all of the personnel, crew, and equipment capabilities into consideration before initiating a specific line-securing technique.

Blacklines

In some instances, the blackline method can be used. The **blackline method**, or blackline concept, consists of creating a buffer of burned fuels between the fire's edge and the rest of the fire. Blacklining is done to reduce heat on holding crews and lessen the chances of spotting. It ensures that fuels and heat remain inside the control line. The most common way to blackline is to use fire to burn small strips or pockets to create the buffer space. Another way of looking at this is that, if there is unburned fuel between the fire and the fireline, a simple firing operation can help to secure the line. It is important to note that this tactic should be used only if you have been trained and credentialed to do this. There are several steps to burning that need to be taken before fire is put on the ground. Once fire has been put down, it cannot be taken back. Chapter 15, *Basic Firing Operations*, discusses blacklining in more detail.

Hoselays

A hoselay is an arrangement of connected fire hose and accessories on the ground, beginning at the first pumping unit and ending at the point of intended water delivery. In most cases, a simple or progressive hoselay will be established to transport water around the perimeter of the fire. Water will significantly help secure the fire by cooling the fuel and bringing its moisture content up. Hoselays are discussed in more detail in Chapter 12, *Water Supplies and Operations*.

Pumping and Rolling (Engine Use)

The pump-and-roll tactic is useful when making a direct attack on fast-moving, low-intensity wildland fires when conditions allow its use. This process utilizes a fire engine that can pump water while moving and putting fire out at the same time. One fire fighter walks in front of an engine using a short section of hose or hose from a live reel. The fire engine or engines will start from the anchor point or heel of the fire and put out fire from inside the black to maintain LCES/LACES. In other instances, the hose and engine can be used to create a water line that other fire fighters can follow up with fire. This establishes a blackline that can be built on. The procedure is called a wet-line burnout and should be attempted only by trained and qualified personnel.

LISTEN UP!

In a mobile attack, good communication, either verbal or visual, is essential. If visual contact is lost, the driver must stop the vehicle. If the driver does not stop, the nozzle operator can hit the windshield with a burst of water to signal the driver to stop. Emergency headlights should be on at all times while operating in the field.

Fireline Improvement Techniques

Holding and improving the fireline are just as important as establishing it. Depending on the staffing for the fire and available resources, some fireline is constructed hastily, meaning that more work needs to be completed to ensure that the fire will not be able to spread any farther. One common improvement required is the breaking up of machine piles. Machine piles are the turf and other organic material piled up on one side or both sides of the fireline that is created by a dozer. Sometimes, fire can be trapped in the pile created by a dozer cutting line in fire, which can continue to burn for several days without any sign. This situation can cause the potential for a fire to escape. Other techniques can include removing limbs on or near the fireline that could be considered ladder fuel. Also, one often used practice is chipping. Chipping reduces the surface area of the fuel and grinds it up so that the fuel is less susceptible to catching fire.

Mop-Up

Mop-up is the action of extinguishing or removing burning material near control lines. It is, in many cases, the final extinguishment process of a fire. Though mop-up is often difficult, dirty, and tiresome, it is one of the most important phases of fire suppression because any remaining burning debris may rekindle the fire, making previous actions ineffective. Complete extinguishment of all material burning within the black may be impractical in large fires and may be left to burn until they are consumed. You will, however, be require to extinguish all smoldering material within a specified distance from the control line as indicated by your supervisor. During mop-up, you are securing the fire scene, and it might require snag felling, trench digging, and widening the control line. Hand tools can assist with scraping, digging, stirring, mixing, separating, and turning logs and heavy materials. There are two types of mop-up: wet and dry.

Systematic mop-up, either wet or dry, should follow these steps:

- Start with the hottest area and progress to the coolest.
- Plan a beginning and an ending point. Keep to the plan and work methodically.
- Work inward from the control line.
- Examine the entire assigned area.
- Make sure instructions are clear. Ask questions.
- For large burns and/or complicated situations, some type of grid or block system should be implemented. Set priorities and number each block.

LISTEN UP!

Watch out for honeybees, hornets, and other stinging insects during mop-up operations!

LISTEN UP!

Each of your four senses can aid in detecting burning materials during mop-up operations:

- *Sight*. Look for smoke, heat waves, white ask, stump holes, steam, and gnats.
- *Touch*. Do not wear gloves, but be careful not to burn yourself. Feel materials with the back of your hand, about 1 in. (2.54 cm) away, then carefully with direct contact.
- *Smell*. Smell for live and old smoke, burning materials, and gases.
- *Hearing*. Listen for cracks, pops, and hisses.

During mop-up, always wear full PPE and remain alert for changes in fire behavior and weather. And watch your footing! Some common mop-up hazards include:

- Overhanging and leaning trees
- Snags
- Broken branches or tree tops
- Trees weakened due to roots burning away
- Trees ready to fall but caught in other trees
- Rolling material on steep slopes
- Slippery footing due to water and foam or gel on the ground
- Fire fighters tripping or falling due to extensive hose lays
- Steam and white ash being released from fire pits and hot stump holes when hit by water
- Stump holes
- Increased wind in an area with numerous trees burning

Dry Mop-Up

Dry mop-up primarily consists of exposing and extinguishing burning materials using hand tools or hand tools and soil, without water or wetting agents. Soil can be used to smother or and cool burning materials. There are several techniques used to perform dry mop-up. **TABLE 5-3** outlines some of these techniques.

Wet Mop-Up

Wet mop-up primarily consists of exposing burning materials using hand tools and then extinguishing the

TABLE 5-3 Dry Mop-Up Techniques

Technique	Description/Actions	Safety Concerns
Boneyarding/ bone piling	In areas with small branches and logs that are not burning, use your hand to check if the material is hot. If cool, move inside the burned area that has been checked and is cool.	Picking up hot materials, missing hot areas, and getting a rekindle in a bone pile.
Chunking and piling	Arrange burning materials to allow them to burn themselves out.	Open fire for an extended time, spotting potential, handling of burning material.
Rearranging fuels	In some instances, a change in fuel arrangement can lower fire intensity and aid in mopup. Spread heavy concentrations of materials near a control line or inside a burned area.	Handling of burning fuels, rolling material, potential for flying embers.
Banking	Cover fuels in soil in order to cool or protect them from ignition. Then mark their location and include in a briefing. To check banked fuels, uncover and scrape out any hot spots.	Banked fuels that are overlooked may cause an ash pit.

burning materials with water, water and soil, or class A foam. When performing wet mop-up, follow these steps:

- Spray, stir, and spray again as necessary.
- Use foam or another wetting agent to penetrate smoldering fuels.
- Break apart higher concentrations of fuel and spread the fuel to lower the fire intensity.
- Apply a fine water spray from the control line inward so as not to push an ember across the control line.
- Use a straight stream to penetrate or reach a distant target.

Wet mop-up is discussed in more detail in Chapter 12, *Water Supplies and Operations*, and Chapter 13, *Class A Foams and Fire-Blocking Gels*.

Patrolling

Patrolling is the action of observing a length of control line during or after its construction in order to locate weak areas, slopover, spot fires, or hot spots. Patrolling ensures that any spots or areas that remain hot after control line completion are located and extinguished. Patrollers should reinforce the line when necessary. If a spot fire is found, report it (via radio, hand signals, mirror, whistle, runner, or other means) and begin initial attack and suppression. Then, flag the area. Patrolling can be accomplished using vehicles or air equipment, or by foot.

Like mop-up operations, an effective technique for patrolling spot fires on the ground includes implementing a grid or block system where each block is numbered and priorities are set. Your supervisor will determine your assigned area coverage.

When patrolling, consider the purpose of the assignment, the coverage of the assigned area, the information to be reported, and the appropriate approach (such as working in pairs or in a system). Always know where your escape routes and safety zones are located.

Decision Points and Safety Considerations

Decision points or *trigger points* are an integral component of situational awareness and the risk management process.

Decision points are the preferred tool to use in the wildland environment because they are objectives based. Objectives are the core element of what we do, from the leader's intent down to the task purpose and end state. Decision points and tactical decision points need to be clearly established and communicated at the beginning of a mission in order for them to be used as an effective risk management tool.

In most cases, decision points are reached as a result of a tactical element's being compromised. Such compromise can occur as a result of the time of day and its relation to the current operational assignment and potential for the operation to be unsuccessful. For example, reevaluating at 5:00 PM might be the decision point for determining whether fire fighters should continue to fight the fire or move off the line if the fire behavior changes from the initial strategy and tactics.

Another decision point might be reached when the dynamics of the fuel, terrain, or topography change in the middle of an assignment. An example includes the humidity dropping or the fuel moisture decreasing as a result of foehn winds. A more common effect that you may face is when a lookout is not able to see you or your crew or the lookout has to leave as a result of the fire overtaking the lookout spot.

It is up to you to make sure you understand the objectives that you have been given and how to implement a change of decision as a result of changing objectives and their relation to the risk management process.

DID YOU KNOW?

Most fire tragedies occur between 2:00 and 4:30 PM. It is important to take a 5-minute pause at 2:00 PM, just before the start of the most dangerous part of the day. This routine is referred to by the Wildland Fire Lessons Learned Center as Take 5@2 **FIGURE 5-12**. During this pause, collect your thoughts; perform a self-assessment; check in with your crew; and reevaluate the weather, fire behavior, and mission objectives. Pausing for just 5 minutes will prepare you for the most dangerous part of the day.

LISTEN UP!

Here are three fireline safety tips:

- Identify escape routes and safety zones, and make them known to everyone.
- Always be aware of changes in fire intensity or a change in direction of the fire.
- Be alert. Keep calm. Think clearly. Act decisively.

FIGURE 5-12 Take 5@2 means take 5 minutes at 2:00 PM, just before the most dangerous part of the day, to collect your thoughts and reevaluate fire behavior and your crew and objectives.
Courtesy of National Wildfire Coordinating Group.

After-Action REVIEW

IN SUMMARY

- Size-up is an ongoing process that involves gathering and evaluating information from a fire, whether by direct observation or derived from details observed by another person or dispatch.
- Crew readiness and available resources need to be constantly evaluated.
- Weather factors are the most changeable factors, so they must be evaluated constantly.
- Incident action plans (IAPs) contain the overall strategy and specific actions for managing an incident. There are three main priorities when developing an IAP:
  - Life safety
  - Incident stability
  - Home and property conservation
- Knowing the parts of the fire is necessary to being successful as a wildland fire fighter. Be familiar with all parts of a wildland fire, including the area of origin, head, heel/rear, flank, finger, pocket, and island.
- Attacking a fire requires in-depth planning. Selecting the best point of attack determines the effectiveness of the attack itself.
  - When attacking an interface fire, the IC decides whether the top priority of the attack is to contain the fire or protect property.

- Before attacking the fire, the escape plan must be known to all fire fighters.
- When the fire cannot be contained properly, the IC must have a backup plan in place.

- The type of attack you will employ on a fire is determined by the types and amounts of resources available as well as the behavior and size of the fire. With more available resources, a more offensive attack can be employed.
  - Direct, indirect, and parallel attacks are all effective under certain circumstances.
- Control lines are natural or constructed barriers and are the primary method of containing the spread of fires. Control lines act to remove fuel from the fire and prevent it from spreading. It is common for firelines to be constructed with hand tools or by using controlled burns to remove fuel that would otherwise be used by the main fire.
  - There are two types of coordinated crew methods for fireline construction: the one-lick method and the leapfrog method. These methods allow for many crew members to rapidly construct a fireline by dividing up the work.
  - Common threats to control lines include:
    - Spotting: Wind can carry embers across a line.
    - Rolling debris: Embers and debris can roll downhill and across a control line.
    - Creeping: Root systems can carry fires across a control line.
    - Radiant heat: Fire inside the line may grow hot enough to ignite fuels outside the line.
- In most cases, a simple or progressive hoselay will be established to move water around the perimeter of the fire.
- Decision points are an objective-based tool in situational awareness and risk management that are best employed to make a choice as a result of a tactical element being compromised.

KEY TERMS

Anchor point An area close to where the fire started and from which vegetation has already been burned.

Area of origin The location where a wildland fire begins.

Blackline method A method in which a buffer of preburned fuels is created between the fire's edge and the rest of the fire.

Control line A constructed or natural barrier used to control a fire.

Detonation line A control line created by using explosives.

Direct attack A method of fire attack in which fire fighters apply a treatment such as wetting or smothering to the fire or by physically separating the burning from unburned fuel

Finger A long, narrow extension of a fire, caused by a change in weather, topography, or fuel.

Fireline A barrier purposefully constructed by fire fighters to control a fire.

Flame length The measurement from the average flame tip to the middle of the flaming zone at the base of the fire; measured on a slant when the flames are tilted due to effects of wind and slope.

Flank The edge between the head and heel of the fire that runs parallel to the direction of fire spread.

Flanking attack A direct fire suppression attack that involves placing a suppression crew along the flank of a fire.

Hand line A fireline that is constructed using hand tools.

Head The most intense and rapidly spreading part of the fire.

Heel The area of a fire close to the area of origin, also called the rear of the fire.

Indirect attack An indirect fire attack that involves building a fire control line along a predetermined route, based on natural fuel breaks or favorable breaks in the topography, and then burning out the intervening fuel

Island An area of land that untouched by fire, but is surrounded by burned land.

Leapfrog method A system to build a fireline in which each crew member is assigned a task at a specific section of line until the task is completed. After the task is completed, the crew member calls out and everyone ahead of them moves up the line.

Machine line A control line constructed using a variety of heavy equipment and machinery.

Mop-up The action of extinguishing or removing burning material near control lines.

Objectives Statements of what is to be accomplished.

One-lick method A progressive system used to build a fireline in which each crew member does one to several licks of work and then moves forward to make room for the person behind them.

Parallel attack A method of fire attack in which a fire control line is built parallel to the fire's edge.

Pincer attack (anchor, flank, and pinch attack) A direct fire suppression attack that involves two or more teams of fire fighters establishing anchor points on each side of the fire and working toward the head of the fire until the fire gets "pinched" between them.

Pocket The unburned area between the finger of a fire and the main body of the fire.

Retardant line A control line created by distributing a chemical solution such as foam that slows the forward progress of the fire.

Saw line A fireline constructed by clearing away trees and brush so that remaining fuel can be removed with hand tools.

Scratch line A temporary, quickly constructed fireline used as an emergency measure to control the spread of fire.

Size-up The process of gathering and evaluating information from a fire to determine a course of action and prepare for response.

Spot fire A new fire that starts outside the perimeter of the main fire.

Spot weather forecast A special forecast, issued upon request, that fits the time, topography, and weather of a specific incident.

Strategy A general plan to meet a set of predetermined objectives.

Tactics The specific steps necessary to support and achieve a given strategy.

Wet line A control line created by spraying water or a mixture of water and foam or retardant along the ground.

REFERENCES

National Wildfire Coordinating Group (NWCG). "Common Denominators of Fire Behavior on Tragedy Fires." Modified September 2019. https://www.nwcg.gov/committee/6mfs/common-denominators-of-fire-behavior-on-tragedy-fires.

National Wildfire Coordinating Group (NWCG). "S-130: Firefighter Training, 2008." NFES 2731. Modified October 23, 2019. https://www.nwcg.gov/publications/training-courses/s-130.

National Wildfire Coordinating Group (NWCG). "S-200, Initial Attack Incident Commander, 2006." Modified October 23, 2019. https://www.nwcg.gov/publications/training-courses/s-200.

National Wildfire Coordinating Group (NWCG), "S-290, Intermediate Wildland Fire Behavior, 2010." Modified October 23, 2019. https://www.nwcg.gov/publications/training-courses/s-290.

National Wildfire Coordinating Group (NWCG). *Wildland Fire Incident Management Field Guide.* PMS 210. NFES 2943. January 2014. https://www.nwcg.gov/sites/default/files/publications/pms210.pdf.

Wildland Fire Fighter in Action

You are the engine boss on a Type 2 engine. You are dispatched to a grass fire next to some homes in the wildland/urban interface. It is 1300 hours, and the fire is at the bottom of a south-facing slope. The angle of the slope is 20 percent. The wind is blowing at 8 mi/h (12.9 km/h) in the direction of the homes. The 1-hour fuel moisture is 10 percent. The homes sit at the top of a long slope with a small 10-ft (3-m) green belt. The fire is about 1 mile (1.6 km) from the homes. You must decide to attack the fire or defend the homes.

1. If you decide to do a frontal attack on the fire, what checklist should you use in the IRPG?
   - **A.** Briefing Checklist
   - **B.** Risk Management Checklist
   - **C.** Downhill Line Construction Checklist
   - **D.** Indicators of Incident Complexity Checklist
2. What guide in the IRPG can you use to help you determine the intensity of the fire?
   - **A.** Probability of Ignition Table
   - **B.** Fire Behavior Hauling Chart
   - **C.** Lightning Activity Level
   - **D.** Relative Humidity Charts
3. In establishing your situational awareness and LCES/LACES, what reference can you use to determine the appropriate size of your safety zone for the homes you are protecting?
   - **A.** IRPG
   - **B.** IAP
   - **C.** Spot weather forecast
   - **D.** IHOG

Access Navigate for flashcards to test your key term knowledge.

CHAPTER 6

Wildland Fire Fighter I

Wildland/Urban Interface Considerations

KNOWLEDGE OBJECTIVES

After studying this chapter, you will be able to:

- Identify wildland/urban interface watch outs. (**NFPA 1051: 4.5.3**, pp. 144–148)
- Identify equipment used to protect improvements. (**NFPA 1051: 4.5.6**, pp. 148–150)
- Describe structure triage and tactical actions as they relate to protecting properties and improvements. (**NFPA 1051: 4.5.6**, p. 148)
- Describe the types of tactical actions for structure and property protection. (**NFPA 1051: 4.5.6, 4.5.8**, p. 148)
- Describe the appropriate tactical actions for low-risk, moderate-risk, high-risk, and extreme-risk structures. (**NFPA 1051: 4.5.6, 4.5.8**, pp. 148–149)
- Describe the differences between surface protection and full preparation of structures. (**NFPA 1051: 4.5.6, 4.5.8**, pp. 148–150)
- Describe why it is important to protect cultural resources. (pp. 150–151)
- Describe the effects of fire management activities on cultural resources. (p. 151)

SKILLS OBJECTIVES

After studying this chapter, you will be able to perform the following skills:

- Use suppression methods to protect improved properties. (**NFPA 1051: 4.5.6, 4.5.8**, pp. 148–150)
- Protect cultural resources. (pp. 150–151)

Additional Standards

- **NWCG S-130 (NFES 2731)**, *Firefighter Training*

You Are the Wildland Fire Fighter

It is 1445 in the afternoon on July 23. The temperature is 98°F (36.7°C), and the wind is at 17 to 22 mi/h (27.4 to 35.4 km/h), gusting to 30 mi/h (48.3 km/h). The humidity is 16 percent, and the Haines Index is 5. There is a fire weather watch in effect. Your pocket danger-rating card is showing the energy release component (ERC) to be in the 85th percentile. This year is turning out to be a year to remember. You are requested to respond to a neighboring town that is experiencing a large and very fast-moving fire, reported to have been started by a train. Your initial assignment is to provide structure protection on Division W. The fuels are light, and 3- to 5-ft (0.9- to 1.5-m) tall flashy grass is mixed with patch brush. Most of the homes are located on a hillside that extends down more than a mile, with a steady, steep slope of 24 percent. As you arrive, you notice that you are parallel to the fire and the fire is burning downhill. Your engine is assigned to operate with several other engines to establish a 20-person handcrew to flank the fire while providing structural protection. You will be constructing a downhill fireline. There are also inbound helicopters and fixed-wing aircraft to assist with the fire effort.

1. What sorts of things would you consider as you formulate your situational awareness and engage the fire?
2. What checklists would you use from the *Incident Response Pocket Guide* (IRPG) before engaging this assignment?
3. What types of firefighting tools might be the most effective in this fuel type and terrain?

Access Navigate for more practice activities.

Introduction

As noted in Chapter 1, the **wildland/urban interface (WUI)** is an area where undeveloped land with vegetative fuels meets with human-made structures **FIGURE 6-1**. Every year the amount of destruction from fires that occur in WUIs continues to rise. California has more fire losses from wildfire than all other states combined. In the 2019/2020 fire season, Australia saw the greatest loss to wildfires in its history.

This chapter covers several topics that will help you tactically, from a planning perspective, and help you understand some of the significant hazards. WUI fires differ from wildland fires. This chapter discusses those differences, including cultural resource protection, WUI watch outs, and the strategy and tactics required to reduce exposure to WUI structures.

FIGURE 6-1 Wildland/urban interface.

Wildland/Urban Interface Watch Outs

Every year the number of incidents that occur in the WUI has contributed to a complexity of fire response that requires a different level of preparation, thought processes, and tactics. As with everything else in the wildland environment, there are specific watch-out situations unique to this urban interface environment. While many of these may seem like common sense, some of these watch-out situations necessitate a closer look as to how one might engage on a particular assignment when encountering specific situations.

Poor Roads

Road conditions need to be at the forefront of your thought process. As you respond to an incident, make sure you're always considering escape routes and what access concerns need to be taken into consideration. Also consider the potential fire behavior you might encounter with heavy fuels and road easements that are not cleared. Use of the roadway may be precluded when the fire arrives. Following are examples of poor roads.

Poor Access and Congestion

With respect to poor access and congestion, some of the things to be aware of include private roads and driveways with only one way in and one way out, driveways and turnarounds with poor access, vehicles blocking egress, and dead-end roads or cul-de-sacs.

Narrow Roads

Narrow roads are notorious for causing problems during times of low visibility or in the early stages of fire when evacuation hampers the initial response. When encountering roads that have a width of less than 16 ft (4.9 m), traffic will usually be a problem. Additional challenges are roads that are winding and when heavy fire equipment or other vehicles are trying to pass one another. One way to mitigate some of the challenges caused by narrow roads would be to secure the road by controlling traffic at each end. If law enforcement officers are available, they are a good resource to utilize to assist you in securing the road.

Poor Traction

Working in areas where the roadbed consists of loose gravel; loose, compacted rock; clay; or sand will often cause difficulties maintaining traction. Additionally, if lots of water is being utilized in the immediate area, the water mixed with the aforementioned soil conditions can prohibit or significantly inhibit movement from wheeled apparatus.

Steep Slopes

According to the International Code Council and the National Fire Protection Association (NFPA), most roads are not supposed to have a slope greater than 10 to 12 percent. However, many of the roads in the wildland/urban interface have slopes that can be as great as 26 percent. If you happen to find yourself on a slope with this percentage of grade, the radiant and convective heat will be exponentially increased, making this a terrible place to be if a fire front is moving toward you.

Adjacent Fuels

If you must utilize a road or a roadbed as a safety zone, keep in mind that some of the fuels adjacent to the road may be easier to burn than others. It is important to pay attention to the characteristics of the fuels adjacent to the road you are on. There have been several situations where fire engines have been compromised or damaged as a result of their proximity to burning fuels next to the roadway. Factors to take into consideration include fuel arrangement, fuel loading, and the fuel moisture content.

Bridge Load Limits

It is common in the interface zone to find bridges that are aged or creatively built. They can include timber, old steel beams, or a combination of both. Understanding the total weight of your vehicle and the limits of the bridge is essential to your safety. Not all bridges are created equal, and many bridges in rural areas as well as culverts could be compromised by the weight of a vehicle. If the bridge load limits are not posted, be very careful and suspicious of the bridge before attempting to cross it. Get out of your vehicle and evaluate the bridge structure before making a decision to use the bridge.

Wooden Construction and Wood Shake Roofs

Depending on the community that has requested protection, there could be several homes or structures that are constructed entirely of combustible material and roofing material that is entirely made of wood. Combustible construction and shake roofs are easy targets for an advancing fire. These structures are vulnerable to radiant heat because of the construction materials used. The following are a few things that can be done to better protect the structure before the fire arrives and prevent considerable damage from fire:

- Move furniture and combustible objects away from openings, windows, and decks. If the house is unlocked, place the furniture inside the house.
- If you have a ladder, clean out the valleys and gutters on the roof if they contain leaves or pine cast.
- Remove woodpiles next to the structure.
- Close all doors and windows, including any garage doors.
- Remove flammable vegetation from around liquified petroleum gas tanks and turn off the gas.
- Cover roof vents.
- Charge garden hoses.
- Under no circumstances should you ever remove anything from a structure. As best you can, try to leave things where they are and do not disturb them. Depending on the circumstances, this could prevent you from unwanted troubles down the road. If you do move something, make sure you document it on your Activity Log (ICS 214) and brief your supervisor.

Power Lines

Power lines and electrical equipment are components of wildland firefighting that must be considered. For some communities, keeping the power on is the only way that water can be used in the interface zone. In other situations, keeping the power on is vital to the local and regional economy. Ultimately, power lines of all types are dangerous. When they end up on the ground, they can create even more of a problem than the fire. It must be assumed that any power line is energized, unless an electrical utility worker informs you directly that they are not. Do not identify safety zones under or near power lines!

Another component of power-line safety is that superheated, ionized air can act as a conductor, causing a power line to arc directly to the ground. Do not identify safety zones under or near power lines!

LISTEN UP!

Following are examples of things not to do around power lines.

Downed power lines (on the ground):

- Do not get within the area of two power poles that contain a downed power line that is on the ground between them.
- Do not approach or attempt to move a downed line.
- Do not spray water at downed power lines or stand in water that may be in contact with downed power lines.
- Do not leave or touch the body of your vehicle if it has come into contact with downed power lines.
- If you must exit your vehicle due to an extenuating circumstance (e.g., vehicle fire), do not make contact with the frame of the vehicle as you exit. Jump away from the vehicle, and keep your feet together when you land. As you move away, shuffle your feet, rather than walking normally. Doing so will prevent you from creating a voltage gradient between your feet and the ground. Voltage gradients allow for the completion of a circuit, which results in electrocution. This can happen within 33 ft (10.1 m) of the downed power line.

Power lines (not on the ground):

- Do not park your vehicle or direct anyone else to park their vehicle under the power lines.
- Do not operate vehicles and pumps under the power lines.
- Do not fuel your vehicle or any other equipment under a power line.
- Do not use any open areas around a power line as a cargo drop.
- Avoid engaging fire beneath power lines if there is dense smoke present.
- Do not locate personnel or equipment in the vicinity of power lines during airborne retardant or bucket drops.

Inadequate Water Supply

It is imperative that personnel maintain situational awareness as it pertains to water flow and water supply. If there are several apparatus using the same water supply, keep an eye out for decreasing flows. This can become critical if operating in an area without power. In some cases, power is required to maintain the water flow for a community In areas where water supplies are inadequate or there are no water systems, water tenders should be used to shuttle water. Portable water tanks can be used to store water from the tenders.

Always make sure to keep a reserve of water for firefighting personnel in the event of an emergency. Also ensure you have enough water to last for the duration of the main heat wave and to protect.

Proximity of Fuels to Structures

Survivability of a structure or outbuilding is significantly reduced if there is fuel present within 30 ft (9.1 m) of the structure. When working to prepare a structure, for added safety make sure that your vehicle is facing in the direction of your escape route. If present, move and pretreat woodpiles with fire-retardant foam if time and opportunity allow.

Structures Located in Steep Terrain

Structures that are located in very steep terrain (e.g., chimneys, box canyons, and steep slopes) have a limited chance of survival. In addition, terrain types like these pose a significant danger to equipment and personnel. In some cases, multiple flame fronts can be created by the terrain influences. In all cases, the rate of spread is increased. This can cause dangerous radiant heat transfer and has the potential to eliminate escape routes and safety zones.

Panic during Evacuation

Panic may occur when evacuations are not conducted in an orderly manner or an evacuation does not go according to plan. Typically, law enforcement will evacuate the public before fire resources engage a fire. However, depending on the fire activity, sometimes this does not occur. When panic sets in and visibility decreases, challenges are exponentially increased. In these cases, one can expect panicked members of the public to disregard the basic rules of the road, not follow posted speed limits, and not drive on the appropriate side of the road. The challenges presented by these behaviors often make it difficult to bring in firefighting equipment to fight the fire. Panicked persons using all lanes of traffic and displaying complete disregard for the legal rules of the road typically lead to traffic accidents, which can block roads or points of access for firefighting equipment. If at all possible, work with law enforcement to help facilitate an orderly evacuation so that panic does not become part of the equation.

Hazardous Materials

A **hazardous material** is a substance capable of causing harm to people, the environment, and property. In the wildland fire environment, there are many types of hazardous materials. It is important to be able to identify key areas where hazardous materials may be present. Hazardous materials present a risk to you and your crew and can result in an increase in fire behavior, smoke-laden plumes containing hazardous materials, or an explosion. If you think there may be a hazardous-materials emergency, follow these general guidelines:

1. Isolate the area or deny entry. Move out of the area and keep people out of the area.
2. Identify the hazards. Check placards, container labels, and shipping papers to identify the hazards.
3. Approach cautiously. Do not rush into the scene. Approach upwind, uphill, and upstream.
4. Obtain help. Advise dispatch of the situation. Let them know where the hazardous material is, what the hazardous material is, how much hazardous material there is, and any potential exposures and safety hazards.
5. Decide on site entry. Determine whether rescue or protection efforts should be made. Compare the risks and benefits.

To think through a hazardous materials situation as it is unfolding, follow the DECIDE decision-making process:

1. Detect hazardous materials present.
2. Estimate how likely harm is.
3. Choose a response objective.
4. Identify an actionable option.
5. Do the best option.
6. Evaluate your progress.

In the event of a hazardous-materials emergency, make sure to keep your supervisor and the rest of your crew informed. Whenever possible, make sure you mark and identify hazardous materials with flagging and mark them on your map to pass along to relief personnel at the end of your shift. Do not attempt to mitigate any hazardous materials without the proper protective gear and training. In some instances, it may be advisable to post a lookout.

There are a plethora of items and materials that present hazardous scenarios to firefighting personnel. Some examples of situations in which firefighting personnel may find themselves exposed to hazardous materials are discussed next.

LISTEN UP!

Avoid inhalation of fumes, smoke, and vapors, even if they are odorless or no hazardous materials are present.

Drug Labs

It is unlikely that you will ever find a sign that says "drug lab." However, there are telltale signs of nefarious activities taking place that personnel can be trained to recognize. The importance of recognizing a drug lab is significant because the chemicals used to make such drugs as methamphetamine can pose a significant health risk and create toxic and explosive conditions. Signs that a scene may contain a drug lab include the presence of discarded propane tanks with corroded valve stems and the overwhelming smell of cat urine. This very recognizable smell will be found in areas where there are no other structures or signs of activity or the observed presence of cats. In some cases, drug labs can be present in vehicles.

Liquified Petroleum Gas Tanks and Compressed Gas Cylinders

Most rural homesites utilize liquified petroleum gas (LPG) for heating and/or cooking. When working around LPG tanks, make sure that there is a 30-ft (9.14-m) vegetation clearance around all storage tanks. If fighting a fire at a rural site containing LPG tanks, attempt to clear the vegetation. Make sure to check outbuildings and barns for flammable liquid storage and LPG tanks.

Abandoned Industrial Sites

It is not uncommon to get called to fight a fire in an abandoned industrial site that is overgrown with brush and tall grass. In some cases, hazardous materials can be hidden among the fuel. It is very important that you do your best to scout the area and try not to be downwind of any smoke.

Sheds and Garages

Sheds and garages are the most common places where hazardous materials associated with residential structures are typically located. Hazardous materials likely to be in these areas include pesticides, gasoline containers, plastics, synthetic materials, vehicles, ammunition, and other explosive and flammable materials.

Protecting Properties and Improvements

It is vital that personnel understand the nomenclature and the various steps necessary to protect properties in the WUI that are threatened by wildfire. The protection of properties can be broken down into three components: structure triage, preparation, and tactical actions. When it comes to the decision points and tactical decision-making process, it is important that you understand the current fire behavior, the available resources, and the time available to protect properties and improvements. In each case, everything begins with triage.

Structure Triage and Tactical Actions

Sometimes, wildland firefighting resources can't protect all structures and properties at once. This is where structure triage comes into play. **Structure triage** is a system of classifying structures based on defensibility and prioritizing protection to the least defensible structures first. Structure triage allows you to protect as many structures as possible. Structures can be divided into four categories, depending on their defensibility: low risk, moderate risk, high risk, and extreme risk. A proper understanding of the categories will help guide personnel in determining what level of preparation can be accomplished and what tactical actions will best fit the assignment.

Tactical actions are the steps taken after triage. Tactical actions must be capable of stopping the fire's advance or preventing the fire from damaging properties while also preventing injuries to fire fighters. Fire behavior should be considered when determining tactical actions.

In most cases, your tactical actions will be decided on in concert with the triage category for the structure or what you are supposed to be protecting:

- Prepare-and-defend tactics involve preparing the structure with defenses, such as gel or foam, and then defending the structure as the fire approaches. These tactics are used when you and your crew are confident that the structure can be saved from the fire. This tactic is used when safety zones or a temporary refuge areas are present and there is time to prepare the structure before the fire arrives.
- Anchor-and-hold tactics involve protecting and defending a structure with large volumes of resources. These tactics utilize control lines and large water streams. This tactic is used to put out structure fires, reduce ember production, and protect structures exposed to fires. These tactics are used when you and your crew are confident that the structure can be saved from the fire.
- Prepare-and-go tactics involve preparing the structure with defenses, such as gel or foam, and then leaving the structure before the fire arrives. This tactic is used when fire intensities are great or when there are no safe zones or temporary refuge areas present.
- Bump-and-run tactics involve putting out spot fires ahead of the flame front to steer the fire around structures to a desired end. Structure defense preparation and perimeter control are secondary considerations when using this tactic. This tactic is often used in the early stages of a fire when resources are light or on fast-moving fires where control efforts are directed toward steering the head of the fire. Constantly identify safety zones and escape routes when using this tactic.
- Check-and-go tactics involve checking inside the structure for occupants, assisting in evacuation as needed, and then leaving the scene. Make note of the structure, its construction features, and how close it is to the wildland fuels. When resources become available, the fire intensity has diminished, or the flaming front has passed, resources can then be assigned to the structure. This tactic is used when there is no safety zone present and the fire is rapidly approaching a structure with inadequate space for defensive tactics.
- Fire front following tactics (also sometimes referred to as retreat-and-reengage tactics) are used when there is no time to employ other tactics. This tactic allows you to disengage and retreat to a safety zone and stay there until the fire has passed and it is safe to reengage (return to the structure).
- Patrol tactics can be used before and after a flame front to monitor an area for potential or remaining threats to structures.

It is important to document your tactical actions on the ICS 214 form. Regardless of the tactical action that you choose, it is important to remember that alternate actions may later be chosen in order to adapt to changes. You can always change your tactics so that they are in line with the current and expected fire behavior.

Low-Risk Structures

Triage: A low-risk structure is a stand-alone, defensible structure that requires little to no action to protect it. It is at low risk for ignition. Safety zones are present and homeowners may or may not have evacuated.

Tactical actions: Prepare and defend. Ensure effective preparation measures are in place, and defend the structure as the fire approaches. A defensible structure will likely require patrolling, but your crew can choose to stay here or move on and check back later.

Moderate-Risk Structures

Triage: A moderate-risk structure is one that can become defensible with some fire-fighter assistance. Without fire-fighter intervention, the structure will have a higher probability of ignition. To defend the structure, safety zones and good escape routes must be present. Homeowners may or may not have evacuated.

Tactical actions: Prepare and defend or anchor and hold. Aggressive structural preparation will be required. As the fire approaches, defend the structure. Fire fighters must remain at the structure as long as it is safe to do so.

High-Risk Structures

Triage: High-risk structures are considered nondefensible. You will be unable to establish a safety zone.

Tactical actions: Prepare and go, bump and run, possibly fire front following (retreat and reengage). Apply simple preventive measures, such as gel or foam, to give the structure a chance at surviving. Encourage homeowners to evacuate. Establish a realistic decision point for retreating, and make it known to all. Post lookouts to notify your crew when the fire has reached your decision point. You should allow adequate time for your resources to reach the safety zone and to avoid the loss of escape routes. If it becomes necessary to disengage, retreat to the closest safety zone, and return after the fire front has passed.

Extreme-Risk Structures

Triage: Extreme-risk structures are nondefensible. You will be unable to establish a safety zone.

Tactical actions: Check and go. There is no time to perform any suppression measures. Attempt to minimize risk to life. Check for occupants requiring removal or rescue. Fire fighters should establish a realistic decision point and make it known. When it becomes necessary to disengage, retreat to the closest safety zone, and return after the fire front has passed.

Preparation

The priority for preparing a structure or an outbuilding will be dictated by one of the four triage categories. It is important to prioritize firefighting actions that will give personnel the greatest ability to preserve the structure or outbuilding. When preparing the structure or outbuilding, there are two preparation tactics: surface preparation and full preparation.

Personnel should attempt to prepare structures without having to physically alter the structure or property. This is especially important in scenarios where a fire front might not affect the area that personnel are attempting to protect. Preparation activities should be completed in the order outlined in the next sections.

Surface Protection

Roof. On the outside of a structure, remove debris from gutters, valleys, and areas where embers may collect. If branches are touching the roof, remove them or any other combustible material that may be affixed to the roof. Plug the downward-facing spouts and fill the gutters with water.

Interior. On the inside of a structure, turn on the lights (this will allow for easier accountability after the fire has come through the area). Ensure all windows and doors are closed. Move the furniture away from the large picture windows on the fire side of the house. Turn off fans in the house. If possible, move furniture on decks into the garage or house. If this is not possible, move the outdoor furniture and other flammable or combustible materials outside of the safe perimeter of the structure.

Full Preparation

Exterior. Additional steps must be taken for full-structure preparation. These steps include:

- Remove all fine, dead fuels; grasses; bark dust; and firewood a minimum of 10 ft (3.0 m) away from the structure.
- Cover all open vents and exterior ducts.
- Cover all underfloor screen vents.
- Using fire wrap, cover areas that can collect embers.
- Shut off natural gas, LPG, or other gases present on-site.
- Clear all flammable vegetation from around LPG tanks to prevent flame impingement on the tank.
- Remove any fencing that is connected to the house, allowing for no less than a 10-ft (3.0-m) gap. This is necessary to allow for fire-fighter access as well as eliminating potential combustibles.

Secondary Priorities. If you have time and it is safe to do so, consider these secondary full-preparation priorities:

- Eliminate any vegetation within 30 ft (9.1 m) of the structure, and scatter the cut material.

- Limb up all trees 5 to 8 ft (1.5 to 2.4 m) from the ground.
- Relocate or scatter woodpiles. If unable to move and if possible, cover piles to protect them from the potential effects of ember cast.
- Construct a fire line around outbuildings, power poles, and fuel tanks.
- If possible, move all vehicles into the garage and leave keys in the ignition or in a visible place. Make sure doors and windows are closed on the vehicle. If the property has no garage, then place the vehicle in a cleared area on the side away from the advancing fire front.
- Consider constructing a quick, unfinished control line, called a scratch line, around the structure to check the fire's rate of spread.
- Plan ahead for burning out (only if qualified to do so and your plan has been communicated to your supervisor).
- Identify whether a sprinkler kit would be useful. If so, order one, or use appliances that may be on-site.
- If available, identify whether the structure might be a candidate for structural protective wrap.
- Consider felling hazard trees or trees that may pose a threat to the house and your escape route (check with a supervisor before you employ this tactic, and make sure to document it on the ICS 214 form).
- Consider additional preparation measures for the road or driveway. An additional 10 ft (3.0 m) on either side is a good place to start. This would include limbing and brushing.
- Identify whether the structure or outbuilding may be a suitable temporary refuge area.

Imminent Flame Front. If the fire is within 30 minutes of the structure or outbuilding that you have been assigned to protect or triage, consider the following tactics if it is safe to do so:

- If you think you can stay and defend, determine whether it is safe to do so.
- Identify safety zones and TRAs.
- Come up with a plan to stretch hose lines around the structure.
- Consider laddering the roof, and predesignate a hose for roof operations.
- Consider whether foam or gel can provide a direct benefit.
- If you have time, plug the downspouts and fill the gutters with water.
- Consider burning out (only if qualified to do so, and your plan has been communicated to your supervisor).

LISTEN UP!

When using fire engines or ground equipment in WUI locations, consider the following:

- Always back in the engine in the event that you need to exit quickly.
- Keep egress routes clear.
- Park behind the structure, placing the structure between equipment and oncoming fire front.
- Do not park next to propane tanks.
- Do not park under power lines.
- Use 1.5-inch (38-mm) hose lines whenever possible.
- Stay mobile and avoid long hoselays.
- Do not park in terrain features such as chimneys or saddles.
- Keep at least 100 gal. (378.54 L) of water in reserve for protection of the engine and crew.
- Always wear full protective clothing.
- Be aware of possible toxic fumes and stay upwind and out of the smoke.
- Never enter a burning structure unless you are properly equipped and trained.
- When working and driving in the fire area, watch for the presence of homeowners, the media, livestock, or other hazards that may be present.

Protecting Cultural Resources

A **cultural resource** is any object relating to past human life, such as architecture, memorials, land reservations, and objects having scientific, historic, religious, or social values. A cultural resource is a potential resource for understanding the past. Cultural resources differ from natural resources in that cultural resources are non-renewable. Once a cultural resource is destroyed, the information the resource could provide is lost forever. Do not remove any artifact regardless of if you think it is significant or not. There may be penalties for removing artifacts from sites.

DID YOU KNOW?

There are several laws that protect historical and cultural resources on public lands:

- National Historic Preservation Act (NHPA) of 1966
- Archaeological Resources Protection Act (ARPA) of 1979

Effects of Fire Activities on Cultural Resources

Fire operations can assist in identifying previously unknown artifacts or sites. However, it is important to be careful when working around cultural resources or suspected cultural resources to avoid destruction or displacement of artifacts. For example, dozer tracks cut deep into soil and can cause damage to cultural resources, water pressure can cause artifact erosion or chipping, and large numbers of people in and around sites can displace artifacts.

Cultural Resource Protection During Suppression

There are several steps that can be taken to protect cultural resources during wildland fire activities:

- Be aware. Incident action plans and briefings should include specific details for suppression activities around cultural resources.
- Avoid significant or unknown sites. Do not disturb large or unknown sites. Instead, mark these sites and make it known to your supervisor, your cultural resource specialist, and the rest of your crew.
- Report any potentially cultural sites and materials to your supervisor and make these sites known to your crew.

LISTEN UP!

Fire fighters should *never* move or collect artifacts.

After-Action REVIEW

IN SUMMARY

- Every year, the destruction from fires that occur in wildland/urban interfaces continues to rise.
- There are specific watch-out situations that are unique to the wildland/urban interface:
  - Poor roads
  - Poor access and congestion (private roads, driveways with one way in/out, vehicles blocking egress)
  - Narrow roads
  - Poor traction (loose gravel/rock/clay/sand)
  - Steep slopes
  - Adjacent fuels
  - Bridge load limits
  - Wooden construction
  - Power lines
  - Inadequate water supply
  - Proximity of fuels to structures
  - Structures located in steep terrain
  - Panic during evacuation
  - Hazardous materials
  - Signs of drug labs, such as discarded propane tanks with corroded valve stems or the smell of cat urine
  - Liquified petroleum gas tanks
  - Abandoned industrial sites
  - Sheds and garages
- Structure triage allows the protection of as many structures and properties as possible by classifying structures based on defensibility and prioritizing protection to the least-defensible structures first.
  - Low risk: Stand-alone, defensible structures; low risk for ignition
  - Moderate risk: Defensible with some fire-fighter assistance; higher probability of ignition without intervention

  - High risk: Considered nondefensible; unable to establish safety zone but time to apply simple protective measures
  - Extreme risk: Nondefensible; unable to establish safety zone and no time to establish any suppression measures
- There are two preparation tactics for structures and outbuildings: surface and full preparation
  - Surface preparation
    - Clear debris from roof, gutters, valleys, areas where embers may collect.
    - Remove fuels from decks, walkways, bases of exterior walls/decks.
    - Turn on lights inside structure, move furniture on decks into the house or to a safe perimeter.
  - Full preparation
    - Remove all fine, dead fuels 10 ft (3.0 m) from structure.
    - Cover all open vents.
    - Cover areas that can collect embers with fire wrap.
    - Shut off natural gas, LPG, etc.
    - Remove fencing connected to house (10-ft [3-m] gap).
    - Additional, secondary preparations can be made if time and resources allow.
- Cultural resources allow us to better understand the past. Be careful when working around cultural resources or suspected cultural resources to avoid destruction or displacement of artifacts. Do not move or collect cultural resources.

KEY TERMS

Cultural resource Any object relating to past human life that can serve as a potential resource for understanding the past.

Hazardous material A substance capable of causing harm to the environment, people, and property.

Structure triage A system of classifying structures based on defensibility and prioritizing protection to the least-defensible structures first.

Tactical actions The steps taken after triage, typically utilizing the PACE (primary, alternate, contingency, and emergency) model.

Wildland/urban interface (WUI) An area where undeveloped land with vegetative fuels meets human-made structures.

REFERENCES

Cowger, Rich. "The Basics of Pre-Incident Structural Triage." *Fire Rescue* 2, no. 7 (2012). https://firerescuemagazine.firefighternation.com/2012/02/01/the-basics-of-pre-incident-structural-triage/#gref.

Federal Emergency Management Agency (FEMA). "Activity Log (ICS 214)." Accessed December 31, 2019. https://training.fema.gov/emiweb/is/icsresource/assets/ics%20forms/ics%20form%20214,%20activity%20log%20(v3).pdf.

FIRESCOPE California. "Wildland Urban Interface (WUI) Structure Defense." Modified October 21, 2013. https://firescope.caloes.ca.gov/ICS%20Documents/WUI-SD.pdf.

National Wildfire Coordinating Group (NWCG). "S-130, Firefighter Training, 2008." Modified October 23, 2019. https://www.nwcg.gov/publications/training-courses/s-130.

Office of the State Fire Marshal (OSFM). "OSFM Structure Protection Plan." Modified July 2017. https://www.oregon.gov/osp/Docs/Structure-Protection-Plan.pdf.

Wildland Fire Lessons Learned Center (LLC). "Annual Incident Review Summaries (2011–2019)." Accessed December 31, 2019. https://www.wildfirelessons.net/viewdocument/annual-incident-review-summaries.

Wildland Fire Fighter in Action

You are called up at 0200 to respond as part of a five-person engine crew to a fire that is threatening multiple structures. It is October 26, and the fire behavior in the region has been identified as extreme by virtue of the Santa Ana winds. The area you are responding to is known for its highfuel load of brush and steep terrain. The roads in the area are not well maintained, and communication is challenging with multiple frequencies being used. The weather includes humidity in the single digits, and the wind speed is between 13 and 31 mi/h (20.9 and 50.0 km/h).

The area you are responding to has a history of 11 large fires in the past 55 years in the same general location. Included in those statistics are three fire-fighter fatalities. The fuel in the area includes manzanita, sage, grass, and mountain mahogany.

By the time you get to the structure, it is about 0600; you have been assigned an octagon-shaped house. The winds are increasing, and the general slope average is 25 percent.

1. What triage category would you classify your structure in, based on the information in the scenario?
   - **A.** Nondefensible—prep and go
   - **B.** Defensible—stand-alone
   - **C.** Nondefensible—check and go
   - **D.** Defensible—prep and hold
2. How many watch outs have been identified? Are you still able to engage in this fire?
   - **A.** Five; Yes
   - **B.** Five; No
   - **C.** It does not matter because LCES is not in place.
   - **D.** Seven; Yes
3. What section in your IRPG would you consult to determine whether you would stay and defend this structure?
   - **A.** Briefing checklist
   - **B.** 10 standard fire orders and 18 watch outs
   - **C.** Risk management checklist
   - **D.** Urban interface watch outs
4. If you determined it was too dangerous to continue defending this structure before the flame front arrived, what job aid would you use to turn down this assignment?
   - **A.** Downhill fireline construction
   - **B.** How to properly refuse risk
   - **C.** Medical incident report (MIR)
   - **D.** Complexity analysis

Access Navigate for flashcards to test your key term knowledge.

Chapter Opener: Courtesy of Joe Lowe; On Scene siren: © Bildgigant/Shutterstock.

CHAPTER 7

Wildland Fire Fighter I

Hand Tools and Equipment

KNOWLEDGE OBJECTIVES

After studying this chapter, you will be able to:

- Identify different types of hand tools and their uses:
  - Scraping and digging tools (pp. 156–157)
  - Swatting tools (pp. 157–159)
  - Cutting tools (p. 159)
  - Power tools (pp. 159–160)
- Identify the proper spacing between fire fighters when using hand tools. (pp. 160–161)
- Identify proper tool handling techniques. (p. 161)
- Identify proper tool care. (**NFPA 1051: 4.3.3**, pp. 161–167)
- Identify proper tool storage. (**NFPA 1051: 4.3.3**, pp. 164–167)

SKILLS OBJECTIVES

After studying this chapter, you will be able to perform the following skills:

- Demonstrate proper use of hand tools. (**NFPA 4.5.4, 4.5.7**, pp. 160–161)
- Inspect hand tools. (**NFPA 1051: 4.3.3**, pp. 161–166)
- Sharpen and maintain hand tools. (**NFPA 1051: 4.3.3**, pp. 161–166)
- Store and prepare hand tools for transport. (**NFPA 1051: 4.3.3**, pp. 164–167)

Additional Standards

- **NWCG S-130 (NFES 2731)**, *Firefighter Training*

You Are the Wildland Fire Fighter

At the morning briefing, your engine boss informs you and the rest of the engine crew that you are being assigned with several other type 6 engines as part of a structural protection strike team. You have been assigned to a subdivision that includes a significant wildland/urban interface challenge with vegetation up against structures, low hanging branches, and structures surrounded by pine trees. The fire that will be impacting the area is not expected to reach the subdivision for 2 days (based upon current weather conditions and predicted forecast). You and your crew have been requested to prep all of the homes in your assigned area, which will include establishing a 30-foot (9.2-m) buffer zone and thinning out to 100 ft (30.5 m) if time allows.

1. Which tools will be most effective in this environment?
2. Which tools would you choose to create a scratch line around a structure?
3. Which job aid will help in determining the life of some hand tools and maintenance?
4. If you were in an area with a lot of grass, which tool might be most effective?

Access Navigate for more practice activities.

Introduction

There are several types of suppression tools to choose from in the wildland environment. Knowing which tools to choose will aid in successfully completing assignments. Not all tools are created equal; if used in the wrong manner or situation, the negative consequences can be considerable. Doing so can cause a significant delay in fire-suppression efforts, place personnel in harm's way, or do more damage to the terrain and area than is necessary. One point to remember is that there is no perfect wildland firefighting tool—that one "tool to have." As you read through this chapter, you should begin to realize that there are many tools and job aids that you have access to that will help reduce the threat of exposure. It is up to you to apply the correct measures, tools, and skills to minimize your potential negative exposure.

Suppression Tools and Equipment

Some of the tools used for wildland firefighting were specifically designed for just that. In other cases, the tools have been adapted from other conventional uses. At the most basic level, suppression tools can be broken down into scraping and digging, swatting, cutting, and powered.

Scraping and Digging Tools

Scraping tools are used to move, spread, or scrape materials across the surface to disperse them and reduce the likelihood of ignition or to create points of fire diversion. Most scraping tools are suited for fireline construction and mop-up operations.

They can be used to clear away small vegetation and assist in removing debris while making a fireline. They are also good for sifting and breaking up debris clumps and rolling over logs that are on fire. Scraping tools are also great for digging and getting down to mineral soil. They are some of the most common tools found in fire suppression and mop-up operations.

Shovel

A fire **shovel** has a long handle and a tapered blade with both edges sharpened **FIGURE 7-1**. A shovel can be used for throwing dirt when a fire is too hot to get close to, cutting light fuels, digging, scraping the final debris off a fireline, and smothering and beating out a fire.

Hoe

A **hoe** is a digging tool with a sharp, flat blade on one or both sides of the head. Hoes are generally used for digging up roots, removing grass, and trenching.

Each agency will have its own preference as to what type of hoes to use for wildland firefighting operations. One of the most common types of hoes you will see is the adze **FIGURE 7-2**, or less commonly, adz, also called the hazel hoe, rhino hoe, or Rinehart. There are many types of adzes, but the original version was made by cutting the tip off a shovel and the blade off the handle and then welding it back together in a 90-degree angle.

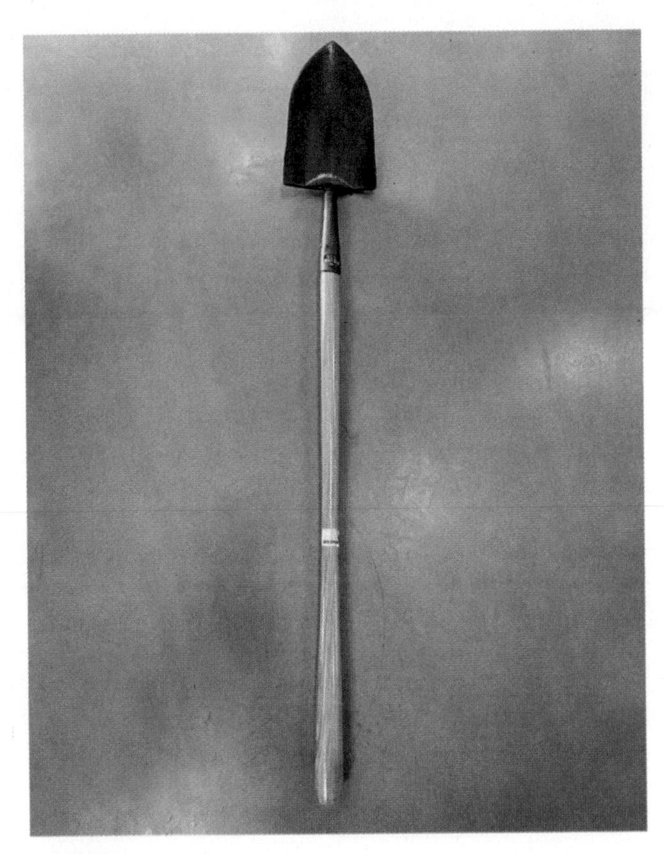

FIGURE 7-1 Shovel.
Courtesy of Jeff Pricher.

FIGURE 7-2 Adze.
© Artproem/Shutterstock.

FIGURE 7-3 McLeod.
Courtesy of Jeff Pricher.

McLeod

The **McLeod** is a combination scraping and raking tool. One side of the head has a blade, which can be used for cutting, grubbing, or trenching. The other side of the head has a rake with five to seven long tines **FIGURE 7-3**. The tines are useful for raking in thick pine needles, duff, or leaf mold, or to dig into a burning log.

Fire Rake

A **fire rake**, also called a council tool, resembles a standard, long-handled garden rake **FIGURE 7-4**. Handcrews like to carry small, one hand–operated rakes for clearing up fine debris on the fireline. Fire rakes are well suited for deciduous leaves and clearing in areas where a McLeod or other scraping tool is ineffective.

Combination Tool

The **combination tool**, also called the combi or Colby tool, is a variation of the military entrenching tool **FIGURE 7-5**. Many fire fighters like this tool because it is very sturdy and lightweight and has several uses: grabbing, digging, smothering, scraping, cutting, and as a pick.

Swatting Tools

In rangeland fires, it is not always practical to try to dig fireline in short grass. In cold climates, such as Alaska's, it is not always feasible to dig down into the tundra, which can be several feet thick. This is where swatting tools come in. Swatting tools are used to

FIGURE 7-4 Fire rake.
Courtesy of Jeff Pricher.

FIGURE 7-5 Combination tool.
Courtesy of Jeff Pricher.

smother fuels, usually in light fuels and sometimes in conjunction with a backpack pump or fire rake.

Fire Swatter

A traditional **fire swatter**, also called a flapper, is a long-handled tool with a thick, flat piece of rubber on the end used to drag over or smother flames. The **Alaska fire swatter** is a version of the traditional fire swatter. This swatter uses several strips of rubber on the end of a long handle, instead of one thick piece of rubber like its traditional counterpart **FIGURE 7-6**. Fire swatters are either purchased commercially or created by an agency to suit the purposes and needs of the agency.

When using a fire swatter, it is important to remember that the swatter is intended to knock down a fire and not to necessarily put out the fire. Personnel will need to work in conjunction with other crews to make sure that water is available to properly extinguish the fire.

When not in use, store a fire swatter so that flap or flaps lay in a flat position.

Gunnysack

Another type of swatting tool is a gunnysack. A gunnysack, also called a wet bag, a burlap, or a Kennedy spanker, is a piece of fabric used to swat at a fire. Usually, a gunnysack is made of burlap or repurposed from damaged fire-resistant clothing that can no longer be worn in the field. The bottom portion of the gunnysack can be wetted before swatting at the fire. When using a gunnysack, it is important to keep it moving to prevent it from burning. Gunnysacks should be dry when stored.

FIGURE 7-6 A traditional fire swatter (top) and an Alaska fire swatter.
Courtesy of Jeff Pricher.

DID YOU KNOW?

Depending on where a fire is located, specific rules apply to what type of suppression tools can be used. In some areas of the country, wilderness fires require that no manmade tools can be used to suppress a fire. The swatting tool of choice may be a pine or a spruce bough. In other nonwilderness situations, if your main tool is a Pulaski, for example, you may want to opt for the natural limb to swat at the fire and speed up suppression.

Cutting Tools

Cutting tools are used to cut brush and small trees in mop-up operations. Cutting tools may also find frequent use in areas of heavy brush and shrubs.

Pulaski

The **Pulaski**, or Pulaski axe, is a chopping and trenching hand tool with a dual-purpose head **FIGURE 7-7**. One side has a blade used for cutting, similar to an axe, and the other side has a flat edge used for grubbing. The Pulaski is very effective for cutting roots and small brush. This tool is generally used in fireline creation and mop-up activities.

Axe

An **axe** is a tool with a short handle and a steel blade used for chopping. Axes are very useful in multiple types of operations, such as felling snags and chopping stumps and logs. Axes can also be used to drive wedges when cutting trees. The two most common types of axes in wildland firefighting are a single-bit axe and a double-bit axe **FIGURE 7-8**.

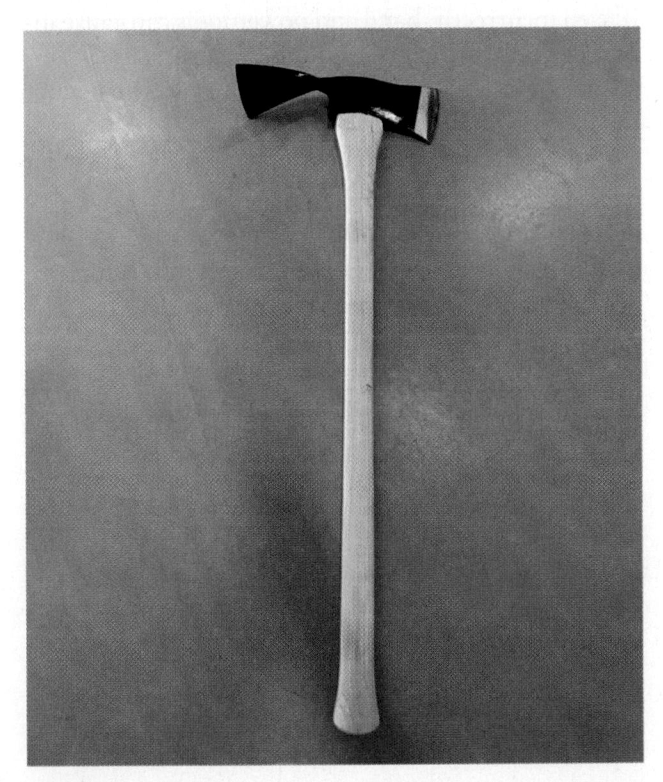

FIGURE 7-7 Pulaski.
Courtesy of Jeff Pricher.

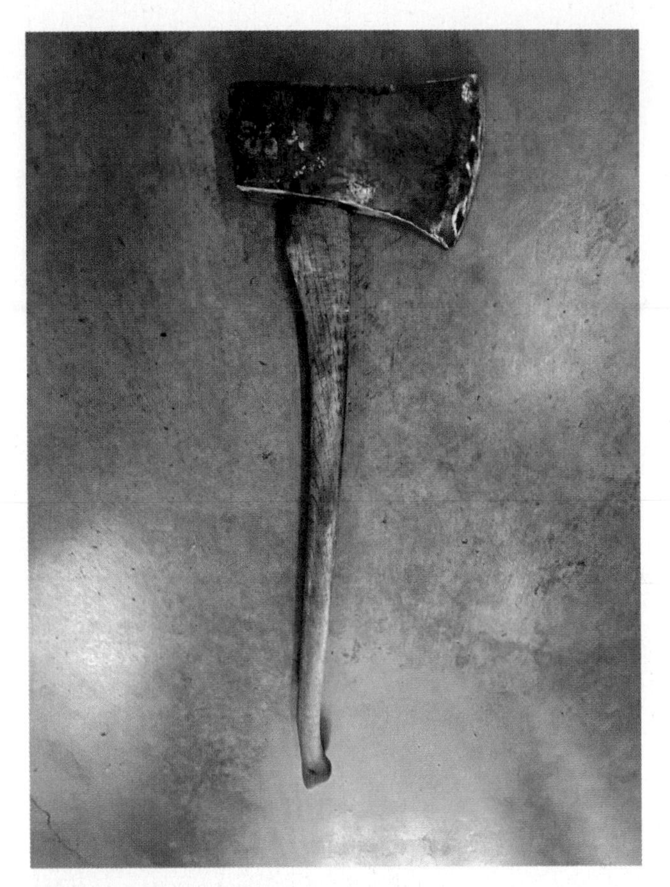

FIGURE 7-8 Single-bit axe.
Courtesy of Jeff Pricher.

Brush Hook

A **brush hook**, also called a brush axe, can be used in a variety of fuel types. Most often they are used for cutting small shrubs or trees, but they can also be used in tall grasses. The main difference between an axe and a brush hook is the shape of the blade head. The brush-hook blade head is curved like the letter *J*. The handle may be curved or straight.

Sandvik

The **Sandvik**, also sometimes referred to as the Swedish brush axe, is a hand tool with a C-shaped blade used to cut small- and medium-sized brush. One drawback of the Sandvik is that it can be very fatiguing because it is designed to be used with one hand.

Power Tools

Power tools are valuable in firefighting efforts to reduce personnel fatigue and provide fast and effective fire clearing. The most common power tools used in fire suppression are powered blowers, string trimmers, and chainsaws. Chainsaws and their operation are covered in the Wildland Fire Fighter II certification. They are discussed in more detail in Chapter 14, *Handcrew Operations*.

Powered Blower

A **powered blower** is a portable tool that uses air to blow leaves and needles away from a structure **FIGURE 7-9**. When there is a lack of time, this tool can do the work of several people. This tool is also good for clearing needles and leaves from gutters during structure protection operations.

In certain parts of the country, blowers are sometimes used to construct a control line. The goal is to blow the fire back on itself and then use a backpack pump to put water on the smoldering fuels. A team of three firefighters—one with a blower, one with a backpack pump, and one with a tool—will accomplish more in less time. Blowers can be effective in fuels shorter than 6 in. (15.2 cm), but often require more fire-fighter assistance. Approach direct fireline applications with caution. Because the powered blower is made of plastic, it can melt under extreme temperatures. Because some issues still exist for the use of powered blowers in fire suppression operations, operate blowers according to your agency standards.

FIGURE 7-9 Powered blower.
Photo by Stephen Strack. Courtesy of AmeriCorps St. Louis.

String Trimmers

A **string trimmer**, also known as a Weedwacker or Weed Eater, is a long metal rod with a spinning blade or string attached to the end **FIGURE 7-10**. String trimmers are used for reducing fuel next to a fireline when a masticator, dozer, or other mechanized equipment is not available. In the West, one of the fuel challenges is blackberry bushes because they are thorny and tough. Chainsaws can be used to clear these bushes but with great effort. Professional-grade string trimmers allow for a saw blade (similar to a circular saw) to be attached, which aids in the removal and clearing of thick brush like this. Make sure you study the manufacturer's guide before using this type of power tool.

Tool Handling

If used incorrectly, hand and power tools can cause injury or even death. It is important that personnel exercise caution and make proper use of tools to reduce the likelihood of injury or death.

Following are some important points to consider when working with hand tools or traversing an area where hand tools are being used.

- Different tools have different spacing requirements, but make sure to leave at least 10 ft

FIGURE 7-10 String trimmer.

(3.0 m) between you and the next fire fighter while working on and/or walking to your assignment.
- When using hand tools in the field, it is best to use short strokes, but, if you have to take a big swing, make sure you call out to prevent a strike on another fire fighter. Usually, the term to yell is "Swinging!"
- When carrying a hand tool, do not run.
- Make sure to hold the tool at its balance point and never place the tool over your shoulder. The only exception to this is if you have a long-handled tool and are using the handle and tool to transport rolls of hose.
- When using a tool, be sure that the cutting edge is facing away from your body. In addition, if you are traversing a steep slope, the tool should be held in the hand that is away from the slope.
- When you need to move around someone, make sure you signal to him or her using a predefined message. The typical verbiage is "Bumping by" or "Coming through." If you hear one of these phrases, that means someone wants to go around you.
- When you are using a hand tool, make sure that you always keep an eye on what you are digging, scraping, or cutting.
- If you have to pass your tool to someone, pass the tool using the handle first. Never pass a tool with the cutting or scraping end first.
- Wear protective eyewear while using hand tools.

Tool Care

Tools will not last if they are not cared for properly. Each tool has its own maintenance requirements. It will be necessary to read the manufacturer's recommendations for sharpening of tools and maintaining the handles. Some basic tool care needs follow:

- Make sure that the heads of tools are securely seated in their respective brackets.
- If you find loose heads, replace them immediately.
- If you find cracked or damaged handles, fix them before use.
- Do not place the wooden end of the tool on the ground and rest your hands on the metal part. This action damages the wooden end of the handle.

LISTEN UP!

Wear gloves, eye protection, and a long-sleeve shirt when sharpening hand tools!

Sharpening

Sharpening your hand tool is an art and takes dedication to be able to consistently bring your tool back to life after each use. Wildland fire tools need to be razor sharp to be most effective in chopping, limbing, and other cutting needs. Even the edges of your shovel need to be sharp **FIGURE 7-11**.

There are many recommendations for how a tool should be sharpened. Every agency has its own standard. For agencies that do not have a standard, the most effective way to sharpen hand tools involves using a sharpening gauge **FIGURE 7-12**. The sharpening gauge is a multipurpose metal gauge that can be used for shovels, Pulaskis, adzes, and combi tools to help determine a tool's health **FIGURE 7-13**. Depending on the tool you are sharpening, the sharpening gauge has specific notches and markings with visual information that can help determine when a tool should be

FIGURE 7-11 Make sure the edges of your shovel are sharp.
Courtesy of Jeff Pricher.

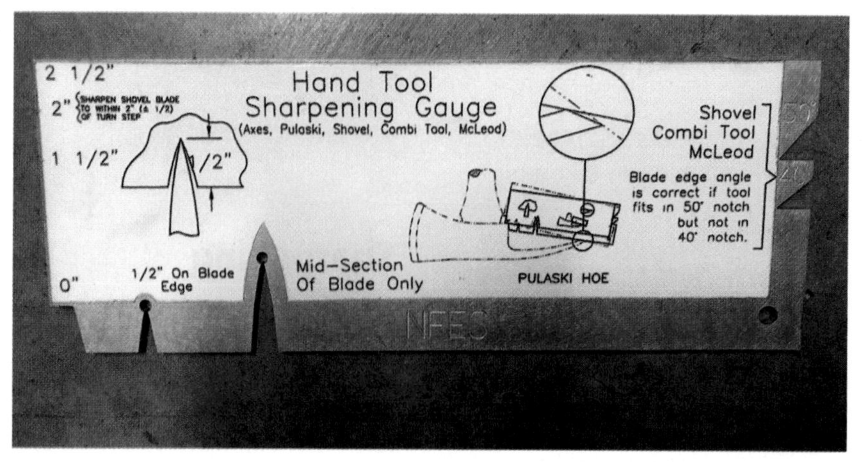

A

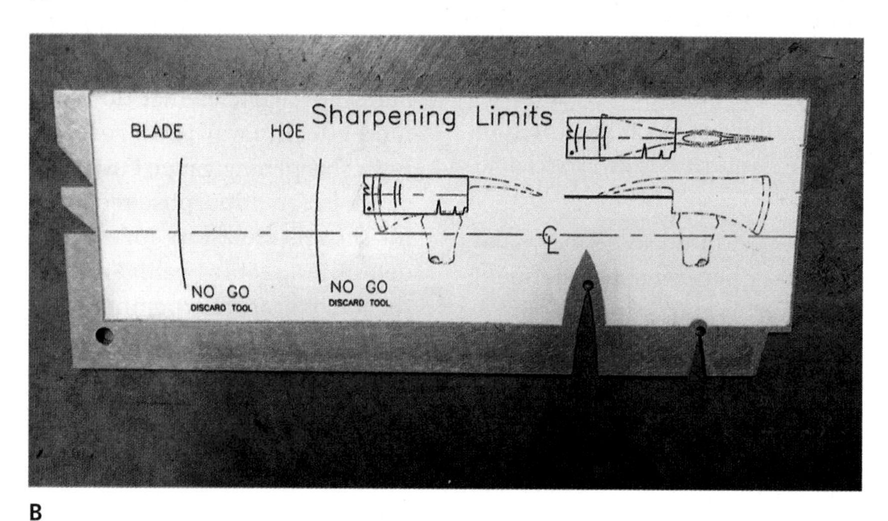

B

FIGURE 7-12 Sharpening gauge. **A.** Front. **B.** Back.
Courtesy of Jeff Pricher.

retired or at what angle a tool should be sharpened. After sharpening or grinding, use the sharpening gauge to determine the health of your hand tool and whether it requires more sharpening. The sharpening gauge is designed for use with several tools. Make sure you are using the right measurements (as seen on the gauge) when determining the life left in the tool or what angle the blades need to be sharpened to. If you have been sharpening for an extended period of time with no progress, it may be time to retire the hand tool.

Sharpening and Maintenance

It is important to use the right type of sharpening tool on your hand tools. The most common sharpening tool is the hand file. Depending on the shape of the hand tool, files with different cuts and levels of coarseness can be used **FIGURE 7-14** and **FIGURE 7-15**. The most common file, which can almost always be found at a fire-camp cache, is a 10-in. (25.4-cm) bastard-cut mill file **FIGURE 7-16**.

When pushing the file across the cutting surface of the hand tool, use even pressure throughout your stroke. At the end of the stroke, make sure to lift the file off the tool. Remember, files are designed to be used in one direction, not back and forth. When sharpening, the tool should be secured in a vise or other comfortable, secured method to prevent injury. When sharpening an adze or a Pulaski in the field, it is helpful to drive the axe head into a stump or a log.

File safety needs to be maintained whenever using the file. Make sure you have a handle and knuckle guard at the end of the file to protect your fingers and hands. Gloves must always be worn (on both hands) as well as a long-sleeve shirt and eye protection.

Files do get clogged and dirty. To clean the file, always clean in the direction or angle of the tool's teeth. It is OK to use a wire brush if the clogging is in deep in the teeth. Do not use oil or water when using or

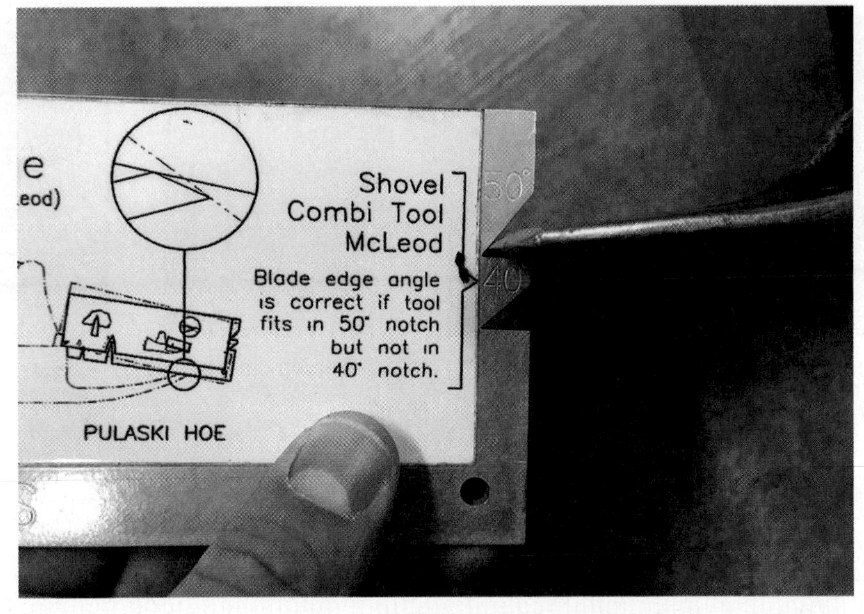

A

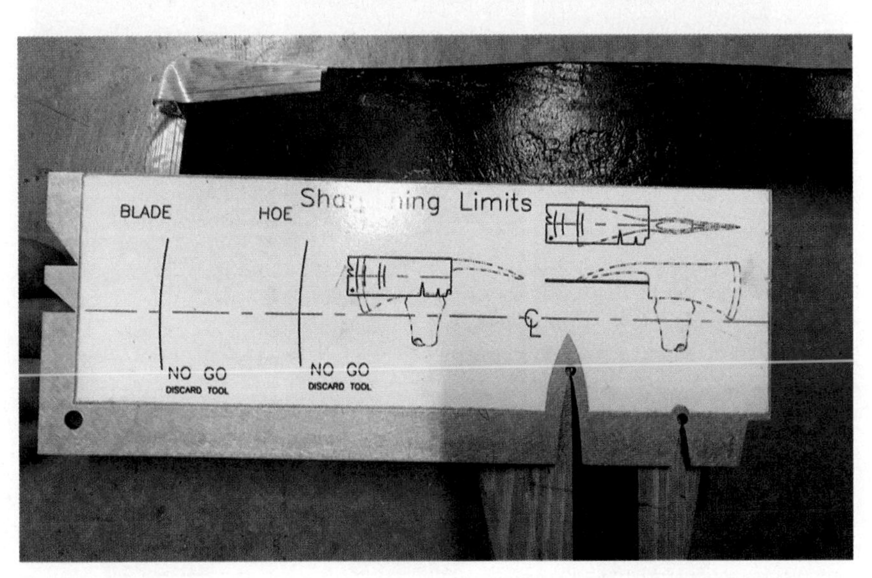

B

C

FIGURE 7-13 The sharpening gauge has several markings and notches that can help you determine the health of a tool.

Courtesy of Jeff Pricher.

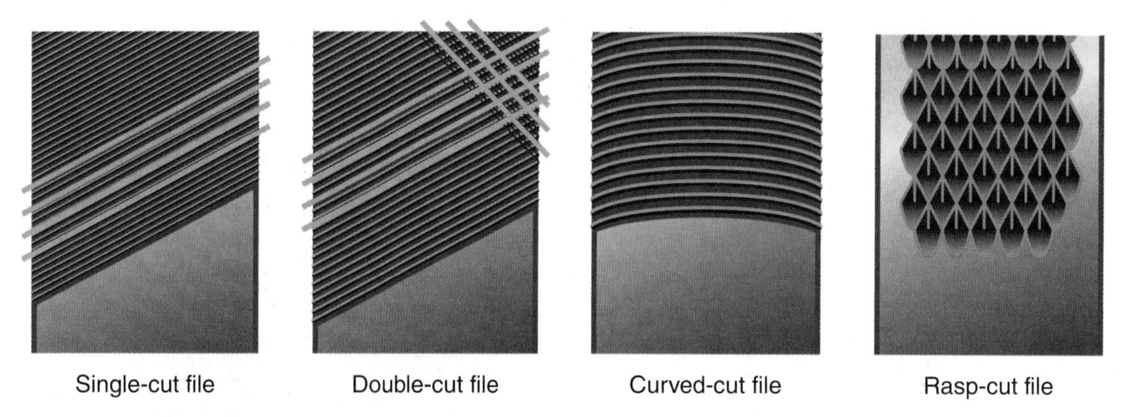

FIGURE 7-14 File cut types.

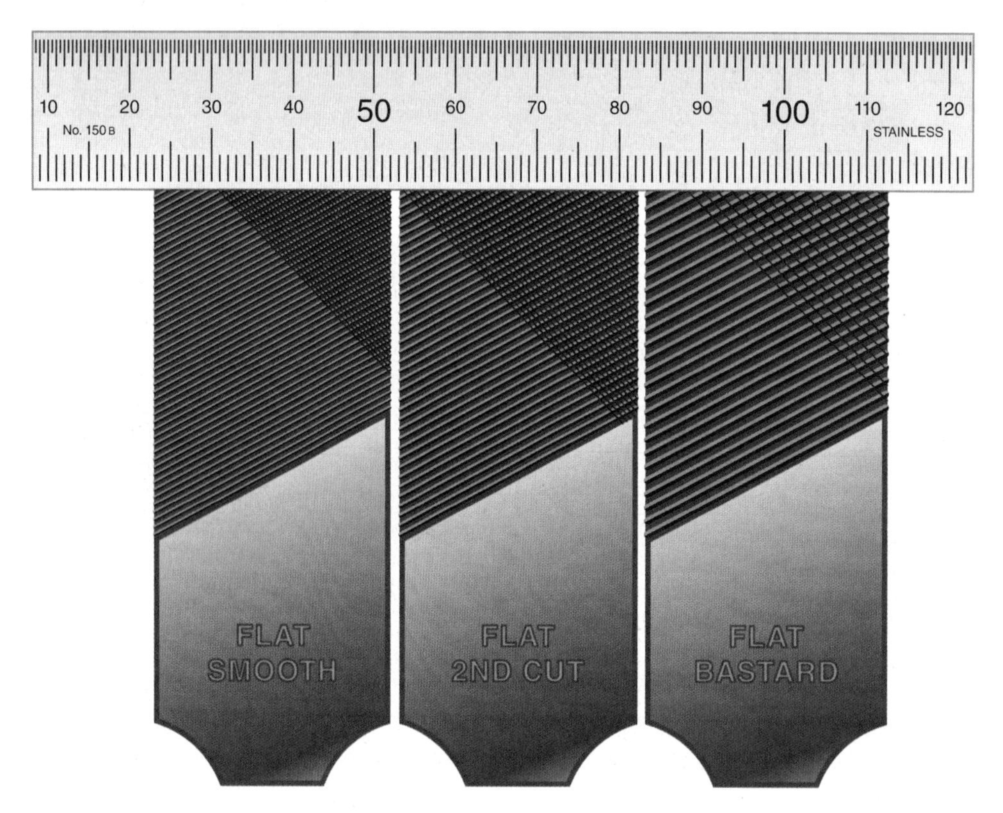

FIGURE 7-15 File coarseness levels.

FIGURE 7-16 10-in. (25.4-cm) bastard-cut mill file.
Courtesy of Mercer Industries.

cleaning the file. Doing so will cause the file to clog faster.

Grinders can be used for heavily damaged tools. However, grinders should be used only by those who are trained to do so. While grinders make light of the work, it is very easy to overheat the metal and make it very difficult to sharpen the tool. In some cases, the overheating can cause the metal to fatigue or become brittle.

Tool handles also need to be maintained. If nicks and burrs are felt or observed, it is recommended to sand down the affected areas with sandpaper and then apply a light coating of linseed oil as necessary. **TABLE 7-1** outlines basic maintenance and sharpening requirements for common hand tools.

Storage and Transport

Once tools have been inspected, cleaned, sharpened, and identified as fire ready, it is important to store them appropriately. Improper storage can lead to unwanted

TABLE 7-1 Hand Tool Inspection and Sharpening Checklist

Tool	Inspection	Sharpening
Shovel	Inspect head for: ■ Damage, such as cracks or wear ■ Rust ■ Safety guards Inspect handle for: ■ Smoothness ■ Alignment ■ Loose hardware	Sharpen edges 1½ in. (3.8 cm) from heel to a point, or per manufacturer's recommendation
Hoe	Inspect head for: ■ Damage, such as cracks or wear ■ Rust ■ Loose hardware Inspect handle for: ■ Smoothness ■ Alignment ■ Loose hardware	Bevel grubbing edge 3/8 in. (1.0 cm) wide on a 45-degree angle, or per manufacturer's recommendation
McLeod	Inspect head for: ■ Damage, such as cracks or wear ■ Rust ■ Safety guards Inspect handle for: ■ Smoothness ■ Alignment ■ Loose hardware	Bevel outside hoe edge to be straight and square with a 12-in. (30.5-cm) file on a 45-degree angle, or per manufacturer's recommendation
Fire rake	Inspect head for: ■ Bent or missing tines ■ Loose hardware Inspect handle for: ■ Smoothness ■ Alignment ■ Security	Sharpening not required

(*Continued*)

TABLE 7-1 Hand Tool Inspection and Sharpening Checklist (*Continued*)

Tool	Inspection	Sharpening
Combination tool	Inspect head for: ■ Damage, such as cracks or wear ■ Rust ■ Loose hardware Inspect handle for: ■ Smoothness ■ Alignment ■ Security	Stick pick end into the ground and sharpen the blade at a 45-degree angle, or per manufacturer's recommendation
Swatting tools	Inspect head for: ■ Damage, such as cracks or wear ■ Loose hardware Inspect handle for: ■ Smoothness ■ Alignment	Sharpening not required
Pulaski	Inspect head for: ■ Damage, such as cracks or wear ■ Rust ■ Loose hardware	Cutting edge: Taper 2 in. (5.1 cm) wide with an even bevel on each side Grubbing edge: 3/8 in. (1.0 cm) wide on a 45-degree angle on one side only, or per manufacturer's recommendation
Axe	Inspect head for: ■ Damage, such as cracks or wear ■ Rust ■ Loose hardware Inspect handle for: ■ Smoothness ■ Alignment	Sharpen back 2½ in. (6.4 cm) on each side with even bevel on both sides, or per manufacturer's recommendation
Brush hook	Inspect head for: ■ Damage, such as cracks or wear ■ Rust ■ Loose hardware Inspect handle for: ■ Smoothness ■ Alignment	Sharpen per manufacturer's recommendation
Power tools (powered blower, Weed Eater, chainsaw)	Follow all recommendations outlined in the user manual	Follow all recommendations outlined in the user manual

FIGURE 7-17 Plastic shovel case.
Courtesy of Jeff Pricher.

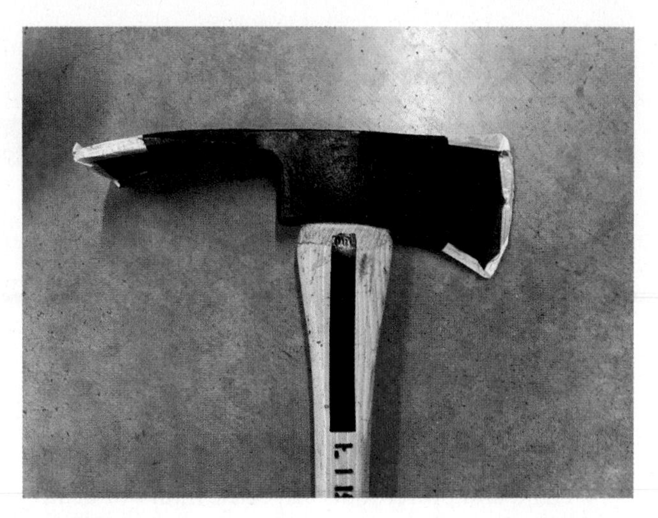

FIGURE 7-18 If a tool does not come with a case, cover the sharp edges with glass tape.
Courtesy of Jeff Pricher.

damage to the tools and equipment on a vehicle or aircraft and in a passenger compartment. It is always your responsibility to properly stow tools according to agency policy and manufacturer's guidelines.

Blades should always be covered so they are protected and unable to injure someone. Some tools are sold with a plastic guard or case that can help protect from cuts and injuries **FIGURE 7-17**. For tools that do not come with a plastic guard, one of the most common ways to protect the sharp edge is to cover it in glass or strapping tape **FIGURE 7-18**. Glass tape helps to prevent inadvertent cuts.

Generally, in wheeled apparatus, there are specific compartments identified for tool storage. Safely secure tools that do not have a dedicated compartment. All tool blades and edges should be covered with their respective guards or with tape. Storing and transporting unprotected tools can lead to tool damage and fire-fighter injury.

In aircraft, generally the aircrew will direct the fire fighters to pool their equipment into specific groups and require you to use straps and/or fiber tape to secure the tools. Like wheeled vehicles, all tools must have an edge guard in place before being placed on an aircraft. With aircraft operations, you must follow all the aircrew's directions. Never stow or place anything on the aircraft, and never open any compartments unless you are specifically directed to do so. Failure to follow instructions could jeopardize the aircraft and everyone on board.

LISTEN UP!

Excess tools near a fireline should be stored a safe distance from the fireline and traffic with sharp cutting edges toward the ground. Cover tool heads with sheaths or cases if available. Excess tools should be visible to crew members, and the area should be flagged if possible.

LISTEN UP!

If your tool doesn't have a case and you don't have any glass tape, you can also protect the sharp edges with any of the following:

- Old hose
- Masking tape
- Old inner tube
- Old conveyor belt

After-Action REVIEW

IN SUMMARY

- There are many types of suppression tools to choose from in the wildland environment. If used in the wrong manner or situation, there can be considerable negative consequences.
- It is important to be familiar with the different types of hand tools.
  - Scraping and digging tools: Shovel, hoe, McLeod, fire rake, combination tool
  - Swatting tools: Fire swatters, gunnysack
  - Cutting tools: Pulaski, axe, brush hook, Sandvik
  - Power tools: Powered blower, string trimmer
- Hand and power tools can cause injury or death if used incorrectly; it is important to be familiar with proper handling procedures for each tool.
- Each tool has its own care and maintenance requirements. Refer to the manufacturer's recommendations for maintaining tool blades and handles.
  - Wildland fire tools need to be sharpened frequently to remain effective at the scene. It is important to use the right type of sharpening tool on your hand tools.
- Improper storage and transport of tools can lead to unwanted damage to tools and equipment, and potentially injure someone if blades are not covered properly.

KEY TERMS

Alaska fire swatter A swatter with several strips of rubber on the end, rather than one thick, flat piece.

Axe A tool with a short handle and steel blade used for chopping.

Brush hook An axe with a curved blade head rather than a standard axe head.

Combination tool A sturdy and lightweight tool with many uses: grabbing, digging, smothering, scraping, cutting, and prying; also called a combi tool or Colby tool.

Fire rake A long-handled rake.

Fire swatter A long-handled tool with a thick, flat piece of rubber on the end used to smother flames.

Hoe A digging tool with a sharp, flat blade on one or both sides of the head.

McLeod A combination scraping and raking tool; one side of the head has a blade and the other side has a rake.

Powered blower A portable tool that uses air to blow leaves and needles away from a structure.

Pulaski A chopping and trenching tool with one side of the head being a blade used for cutting, and the other side being a flat edge used for grabbing.

Sandvik A cutting tool with a C-shaped blade used to cut small- and medium-sized brush.

Shovel A tool with a long handle and a tapered blade with both edges sharpened.

String trimmer A long metal rod with a spinning blade or string, used for reducing fuel next to a fireline.

REFERENCES

National Fire Protection Association (NFPA). "NFPA 1977, Standard on Protective Clothing and Equipment for Wildland Fire Fighting." 2016 ed. https://www.nfpa.org/codes-and-standards/all-codes-and-standards/list-of-codes-and-standards/detail?code=1977.

National Wildfire Coordinating Group (NWCG). "NWCG Standards for Fire Equipment Storage and Refurbishing, PMS 448." April 2018. https://www.nwcg.gov/publications/pms448.

National Wildfire Coordinating Group (NWCG). "S-130, Firefighter Training, 2008. (NFES 2731). Modified October 23, 2019. https://www.nwcg.gov/publications/training-courses/s-130.

Wildland Fire Fighter in Action

It is July 11, and you and your crew are returning from a fire assignment that had you up most of the night. Your last shift was scheduled to be on your twelfth day of the assignment, but you and your crew have not had a day off in 17 days. Your crew is in a convoy of one 10-passenger van and three 4-door pickup trucks. The van contains all of the personal gear bags for the crew and all of the tools and line gear are in the bed of the three pickups. The route from the fire has you travelling on an interstate for about 18 miles. On this particular day, the weather is not optimal and the wind is howling at over 50 mi/h (29 km/h). The wind effect is so significant that, about halfway into your drive to fire camp, some of the hand tools and a few sets of line gear are blown out of the truck and land on the interstate.

1. How could this incident have been avoided?
   - **A.** Tools should have been secured in holders.
   - **B.** Tools should not have been on the top of the line gear.
   - **C.** Tools could have been grouped together and bungeed to the vehicle.
   - **D.** All of the above
2. Why is it important to keep edge guards and sheaths on tools and equipment?
   - **A.** To prevent accidental damage to others and equipment
   - **B.** To protect the sharp edges from damage
   - **C.** Edge guards are not really needed.
   - **D.** Both A and B
3. Assuming the tools were not badly damaged, what type of personal protective equipment would need to be worn when sharpening the tools before their next use?
   - **A.** Chaps
   - **B.** Eye protection, long sleeves, gloves
   - **C.** Eye protection, hearing protection, gloves, and long sleeves
   - **D.** Gloves and eye protection
4. What was probably a significant contributing factor to this incident?
   - **A.** Speed of the vehicles
   - **B.** Higher than normal wind
   - **C.** Fatigue and not properly securing the tools
   - **D.** Both A and B

Access Navigate for flashcards to test your key term knowledge.

CHAPTER

Wildland Fire Fighter I

Ground and Air Equipment

KNOWLEDGE OBJECTIVES

After studying this chapter, you will be able to:

- Identify safety guidelines and weight limitations for various transportation modes. (**NFPA 1051: 4.5.2**, p. 172)
- Identify the types of mobile fire apparatus used on wildland fires and the proper use of each:
  - Type 1 engines (p. 173)
  - Type 2 engines (p. 173)
  - Type 3 engines (pp. 173–175)
  - Type 4, 5, 6 and 7 engines (pp. 175–176)
  - Water tenders (p. 176)
- Identify the types of mechanized ground equipment used on wildland fires and explain the proper use of each:
  - Dozers (pp. 177–179)
  - Tractor plows (pp. 179–180)
  - Excavators (p. 180)
  - Feller bunchers (pp. 180–181)
  - Other types of logging equipment (pp. 181–182)
  - Skidgens (p. 182)
  - Masticators (pp. 182–183)
  - Terrain vehicles (p. 183)
- Identify the types of air equipment used on wildland fires and explain the proper use of each:
  - Helicopters (pp. 183–184)
  - Fixed-Wing Air Tankers (pp. 184–185)
  - Unmanned Aircraft Systems (pp. 185–187)

SKILLS OBJECTIVES

This chapter contains no skills objectives for Wildland Fire Fighter I candidates.

Additional Standards

- **NFPA 1906**, *Standard for Wildland Fire Apparatus*
- **NWCG S-130**, *Firefighter Training*
- **NWCG PMS 310-1**, *Standards for Wildland Fire Position Qualifications*
- **NWCG PMS 510**, *Standards for Helicopter Operations*

You Are the Wildland Fire Fighter

It's mid-August, and you are in your third season as a wildland fire fighter. The assignment you have on the Type 1 fire you are assigned to is to scout the western flank on the fire in one of two utility terrain vehicles (UTVs) with a total of six people. You are on a decommissioned road that has not been used in over 15 years. The road is rutted out and in significant disrepair.

As you come around a bend, the road turns into what amounts to be a very steep trail. As you slowly roll down the hill, the undercarriage scrapes the dirt. Almost immediately, the UTV stops. As you start to exit the vehicle, you feel as though the vehicle is about to roll over. In addition, you smell and see smoke coming from the floorboard area.

1. What personal protective equipment (PPE) should you be wearing for this assignment?
2. What is most important about UTV use?

Access Navigate for more practice activities.

Introduction

Wildland fires are dynamic events that require a variety of resources and staffing levels. Before you can develop an overall plan for fighting a wildland fire, you must understand the types of ground and air resources available and their capabilities. Some engines are capable of operating off road, whereas others are not.

Resource typing is the categorization of resources that are commonly exchanged in disasters via mutual aid by capacity or by capability. Resource typing allows for ease of ordering and tracking of fire resources. Resources are classified (typed) in general categories: engines and water tenders, heavy or mechanized equipment, and aircraft. Classifying resources enables incident commanders to order the appropriate resources to match the incident objectives.

The classification system used in this book is based on the following sources:

- National Interagency Fire Center's *Interagency Standards for Fire and Fire Aviation Operations* (Red Book)
- Federal Emergency Management Agency's *Typed Resource Definitions: Fire and Hazardous Materials Resources*

Both of these sources are intended to serve as field reference guides for wildland fire agencies using the Incident Command System in the control of wildland forest and range fires.

DID YOU KNOW?

When traveling from your home station to an incident, there are safety guidelines that need to be followed. While each type of transportation equipment may have more specific requirements, here are a few considerations and safe practices to utilize. If you see someone who is not following safe practices, you should intervene to prevent an accident from occurring. Consider the safety concepts for different types of transportation.

When traveling by vehicle:

- Wear your seat belt.
- Do not distract the driver (music, cell phone, or other media devices can all be distractions).
- Everyone should be acting as a lookout.
- Do not ride in the back bed of pickups.

When traveling by utility vehicle (side by side):

- Always wear a helmet.
- Always wear a seat belt.
- Know the limitations of the vehicle; do not push beyond them or your skill ability.

When traveling by boat or raft:

- Always wear a personal flotation device.
- Be prepared to swim.
- Follow the commands of the boat pilot at all times.

When traveling by helicopter or other aircraft:

- Wear full PPE (including chin strap on helmet).
- Know the weights of your gear and yourself. The helicopter crewmember will ensure the aircraft does not exceed its weight limit for cargo and passengers.
- Always receive a briefing from an aircraft crew member.
- Follow all commands of the aircraft crew member or pilot.

When traveling by foot:

- Follow the pace and route determined by your supervisor
- Stay with your crew
- Maintain proper distance between crew members
- Watch out for hazards

Mobile Fire Apparatus

A fire engine or apparatus is one of the most versatile pieces of fire equipment in use on a wildland fire. An

apparatus is a motor-driven vehicle designed for the purpose of fighting fires. Wildland fire apparatus range from small pickup trucks or Jeep-type vehicles to large trucks. Most wildland apparatus is designated with all-wheel drive; underside protection; and high road clearance, which enables vehicles to travel off road.

Apparatus carry personnel and equipment—hoses, water, hand tools, and sometimes Class A foam—to a wildland incident. These tools and equipment allow fire crews to attack fires by direct or indirect attack methods. They can mobile pump fires where terrain allows this type of operation. Many types of apparatus can be found on almost any brush or interface fire. There are typically seven types of engines, two of which are structural. **TABLE 8-1** compares the different types of engines and their components.

Type 1 Engines

The **Type 1 engine** is a typical structure-fire apparatus for use in urban settings **FIGURE 8-1**. These apparatus are not typically used off of roadways due to a lack of a mobile pump, all-wheel drive, and underside protection. Type 1 apparatus are used primarily for the protection of structures; however, this does not preclude their use along the side of a roadway to supply water to wildland hoselays.

According to the National Wildfire Coordinating Group (NWCG), Type 1 engines must meet the following minimum standards (as outlined in Table 8-1):

- Have a minimum pump capacity of 1000 gallons (3785.4 L)
- Have a minimum tank capacity of 300 gallons (1135.6 L)
- Carry 1200 ft (365.8 m) of 2.5-in. (63.5-mm) hose
- Carry 500 ft (152.4 m) of 1.5-in. (38.1-mm) hose
- Be able to produce a 500 gallon per minute (1892.7 L/min) master stream
- Have ladders that meet National Fire Protection Association (NFPA) 1931, *Standard for Manufacturer's Design of Fire Department Ground Ladders*
- Have a minimum of four personnel

LISTEN UP!

It can be very enticing for Type 1 engine operators to take the engine off the pavement. Do so only if you are confident that the engine will not get stuck or damaged.

Type 2 Engines

The **Type 2 engine** is typically used for structure protection but can also be used for wildland-fire suppression **FIGURE 8-2**. They are versatile and can be used to respond off road on a wildland fire if needed. Unlike Type 1 engines, many agencies will require their Type 2 engines to have the capability to **mobile pump**, or operate a water pump while moving. This pump-and-roll capacity is achieved by powering the pump with a separate engine or powering the pump from a power takeoff shaft. In some cases, Type 2 engines have a higher ground clearance and are built with a locking differential to maximize traction.

Type 2 engines generally have a smaller wheel base than Type 1 engines and must meet the following minimum standards according to the NWCG and as listed in Table 8-1:

- Have a minimum pump capacity of 500 gallons per minute (1892.7 L/min)
- Have a minimum tank capacity of 300 gallons (1135.6)
- Carry 1000 ft (304.8 m) of 2.5-in. (63.5-mm) hose
- Carry 500 ft of 1.5-in. (38.1-mm) hose
- Meet NFPA 1931, *Standard for Ladder Complements for Engines*
- Have a minimum of three personnel

LISTEN UP!

There are many Type 2 engine configurations that will provide increased capabilities to do more when off paved roads. In areas that have big dips in gravel or dirt roads, make sure a crew member inspects and walks in front of the apparatus to verify that the engine is capable of the operation.

Type 3 Engines

The **Type 3 engine** is a typical brush engine used in wildland areas **FIGURE 8-3**. It is a smaller apparatus, compared to type 1 and 2 engines, with a short wheelbase and high clearance that can be used off road. Type 3 engines have the ability to mobile pump and do not carry larger-diameter hose complements. The versatility of this engine cannot be understated. With its water- and equipment-carrying capacity and nimbleness, this type of engine is quickly becoming a favorite choice for interface firefighting in rural areas.

TABLE 8-1 Firefighting Engines

Engine Type	Structural Engines		Wildland Engines				
Components	**Type 1**	**Type 2**	**Type 3**	**Type 4**	**Type 5**	**Type 6**	**Type 7**
Minimum tank capacity in gal (L)	300 (1135.6)	300 (1135.6)	500 (1892.7)	750 (2839.1)	400 (1514.2)	150 (567.8)	50 (189.3)
Minimum pump flow in GPM (L/min)	1000 (3785.4)	500 (1892.7)	150 (567.8)	50 (189.3)	50 (189.3)	50 (189.3)	10 (37.9)
2.5-in. (63.5-mm) hose in ft (m)	1200 (365.8)	1000 (304.8)	N/A	N/A	N/A	N/A	N/A
1.5-in. (38.1-mm) hose in ft (m)	500 (152.4)	500 (152.4)	1000 (304.8)	300 (91.4)	300 (91.4)	300 (91.4)	N/A
1-in. hose (25.4-mm) in ft (m)	N/A	N/A	500 (152.4)	300 (91.4)	300 (91.4)	300 (91.4)	200 (61.0)
Ladders	Yes	Yes	No	No	No	No	No
Pump-and-roll capacity	No	No	Yes	Yes	Yes	Yes	Yes
Maximum operating weight in lb (kg)	N/A	N/A	N/A	N/A	26,000 (11,793.4)	19,500 (8845.1)	14,000 (6350.3)
Minimum personnel	4	3	3	2	2	2	2

NWCG/Red Book. Interagency Standards for Fire and Fire Aviation Operations. Table from page 304: https://www.nifc.gov/PUBLICATIONS/redbook/2019/RedBookAll.pdf.

FIGURE 8-1 Type 1 engine.
© ND700/Shutterstock.

FIGURE 8-2 Type 2 engine.
Courtesy of Eric Smythe.

FIGURE 8-3 Type 3 engine.
Courtesy of Jeff Pricher.

Type 3 engines must meet the following minimum standards per NWCG and as noted in Table 8-1:

- Have a minimum pump capacity of 150 gallons per minute (567.8 L/min)
- Have a minimum water tank size of 500 gallons (1892.7 L)
- Carry 1000 ft (304.8 m) of 1.5-in. (38.1-mm) hose
- Have 500 ft (152.4 m) of 1-in. (25.4-mm) hose as minimum complement
- Must be able to pump and roll (mobile pump)
- Have a minimum of three personnel

Type 4, 5, 6, and 7 Engines

Types 4, 5, 6, and 7 engines are small pickups, or **patrol vehicles**, found in many rural areas **FIGURE 8-4**. Many of these engines have a four-wheel drive capability.

A

B

FIGURE 8-4 Patrol engines. **A.** Type 4 engine. **B.** Type 6 engine.
A: Courtesy of Jeff Pricher; **B:** Courtesy of Joe Lowe.

These engines are more effective than Types 1, 2, and 3 in rocky or steep slopes due to their smaller size and shorter turning radius. Types 4, 5, 6, and 7 engines must meet minimum standards per NWCG and as noted in Table 8-1.

Water Tenders

A **water tender** is used to support engine operations by transporting water to the scene of a wildland fire **FIGURE 8-5**. They are usually equipped with a pump that enables them to fill from a drafting site and to off-load water when needed. Many tenders are equipped with a dump valve for rapidly unloading water into a portable tank.

They fall into two classifications on an incident: support tenders and tactical tenders. Support tenders are used, as the name implies, to support operations that require water. Tactical tenders are used on the fire for operational purposes. Tactical water tenders have large tank capacities and mobile-pump capabilities. Water tenders must meet the minimum standards shown in **TABLE 8-2**.

FIGURE 8-5 Water tender.
Courtesy of Jeff Pricher.

Ground Equipment

Heavy or mechanized **ground equipment** are effective initial attack tools on wildland fires. These powerful heavy pieces of mechanical equipment are used to construct firelines within a variety of fuel and soil types. They must be closely supervised and integrated into your plan.

Heavy or mechanized equipment used in wildland firefighting includes ground vehicles such as dozers, tractor plows, feller bunchers, skidgens, and masticators.

While you may not be directly responsible for the operation or direction of the heavy equipment, it is very important to understand some of the basic operations. Most of this equipment will be transported to the fireline by a tractor or dump truck. These big trucks will be towing a trailer that will need a space to park or turn around. As you work off a road, pay attention to areas that may be a good spot for the transport equipment to park or stage for loading and unloading. Also, make sure you communicate road conditions to your supervisor or the transport driver. This will help him or her determine whether the equipment will be able to make it up a road.

TABLE 8-2 Minimum Standards for Water Tenders

Requirements	Support 1	Support 2	Support 3	Tactical 1	Tactical 2
Tank capacity in gal (L)	4000 (15,141.6)	2500 (9463.5)	1000 (3785.4)	2000 (7570.8)	1000 (3785.4)
Pump capacity in GPM (L/min)	300 (1135.6)	200 (757.1)	200 (757.1)	250 (946.4)	250 (946.4)
Maximum refill time in minutes	30	20	15	N/A	N/A
Pump-and-roll capacity	No	No	No	Yes	Yes
Minimum personnel	1	1	1	2	2

NWCG/Red Book. *Interagency Standards for Fire and Fire Aviation Operations.* Table from page 304: https://www.nifc.gov/PUBLICATIONS/redbook/2019/RedBookAll.pdf.

Dozers

A bulldozer, or **dozer**, is a tracked vehicle with a front-mounted blade used for moving fuels and exposing soil. Dozers are a great resource for constructing firelines and safety zones, opening up roads, and rapidly clearing fuels in advance of the fire. Dozers are best used in pairs, with the lead dozer pioneering the line and the second dozer widening, strengthening, and completing the line. If one gets stuck, the other can assist. However, dozer progress can be slowed in rocky soils or areas of dense timber. Some conditions, such as steep slopes or sidehill operations, can limit their use.

Use of dozers must be managed closely because they can damage fragile soils and leave scars that remain for decades. Some jurisdictions require special permission before dozers can be utilized for fireline construction. Generally speaking, state and national parks and national forest wildernesses require special permission before using dozers in fire suppression. With proper management of this resource, the damage caused by their use can be minimized.

Dozers are typed by size and horsepower. **TABLE 8-3** outlines these characteristics. On a fire, dozer organization and the types of dozers used will vary with the size of the fire, amount and type of fuels, resource availability, and local practices.

Heavy Dozers

Heavy dozers (Type 1) do not maneuver well in tight situations, especially in steep terrain. Generally, they are too big to be used to construct firelines in average situations **FIGURE 8-6**. They are best used as lead dozers in heavy fuels on moderate slopes.

TABLE 8-3 Wildland Dozer Types and Characteristics

Dozer Type	Description	Example	Minimum Horsepower
Type 1	Heavy	D7, D8, D9, or equivalent	200 HP
Type 2	Medium	D5N, D6N, or equivalent	100 HP
Type 3	Light	D4 or equivalent	50 HP

Medium Dozers

Medium dozers (Type 2) perform well on moderately steep slopes and are maneuverable and usually considered the best size for all-around use on wildland fires **FIGURE 8-7**. They are well suited for the average fuel and terrain features found in mountainous areas. They can be fitted with wider tracks, which enable them to work well in wet or boggy areas.

Light Dozers

Light dozers (Type 3) are very maneuverable in tight situations and best used in soil types that have few rocks and light fuels **FIGURE 8-8**. Light dozers work well in wet soil if they are equipped with wide tracks and are effectively used on level ground and moderate slopes.

FIGURE 8-6 Heavy dozer.
Courtesy of Jeff Pricher.

FIGURE 8-7 Medium dozer.
Courtesy of Jeff Pricher.

FIGURE 8-8 Light dozer.
Courtesy of Jeff Pricher.

Dozers and Slope Operations

Consider the maximum percentage of slope that a dozer can operate on. It is not only important to consider uphill and downhill operations but also sidehill operations. Maximum percentages of slope for uphill and downhill operations vary by dozer type. Generally, dozers should not be operated across sidehills greater than 45 percent, uphill slopes greater than 55 percent, or downhill slopes greater than 75 percent. These percentages are dependent on soils, fuels, and other factors. If these percentages are exceeded, then consider the use of handcrews. Handcrews are discussed in Chapter 14, *Handcrew Operations*.

LISTEN UP!

Avoid working downslope from a dozer! Dozers are a powerful piece of equipment that will knock over trees, cause logs to roll, and dislodge rocks that can damage equipment or injure you **FIGURE 8-9**. It is difficult for operators to communicate; if they cause something to break loose, they may not be able to warn you fast enough.

Dozer Control Systems

Dozers have hydraulic systems that control blade operations. The advantage to a hydraulic blade-control system is that pressure can be exerted either up or down on the blade, and it can exert downward blade pressure. Dozers with hydraulic systems are best used on firelines and are a versatile tool when used to construct a fireline.

The hydraulic system of blade control enables the bulldozer to dig in hard ground and can be used to brake while going downslope. If the dozer gets stuck or high centered, then downward pressure on the blade can be exerted and used to raise the dozer; then materials can be placed underneath the bulldozer to free it.

FIGURE 8-9 This rock fell off a hill that a dozer was working on. Avoid working downslope from a dozer.
Courtesy of Jeff Pricher.

LISTEN UP!

Dozer management is vital. Remember that they can produce a lot of finished fireline but, if not properly managed, they can cause considerable damage to the environment. The dozer operator must understand the plan because he or she is the vital link between the dozer and what you are trying to accomplish.

Dozer and Tractor Plow Safety

Dozers and tractor plows (discussed in the next section) have similar safety considerations. When working with or around dozers or tractor plows, follow these NWCG safety guidelines:

- Load and unload equipment from the transport in a safe manner on a level, stable surface.
- Park transport in an area free of fuel. Clear an area if needed to protect parked equipment.

- Do not sit or bed down near equipment. Walk around the equipment before starting or moving it.
- Lower the dozer blade and/or plow to the ground when the equipment is idling or stopped.
- When operating a dozer or tractor plow, stay at least 100 ft (30.5 m) in front or 50 ft (15.2 m) behind the equipment.
- Only the operator should ride on the equipment.
- Never get on or off equipment while it is moving.
- Provide front and rear lights for equipment working at night or in heavy smoke.
- Provide lights and fluorescent vests to personnel working with dozer and tractor plow units.
- Use hand signals for direction and safety.
- Do not use a dozer or tractor plow without a canopy or brush guard, and radio communications.
- Dozer and tractor plow operators must wear required PPE and carry a fire shelter. Tractor plow operators must wear protection for the head, face, eyes, and ears while also providing radio reception and ventilation capabilities.
- Be aware of different fuel types, rates of spread, and flammability.
- Watch for wetlands, steep slopes, rocks, ditches, and other obstacles that might stop the equipment. Bogging down in swampy areas is common.
- Do not get too far ahead of a firing crew during firing operations.
- Anchor the line to a secure firebreak and create a black line (burn out) until the fire is completely enclosed.
- When the dozer or tractor plow is equipped with a hand-clutch lever, always take equipment out of gear when mounting and dismounting.

Other safety concerns when operating a dozer or tractor plow are as follows:

- Establishing lookouts for dozers and tractor plows proves nearly impossible in flat, brushy terrain without the use of aircraft.
- Coordination and communication with other fire fighters is critical.
- Hearing the radio on an open-cab tractor plow is sometimes difficult when a dozer is operating, unless the dozer has an environmental cab.
- In thick, swampy terrain, dozer and tractor plow operators usually have only one good escape route: back down the line they just plowed. Few options remain if that escape route is compromised.
- Safety zones for tractor plow operators are typically in the black as the backfire is carried behind the dozer. Experience and good communications remain key.

Tractor Plows

A **tractor plow** is best described as a dozer pulling a plow unit. The blade on the dozer is used to clean downed material, slash, and debris out of the way while the plow unit, being pulled, constructs the fireline **FIGURE 8-10**.

Blades can determine a tractor plow's operations. A tractor plow with a straight blade can be useful for clearing a pathway for a small brush engine. Some tractor plows have C-blades, which do not disturb the soil and are used to push small trees out of the way. In some parts of the country, tractor plows have V-blades, which cut and push small trees out of the way.

Tractor plows are effective fireline tools used to build a control line. They are used in the flat woods and coastal plains. Tractor plows are best used in topography that has limited slope and with soils that have minimal rock. Rocky soils and increased slopes are a problem for plows. Plows are pulled by a dozer or a four-wheeled vehicle.

Tractor plows are best used in areas that are relatively flat or have rolling hills and in soils that crumble easily or are sandy. Plows do not work well in rocky areas. In areas of standing timber, tractor plows should not be used unless the spacing of the trees is such that they do not interfere too much with the plow unit.

FIGURE 8-10 A tractor plow.
Ross Frid/Alamy Stock Photo.

Tractor Plow Operations

Generally, a tractor plow can construct lines at about 3 mi/h (4.8 km/h), providing that the fuels, soils, and terrain are suitable for their use. The speed drops off as the slope increases or the plow unit encounters boggy sections, stumps, trees, and poor soil conditions.

In areas of trees and heavy brush, one tactic to consider is to have a dozer take the lead with a V-blade and follow it with a tractor plow. The dozer with the V-blade takes out the heavy brush and trees and pushes them to the outside of the fireline.

When using tractor plows, be aware of the rate of fire spread, the plow rate of line construction, the distance the unit is to the fire, the fire's intensity levels, and the time it is taking to complete the burnout operations. Tractor plow crews should consist of a minimum of two people.

Tractor Plow Safety

Tractor plows and dozers have similar safety considerations. NWCG safety guidelines for tractor plow and dozer operations were previously discussed in the Dozer and Tractor Plow Safety section.

LISTEN UP!

Dozers and tractor plows must be staffed with personnel who meet the training and experience standards for Dozer Operator (DZOP) or Dozer Operator Initial Attack (DZIA). Dozer and tractor-plow operators must meet the local department of transportation requirements.

Excavators

An **excavator** is a tracked vehicle with a rotating cab; a long, bent neck with a bucket; and, usually, a thumb at the end **FIGURE 8-11**. This piece of heavy equipment is effective for fireline construction; rehab; and, in some instances, assisting with road improvements.

There are four categories of excavators. This range covers the needs from very large (Type 1) to very small (Type 4). If ordered correctly from the incident staff, the excavator will arrive with a bucket and a thumb. The thumb allows the operator to be able to pick up logs, trees, boulders, and other large items. One of the most common uses for the excavator is rehab after the fire. The excavator is very effective in pulling berms and scattering trees and logs all over the fireline. It is not uncommon for excavators to also assist with road repair and replacing damaged culverts.

A

B

FIGURE 8-11 Excavators. **A.** Type 2. **B.** Type 3.
Courtesy of Jeff Pricher.

Feller Bunchers

A **feller buncher**, or harvester, is a type of mechanized logging equipment that allows for accelerated line construction by cutting down and removing trees in light to moderate timber. The equipment is very similar in visual appearance to an excavator. Feller bunchers may be tracked or wheeled.

There are two types of feller bunchers that are commonly used: self-leveling and nonleveling. Self-leveling feller bunchers are used on very steep slopes because the cab is able to level against the slope **FIGURE 8-12**. Nonleveling feller bunchers can be used only on flatter terrain.

Feller bunchers are able to do the work of several saw teams with more protection and safety than hand felling. When a fireline needs to be established through dense canopies, this machine is able to grapple onto a tree, cut it, and stack it away from the fireline. Generally, these types of equipment are not able to work in timber of 34 in. (86.4 cm) in diameter or larger.

FIGURE 8-12 Self-leveling feller buncher.
Courtesy of Paul Jones.

> **LISTEN UP!**
>
> Feller buncher–operation requirements vary from region to region. These types of equipment can cost anywhere from $750K to over $1 million, so it is best to follow your agency's operation requirements to avoid damaging the equipment.

Other Types of Logging Equipment

With the intensity of fires and the increasing shortage of resources during busy fire years, wildland fire fighters have long been requesting logging equipment on fire scenes. Logging equipment can increase productivity and safety measures by decreasing fire-fighter exposure to felling and logging operations during fireline construction.

During the felling and logging process, as the feller bunchers clear the way, other equipment is needed to completely remove and process the fuel. These other pieces of equipment include logging processers, skidders, log loaders, and forwarders.

Logging Processors

Logging processors are a type of heavy ground equipment used to remove all the limbs from trees and cut them into merchantable lengths for sale. The grapple picks up the log and runs it through a grinder built into the head, which takes off all the limbs, measures the log for board feet, and then cuts the tree into appropriately sized logs **FIGURE 8-13**.

FIGURE 8-13 Logging processor.
Courtesy of Jeff Pricher.

Skidder

A skidder is a piece of logging equipment used to grab logs and haul them from one location to another. Skidders can be rubber tired or tracked **FIGURE 8-14**.

FIGURE 8-14 A tracked skidder (left) and a rubber-tired skidder.
Courtesy of Jeff Pricher.

Log Loaders

Once the logs are hauled off the cutting block to an area that will become a log deck, log loaders come into play. A log, or logging, loader is a type of logging equipment with an open/close gripping end used to stack logs and load them onto trucks **FIGURE 8-15**.

Forwarders

Forwarders are heavy equipment that can move large loads of logs from where they are harvested and processed to the log deck **FIGURE 8-16**. There are many variations of this equipment type. If many trees need

> **LISTEN UP!**
>
> Each of these types of logging equipment have a use and several variations. To understand all the variations, consider taking a Heavy Equipment Boss course (NWCG S-236). In this class, you will learn about inspection, loading and unloading, equipment uses, and other specialized information related to heavy logging equipment.

FIGURE 8-15 Log loader.
Courtesy of Jeff Pricher.

FIGURE 8-17 Skidgen.
Courtesy of Jeff Pricher.

FIGURE 8-16 Forwarder.
© Esemelwe/E+/Getty Images.

FIGURE 8-18 Masticator.
Courtesy of ASV Holdings Inc.

to be moved and skidders are not available, a forwarder can be used.

Skidgens

Skidgens, or skidgines, are a very versatile component of the heavy-equipment category. A **skidgen** is a hybrid vehicle that got its start as part logging skidder and part fire engine **FIGURE 8-17**. Generally equipped with a plow on the front and a water tank and pump on the back, this vehicle can traverse terrain that no fire engine ever could. For initial attack and mop-up, this equipment is the king of versatility. Some of the newer skidgen variations include multiple wheels or tracks, like a dozer/tank.

Masticators

A **masticator**, also referred to as a mulcher, is a type of mechanized equipment used to grind brush and debris while driving through it **FIGURE 8-18**. It can also clear brush alongside roads. Masticators are ideal for use in areas with delicate soil conditions. This piece of equipment can do the work of two or three crews in minutes when it comes to thinning and reducing fuel along a road or fireline.

Masticators come in a few varieties: boomed, wheeled, or track mounted. Just like all the other equipment discussed in this chapter, having the right masticator is paramount to accomplishing a task. Depending on the slope, a tracked masticator might work better than a wheeled masticator. Masticators are a fantastic tool for reducing fuel around a timber stand or brush along a road or mulching a small timber stand.

It is a good idea to stay at least 150 to 200 yards (137.2 to 182.9 m) away from the masticator. The grinder heads are incredibly powerful and spin at high revolutions per minute. When they start to grind larger logs or trees, the wood chips can be thrown a considerable distance. It is not uncommon for vehicle windows to be broken if they are in close proximity to an operation that is throwing a lot of wood chips.

Terrain Vehicles

All-terrain vehicles (ATVs) and utility terrain vehicles (UTVs), also called side-by-sides, are becoming more commonplace on wildland fire scenes **FIGURE 8-19**. Towed behind engines or trucks, these types of versatile equipment are used to shuttle people and gear in areas where engines and other standard vehicles can't go. The configurations for UTVs are in one, two, or four passengers and, in some cases, are set up for medical transport. ATVs are one-person vehicles only, even if they have the capacity for two people.

Special training is needed to operate terrain vehicles on federal fires. In addition, full-face helmets and seat belts are required. The use of these vehicles aids in the overall operational environment, but it is important to remember that the use of these vehicles is considered high risk. Terrain vehicles should be used only when they are essential to completing an assignment, rather than for convenience.

A

B

FIGURE 8-19 UTVs.

Courtesy of Jeff Pricher.

LISTEN UP!

ATV/UTV operation qualifications vary by agency. Typically, operators will be required to attend a training course every 3 years. Regardless, eye protection, helmets, and seat belts must be worn when operating utility vehicles. When working with a utility vehicle on the fireline, standard wildland PPE is required.

Air Equipment

The use of aircraft during initial attack helps keep fires small. In California, where there are some of the busiest air-attack bases, an air tanker can be on a wildland fire within 20 minutes. Aircraft support the ground attack forces; however, ground forces ultimately have to go in and extinguish the fire.

Costs must always be analyzed versus the benefits gained. Aircraft are a very costly resource. There is an hourly fee as well as a per gallon cost for the retardant. Incident commanders must therefore know how to use this tool effectively and to its full potential.

Two broad categories of aircraft are found on a wildland incident: helicopters and fixed-wing air tankers. A third type of aircraft, unmanned aircraft systems, are growing in popularity on wildland fires.

Helicopters

Helicopters, also called rotary-wing aircraft, are aircraft with revolving head rotors. Helicopters are usually used on the fireline for direct-attack applications. They normally carry water, although they can be loaded with Class A foam or retardant. Water is applied directly on the flames to extinguish them **FIGURE 8-20**. Retardant and its use is discussed in more detail in Chapter 13, *Class A Foams and Fire-Blocking Gels*.

Helicopters are also a valuable resource for transporting supplies to and from remote locations on the fireline, such as a **helispot**, which is a temporary location where supplies, equipment, or personnel are transported or picked up. Helispots are numbered and close to the fireline. Helicopters also fly other logistical or support needs. For instance, they may be used to establish a cellular phone site on a mountain peak, and they can provide aerial reconnaissance on wildland fires.

Helicopters can also carry a helitack crew to the fireline. These handcrews are dropped off to construct hand lines near the head of the fire and are supported by water drops from the helicopter that transported them. Helitack crews are discussed further in Chapter 14, *Handcrew Operations*.

On many large fires, a helicopter is dedicated for medical evacuation purposes. If a fire fighter is injured, he or she is immediately flown to the local hospital.

Helicopters are typed by their seating capacity, the carrying-weight capacity, and the water or retardant capacity **TABLE 8-4 FIGURE 8-21**.

FIGURE 8-20 Type 1 helicopter dropping water on a wildland fire.
Courtesy of Jeff Pricher.

Helicopters are advantageous because they can work in terrain that is too difficult for safe and accurate air-tanker drops, and they have a short turnaround time between water drops. After they drop water on the fire, vertical takeoffs and landings allow them to utilize refill sites close to the fire. Lakes, ponds, the ocean, portable tanks, and engines placed next to conventional water sources are used as refill sites.

Helicopters drop water from either fixed tanks or portable buckets. They can carry as little as 75 gallons (283.9 L) and as much as just over 2600 gallons (9842 L) of water.

As with any aerial resource, it is very important to consider safety when this equipment is operating nearby. Specific to helicopters, do not operate directly beneath the aircraft. During bucket operations, there is always a potential for the bucket to strike the crown of a tree and break it off. Should that occur and you are below it, the chance for accident is very real. The same can be said for when water is being released. One other consideration is the fire behavior and negative effects of the rotor wash on spotting and changing the expected fire behavior. If the aircraft affects the fire, make sure you or your supervisor communicates this to the pilot. Make sure you have the ability to monitor the air-to-ground frequency so that you can be kept abreast of what is going on above you.

Fixed-Wing Air Tankers

Fixed-wing aircraft, also called **fixed-wing air tankers**, are generally used to drop long-term retardant, which slows the spread of the fire just ahead of the flame front **FIGURE 8-22**. They can also be used to drop water or short-term retardant. Fixed-wing water delivery aircraft can deliver large volumes of retardant or fire suppressants very quickly on a wildland fire. They

TABLE 8-4 Helicopter Standards

Components	Type 1	Type 2	Type 3	Type 4
Seats, including pilot	15 or more	9–14	4–8	3
Carrying-weight capacity in lb (kg)	5000 (2268)	2500 (1134)	1200 (544.3)	600 (272.2)
Tank/bucket water or retardant capacity in gal (L)	700 (2650)	300 (1135.6)	100 (378.5)	75 (283.9)
Examples of helicopters	Boeing 234, S-64, S-61, AS-332; Bell 214, K-MAX	Bell 204, 205, 212, BK-117, S-58T	Bell 206B-III and 206L-3, Lama, MD 500	Bell 47, Hiller 12E

A

B

C

FIGURE 8-21 Helicopters. **A.** Type 1. **B.** Type 2. **C.** Type 3.
Courtesy of Jeff Pricher.

work well as an initial attack tool, keeping the fires small and easy to extinguish. Some of the disadvantages to their use include long load and return times to retardant bases and their unsuitability for steep slopes, deep canyons, and heavy timber canopies with

FIGURE 8-22 Fixed-wing air tanker.
Courtesy of Jeff Pricher.

FIGURE 8-23 VLAT.
Courtesy of Joe Lowe.

a surface fire. Fixed-wing air tankers are typed by their load capacity:

- VLAT (Very Large Air Tanker) **FIGURE 8-23**
- LAT (Large Air Tanker/Types 1, 2, and 3)
- SEAT (Single-Engine Air Tanker/Type 4)

Types of fixed-wing air tankers are detailed in **TABLE 8-5**.

LISTEN UP!

Vehicles that have wheels and carry water are called *tenders*. Vehicles with wings are called *tankers*.

Unmanned Aircraft Systems

Unmanned aircraft systems (UASs), also called unmanned aerial vehicles or drones, are air vehicles without a human pilot on board **FIGURE 8-24**. UASs

TABLE 8-5 Types of Fixed-Wing Air Tankers

Tanker	Maximum Gallons of Retardant
Very Large Air Tanker (VLAT), 8000+	
Boeing 747	24,000
DC-10	12,000
Type 1 Tankers, 3000+	
Martin Mars	7200
KC-97	4500
C-130	3000
P3	3000
DC-7	3000
Type 2 Tankers, 1800–2999	
DC-6	2450
P2V	2700
SP2H	2000
PB4Y2	2200
DC-4	2000
DC-4 Super	2200
Type 3 Tankers, 600–1799	
B-26	1200
Super PBY	1400
CL-215	1400
CL-415	1600
AT802	800
S2	800
S2T	1200
Type 4 Tankers, 100–599	
Turbine Thrush	350
Ag-Cat	300
Dromadear	400
Beaver	108

LISTEN UP!

When both fixed-wing aircraft and helicopters are working on the same fire (typically, larger fires), they are separated by altitude differences and, possibly, geographically. Helicopters orbit in a clockwise, or right-hand pattern, whereas fixed-wing aircraft orbit in a counterclockwise direction. The air tactical supervisor orbits above both of these resources.

A

B

FIGURE 8-24 UASs. **A.** Type 4. **B.** Type 2.

Courtesy of Jeff Pricher.

TABLE 8-6 UAS Characteristics

UAS Type	Configuration	Endurance	Data Collection Altitude (Above Ground Level) (ft)	Equipment Weight (lb)
Type 1	Fixed-wing rotorcraft	6–24 hours	3000–5000	>55
Type 2	Fixed-wing rotorcraft	1–6 hours	1200–3000	15–55
		20–60 min	400–1200	
Type 3	Fixed-wing rotorcraft	20–60 min	<1200	5–14
Type 4	Fixed-wing rotorcraft	Up to 30 min	401–1200	<5
		Up to 20 min	<400	

Modified from PMS 515, NWCG Standards for Fire Unmanned Aircraft System Operations. URL: https://www.nwcg.gov/sites/default/files/publications/pms515.pdf.

are more frequently being integrated into the aviation capabilities of wildfire operations. During the 2018 fire season, UASs were used in several instances where manned aircraft could not fly, such as night operations, to obtain situational awareness, map an area, and determine aerial ignitions. This technology will continue to evolve, and new uses will be developed in the coming years. Currently, the Bureau of Land Management is the pioneering agency with UASs; it has several waivers from the Federal Aviation Administration that enable UASs to fly in ways that typical aircraft cannot.

Coordination is essential with UASs; these tools should never be utilized without the proper licensing, approvals, and an established operations manual. While it is very easy to obtain an off-the-shelf UAS, improper use in the national airspace could lead to accidents or interrupt current firefighting operations.

Consult *NWCG Standards for Fire Unmanned Aircraft Systems Operations* (PMS 515) for more information. Keep in mind that not all commercially available UASs are certified for use. **TABLE 8-6** outlines UAS types and characteristics.

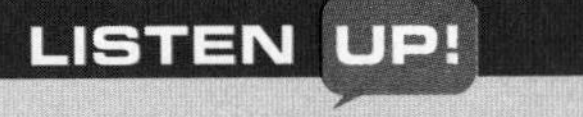

Know resource capabilities and use them wisely.

After-Action REVIEW

IN SUMMARY

- Ground and air resources are classified into three categories: engines and water tenders, heavy or mechanized equipment, and aircraft.
- Wildland fire apparatus are designed with all-wheel drive, high road clearance, and underside protection, which enables them to travel off road to fight fires. Typical wildland fire apparatus range from small pickup trucks to large trucks.
  - Apparatus can be of several types: Types 1, 2, 3, 4, 5, 6, and 7. Each type of apparatus is most effective in certain terrains. For example, Type 1 engines are typically used for urban structure firefighting, while Type 3 engines are smaller apparatus with high road clearance and are best suited for wildland areas.
- Water tenders transport water to the scene of a wildland fire and fall into two categories: support tenders and tactical tenders.
- Heavy or mechanized ground equipment used in wildland firefighting includes ground vehicles, such as all-terrain vehicles, dozers, excavators, feller bunchers, masticators, skidgens, and tractor plows. It is important to understand the basic operations of various types of heavy equipment.

    - All-terrain and utility-terrain vehicles are used to shuttle people and gear in areas where engines and standard vehicles can't go. UTVs can be configured to carry 1, 2, or 4 passengers and, in some cases, can be set up for medical transport. ATVs are one-person vehicles only.
    - Dozers can be of various grades (heavy, medium, or light), with each grade being best suited for certain terrains. Heavy dozers are large vehicles best used as lead dozers in heavy fuels on moderate slopes. Medium dozers perform and maneuver well on moderately steep slopes. Light dozers are best used in terrains that demand tight maneuverability and contain few rocks and light fuels.
        - Maximum percentages of slope for uphill and downhill operations vary by dozer type. If slope exceeds your dozer type maximum, use handcrews instead.
    - Excavators are heavy equipment that can effectively construct firelines. They have a rotating cab and a long, hinged neck with a bucket at the end to dig into the ground. Excavators come in four categories, which cover needs from very large (Type 1) to very small (Type 4).
    - Feller bunchers, also called harvesters, are a type of mechanized logging equipment that allows for accelerated line construction by cutting down and removing trees in light to moderate timber.
        - Self-leveling feller bunchers are used on very steep slopes, and nonleveling feller bunchers can be used only on flatter terrain.
        - Other equipment is necessary after feller bunchers start clearing the way. These other pieces of equipment include logging processors, skidders, log loaders, and forwarders.
    - Masticators, also called mulchers, are pieces of mechanized equipment used to grind brush and debris while driving through them. Depending on the terrain, different varieties of masticators should be used: boomed, wheeled, or track mounted.
        - Depending on the slope, a tracked masticator will work better than a wheeled masticator.
    - Skidgens are versatile pieces of heavy equipment because they are part logging skidder and part fire engine. They are generally equipped with a plow on the front and a water tank and pump on the back.
    - Tractor plows are dozers that pull a plow unit and are used to clear downed materials out of the way while simultaneously constructing a fireline. There are different blades a dozer can be equipped with that are useful for different situations.
- The use of aircraft during initial attack helps keep fires small. Aircraft can support ground attack forces; however, ground forces ultimately have to go in and extinguish the fire. Two broad categories of aircraft are usually found on a wildland incident: fixed-wing air tankers and helicopters. A third category, unmanned aircraft systems, are growing in popularity on wildland fires.
    - Helicopters are aircraft with revolving head rotors and are usually used for direct-attack applications. They normally carry water but can be loaded with Class A foam or retardant. Helicopters are also useful for transporting supplies and personnel from locations called helispots.
    - Fixed-wing aircraft are generally used to drop long-term retardant just ahead of the flame front, which slows the spread of the fire. They can also be used to drop water or short-term retardant.
        - Fixed-wing aircraft can have long load and return times, and are unsuited for steep slopes, deep canyons, and heavy timber canopies.
    - Unmanned aircraft systems (UASs) are more frequently being integrated into the aviation capabilities of wildfire operations because they do not have a human pilot on board. They can be used in instances where manned aircraft cannot fly, such as night operations, mapping, obtaining situational awareness, and aerial ignitions.

KEY TERMS

Apparatus A motor-driven vehicle designed for the purpose of fighting fires.

Dozer A tracked vehicle with a front-mounted blade used for moving fuels and exposing soil.

Excavator A tracked vehicle with a rotating cab and a long, jointed neck with a bucket.

Feller bunchers Mechanized logging equipment used for accelerated fireline construction by cutting down and removing trees. Also called harvesters.

Fixed-wing air tankers Fixed-wing aircraft generally used to drop long-term fire retardant to slow the spread of fire.

Ground equipment Heavy, mechanized equipment used to construct firelines.

Helicopters Aircraft with revolving head rotors.

Helispot A temporary location where supplies, equipment, or personnel are transported or picked up by helicopters.

Masticator A type of mechanized equipment used to grind brush and debris while driving through it. Also referred to as a mulcher.

Mobile pump The operation of a water pump while in motion.

Patrol vehicles Small pickup trucks, found in many rural areas, that are effective in rocky or steep terrain.

Skidgen A vehicle equipped with a plow on the front and a water tank and pump on the back, designed to traverse terrain fire engines cannot. Typically, used for initial attack and mop-up.

Tractor plow A dozer with a plow unit behind it.

Type 1 engine A fire apparatus used primarily for fighting urban structural fires.

Type 2 engine A fire apparatus that can be used for both structural and wildland firefighting.

Type 3 engine A smaller apparatus with the ability to mobile pump, used typically in wildland areas.

Unmanned aircraft systems (UASs) Air vehicles without a human pilot on board. Also called unmanned aerial vehicles (UAVs) or drones.

Water tender A vehicle used for transporting water to the scene of a wildland fire; it is usually equipped with a pump-and-dump valve.

REFERENCES

National Fire Protection Association (NFPA). "NFPA 1906, Standard for Wildland Fire Apparatus." 2016 ed. https://www.nfpa.org/codes-and-standards/all-codes-and-standards/list-of-codes-and-standards/detail?code=1906.

National Interagency Fire Center. *Interagency Standards for Fire and Fire Aviation Operations* (Red Book). (NFES 2724). February 2019. https://www.nifc.gov/PUBLICATIONS/redbook/2019/RedBookAll.pdf.

National Wildfire Coordinating Group (NWCG). *Dozer/Plow Operations.* October 2019. https://www.nwcg.gov/committee/6mfs/dozerplow-operations.

National Wildfire Coordinating Group (NWCG). *Incident Response Pocket Guide.* PMS 461. NFES 001077. April 2018. https://www.nwcg.gov/sites/default/files/publications/pms461.pdf.

National Wildfire Coordinating Group (NWCG). "NWCG Standards for Fire Unmanned Aircraft Systems Operations." PMS 515. February 2019. https://www.nwcg.gov/sites/default/files/publications/pms515.pdf.

National Wildfire Coordinating Group (NWCG). *NWCG Standards for Helicopter Operations.* PMS 510. May 2019. https://www.nwcg.gov/sites/default/files/publications/pms510.pdf.

National Wildfire Coordinating Group (NWCG). "NWCG Standards for Wildland Fire Position Qualifications, PMS 310-1." Modified December 5, 2019. https://www.nwcg.gov/publications/pms310-1.

National Wildfire Coordinating Group (NWCG). "S270, Basic Air Operations, 2011." Modified October 23, 2019. https://www.nwcg.gov/publications/training-courses/s-270.

National Wildfire Coordinating Group (NWCG). *Wildland Fire Incident Management Field Guide.* PMS 210. NFES 2943. January 2014. https://www.nwcg.gov/sites/default/files/publications/pms210.pdf.

United States Department of Homeland Security Federal Emergency Management Agency (FEMA). *Typed Resource Definitions: Fire and Hazardous Materials Resources.* FEMA 508-4. July 2005. https://www.fema.gov/pdf/emergency/nims/fire_haz_mat.pdf.

United States Forest Service. "Appendix D. Pre-Course Work." Accessed January 2, 2020. https://www.fs.usda.gov/Internet/FSE_DOCUMENTS/fseprd566744.pdf.

United States Forest Service. "Mulchers." Accessed January 2, 2020. https://www.fs.fed.us/forestmanagement/equipment-catalog/mulchers.shtml.

Wildland Fire Lessons Learned Center. "Dozer and Tractor Plow Lessons Learned." Accessed January 2, 2020. https://www.wildfirelessons.net/HigherLogic/System/DownloadDocumentFile.ashx?DocumentFileKey=b4ad3135-44ba-c3f2-fc4b-ffa2a329e545&forceDialog=0.

Wildland Fire Fighter in Action

Because the training class you provided was such a success, the battalion chief has asked you to be in charge of a fully functional wildland fire exercise. You found a perfect interface area for the exercise. Numerous homes are adjacent to the interface and the drill site. The drill will begin at 1500 hours and fire fighter Blake will be assigned as the structure protection specialist for the exercise. In the staging area, the following resources are available for use:

- Type 1 Engine Strike Team with Leader
- Type 2 Initial Attack Handcrew
- Type 2 Water Tenders

The wind is blowing toward the structures at 5 mi/h (8 km/h), and the fuels are grass and brush. At 0700 hours, the temperature is 62°F (16.7°C). The fuel moistures are low. It is mid-July. The homes sit at the top of a slight slope. The fire is 2 miles (3.2 km) away and moving slowly toward the homes. The afternoon temperature will be 85°F (29.4°C), and the winds will increase to 10 mi/h (16.1 km/h). Some of the houses have defensible space, whereas others need better clearance. Water supplies in some of the homes will support larger fire flows. Sections with older homes have limited water supplies. There are 20 homes that need defending, and you are responsible for triaging them.

1. Where would you use your handcrews?

A. Wait until the fire arrives, and use them to cut firelines on the fire's flanks.

B. Use them to improve fuel clearances around homes that have fuels too close to the structures.

C. Use them to check the fire hydrants before the fire arrives.

D. Use them to enter houses and move furniture away from the windows.

2. What do you feel is the best use for your water tenders?

A. Use them to supply engines in areas where water supplies are inadequate.

B. Use them to extinguish homes that have caught fire.

C. Use them to pump wildland hoselays.

D. Use them to water the road for dust abatement.

3. According to the firefighting equipment chapter of the *Interagency Standards for Fire and Fire Aviation Operations* (Red Book), Type 1 engines carry how many feet of 2.5-in. (64-mm) hose?

A. 800 ft (243.8 m)

B. 2000 ft (609.6 m)

C. 1200 ft (365.8 m)

D. 1800 ft (548.6 m)

4. Based on the time of day, fuel types, exposures, and threat potential, what might be the best thing to have in place?

A. Spot weather forecasts

B. Escape route

C. Safety zone

D. Lookouts

Access Navigate for flashcards to test your key term knowledge.

SECTION 2

Wildland Fire Fighter II

CHAPTER 9

Wildland Fire Fighter II

Human Resources

KNOWLEDGE OBJECTIVES

After studying this chapter, you will be able to:

- Inspect crew members' personal protective equipment, tools, and supplies for state of readiness. (**NFPA 1051: 5.2.1, 5.3.1, 5.5.2**, pp. 195–200)
- Inspect crew members' qualifications and physical fitness levels for state of readiness. (**NFPA 1051: 5.2.1, 5.5.2**, pp. 200–201)
- Identify the five communication responsibilities for fire fighters. (p. 201)
- Describe operational leadership techniques. (**NFPA 1051: 5.1.1, 5.2.3, 5.5.3**, pp. 202–203)
- Describe the risk management process and how it relates to decision making and task management. (p. 203)
- Describe the importance of team cohesion. (**NFPA 1051: 5.1.1**, p. 203)
- Brief assigned personnel on incident and task information necessary to carry out assignments. (**NFPA 1051: 5.2.2**, pp. 203–204)
- Describe how to serve as a lookout and update fire fighters when conditions change. (**NFPA 1051: 5.5.3, 5.5.8**, pp. 204–207)
- Describe types of evidence. (**NFPA 1051: 5.5.7**, pp. 208–209)
- Describe different ways to secure a site and evidence. (**NFPA 1051: 5.1.1, 5.5.7**, pp. 208–209)
- Describe the importance of site security and evidence preservation. (**NFPA 1051: 5.5.7**, p. 209)
- Describe how to conduct an after–action review. (pp. 211–212)
- Describe typical administrative duties and other helpful references. (pp. 212–214)

SKILLS OBJECTIVES

After studying this chapter, you will be able to perform the following skills:

- Inspect members' personal protective equipment. (**NFPA 1051: 5.2.1**, pp. 195–200)
- Inspect members' qualifications. (**NFPA 1051: 5.2.1**, p. 200)
- Inspect members' physical fitness level. (**NFPA 1051: 5.2.1**, pp. 200–201)
- Lead fire fighters through operational leadership. (**NFPA 1051: 5.2.3**, pp. 202–203)
- Provide an incident briefing. (**NFPA 1051: 5.2.2**, pp. 203–204)
- Verbally communicate fire behavior. (**NFPA 1051: 5.5.8**, pp. 206–207)

- Communicate fire behavior through communication equipment. (**NFPA 1051: 5.5.8**, pp. 206–207)
- Communicate fire behavior through hand signals. (**NFPA 1051: 5.5.8**, pp. 207)
- Demonstrate evidence preservation techniques. (**NFPA 1051: 5.5.7**, pp. 209)
- Secure a fire origin site with marking devices. (**NFPA 1051: 5.5.7**, pp. 209)
- Document the details of a fire origin site. (**NFPA 1051: 5.1.2**, pp. 209–210)

Additional Standards

- **NFPA 921**, *Guide for Fire and Explosion Investigations*
- **NFPA 1033**, *Standard for Professional Qualifications for Fire Investigator*
- **NFPA 1977**, *Standard on Protective Clothing and Equipment for Wildland Fire Fighting*
- **NWCG S-130 (NFES 2731)**, *Firefighter Training*
- **NWCG L-180 (NFES 2985)**, *Human Factors in the Wildland Fire Service*
- **NWCG S-215**, *Fire Operations in the Wildland/Urban Interface*
- **NWCG L-280 (NFES 2992)**, *Followership to Leadership*
- **NWCG PMS 412-1 (NFES 1874)**, *Guide to Wildfire Origin and Cause Determination*

You Are the Wildland Fire Fighter

It is mid-July, and you are dispatched as part of a group of engines to a lightning-caused fire from the night before. Several other fires are in the area from the same storm, which means resources are very limited. The fire you are assigned can be accessed only by foot. As you arrive at the staging area, you are given a quick briefing that covers LCES (lookouts, communication, escape routes, and safety zones)/LACES (lookouts, awareness, communication, escape routes, and safety zones). You and your crew are to join several other engines to form a handcrew. As a Wildland Fire Fighter II (National Wildfire Coordinating Group [NWCG] Firefighter Type 1 [FFT1]), you are responsible for a squad of fire fighters who have to hike to the fire to begin your assignment. Before your arrival and during the briefing, a dozer was pushing line up from the staging area, which was at an elevation of 4600 ft (1402 m) and followed a ridgeline to 5100 ft (1554.5 m). The fighters assigned to you are the only ones qualified to run chainsaws. Your assignment is to brush line and fell trees ahead of the fire fighters digging the hand line.

After a 20-minute hike, you crest the ridge, follow the flagging route established by the fire officer, and begin brushing around the fire. The technique that you use is 1 ft (0.3 m) in the green and 1 ft (0.3 m) in the black. At about 1430, you pause and realize all the fire fighters have been engaged for at least 5 hours in a very active firefight. Suddenly, you hear one of the fire fighters shouting that he has been cut by a chainsaw.

As you reach for your *Incident Response Pocket Guide* (IRPG), you flip to the pink pages that cover the medical incident report (MIR). Almost at the same time, you call dispatch on the command channel and state that you have emergency traffic and need to establish an incident within an incident (IWI).

1. How many times have you practiced an MIR?
2. What type of paperwork will you need to complete?
3. How familiar are you with the green section of the IRPG, and do you have a plan?

Access Navigate for more practice activities.

Introduction

In this chapter, you will learn many facets of the human resource management principles that are required for you to be successful as a wildland fire fighter. This chapter discusses fire-fighter preparedness and the relationship between operational leadership, the leader's intent, and team cohesion. It also discusses the crew readiness evaluation, the personal protective equipment (PPE) inspection, incident briefings, incident lookouts, securing evidence, after-action reviews, and administrative duties.

Evaluating Crew Readiness

A **crew** is a team of two or more fire fighters. The most important aspects to consider when preparing for any assignment include the mental and physical state of crew members, the crew's workload, and team dynamics. Preparation starts on day one if you are a new supervisor, a new crew member, or part of a rapidly assembled squad or crew. The key to crew readiness and good team dynamics is good leadership and leadership by example. The readiness of the crew starts with the steps that you or your supervisor takes.

Without good leadership, people will get hurt, accidents will happen, and assignments will not be executed in concert with the incident objectives and timelines. It is imperative that all wildland fire fighters and supervisors prescribe to the same ideals before responding to incidents.

There are several exercises that can be employed (such as pack tests) and classes that can be taken (such as NWCG L-180 and L-280) that prepare fire fighters for the rigors of team dynamics and teach the human factor skills required in order to be successful. The principles described in this chapter are considered the foundation for development. It will take every fire fighter, squad leader, and supervisor watching out for each other to be successful in assignments.

All fire fighters and first-line supervisors should be well acquainted with the inspections of PPE and the tools they have been issued or are equipped with. The tools vary based on whether you are assigned to a felling module, handcrew, helitack crew, rappel crew, smokejumper crew, or engine company. Each type of firefighting group will have to prepare for its assigned area of operation and make sure that everything is in pristine condition, working order, and always fire ready. Each agency (federal, state, county, or local) will have its own set of standards that need to be followed. Make yourself as familiar with these documents as soon as they are available. The distinct responsibilities and skills of each agency and crew are what make them strong when given an assignment.

LISTEN UP!

Your personal gear should weigh no more than 45 pounds (20.4 kg), and your web gear/fire pack should weigh no more than 20 pounds (9.1 kg). If your total gear weighs more than 65 pounds (29.5 kg), reevaluate your pack.

Inspecting Personal Protective Equipment

PPE is an essential component of a fire fighter's safety system. At the beginning of every shift and as you put on your duty clothes for the day, your PPE must be inspected to ensure each piece is in good working order. PPE improves performance and physical capabilities. Your PPE could be the difference between getting unnecessary cuts, scratches, and burns.

LISTEN UP!

Regardless of your title, remember that it is essential to make sure that everyone's PPE is in good working order. When the unthinkable event happens, your gear will need to perform with 100 percent reliability. If something looks off, say something! Your observation of your crew and other team members might save their life or yours.

There are several components that make up the PPE ensemble, and it is incumbent on every fire fighter to understand how each piece works and must be maintained. Whatever PPE is issued to you, you are responsible for ensuring it is always fire ready. If you note a defect or that performance has been compromised, you are responsible for taking it out of service and requesting a replacement or exchanging it. In some cases, depending on what part of the country you are in, you may be required to immediately exchange it upon return from the fireline if you have come into significant contact with poison oak or poison ivy.

Additionally, dirty gear is not a badge of honor. It is very dangerous. Carbon by-products that collect in fabric or PPE that is embedded with fuel or oil can become flammable and will cause immediate burns and increased risk of cancer with continual use.

Starting from the head down, the following are the critical items that you need to inspect and maintain.

Helmet

Your fire helmet (which can come in a variety of styles) needs to be inspected for dents, dings, or significant scratches. If any of these blemishes are identified, the helmet must be removed from service and replaced. Helmets over 10 years old need to be replaced regardless their condition **FIGURE 9-1**.

Check the inner suspension for tightness of the straps, and check the securing devices of the suspension to the helmet itself. Replace the sweatband annually if it becomes too soiled or damaged to clean. If the

FIGURE 9-1 To determine the age of a helmet, look for a manufacturer's stamp. It is usually located on the brim and may be covered by a sticker. The year is stamped as a two-digit number inside a ring of numbers from 1 to 12. The arrow inside the ring points to the month. In this photo, the helmet was made in April 1995.
Courtesy of Jeff Pricher.

helmet absorbs a significant strike, remove the helmet from service. Check to make sure the chin strap is in good working order and is attached. Some fire fighters like to remove the chin strap; however, when working in high winds and around aircraft, the helmet will blow away in the right conditions. Make sure that the helmet shroud is attached. The helmet shroud provides incredible insulating properties when working a fireline and prevents embers from getting around your collar.

Another component you should become familiar with is the manufacturer's recommended retirement date. Most helmets are made with plastics. Because of this material, helmets that spend a significant amount of time in the sunlight and are exposed to ultraviolet light will lose strength over time, which can cause a helmet to fail. Test each helmet annually by compressing each side of the helmet inward about 1 in. (2.5 cm) and then immediately releasing the pressure. The helmet should return to its original form. If it does not, it should be removed from service.

Eye Protection

Eye protection can be in the form of clear American National Standards Institute (ANSI)-approved lenses or, in many cases, ANSI-approved sunglasses. Protecting your eyes is just as important as protecting your head. Eye injuries during wildland firefighting are common. Factors contributing to these injuries include dust blowing in the wind; bark and other debris falling from trees or other objects; dirt and rocks flying back at you during mop-up; and other high-risk activities, such as chainsaw work. There are several styles of eye protection, so make sure you use the style that meets your authority having jurisdiction (AHJ) equipment standards.

Flame-Resistant Shirt

The flame-resistant shirt issued to you must be kept in pristine order. One of the common flame-resistant materials used in shirts is Nomex. The shirt protects your torso, which, in most cases, has the highest surface area exposure of your body. Your shirt must be free of holes and tears. Just because a shirt is the "old cool style," some fire fighters like to delay its retirement by adding patches. Such a shirt has been compromised and may fail at the worst possible time. Shirts must fit in a way in which it is not too tight or too loose. You need a little space for your upper body to move around in. Also, the small air pockets provide a bit of insulation from radiant heat when you are engaged in hot-line construction.

Check all zippers, buttons, and Velcro to make sure that pockets are not open so they do not collect embers **FIGURE 9-2**. When wearing the shirt, do not roll up your sleeves. In an emergency, trying to roll down your sleeves may take too long and put you in a situation where you may get burned. Though it may look cool and you may see others doing it, rolled-up sleeves are very dangerous and unprofessional. In general, when the shirt becomes soiled, it can be washed in a washing machine and hung to dry. Check the manufacturer's recommendations for best practices on maintenance, which will be on a tag located on the inside of the garment.

LISTEN UP!

Always wear a cotton shirt as your base layer. Do not choose a synthetic material! For undergarments and socks, consider cotton or wool.

LISTEN UP!

Do not wash your PPE with personal items or in a personal washing machine. Doing so can cause contamination of your personal items and machine.

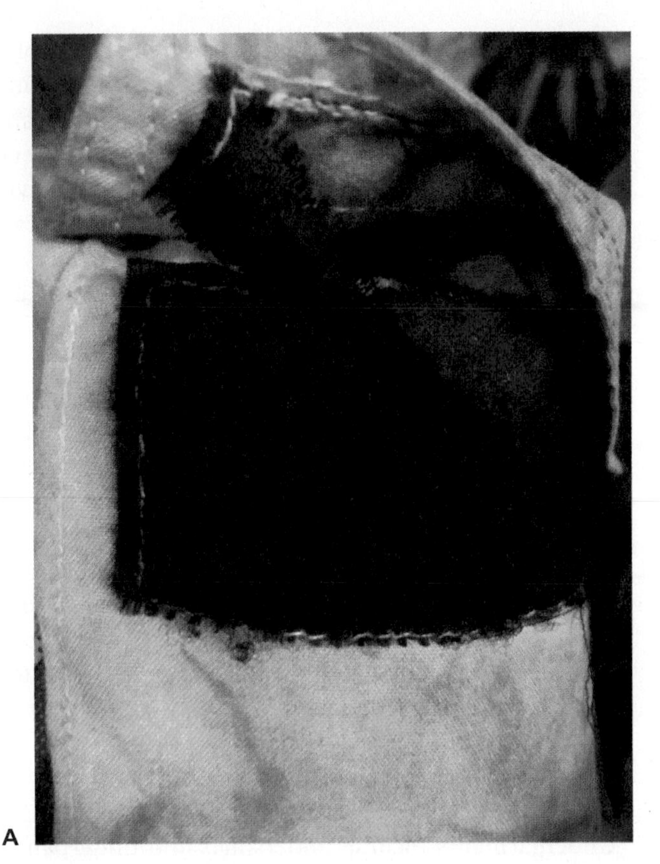
A

B

FIGURE 9-2 Damaged Velcro. **A.** Jacket pocket. **B.** Sleeve cuff.
Courtesy of Jeff Pricher.

Flame-Resistant Pants

There are many different types of flame-resistant pants, but they all require that you keep them in pristine order. They are made of advanced fabrics, including Nomex/Kevlar blend, Nomex, or, in some areas, fire-resistant cotton. While the latter is an option, it is not typically recommended due to its poor abrasion resistance and overall performance. Your organization will determine which materials you should wear.

Check zippers, Velcro, and seams for failure or other defects. If the pants have burn marks or holes, they will need to be taken out of service **FIGURE 9-3**.

An activity that may require you to immediately change your pants is fuel or mixed gas soaking into the pant leg during drip-torch operations or work with a chainsaw. In general, when the pants become soiled, they can be washed in a washing machine and should be hung to dry. Check the manufacturer's recommendations for best practices on maintenance, which will be on a tag located on the inside of the garment.

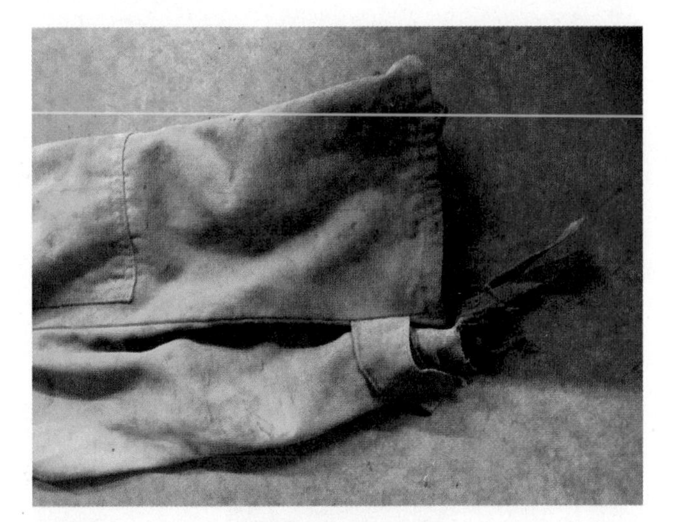

FIGURE 9-3 Damaged pants.
Courtesy of Jeff Pricher.

Boots

Depending on the wearer's use, boots can have a very short life. They should be formfitting, not too tight nor too loose, and the uppers should be free of defects. The most common boot type is the leather boot. Leather boots shall have a height of no less than 8 in. (20.3 cm) and should not have a steel toe or midsole. Boots with metal in them can absorb heat and burn your feet.

Boots require a rubber or Vibram sole. Make sure that the midsole is made entirely of rubber and does not contain any ethylene-vinyl acetate (EVA) foam. EVA foam (similar to what you would see in running shoes or lightweight hiking boots) will melt with any sort of heat or flame impingement. This will cause the

Vibram sole to fall off, exposing your feet to only the leather upper and stiches.

Boots should be regularly cleaned and, if they are leather, oiled. There are several products that can be applied to condition the leather. Check with your boot manufacturer to make sure you choose the appropriate product.

Boot laces often have a high failure rate. It is a good idea to carry an extra set of laces in the event the ones you are using break.

Some boots can be rebuilt or repaired depending on the vendor. If you notice that your soles are worn out, check to see whether you can send them in to be rebuilt or repaired. If your boots don't seem to fit right, take them to a cobbler and see whether they can be adjusted. Proper-fitting boots can help to avoid blisters.

Not all boots are created equal. Just because a leather boot meets some of the requirements does not mean it meets all of them **FIGURE 9-4**.

Socks

Even though socks are not typically thought of as PPE, they are an important component of taking care of your feet. Change your socks as often as feasible. If you are fighting fire in environments that put your feet in a constant state of moisture, such as in rain and snow or near streams, change your socks every 8 to 12 hours. Frequent sock changing in these environments should be a routine. Failing to do so can result in trench foot.

FIGURE 9-4 This pair of leather boots does not meet some of the requirements. These boots have an EVA foam midsole. EVA foam has a very high failure rate when exposed to heat. These boots also have a metal toe cap. Metal toe caps can cause severe burns to fire fighters.
Courtesy of Jeff Pricher.

Firefighting is hard on feet and socks. Make sure your socks have plenty of cushion to prevent blisters on your feet. If you notice the cushioning on your socks is starting to wear away or is missing or if your socks start to wear thin, replace them. Some fire fighters wear multiple pairs of socks together, such as a liner sock and an outer sock or a running sock and an outer sock. Employ a strategy that works for you.

A good wool or wool-blend sock will be one of the best investments you can make. While wool socks may be more expensive than socks made from other materials, they will last longer and have better moisture-wicking properties.

Undergarments

Undergarments include undershirts and underpants. One general principle to remember is to stay away from synthetic materials, such as acrylic, polyester, and nylon. Synthetics are great for use during physical fitness, but, in a fire setting, they can be detrimental to the fire fighter if a heat injury is experienced. Synthetic materials tend to melt onto the wearer in extreme temperatures. If an undergarment melts to your body, the aftermath and process of treating burns should be a reminder to choose wisely.

The most common protective undergarment material is cotton. Depending on what part of the country you work in, you may be required to wear a long- or short-sleeve cotton T-shirt. In any event, the undergarment functions as another layer of protection, and, on hot days after a lot of sweating, the saturated cotton can aid in the cooling process.

Gloves

Use of gloves should be a no-brainer. However, many fire fighters opt not to wear them or are not dutiful about using them because they add to the heat stress on the body, can limit dexterity, and easily get contaminated. However, gloves are essential in the total PPE ensemble. Gloves help prevent blisters, make it easier to handle hot debris during mop-up and cold trailing, protect from cuts when maintaining hand tools or sharpening chainsaw chains, and prevent burns. A significant burn on the palm of your hand equates to a 9 percent total body surface area burn.

Not all gloves are able to withstand the rigors of firefighting. Wildland gloves and their attached wristlets should have a thermal protection performance greater than 20 (this number will be indicated by the manufacturer). Though there is no specific thickness requirement, gloves should be thicker than the standard garden glove and need to meet cut and puncture-resistance standards. When fire fighters are sharpening a Pulaski or working with a chainsaw chain, the gloves must be able to protect the fire fighter's hand from a cut.

During your daily PPE examination, check the stitching on the glove to make sure that everything is sewn shut and check for holes or punctures. If you find any issues with your gloves, you should replace them immediately.

Hearing Protection

Ear protection is key to preventing you from losing your ability to hear. The Occupational Safety and Health Administration (OSHA) requires hearing protection around power equipment, including helicopters, dozers, pumps running on engines or portable pumps, and chainsaws.

The most common form of hearing protection available to fire fighters is foam earplugs. They need to have a decibel rating of at least 30 for maximum protection. Get in the habit of having several pairs at your disposal.

To use the foam earplugs, first, make sure your hands are clean and free of bar oil, fuel, or dirt. Then, squish them lengthwise, and roll them between your fingers so that the foam is condensed and looks like a short stick. Next, pull up on the upper part of the ear, and slide the foam into the ear canal. Do not push it in too far, or it may get stuck. After several seconds, the foam will expand and occlude the ear canal, which will prevent loud noises from damaging your eardrum.

Foam earplugs should be replaced every 3 to 4 months after heavy use or when they appear dirty, damaged, or misshapen.

LISTEN UP!

If you are working in high temperatures close to a fire, do not use foam ear plugs! The high temperatures could cause them to melt to your body.

Web Gear/Fire Pack

Not all agencies issue web gear/fire packs as part of the PPE ensemble, but for those that do, it is crucial that all parts of the web gear/fire pack are fire ready. A web gear/fire pack is a belt and harness used to carry the fire shelter and canteens **FIGURE 9-5**. Fire shelters are discussed later.

FIGURE 9-5 Fire pack.
Courtesy of Jeff Pricher.

Check all zippers and Velcro (assuming that the Velcro may be part of the deployment system for the fire shelter) to make sure they are in working condition. Any pack that is missing buckles, has burn holes, or does not meet performance standards should be removed from service. Consult the manufacturer's instructions for inspections or agency guidelines for removing line gear from service. One common problem is the failure of fluid fire shelter deployment. If you are not able to deploy your shelter because it is stuck or the pull tab does not perform during inspection, the line gear must be taken out of service and tagged so no one else uses it.

LISTEN UP!

If your web gear/fire pack has added capacity, some items to consider carrying are at least 1 gallon (4 L) of water and some extra food items (energy bar or other snack). If you have room, consider a signal mirror, a whistle, a knife, a lighter, extra socks, toilet paper, fusees, safety glasses, warm clothing for night responses or late-season fires, a pen, and a global positioning system (GPS) device or compass.

Fire Shelter

A fire shelter is a required device in the United States but is not required in Canada or Australia. The fire shelter itself has an indefinite shelf life, and the materials used in its construction do not have an expiration date. Firefighting, though, is very rough on equipment and, as a result, necessitates that you regularly inspect your shelter. Depending on your agency policy, this could be a daily or a weekly check. You should never go longer than 2 weeks between inspections.

The fire shelter is packaged in a permanent clear plastic wrap and surrounded by a hard-plastic case. The shelter should always be kept in that hard case to protect it from impacts. The most common damage that occurs to the fire shelter is abrasion. While you will never take the shelter out of the clear plastic wrap, signs of abrasion will be very apparent. If abrasion occurs, the clear plastic will turn a black or grayish color. When this happens, your shelter needs to be taken out of service. Dirt or other debris inside the clear wrap is another indicator that the shelter needs to be taken out of service.

A little prevention and maintenance can extend the life of the shelter. Avoid any rough handling of your gear. Don't throw it on the ground or into a compartment. Don't place heavy objects on your fire shelter. Protect your shelter from sharp objects, and do not use it as a pillow **FIGURE 9-6**.

LISTEN UP!

It is a good idea to develop muscle memory of fire shelter deployment.

FIGURE 9-6 Damaged fire shelter.

Photo from Tony Petrilli (2011). Fire Shelter Inspection Guide and Rebag Direction (1151 2301P). Missoula, MT: U.S. Department of Agriculture, Forest Service, Missoula Technology and Development Center.

Training Requirements

To successfully manage and lead other fire fighters, you must be familiar with their qualifications. With all the many jobs that are required as part of the firefighting effort, it is important to remember to place fire fighters in appropriate positions based on their qualifications. Wildland fire-fighter qualifications are not limited to just Firefighter I and Firefighter II skill levels. Other certifications include chainsaw operation, terrain-vehicle operation, emergency medical technician (EMT), apparatus driver/operator, helicopter crew member, and more. Placing fire fighters in positions that they are not qualified for can put their safety at risk and the operation in jeopardy.

Another important evaluation of crew readiness is to have a good understanding of both past and current training. Every year, there are changes to the standards. It is required that everyone receive an annual refresher. Refreshers cover specific topics that are designed to bring to light concepts that may have been forgotten or skills that are not often used. This helps to prepare fire fighters for the upcoming season, as well as share new information and concepts.

A good place to learn more on these topics is through the NWCG RT-130, Wildland Fire Safety Training Annual Refresher (WFSTAR) course. Every year the topics are changed and refreshed with new and relevant content. Most qualifications are performance based and can be documented using a position task book. Make sure you are familiar with all the components of the position task book. The position task book is discussed in more detail in Chapter 1, *The Wildland Fire Service*.

Physical Fitness Testing

Firefighting is a very difficult job that requires a high degree of physical fitness (see Chapter 3, *Safety on Wildland Fires*). Fire fighters are expected to work long hours in uncomfortable environments and rough terrains.

One of the challenges in evaluating physical fitness is trying to be fair with all of the different fitness levels of the crew you may be working with. Teach the whole crew or team to motivate each other to be at the same level. At the beginning of the season, pay close attention to complaints among your crew members of muscle cramps or pain that doesn't seem to go away. On your first fire assignment you will have a chance to reevaluate the crew's fitness level. Note any problems encountered by crew members and work with them to correct the issues.

One of the national standards for evaluating physical fitness of fire fighters is the **arduous pack test** (also known as the work capacity test). The arduous pack test is completed on a flat course that is 3 miles (4.8 km) long. The fire fighter must walk, jog, or run through this course while wearing a 45-pound (20.4-kg) pack. The fire fighter must cross the finish line in 45 minutes or less. This test is pass/fail, and there are no points for getting a lower time. Other ways to evaluate physical fitness include a 1.5-mile (2.4-km) run. And, those who are trying out for the smokejumper's propellers or the hotshots can expect to also have to complete a series of push-ups and pull-ups.

It is important to make sure that all fire fighters are physically and mentally ready for a fire or assignment.

Individual fitness responsibilities, nutrition, and hydration are discussed in more detail in Chapter 3, *Safety on Wildland Fires*, as well as social support, mental stress, and hazardous attitude barriers.

LISTEN UP!

Fire fighters have a responsibility to recognize and minimize stress barriers. It is important to look out for yourself and your team.

Communications

Effective communication is essential for effective briefings and successful assignment completion. When things go wrong during an assignment, improper communication is frequently an issue. Communication can be verbal or nonverbal and can be through technology or, the most trusted method, face-to-face. Most people think that all the responsibility of communication rests with the supervisor. Ultimately, the responsibility of communication is required by all fire fighters at all levels. Fire fighters should adhere to the five communication responsibilities:

- Brief others.
- Debrief your actions.
- Communicate hazards.
- Acknowledge messages.
- Ask when you don't know.

Effective communications can be achieved when senders and receivers are careful of errors, when direct statements are made, and when standard communication procedures are followed.

Sender and Receiver Roles

The sender is the person speaking. The receiver is the person listening to the speaker. These roles can be established in person or over communication devices, such as a radio. When working in high-risk situations, communication errors are likely to occur. **TABLE 9-1** outlines common sender and receiver errors to be aware of.

Direct Communication

Direct communication, or direct language, is an important part of effective communication. Direct communication is language that is concise and factual and provides closure. Indirect language is passive, lacks details, and sometimes used out of politeness. There are six components of direct communication:

- Use the person's name.
- Use "I"—"I think" or "I feel."
- State a clear message.
- Use appropriate emotion—don't act casually during an emergency, for example.
- Phrase your statement so the receiver is obligated to respond—"What do you think?" or "Do you agree?"
- Don't let the receiver disengage until you have a clear understanding.

TABLE 9-1 Common Sender and Receiver Errors

Common Sender Errors	Common Receiver Errors
Omitting information	Only able to focus on a previous perception
Providing biased information	Not recognizing perceptions or not asking for facts
Verbally saying one thing but nonverbally suggesting another	Not paying attention to inconsistencies between verbal and nonverbal cues
Unwilling to repeat information	Not asking for clarification
Using indirect statements	Inability to construct a direct statement from an indirect statement
Disrespectful communication	Disrespectful communication

"It feels like the wind is stronger than predicted and coming from the south now. Do you think we'll have any problems down below?" is a direct statement. "Does it seem like the wind is picking up?" is an indirect statement. Direct communication should be used even if it feels rude or bossy.

Direct language should always be used, especially when you are unsure of something, when you see a problem, when you have an answer to a question, or when a receiver is not understanding you.

Standard Communication Procedures

Standard communications establish a common understanding among team members. Standard communications help clarify and save time because certain terms mean certain things or actions. This allows you to quickly comprehend your task and your team members' tasks. Common standard communication procedures include helicopter and dozer hand signals, radio calls (see Chapter 10, *Radio Communications*), and response protocol. Your department may have standard communication procedures specific to it.

Task Management

Task management can be a significant challenge for anyone involved in firefighting. There are many dynamics that affect the management of different tasks and assignments on incident. Fire-fighters' dynamics are influenced by their leader or supervisor. Other factors can be influenced by the fire itself.

Leader's Intent

Leader's intent is an important practice of successful wildland firefighting. It is a clear, concise statement that outlines what individuals must know in order to be successful for a given assignment. It is an extension of number 8 of the 10 standard fire orders (give clear instructions and ensure they are understood, as noted in Chapter 3, *Safety on Wildland Fires*). It encourages communication among crew members and ensures that everyone's goals and responsibilities are understood.

Leader's intent communicates three pieces of information: task, purpose, and end state. (Task, purpose, and end state can be found in the IRPG.)

Task

For a fire fighter or a group of fire fighters to have a clear understanding of the assignment, it is essential for the leader to make sure that everyone understands the task, that is, the goal or objective. If a task is poorly explained, crew members might deviate from the leader's plan. When explaining the task, make sure that your crew is able to clearly communicate the task and the goal back to you.

Purpose

The next step is to explain the purpose, or why the task needs to be done. If crew members have a solid understanding of why they need to do what they're being asked to do, they will be able to clearly visualize the situation and the task. Understanding the purpose of the task helps to establish a sense of ownership among your crew members and validates how their work will fit into the big picture.

End State

The final component of the leader's intent is to explain what the end state, or the end product, should look like upon successful completion. When crew members are able to understand the task, purpose, and end state, everyone will be on the same page.

LISTEN UP!

It is important to encourage fire fighters, especially new recruits, to ask questions.

Operational Leadership

Another component of task management includes operational leadership. While leader's intent is specific to the big picture, **operational leadership** takes human factors and processes into consideration. Operational leadership includes improving systems and processes, taking charge, motivating fire fighters, demonstrating initiative, and using direct communication.

Ultimately, if you are responsible for making assignments, you need to make sure that you set the example with the fire fighter or group of fire fighters assigned to you. When you are faced with difficult decisions, make sure you choose the right decision, even if it is the more difficult one.

Once your crews have been given an assignment and briefed on the leader's intent, it will be equally important for you to continually evaluate the assignment. Questions to ask yourself during evaluation include:

- Is the assignment going as planned?
- Are the fire fighters task-saturated, or are they looking for work?
- Do the crews need support, or do you need to provide guidance?

Additionally, it is important for leaders (regardless of their actual title or rank) to promote positivity whenever possible. It is easier to work for someone who has a lot of energy and is excited and knowledgeable about the job. An example of this would be during mop-up (generally, a dirty and less than fun assignment when compared to other tasks such as burning). A leader who helps to excite and challenge the crew will be more successful than one who is less than enthusiastic.

As a squad or crew boss, it will be up to you to make sure that you are constantly evaluating your environment. This will aid in successful completion of assignments and allowing you to keep abreast of any safety concerns that may come up.

Decision-Making

Decision-making is an important part of task management. The ability to make sound, justified decisions in high-risk situations depends on experience, situational awareness, and preplanning.

Fire fighters have a responsibility to minimize risk. The **risk management process**, sometimes called the decision cycle or the situational awareness self-check tool, is a series of steps used to make decisions and observe their outcomes. This process includes the following steps:

- Situational awareness. Know what is going on around you at all times.
- Hazard assessment. Recognize and identify hazards.
- Hazard control. Determine how to minimize or omit the hazards that will prevent you from successful task completion.
- Decision point/implement controls. If hazard controls are in place, tactics have been selected, and instructions have been provided and understood, then action can be initiated. If one or more of these items are not met, reassess the situation and start the risk management process over again.
- Evaluate. Assess changes in the situation.

Preplanning supports the risk management process. It helps organize thoughts, observations, and plans before they are put into action. Preplanning can help to identify potential problems, hazards, and decision points before they happen. An *if A then B, if C then D* tactic can be developed and incorporated into the preplanning stage in order to put several decisions in place in the event that the main course of action needs to change.

Team Cohesion

Teamwork is an important part of performance and task management. Fire fighters have a responsibility to work as a member of a team. Good teamwork skills are the foundation for good leadership skills. Good leadership skills are a foundation for a successful assignment. Remember that all team members bring something different to the table and diversity is key.

The following items are traits of good team members:

- Strong communication skills
- Commitment to results
- Accountability
- Job proficiency and work ethic

The team leader must bring the team together through team building, physical training, and other exercises. Team members need to understand, accept, and learn from the strengths and weaknesses of every member on their team. A good team leader understands where to place individuals based on where they will serve the team to the best of their abilities. This leads to a successful team that will work, learn, and grow together as a unit.

LISTEN UP!

Those looking to improve on their leadership skills in wildland/urban interface should consider taking a NWCG S-215, *Fire Operations in the Wildland Urban Interface* course.

Briefings

The firefighting environment is very dynamic and includes risk and benefit. To balance those and prevent the possibility of accidents or mistakes, it is necessary for all fire fighters to communicate critical information. One of the best ways to communicate information is through briefing. A **briefing** is a meeting designed to provide information and instruction. There are several types of briefing.

Shift Briefing

At the beginning of every shift, it is important for all crew members and fire fighters to participate in a daily shift briefing. At this briefing, everyone can communicate about the previous day's actions and the day's current assignments. If there is equipment that needs to be repaired or maintained, everyone can be made aware of it at one time. If there is a training assignment for the day, the briefing is the perfect opportunity to explain to everyone what will be expected of them. Another important component of the shift briefing is to go over the day's weather report and into the next 24 hours. If you cover a large geographic area, it is a good idea to look at specific weather forecasts for the different areas you're responsible for. Make sure you cover:

- Precipitation
- Lightning activity level
- Temperature
- Humidity
- Probability of ignition
- Haines Index
- Local energy release component
- Wind direction and wind speeds

Incident Briefing

Successful fire assignments result from proper incident briefings. The style of the briefing isn't as important as the content of the briefing and making sure all the crew members understand their personal responsibilities and the leader's intent. A common misconception is that briefings occur only at the beginning of an incident or at the beginning of a shift. Briefings can occur at any point in time during the fire assignment **FIGURE 9-7**.

To deliver a complete and organized briefing, use the briefing checklist on the inside back cover of the IRPG. The core components of the briefing checklist include the situation, mission/execution, communications, service/support, risk management, and any questions or concerns that you or others may have.

While the briefing checklist is fairly comprehensive, it does not cover all elements that may need to be covered for a particular briefing, such as incident-specific information. Make sure to include those elements, depending on what your fire problem is.

FIGURE 9-7 Crew briefing at a fire scene.
Courtesy of Jeff Pricher.

Large-Incident Briefing

If you are on a large incident, project fire, or a state mobilization, you can expect that there will be two briefings per day if the incident has been divided into two operational periods. These briefings are the culmination of information gathered throughout the day or night and input into an incident action plan (IAP) during several meetings of the 12-hour planning cycle. They are called the operational period briefings. Day shift crew members attend the morning briefing and night shift crew members attend the night briefing before being deployed to their fireline assignment.

Large-incident briefings will include new components of information that may not be pertinent to smaller incidents. An example of this is human resources. In large incidents, where the number of personnel may span from 1000 to 5000, other elements of the Incident Command System (ICS) will be activated to manage span of control and support necessary for the volume of people in a small living and working environment.

Make sure to arrive at the briefing early. Obtain a map and an IAP so that you can follow along as the briefing occurs and take notes on important information. Generally, the briefing is run by the planning section chief. The planning chief will invite other members of the incident management team to provide information for the incident personnel. After the briefing, crew members will break out with their respective supervisors for further information. This process happens relatively quickly, so make sure you pay attention. If you are the supervisor, delegate the morning assignments to your crew members before the briefing. Assign people to obtain lunches, obtain supplies from the cache, and inspect the vehicles.

LISTEN UP!

Remember that the ICS is adaptable to the size of an incident. Only necessary positions need to be filled. As an incident grows, more positions can be filled. ICS expansions will be included in the incident briefings.

Lookouts

LCES/LACES is an important part of wildland fire safety and communications. The first letter of the acronym is *lookouts*. After a fire as been sized up and before fire fighters engage in suppression actions, establishing a lookout needs to be the first priority. A **lookout** is a person designated to detect and report fires from a vantage point **FIGURE 9-8**. The lookout will observe the fire and communicate environmental and fire

FIGURE 9-8 Lookouts are necessary for almost every type of fire.
Courtesy of National Park Service.

behavior changes to the crew either verbally, through communication equipment, or through hand signals.

Based on the fire's location, size, fuel type, or position on a slope, getting a lookout in place may take awhile and may be needed to maintain communications for the incident. Lookouts are necessary at just about every type of fire.

Every crew member must be familiar with the requirements of being a lookout and prepare for this type of assignment. Filling this role can be very demanding and is not the job for slacking off. There are many requirements and skills associated with being a lookout. The lookout will need to know how to use a variety of equipment and tools and how to read tables and charts in order to calculate information to be shared with your crew, other crews, and, in some cases, aircraft. There are many instances when a lookout may be serving more than one division or crew, based on their view of the fire.

Lookout Equipment and Preparedness

While there are many types of equipment a lookout may be required to use, following is a basic list of tools and communication equipment that you should gain access to as a lookout:

- Handheld and mobile two-way radio
- Signal mirror
- Whistles or air horns
- Binoculars
- IRPG (tables in the IRPG for calculating the probability of ignition, or PIG)
- Belt weather kit or other digital weather-gathering instrument
- Extra batteries for electronic devices and portable radio
- Sunscreen
- Food for the shift
- Water for the shift
- GPS device and familiarity with that device (such as how to read the device's latitude and longitude format; GPS is discussed in more detail in Chapter 11, *Orienteering and Global Positioning Systems*)

In general, the person assigned to this role needs to be one of the more experienced fire fighters on the crew. The lookout needs to be able to:

- Find a good vantage point where he or she can see both the fire and the crew(s).
- Identify a safety zone.
- Choose an escape route.
- Have a solid understanding of fire behavior and fuel types.
- Monitor and track weather changes.
- Have exceptional communication skills.
- Retain and extrapolate information from an IAP.
- Estimate decision points and communicate the possibility of needing a new safety zone.
- Have the common sense to communicate the right observations at the appropriate time, using the right communication method and tools.

As you prepare for your assignment, consider where you will be posted as the lookout. Will it take you a long time to get there? If so, is there a better location that you can get to faster? Once there, are you able to see the whole fire or just part of the fire? Make sure you are able to meet the objectives outlined in your briefing as you plan ahead for what the crew, engine, or other fire fighters will be doing throughout the shift.

Be prepared. In certain situations, you may be dropped off at your lookout point by aircraft and be expected to remain there overnight or for a few days.

In other lookout settings, you may not be in a faraway location. You may be assigned to watch out for rockfalls or other vehicles (if your fire is on a road or highway), or you may be assigned to a big snag patch. In the last instance, you may just need to look up to observe trees.

LISTEN UP!

Most lookouts are in radio communication with their crew. Radios are discussed in more detail in Chapter 10, *Radio Communications*.

Communicating Safety Zones

One aspect of being a lookout is having good situational awareness regarding the location of the escape routes to the safety zones. As a lookout, you could be responsible for having to direct your crew or others to the safety zone. You may also be required to gauge the behavior of the fire and let others know that they may have to go to a secondary escape route as a result of the primary safety zone being cut off. Additionally, you may need to identify or pre-identify other safety zones in the event that the primary safety zone is compromised.

Determining Fire Behavior

In developing your situational awareness as you get your lookout assignment, it can be very helpful to look at local Remote Automatic Weather Stations (RAWS); your IAP; and, in general, the available fuel surrounding you and what your crew is working in. Try to determine the spotting potential and expected fire behavior. Much of this will be based on past experience. As a result, remember that the lookout needs to have this understanding.

Communicating Weather

As you arrive at your lookout destination, you should immediately begin to start tracking the current weather and PIG and track the trend throughout the day. Establishing this baseline will help you to determine whether the weather is getting hotter and drier. In general, if you notice anything that might affect the crews or others in the area that you can see, speak up. Timing is everything. Waiting too long could be detrimental to others.

Communicating weather is a very important skill that all fire fighters need to know. As the lookout, you are the conduit to assisting everyone in your area to keep safety in the forefront of all tactical decisions. There have been many occasions when a tactical operation has been halted as a result of good observations and weather reporting.

Weather-Taking Instruments

Be very comfortable with your weather-taking instruments. The most common is the belt weather kit. The contents include deionized water, a sling psychrometer (used to determine wet and dry bulb temperatures), an anemometer, a compass, a weather notebook, a clipboard, a pencil, a blank notepad, and a conversion chart. The conversion chart uses your known altitude for determining humidity and dew point. When using the chart, make sure you are using the page associated with your current altitude. Using the wrong altitude can throw off your weather readings. One easy tip for remembering the difference between the humidity and dew point (they are in the same box) is to look elsewhere on the chart for negative numbers. The negative number is important because it is impossible to have a negative humidity. So the top number in the chart is the dew point (can have a negative number), and the bottom number is the humidity.

Another common weather instrument is an electronic weather device. There are several manufacturers that make these, but there are a couple that are specifically made for wildland fire. These specific wildland fire weather instruments will do all the calculations for you. And, yes, there are also several weather apps that can do the same functions on a smartphone or tablet.

Determining Probability of Ignition

The next step in your weather baseline is to determine the PIG. To do this you will need the 2018 IRPG or other reference provided to you by your supervisor. Table A will help you determine your reference fuel moisture (RFM). To determine PIG, you will need the current relative humidity (RH). Depending on what month you are on your assignment, you will need to refine the RFM or adjust it by using Table B, C, or D. Factors to consider will require you to determine whether the fuels are greater or less than 50 percent shaded, the percentage of slope, and the aspect the fuel is facing (N, S, E, W). Time is also part of the chart and calculation, as well as the area of concern. What this means, based on your weather readings, is whether the area that you are calculating the PIG is above or below you or at the same elevation. When you intersect all that information on the tables, you will end up with a number.

The last step in calculating the PIG is to use the number obtained in the last chart and correlate it with your current temperature and shading of the fuel (greater than or less than 50 percent). When you intersect all this information, you will have the PIG.

Broadcasting Weather Findings

Now that you have carefully referenced and recorded all this information, you need to make sure that you get this information out. If you are working on a small incident, you may be asked to broadcast to everyone on the fire but, specifically, call in the observations to the incident commander. On a larger incident, you will most likely be calling a task force leader or division supervisor. In most cases, weather will be expected to be transmitted every hour on the hour throughout the entire shift. Here is an example of what you would need to transmit over the radio:

"Division Foxtrot from Foxtrot Lookout. Break."

"Go ahead, Foxtrot Lookout. Break."

"I have the 1530 weather report. Are you prepared to copy? Break."

"Foxtrot Lookout from Foxtrot, go ahead with the weather. Break."

It is very important that you ask whether it is OK to transmit the weather. There may be critical tactical operations going on that may require clear radios, or there could be a medical incident that requires minimal radio traffic.

"Foxtrot copies dry bulb 75, wet bulb 55, RH 12 percent, PIG of 95 percent. Break.

"Continuing, wind is out of the east at 15 miles per hour with gusts of 22. There are scattered clouds, and the visibility is less than 2 miles due to the smoke. Break.

"Continuing, weather was taken on a south aspect at an elevation of 6200 feet on the top of Bald Butte. Break."

Another good practice when describing the temperature, humidity, and PIG is to notate whether there was a change. Specifically, if the temperature was previously 68°F (20°C) and changed to 75°F (23.9°C), notating that the temperature is "up 7" could alert all fire fighters that a significant change has occurred in the past hour.

Additional radio communication skills and techniques are discussed in Chapter 10, *Radio Communications*.

Communicating with Hand Signals

It is important that you understand how to communicate emergency fire behavior to air and ground crews when other forms of communication are unable to be used. This can be done through basic emergency hand signals **FIGURE 9-9**. It is important that emergency hand signals be clear and full-arm motions used. Both air and ground crews should be familiar with emergency hand signals.

FIGURE 9-9 Basic emergency hand signals. **A.** Fire location. **B.** Evacuate. **C.** Stop all activity. **D.** Fire or emergency is contained.

Securing Potential Evidence

Arrival upon the Scene

One of the greatest challenges upon arrival of a fire scene is to make sure you and your crews do not accidently disrupt the area where the fire started, called the **point of ignition (POI)**, or area of origin. Locating and identifying the POI can be a significant challenge.

Upon arriving at a fire scene, take a minute to develop your situational awareness and visualize where you think the fire may have started and what started it **FIGURE 9-10**. Once the general origin area is identified, mark off the area to prevent people from damaging the site. Sometimes, pinpointing the exact point of origin is incredibly difficult because of damage caused from suppression efforts (tire tracks, skid marks, removed brush or trees, etc.).

Upon arriving at a fire scene, make sure that you are wearing all of your PPE. Many exposures need to be identified before collecting information related to the fire. Remember your safety first. Post a lookout in areas near steep hillsides or when heavy equipment is still working above you. To preserve the scene, request that work around you cease, until you obtain the necessary information needed to determine the fire's point of origin. Such work would include dozers or excavators, felling operations, and water delivery. Look up, look on the ground, and look around to sustain your safety and anyone else involved in securing the scene or evidence.

Protecting the Scene and Evidence

Protecting the fire scene is paramount to providing the fire investigators the clearest picture of what may have started a fire. The **fire investigator** is an individual trained to examine a scene and determine a POI. Any burn indicators and patterns, debris, discarded items, footprints, and tire tracks will be documented as information or evidence. Disturbing any of these items makes it almost impossible for the fire investigator to successfully complete an investigation. From the moment you arrive on a fire scene to when you start mop-up, make sure you are not operating in the suspected area of origin.

If you are able to identify a POI, in general, everything in that area is considered evidence. This can include soda bottles, receipts, cups, cigarettes, shell casings, or other items that appear to be garbage or out of place **FIGURE 9-11**. In many cases, based on what investigators find, they can backtrack to a store or other location to get more information about a person who may have been the last one at a fire scene. That is why it is very important for all fire fighters to identify the POI and protect it.

In most cases, the area of origin will be next to a road or adjacent to a power pole. Depending on the dynamics of a given fire (wind, terrain, fuel, and other influences), there are common burn patterns that can be identified at many fire scenes. In some cases, it could be an unusual-looking blob that ends up being the result of a road flare. In most cases, the POI will be located at the heel of the fire and will look like a *V*- or *U*-shaped pattern on the ground. This pattern is best identified when the point of the *V* is located at the POI. The two sides of the *V* extend out and away, making the overall shape that looks like a *V*. In some cases, there will be no *V* pattern, but eyewitnesses will provide statements or photos that will direct you to where the fire may have started.

Fires can be caused by many sources. Some of the more common ones include lightning, campfires, burn piles, utilities, equipment use (e.g., mowers and vehicle exhaust in tall grass), railroad tracks, children, fireworks, firearms and exploding targets, welding

FIGURE 9-10 Aerial view of a fire scene.
Courtesy of Jeff Pricher.

FIGURE 9-11 Everything in a general origin area is considered evidence.
Courtesy of Jeff Pricher.

and grinding, electric fences, reflections (e.g., glass or other objects), flares, blasting, candle lanterns, wind turbines, embers from a structure fire, and, in some cases, spontaneous ignition.

Identifying the area of ignition will help you start protecting the fire scene for the fire investigators. There are many strategies that you can employ to protect the scene, but the most effective, if you have the available manpower, is to have a fire fighter stand guard on the scene. A person fulfilling this role can warn others from entering the area or advise them not to take heavy equipment or tools through the area until the fire investigator arrives. A scene guard can also assist with the initial scene documentation of the area (discussed later).

If manpower is not available, using apparatus or other vehicles to block off an area is advisable. The vehicle or apparatus provides a visual indicator that no one should be in a specific area.

When apparatus, vehicles, and manpower are sparse (which is a common occurrence), the best way to protect the scene is to flag it off with high-visibility surveyor's tape or fireline tape, printed with the words "Fireline Do Not Cross" **FIGURE 9-12**. If you see flagging anywhere on an incident, specifically at the heel or POI, stay out of the area **FIGURE 9-13**. Fire investigators will be scrutinizing this area for shoe prints, tire tracks, metallic objects, and anything else that could have contributed to the fire. It is everyone's job to protect these areas.

If you are able to identify the area of origin and the POI, make sure it is marked off and preserved. Under no circumstances should you ever remove anything from a fire scene or area of origin. Disturbing the integrity of the scene might make or break a case for the investigator. There may be rare instances when you may need to collect something that could be considered evidence or remove an item to protect it. If that needs to happen, immediately notify a supervisor, and take the extra time to document everything you or your crew did and take multiple pictures.

FIGURE 9-12 It is sometimes necessary to mark a scene with tape to secure evidence.
Courtesy of New Jersey State Forest Service.

FIGURE 9-13 Evidence flagging.
Courtesy of Jeff Pricher.

LISTEN UP!

Don't walk or drive through a POI or suspected POI area! Protecting the fire scene is essential for the fire investigator's efforts.

LISTEN UP!

Guarding the fire scene is essential for maintaining scene integrity. If the fire department leaves the scene at any point, in most cases, any evidence collected cannot be used in court. Leaving the scene could allow for the potential of evidence tampering or allow someone to add something that was not there originally.

Documenting the Scene

The next step in the process for protecting the scene will require a methodical and systematic approach to document everything you have observed and led you to believe that you have arrived at the area of origin or POI. Accurate and detailed record keeping will be crucial for the fire investigator to use as he or she establishes what happened.

LISTEN UP!

If a fire fighter or civilian is injured as a result of a fire, the initial documentation and scene preservation could make or break a criminal case if one is brought against a suspect.

Tools needed for scene documentation include, but are not limited to, cell phone camera, camera, paper to sketch or draw what you see, GPS coordinates to identify specific areas or locations, tablet device that allows you to establish markers with mapping programs, and written and communicated statements from the public or your crew.

Photograph the Area

Take some time to ponder once things have calmed down from the initial attack. It is wise to logically process the event to prevent loss of critical memories or information. Take photos of the area. If you identify footprints, tire marks, or other items that seem out of place, capture those images. When taking photos, think about sequencing them with the intention of starting with an overall wide area view or general overview of the fire area and then slowly working your way into a specific detail. This will allow you to maintain a systematic approach that can easily be followed. If you identify an item or area of importance, make sure to take photos from several angles.

Log GPS Coordinates

GPS has been established as a go-to source for accurate cataloging of all sorts of information on fires. In terms of protecting the scene, you can use your GPS device to start a log. Remember, as you identify items or areas in the area of origin, create a systematic and consistent method to naming the items.

Additionally, there are several mapping products available for GPS-enabled tablets that can be used to show items on a topographic map. Once the data points have been created, they can be shared with the investigators or graphic information specialists, who can then add the data to larger maps for an investigation.

If you do use GPS devices, make sure that you have the right data set formatted for your latitude and longitude before collecting data. One of the more common formats for designating latitude and longitude in wildland firefighting is the degrees decimal minutes (DDM) format (discussed later in Chapter 11, *Orienteering and Global Positioning Systems*).

Document the Details

Make sure you document what you and your crew did when you arrived. Spend time talking about where the fire was burning and what direction it was heading. Explain the fire behavior, and estimate the number of chains per hour (1 chain is 66 ft). (Calculating chains is discussed in more detail in Chapter 11, *Orienteering and Global Positioning Systems*.) Explain the fuel type you are in and whether the fuel was continuous or patchy. If there are elements that affected the fire behavior, be as specific as possible when describing them. Examples of these elements include topography, aspect of the slope, time of day, and weather elements that you or your crew documented.

Once you have compiled this information, verify the information with each of your crew members who were working with you. Fire investigators need to be able to obtain an understanding of what the crew members saw and what they did when they arrived on scene. Sometimes, crew members might be able to remember details that you didn't. They may remember the color of a vehicle, how many people were in it, or what the license plate was. While you were focused on the fire as you arrived, they may have noticed that there was a vehicle leaving the scene at the same time. This information will be very valuable for fire investigators and law enforcement.

It is also important to recognize the possibility of arson. If you observe a crater, smell a strong odor of flammable liquids, or notice bizarre burn patterns, or someone said he or she saw a flash or heard a bang before the fire started, then documenting this needs to be a priority.

Overall, protecting and documenting the scene are requirements for all fire fighters. Remember to use the IRPG if you forget some of the steps that are required for this process. There is a checklist in the white pages of the IRPG, titled Fire Cause Determination Checklist.

DID YOU KNOW?

Wildfire investigation should be conducted by qualified fire investigators. A fire investigator will observe when traveling to a fire scene, after arriving at the scene, and during initial attack.

While arriving at the scene, a fire investigator will take note of vehicles or people leaving an area, vehicles without lights, and damaged or downed gates, fences, and power lines.

After arriving at a scene, a fire investigator will take note of tire marks or footprints leading to the scene, suspicious people, onlooker conversations, foreign objects in the scene, and burning campfire or debris.

During initial attack, a fire investigator will note and report ignition devices, such as matchbooks, candles, and road flares; license plate and vehicle descriptions; descriptions of people at the scene; and descriptions of items that he or she believes may be important in determining the cause of the fire, as well as times when all this information was recorded.

FIGURE 9-14 The after-action review occurs after the incident.
Courtesy of Jeff Pricher

After-Action Review

The **after-action review (AAR)** is a debrief discussion after the incident that encourages self-evaluation. It enables participants to discover for themselves what happened, why it happened, and how to sustain strengths and improve on weaknesses. AARs are intended to reflect on and improve fire-fighter performance **FIGURE 9-14**. Remember that every assignment is a chance to learn and improve.

The military initially developed AARs to create an environment that builds a team and improves the team dynamics. It is important to remember that, during this process, finger-pointing or blaming accomplishes nothing. Because AARs are conducted in a group setting, it is very important to be respectful of all participants. These types of reviews are not intended to last for an extended period of time. Instead, they should be focused and organized so that everyone has an opportunity to participate.

AARs can be conducted in several ways. They can be adapted based on the audience or the type of incident that needs to be reviewed. A good habit to develop would be to host an AAR immediately after an event and at a time when everyone is present. The most common AAR formats include the standard four-question format, the chainsaw format, the "what" format, and the PLOWS format.

LISTEN UP!

Be accountable for your actions. Acknowledge your shortcomings as well as your accomplishments.

Standard Four-Question Format

The four-question format is included in the IRPG. The military initially developed this format. It summarizes what was planned, what actually happened, why it happened, and what we going to do the next time.

Chainsaw Format

The chainsaw format is great to use when time is of the essence. When you have worked a long shift or your crew is tired, this AAR is designed for speed.

1. Gather everyone in a circle.
2. Identify a starting point in the circle.
3. Ask each of the following four questions, one at a time, and make sure everyone answers.
   - What is one thing that went well on this shift?
   - What is one thing that did not go well on this shift?
   - What is one thing you would do differently next time?
   - What is one thing you learned today?
4. Make sure everyone has a chance to answer the questions, and do your best to limit each question to 30 to 45 seconds.
5. Make sure fire fighters stay on track and do not get too specific on any one topic.
6. Document important comments for future discussion.

"What" Format

The "what" format works well in a setting when you are working with a group of people not familiar with the AAR process. It asks the following "what" questions:

1. What?
2. So what?
3. Now what?

PLOWS Format

The PLOWS (plan, leadership, obstacles, weaknesses, and strengths) format is a great alternative to the standard AAR format because it facilitates a discussion in a format more akin to how we take on an assignment.

1. Plan:
   - Did everybody know what the plan was?
   - Was the plan sufficient to complete the objectives?
2. Leadership:
   - What leadership was in place?
   - Was the chain of command clear?

- Was the leader's intent communicated and sufficient?

3. Obstacles:
   - What obstacles were encountered, and how were they mitigated?
4. Weaknesses:
   - What were the weaknesses that should be improved upon?
   - How will they be improved?
   - Is follow-up action required?
5. Strengths:
   - What were the strengths that should be sustained?
   - How will they be sustained?
6. Are there additional questions or topics that need to be discussed?

AARs are the perfect time to assess abilities that need improvement by looking at every incident response. The only way to grow and progress is to learn from our mistakes. This process has had great success in the aviation industry, the military, big business, and the fire service. Remember, when you implement your AAR format, make sure you explain the rules so that everyone is on the same page, and keep the pace consistent. If you dwell too much on one topic, you may lose your window for growth.

Administrative Duties and References

Taking care of your crew and personnel is a skill that takes time to develop. You will be responsible for an incredible amount of information, so the process will not happen overnight.

In addition to all the information explained in this chapter, one other important part of your job will be to complete the required administrative components. Some of the important administrative tasks are listed next, but every agency has its own requirements that will need to be learned. The administrative side of your responsibilities help you by establishing statistics that can be relied on in the future to protect your employees and your organization.

Additionally, if you will be responding to interagency incidents, it is equally important that you understand your neighbors' standards to prevent confusion or asking them to violate other agency standards.

There are several administrative references that you should familiarize yourself with that you will use as a first-line supervisor.

SAFENET

SAFENET is an online portal designed for frontline fire fighters and support staff to submit near misses, unsafe situations, or unhealthy situations. This reporting is done anonymously and is designed to give supervisors insight into situations that they may be unaware of. SAFENET is administered by the National Interagency Fire Center and can be found at safenet.nifc.gov **FIGURE 9-15**.

SAFECOM

SAFECOM, which is short for Aviation Safety Communiqué, is similar to SAFENET but established specifically for aviation. The Federal Aviation Administration (FAA) has very strict guidelines on all things aviation. One of the reasons that the aviation industry is as successful as it is, is in large part due to the reporting requirements and standards that have been established. The SAFECOM portal can be accessed at safecom.gov.

Red Book

The Red Book is a guide that provides interagency standards for fire and fire aviation operations. Every year, this document is updated to reflect the changes or best practices from the previous fire season or new standards that have been developed. Most of the large incidents that resources are sent to include more than just state and local resources. It is important to understand that every agency has a specific way of doing things but that all of them have agreed to perform as one cohesive unit to maintain consistency.

ICS-214

This document is commonly referred to as the unit log. The unit log is designed to capture all the daily activities of you and your crew. The challenge with this is making sure you place information into it. Some agencies require that the highlights of the day be entered, and others require that all movements be logged as entries. You will need to check with your agency to determine exactly what needs to be entered. Think of this of as a way to justify what you have done, struggles you have encountered, and equipment problems that have come to light. This will be key when supervisors are trying to track multiday efficiencies and establish a timeline of what has occurred.

SAFENET

Wildland Fire Safety & Health Reporting Network

SAFENET Event Information

Event Start Date:	* *Format mm/dd/yyyy*
Event Start Time:	* *Format hhmm Military Time*
Event Stop Date:	*Optional - if used, Format mm/dd/yyyy*
Event Stop Time:	*Optional - if used, Format hhmm Military Time*
Incident Name:	*
Fire Number:	*P-Code or 4 digit Fire Code*
State:	* -- Select State --
Jurisdiction:	* -- Select Jurisdiction --
Local Unit:	*Administrative unit (district, park, reserve, etc.) where the event occurred.*
Incident Type:	* ☐ Wildland ☐ Prescribed/Fuels Treatment ☐ All Hazard ☐ Training ☐ Work Capacity Test
Incident Activity:	* ☐ Line ☐ Support ☐ Transport to/from ☐ Readiness/Preparedness
Stage of Incident:	* ☐ Initial Attack ☐ Extended Attack ☐ Transfer of Command ☐ Mop Up ☐ Demobe ☐ Non-incident ☐ Other
Position Title:	*Firefighter, Division Supervisor, Facilities Unit Lead, etc*
Task:	*Line Construction, Structure Protection, Camp Activities, etc*
Management Level:	* ○ 1 ○ 2 ○ 3 ○ 4 ○ 5 ○ N/A *Incident Types 1, 2, 3, 4, 5*
Resources Involved:	*Crew, Equipment, Overhead, etc*

FIGURE 9-15 SAFENET online portal.

SAFENET, Wildland Fire Safety & Health Reporting Network, U.S. Department of the Interior. Retrived from https://safenet.nifc.gov/safenet.cfm.

CA-1 and CA-16

If you are on an incident that involves federal employees and you happen to be assigned to supervise them, make sure you are familiar with the CA-1 (Federal Employee's Notice of Traumatic Injury and Claim for Continuation of Pay/Compensation) and the CA-16 (Authorization for Examination and/or Treatment) documents. If an injury or accident occurs with a federal employee, familiarizing yourself with these documents can be beneficial. Both of these documents can be downloaded from the Internet.

Crew Reports and Individual Performance Ratings

Interagency training standards require that persons completing trainee assignments or contractors assigned to a fire need to have a performance review completed at the end of every assignment. This document helps to reinforce the performance of the crew or individual that pertains to the assignment just completed. When filling out these documents, it is essential that ratings or reviews be pertinent to the incident and related to specific performance-related operations. Deviating from this may cause unnecessary additional paperwork and a meeting with human resources.

Depending on what your agency requires, make sure you have the appropriate paperwork at your disposal. This could include department-specific forms or OSHA-required paperwork, such as the 801 form.

Position Task Book

Most agencies utilize the Position Task Book (PTB) for evaluating the performance of personnel who are acting in the capacity of trainees. The PTB is a vital document that needs to be taken seriously. Any entry in this book needs to be realistic and verifiable. In most areas of the country, for a fire fighter to acquire more qualifications, the completed PTB will be sent to a committee for review before being approved. The following are some key requirements for filling out this document:

1. PTB initiation must occur by an authorized individual from your agency. This person is usually the training chief or training officer. Not all PTBs can be initiated without the prerequisite information and training documented.
2. A PTB has a 3-year time frame during which the first assignment must be entered. If no assignments have been entered, the PTB becomes invalid.
3. After the first assignment has been entered into the PTB, you will have an additional 3 years to complete it.
4. New standards for the PTB include documenting the specific day that a skill or task was completed. Entering a blanket date, if challenged in court or during an investigation, would make it seem as though everything had been completed in 1 day.
5. In general, no more than two PTBs can be opened in a given area at the same time. An example of this would be two in operations, two in planning, etc.
6. Multiple PTBs cannot have signatures in them from the same assignment with the same dates. What this means is that, if you are trying to complete your squad boss PTB, you cannot have signatures in a crew boss PTB from the same fire using the same dates. When you are on an incident, you are generally given only one assignment.

Other reference materials, such as the IRPG and the Fire Danger Pocket Cards, are discussed in Chapter 1, *The Wildland Fire Service*.

After-Action REVIEW

IN SUMMARY

- Personal protective equipment (PPE) must be inspected for damage and cleanliness at the beginning of every shift to ensure each piece is in good working order.
  - This includes your helmet, eye and hearing protection, flame-resistant shirt and pants, boots, socks, undergarments, gloves, web gear/fire pack, and fire shelter.
- To successfully manage and lead fire fighters, you must be familiar with their training and physical fitness levels. Placing fire fighters in appropriate positions based on their qualifications is important.
  - Qualifications extend beyond Firefighter I and II and can include chainsaw operations, terrain-vehicle operation, EMT, apparatus driver/operator, and helicopter crew.
- The team leader or supervisor influences the dynamics of task management.
  - Leader's intent communicates the task, purpose, and end state.
  - Operational leadership takes human factors into consideration, such as communications and motivation.
- All fire fighters at all levels are responsible for communicating with other members of the team and with team leaders. Communication responsibilities can be broken into five areas:
  - Briefing others
  - Debriefing your actions
  - Communicating hazards

- Acknowledging messages
- Asking when you do not know
- Daily shift briefing is important for all crew members to participate in to discuss the previous day's actions and current day's assignments and training expectations.
  - Necessary equipment maintenance is communicated to everybody at once.
  - Expectations for training assignments are explained to the team.
  - Weather reports for your jurisdiction can be covered.
- Incident briefing can occur at any time during the fire assignment.
- Lookouts are necessary at every type of fire for communicating environmental changes, traffic safety, fall hazards, fire behavior, and trigger points.
- Protecting a fire scene the moment you arrive provides the fire investigator with the clearest picture of what may have started a fire and avoids damage to potential evidence:
  - Protecting the scene can be done with a scene guard or by blocking off the scene with apparatus and other vehicles or with tape.
- Documenting the scene is a methodical and systematic approach to recording everything you observed at the scene and suspected evidence of the origin of the fire.
- After-action reviews are post-incident debriefs that encourage fire fighters to reflect on their actions and improve their performance after discussing an incident.
- An important part of the job is to complete required administrative components, which establish statistics that can be relied upon in the future.
  - These components include SAFENET, SAFECOM, the Red Book, ICS-214, CA-1 and CA-16, crew reports and individual performance ratings, and the Position Task Book.

KEY TERMS

After-action review (AAR) A post-incident debrief that encourages self-evaluation.

Arduous pack test A physical fitness test for wildland fire fighters; also called the work capacity test.

Briefing A meeting designed to provide information and instruction on an incident.

Crew A team of two or more fire fighters.

Direct communication Language that is concise and factual and provides closure.

Fire investigator An individual trained to examine a scene and determine a point of ignition.

Leader's intent A clear, concise statement that outlines what individuals must know in order to be successful for a given assignment.

Lookout An individual designated to detect and report fires from a vantage point.

Operational leadership Management technique that takes human factors and processes into consideration; includes improving systems and processes, taking charge, motivating fire fighters, demonstrating initiative, and directly communicating.

Point of ignition (POI) The area where a fire started; also called area of origin.

Risk management process A series of steps used to make decisions and observe their outcomes.

REFERENCES

American National Standards Institute (ANSI) and International Safety Equipment Association (ISEA). ANSI/ISEA Z87. 1-2015, *American National Standard for Occupational and Educational Personal Eye and Face Protection Devices*. 2017.

Aviation Safety Communiqué (SAFECOM). "Safety Communiqué Form." https://www.safecom.gov/SAFECOM_doi.pdf.

Avsec, Robert P. "Best Undergarments to Protect Firefighter from Burns." Fire Chief by Fire Rescue. Published

September 16, 2016. https://www.firechief.com/fire-products/personal-protective-equipment-ppe/articles/best-undergarments-to-protect-firefighter-from-burns-lVlzkxVjEFn7kAlu/.

Federal Emergency Management Agency (FEMA). "Activity Log (ICS 214)." Accessed October 14, 2019. https://www.fema.gov/media-library-data/20130726-1922-25045-6289/ics_forms_214.pdf.

Flight Safety Foundation. "Emergency Hand Signals." Skybrary. Modified June 22, 2019. https://www.skybrary.aero/index.php/Emergency_Hand_Signals.

National Fire Protection Association (NFPA). "NFPA 921, Guide for Fire and Explosion Investigations." 2017 ed. https://www.nfpa.org/codes-and-standards/all-codes-and-standards/list-of-codes-and-standards/detail?code=921.

National Fire Protection Association (NFPA). "NFPA 1033, Standard for Professional Qualifications for Fire Investigator." 2014 ed. https://www.nfpa.org/codes-and-standards/all-codes-and-standards/list-of-codes-and-standards/detail?code=1033.

National Fire Protection Association (NFPA). "NFPA 1977, Standard on Protective Clothing and Equipment for Wildland Fire Fighting." 2016 ed. https://www.nfpa.org/codes-and-standards/all-codes-and-standards/list-of-codes-and-standards/detail?code=1977.

National Wildfire Coordinating Group (NWCG). *Guide to Wildland Fire Origin and Cause Determination.* PMS 412-1. NFES 1874. April 2016. https://www.nwcg.gov/sites/default/files/publications/pms412.pdf.

National Wildfire Coordinating Group (NWCG). "L-180, Human Factors in the Wildland Fire Service, 2014." NFES 2985. Modified July 31, 2019. https://www.nwcg.gov/publications/training-courses/l-180.

National Wildfire Coordinating Group (NWCG). "L-280 Followership to Leadership, 2008." NFES 2992. Modified August 21, 2019. https://www.nwcg.gov/publications/training-courses/l-280.

National Wildfire Coordinating Group (NWCG). "RT-130, Wildland Fire Safety Training Annual Refresher (WFSTAR)." Modified July 26, 2019. https://www.nwcg.gov/publications/training-courses/rt-130.

National Wildfire Coordinating Group (NWCG). "S-130, Firefighter Training, 2008." NFES 2731. Modified August 21, 2019. https://www.nwcg.gov/publications/training-courses/s-130.

Petrilli, Tony. "Fire Shelter Inspection Guide and Rebag Direction." United States Forest Service. Published January 2011. https://www.fs.fed.us/t-d/pubs/htmlpubs/htm11512301/index.htm.

SAFENET. "Wildland Fire Safety & Health Reporting Network." National Interagency Fire Center (NIFC). Accessed October 14, 2019. https://safenet.nifc.gov/safenet.cfm.

U.S. Department of Labor. "Authorization for Examination and/or Treatment." Form CA-16. Revised February 2005. https://apwu.org/sites/apwu/files/resource-files/hr-injurycomp-wna-align-120126-ca-16.pdf.

U.S. Department of Labor. "Federal Employee's Notice of Traumatic Injury and Claim for Continuation of Pay/Compensation." Form CA-1. Revised October 2018. https://www.dol.gov/owcp/regs/compliance/ca-1.pdf.

Wildland Fire Fighter in Action

It is July 17, and you are on your second day of your fire assignment. Today, your crew boss has assigned you to be the lookout for your crew. Your crew boss advises you that where your crew will be working will require you to drive around to the far side of the fire and then hike up a hill to 7200 ft (2194.6 m). It will take you 3 hours to complete the drive and hike to the ridgetop. You will be spending the night at the lookout spot due to the location of the fire and how long it takes to get to it. From the lookout spot, you can see the whole fire, all the crew, and the additional resources that have been assigned to the fire. You were told to bring enough food and water for 2 days, as well as stock up on batteries for your handheld radio. Your crew boss also advises you that you will be the lookout for Division Foxtrot, which is your neighboring division.

1. To make sure you thoroughly understand your assignment, you will want to make sure the crew boss covered which of these?
   - **A.** Responsibilities, estimated time of arrival, and weather taking
   - **B.** Task, purpose, and end state
   - **C.** Safety zones, escape routes, and radio frequencies
   - **D.** What is in the IAP, map features, and communications channels
2. What types of equipment will you need to have as a lookout?
   - **A.** IRPG, radio, belt weather kit, and signal mirror
   - **B.** IRPG, signal mirror, radio, lunch, and a book
   - **C.** Binoculars, signal mirror, IRPG, and a book
   - **D.** Signal mirror, IRPG, maps, and lunch
3. Where are the tables for calculating the probability of ignition located?
   - **A.** Yellow Book
   - **B.** Red Book
   - **C.** IRPG
   - **D.** IHOG

Access Navigate for flashcards to test your key term knowledge.

CHAPTER 10

Wildland Fire Fighter II

Radio Communications

KNOWLEDGE OBJECTIVES

After studying this chapter, you will be able to:

- Describe the types of radios available to wildland fire fighters. (pp. 220–221)
- Describe radio frequencies and how they affect radio communications. (pp. 221–224)
- Operate a portable radio so that communication with assigned crew members is clear, concise, and accurate. (**NFPA 1051: 5.1.1, 5.1.2, 5.5.6**, pp. 225–227)
- Describe the types of radio traffic. (p. 225)
- Describe basic radio etiquette. (pp. 225–227)
- Describe basic radio precautions, care, and maintenance. (pp. 228–229)

SKILLS OBJECTIVES

After studying this chapter, you will be able to perform the following skill:

- Operate a portable radio. (**NFPA 1051: 5.1.1, 5.1.2, 5.5.6**, p. 228)

Additional Standards

- **NWCG S-130 (NFES 2731)**, *Firefighter Training*

You Are the Wildland Fire Fighter

You are on your way to patrol an area of your district as a result of a report from dispatch of smoke in the area. The engine has just started to pull away from the station when the radio squawks with the dispatcher advising that you are being sent to a neighboring county for a fast-moving fire that is threatening many structures. During the 20-minute response time, your captain advises that you will need to switch radio groups or zones so that you will be able to communicate with the other resources that are responding. You then change your radio to the new assigned radio group or zone for the incident.

1. Do you remember the steps required to change groups or zones?
2. Do you have enough batteries and clamshells?
3. If you do not have the group or zone in your radio, what can you do?
4. Are you prepared to broadcast an emergency if needed?

Access Navigate for more practice activities.

Introduction

Radios are an integral part of the Incident Command System (ICS) because they link all the units responding to an incident. This chapter discusses radio types, radio frequencies and channels, radio traffic, radio communications, and basic radio care.

Radio Types

There are three main types of fire service radios: portable, or handheld, radio; mobile radio; and base station radio.

Portable Radios

A **portable radio** is a two-way radio that is small enough for a fire fighter to carry at all times **FIGURE 10-1**. This is the most important radio for the wildland fire fighter. The portable radio body contains an integrated speaker and a microphone, an on/off switch or knob, a volume control, a channel select switch, and a push-to-talk (PTT) button. Portable radios also require batteries. The signal from a battery-operated portable radio can be heard within only a certain range and is easily blocked or overpowered by a stronger signal.

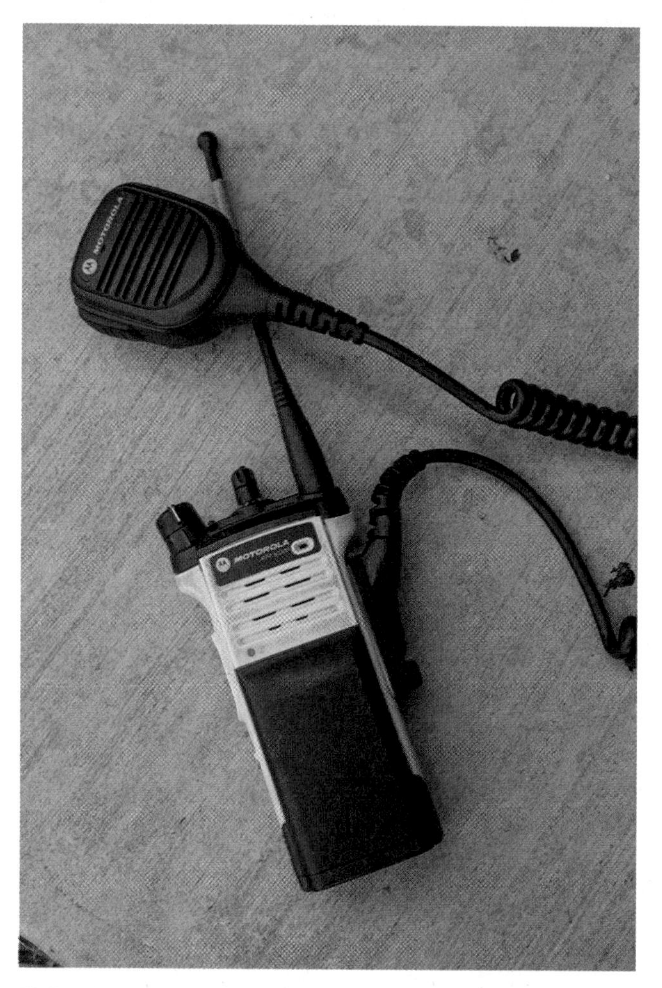

FIGURE 10-1 Portable radio.

Portable Radio Batteries

Portable radios require charged batteries to operate. There are many varieties of batteries. Become familiar with the batteries your agency uses, and follow the manufacturer's recommendations for charging.

If you are in a situation when you do not have the ability to charge a battery, there are add-on devices, called clamshells, that incorporate AA batteries as the power source for portable radios **FIGURE 10-2**. Usually, when a portable radio starts to chirp during an attempt at transmitting or the receiver provides feedback

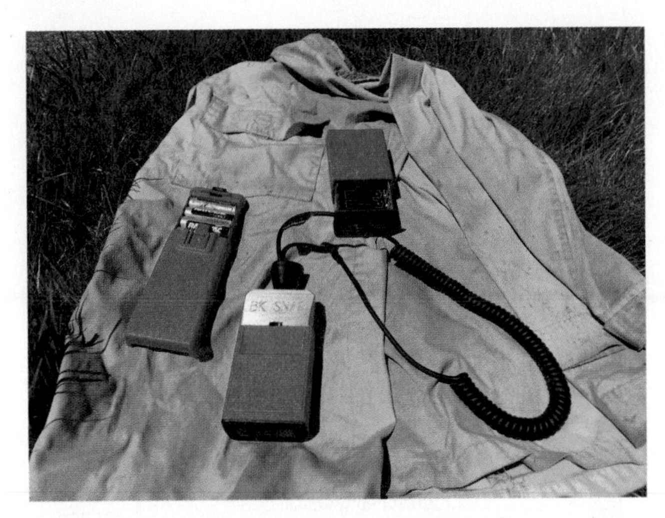

FIGURE 10-2 Clamshell power sources.
Courtesy of Jeff Pricher.

FIGURE 10-3 Bendix King Radio.
Courtesy of BK technologies.

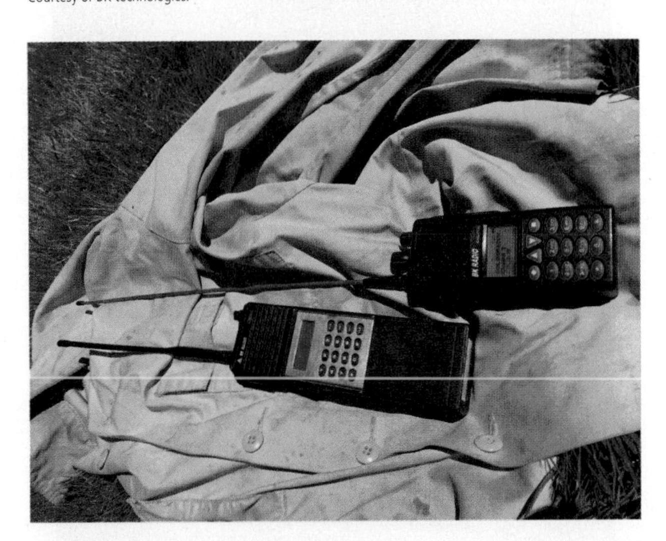

FIGURE 10-4 Old and new radios.
Courtesy of Jeff Pricher.

that you seem to be cutting out on the radio, chances are the battery on the radio is out of power. Having two clamshells is a good idea so that the time between batteries dying and replacement is kept to a minimum.

The older-style Bendix King Radio (discussed later) clamshells have aftermarket adapters that allow you to utilize a battery to charge other equipment or power a device that uses a cigarette lighter adapter.

LISTEN UP!

Make sure your portable radio battery is charged before your work shift begins!

Bendix King Portable Radio

Though there are many types of portable radios, the most common in the wildland environment is the Bendix King Radio **FIGURE 10-3**. This radio is a favorite because it can be programmed in the field by a fire fighter or can be cloned from mobile to handheld or handheld to handheld **FIGURE 10-4**. (Radio cloning is discussed in more detail later in this chapter.)

Mobile Radios

A **mobile radio** is a more powerful two-way radio that is permanently mounted in vehicles and powered by the vehicle's electrical system **FIGURE 10-5**. Mobile and portable radios share similar features, but mobile radios usually have a fixed speaker and an attached microphone.

Base Station Radios

Base station radios are permanently mounted in a building, such as a fire station or remote transmitter site. Base radio transmissions have a wide coverage area thanks to antennas mounted on radio towers **FIGURE 10-6**.

Radio Frequencies and Channels

Frequencies

In the United States, the design, installation, and operation of two-way radio systems are regulated by the Federal Communications Commission (FCC). The FCC has established strict limitations governing the assignment of frequencies to ensure that all users have adequate access. Every system must be licensed and operated within these guidelines. Some frequencies are specifically allocated for emergency services **FIGURE 10-7**. Of those frequencies that are allocated

FIGURE 10-5 Mobile radio.
© Jones & Bartlett Learning. Photographed by Glen E. Ellman.

FIGURE 10-6 Base station radio.
© Jones & Bartlett Learning. Photographed by Glen E. Ellman.

Allocation		Frequency (MHz)
RADIONAVIGATION-SATELLITE	MOBILE-SATELLITE (Earth-to-space)	149.9
FIXED	MOBILE	150.05
FIXED	LAND MOBILE	150.8
LAND MOBILE		152.855
FIXED	LAND MOBILE	154.0
MARITIME MOBILE		156.2475
MARITIME MOBILE (distress, urgency, safety and calling)		156.7625
MARITIME MOBILE		156.8375
MARITIME MOBILE		157.0375
MOBILE except aeronautical mobile		157.1875
FIXED	LAND MOBILE	157.45
MARITIME MOBILE		161.575
LAND MOBILE		161.625
MOBILE except aeronautical mobile		161.775
MARITIME MOBILE (AIS)		161.9625
MOBILE except aeronautical mobile		161.9875
MARITIME MOBILE (AIS)		162.0125
FIXED	MOBILE	163.0375
FIXED	Land mobile	173.2

FIGURE 10-7 Radios are the most common form of communication in large wildland incidents.
Courtesy of National Telecommunications and Information Administration, U.S. Department of Commerce.

for public safety, several are used by zone dispatch centers. Initial attack radio frequencies on wildland fires that involve state and federal partners are developed and approved by a zone board of directors. Responding initial attack resources are given a radio frequency common to all agencies that are responding to the incident, by the zone dispatch center.

Radios work by broadcasting electronic signals on certain frequencies, which are designated in units of megahertz (MHz). A **frequency** is the number of cycles per second of a radio signal; only those radios tuned to that specific frequency can hear the message.

During incidents that involve several fire departments and emergency service crews, radios can be programmed with over 20 different frequencies. This allows fire fighters to easily switch between mutual aid resource frequencies as needed. It is very important to be familiar with your radios. There will be many situations when you will need to switch from one group or zone to another.

Radio Cloning

Radio **cloning** is a programming process that allows for the copying of frequency information from one radio and sending that copy to another radio to avoid manual programming. A specific set of cables and programming sequences are needed for successful radio cloning **FIGURE 10-8**.

Radio cloning is a skill that is needed at the Wildland Fire Fighter II (National Wildfire Coordinating

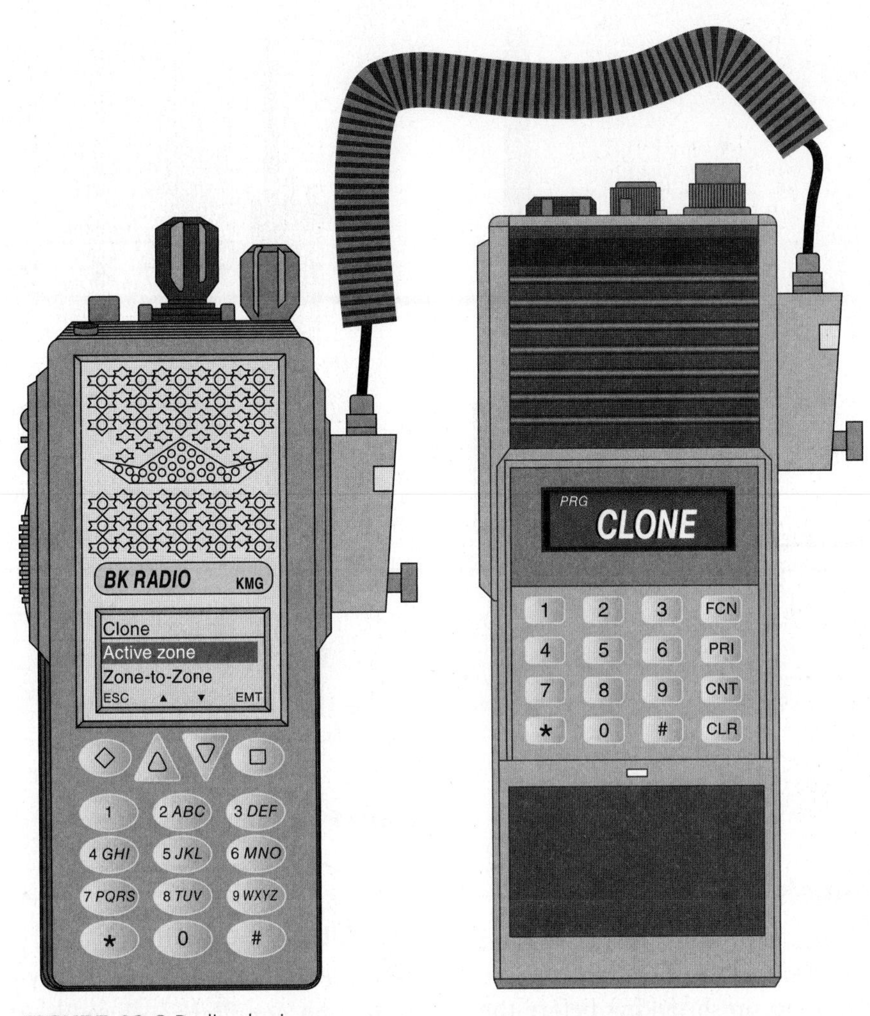

FIGURE 10-8 Radio cloning.

Group [NWCG] Firefighter Type 1 [FFT1]) level. It is not uncommon to have to program a new frequency into a radio on a fire when multiple aircraft are being used or fire fighters at another fire in close proximity may be sharing the same frequency. With respect to cloning, depending on the expanding incident, clones may be required before the beginning of every shift. Radio cloning can be accomplished by following the instructions that accompany the radio cloning cables.

Radio Channels

A radio **channel** is an assigned frequency or frequencies used to carry voice and data communications. Assigned radio frequencies may be used in a variety of systems. The specific programming of your radio will dictate how to select the different radio channels. Most commonly, there is a knob on the top of the radio with numbers from 1 to 16. Each one of the 16 numbers is associated with a frequency (and assigned a channel name). Turn the knob clockwise to go from 1 to 16. Some of the newer radios do not have 16 numbers but, instead, have an infinity knob. The radio screen will tell you which channel you are on.

Simplex Channels

Simplex channels (push to talk, release to listen) send information on the same frequency and in one direction only **FIGURE 10-9**. This type of communication is often referred to as car-to-car or line-of-sight communication because a message goes directly from one radio to every other radio that is set to receive that frequency. In general, the radios have no obstruction or are being used within a small geographic area. Transmissions can occur in either direction but not simultaneously.

One of the benefits of a simplex channel is that there is no delay from the time the PTT button is pressed and the speaker talks. A simplex channel does not rely on sophisticated equipment to make the transmitting signal reach the receiving devices. The transmitting and receiving radios simply need to be programmed to the same frequency number.

On the technical side, a simplex frequency would look like this: 169.925. When transmitting a message or a signal, both the transmitting and the frequency sides of the radio would be programmed with 169.925. This is known as a simplex frequency.

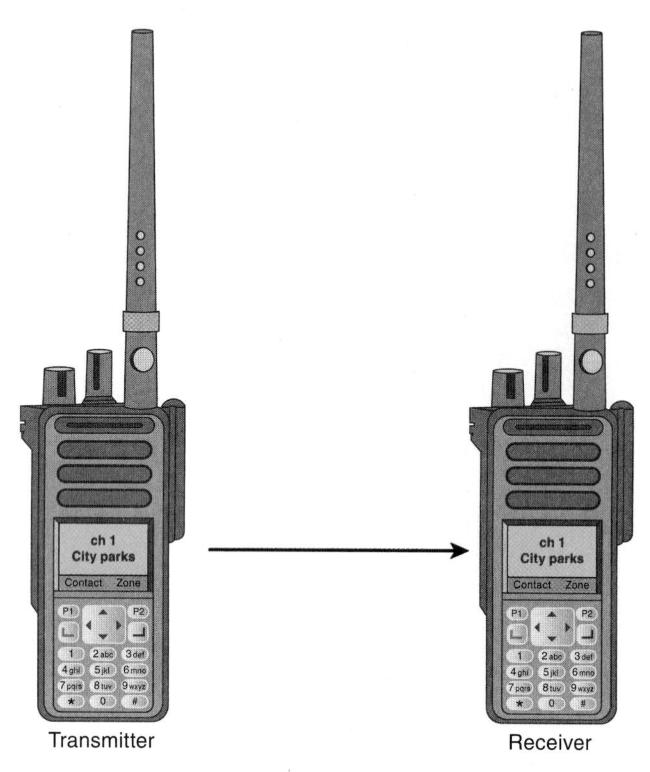

FIGURE 10-9 Simplex radio communications.

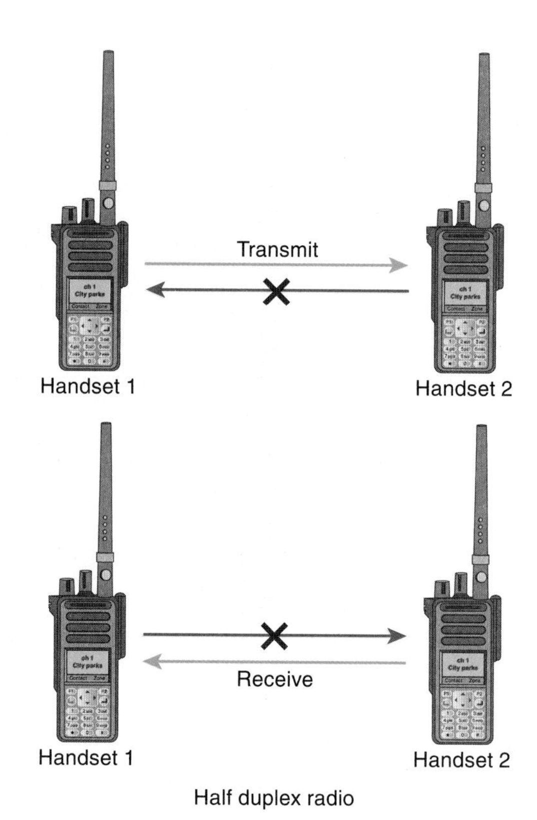

FIGURE 10-10 Half duplex channel.

Half Duplex Channels and Repeaters

Half duplex channels also allow both parties to send information but not simultaneously. The receiver must wait for the sender to finish talking before the receiver can transmit a reply **FIGURE 10-10**. Half duplex channels are used with a repeater system.

In some instances, it is important to be able to talk from one side of a mountain ridge to the other. When distance is part of the equation, simplex frequencies do not have the range or the ability to go through hard objects. This is where repeaters come in handy.

A **repeater** is a special base station radio that uses two separate frequencies—one to receive and one to transmit messages. Repeaters are an important tool because they extend our communication capabilities. A repeater has a large antenna that is able to receive lower-power signals, such as those from a portable radio, from a long distance away. The signal is then rebroadcast to all the radios set on the designated channel with all the power of the base station.

The radio would need to be programmed with a specific frequency on the transmitting side (an example being 169.925) and another frequency programmed for the receiver (an example being 164.875). The challenge with this type of communication is that the sender and the receiver are not able to talk at the same time. If they did, no one would be able to hear each other. It is very important to make sure that no one is speaking on the channel you intend to use before you talk.

Full Duplex Channels

Full duplex channels allow both parties to send information simultaneously without any interference, such as when using a cell phone. Full duplex channels are not currently an option in the wildland firefighting environment. While some fire fighters do use full duplex, it is not ideal because generally only two people are able to hear the communication as it is occurring.

Trunked (700/800 MHz) Channels

Many public safety agencies utilize trunked, or 700/800 MHz, channels. In general, this frequency is considered ultra high frequency (UHF). **Trunked channels** automatically group users to available radio channels as needed. While there are several benefits of using this type of system, it is not able to function well in mountainous terrain without the aid of several repeaters and specialized computer equipment.

Radio Traffic

Communication equipment, like any other firefighting equipment, has very specific words or phrases that are important to know. Depending on the region or part of the country you are from, there will be area-specific phrases you will need to be familiar with.

First, make sure you know the dispatch center or centers you will be talking with. Knowing their names before using the radio or traveling to another jurisdiction will allow you to be able to communicate when it is most important. Examples include Y-COMM, C-COMM, dispatch, or other area-specific names.

It is also important to understand and practice the proper ways to identify important information and emergency information. It is imperative for all fire fighters to be able to communicate critical information in a way that lets all users know they need to keep communication to a minimum or not at all.

Routine Traffic

Routine traffic is the normal communication that is required in response to an incident, during an incident, and returning from an incident. Generally, critical information is kept to a minimum or is nonexistent. Routine traffic is akin to having a casual conversation with another person in a way that keeps the conversation short and concise.

Priority Traffic

Priority traffic is a message that contains critical information that needs to be shared over a single channel or frequency. Priority traffic restricts all radio communications to information that is only mission critical. An example of priority traffic could be needing to clear the air to allow a supervisor to guide an ambulance to a specific location or to share important information with an aircraft that is about to drop water or retardant. In these types of situations, the procedure would be to identify yourself over the radio and state that you have priority traffic. In practice, once this message has been transmitted, everyone should cease communication until you get your message out. In pauses of information sharing, it is OK for others to share mission-critical information to continue with the current operation. When you have completed your priority traffic, it is imperative that you transmit over the air that the channel is clear and priority traffic has been completed.

Emergency Traffic

Emergency traffic is an urgent message that takes priority over all other radio communications. When an emergency has occurred (incident within an incident or other emergency) and you need the channel or frequency to get life-threatening or life-saving information out, you will want to use this type of communication. While emergency traffic is similar to priority traffic, with emergency traffic, everyone on that channel or frequency must stop communicating unless you are calling them.

To declare an emergency, break into the radio traffic, and ask for the channel to be cleared for emergency traffic. Most fire agencies use the term **mayday** to break into the radio traffic and communicate an emergency in which a fire fighter is lost, missing, injured, or requires immediate assistance. Some agencies simply use the terms *lost*, *hurt*, or *injured*. In essence, you are using the equivalent of lights and sirens on the radio to get the important information out to people who need to know. Use a calm, normal tone of voice while delivering the emergency message.

LUNAR

The fire fighter making the mayday, or emergency, call should describe the situation, location, and help needed. Some agencies utilize the acronym **LUNAR** to help remember all the information needed when reporting a mayday. LUNAR stands for *location*, your location in the incident; *unit*, the unit you are assigned to; *name*, who you are; *assignment*, where you were last assigned; and *resources*, what you need to get you out of the mayday situation.

It may take a bit of time to get all of this information out, but, if you are in a situation where you have only a short time to talk, it is important to let others know who and where you are. If you have a medical emergency or are reporting an accident that involves an injury, make sure you follow the procedures in the *Incident Response Pocket Guide* (IRPG) for calling in a medical incident report (MIR). Finally, like priority traffic, when you are finished, you will need to state that the emergency traffic has been completed and you are releasing the channel or frequency back to routine traffic.

Radio Communications and Etiquette

It is very important to make sure you are familiar with your communication devices **FIGURE 10-11**. This includes how to change a channel, change a group or zone, scan, and transmit.

Radio Sequence

One of the most important things to remember when operating a radio is this sequence: push, pause, talk, pause, release. This sequence is important because it will prevent your messages from being cut off. When you talk on a repeater, generally, there is a delay of about

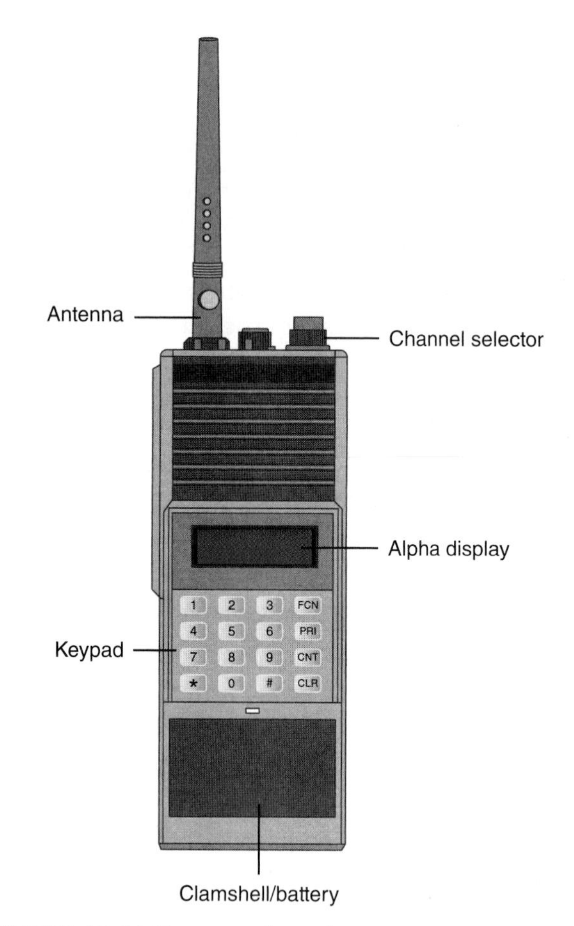

FIGURE 10-11 The parts of a radio.

1 second. That delay is caused by the signal leaving the radio and making contact with the repeater. If you talk during that delay time, that portion of your message will be cut off. To prevent this, pause for 2 seconds before and after you are finished speaking. Pausing before you speak will give the radio the time it needs to activate the repeater. That second is the time it takes to turn the repeater switch in the on position. Pausing after you are finished speaking and before letting go of the PTT button prevents your message from being cut off. That last second is the time it takes to turn off the repeater.

Speaking over a Radio

Another area where fire fighters have a challenge is not speaking coherent thoughts during their transmission. This allows for confusion and wasted time. To avoid this situation, it is imperative that you think about what you are going to say before you say it. If you have time, run through what you intend to say quietly aloud or in your head to make sure you cover all of your message. This process will allow for complete, accurate, and concise messaging.

Your messages should be kept to 15 to 20 seconds maximum. Transmissions longer than this will potentially prevent someone from being able to get an important or time-sensitive message out. If you do have a long message to transmit, pause for 5 to 10 seconds after every 15 seconds of transmission. Announce "break" between each transmission. This break will allow someone to chime in with emergency or priority traffic.

Always use clear language. In the early use of radios, it was common for fire fighters to use 10 or 12 different codes. These codes were specific messages that shortened transmission times. Unfortunately, there was never a nationwide or regional standard. These codes caused several communication problems on large fires and created several situations where misunderstandings were common. One of the recommendations that came out of the 9/11 Commission Report was for emergency responders to use plain language without coded messages.

LISTEN UP!

Whenever possible, be brief when speaking over a radio. Radio channels are busy!

Communication Styles

There are two common ways to communicate on a radio when talking to another user. They are generally "Hey you, it's me" and "It's me, hey you." Depending on where you are, it will be easy to differentiate between the two.

While not all users of communication devices have adopted the "Hey you, it's me" style, in most cases, it is the preferred method. With this style, the sender of information is hailing the person he or she needs to talk to. The benefit of this is that the receiver is more apt to hear his or her name or unit called and immediately pay attention or recognize that someone is trying to get his or her attention. When mission-critical information needs to be communicated, getting the receiver's attention makes for a more successful transmission of information. An example of this would be "Command from Division Bravo." In this instance, command is alerted and then, in theory, would understand that Division Bravo is trying to get a message across.

There are still some places that use the "It's me, hey you" style to communicate. This style is not preferred because, in most cases, the receiver will have to ask who is calling him or her. If you identify yourself first and then request the attention of the receiver, he or she might not be listening and, therefore, will probably miss the message. An example of this would be "Division Bravo to command." In this instance, by the time command processes that someone is trying to contact him or her, he or she has missed the first part of the transmission.

Your best success for communicating will be to use "Hey you, it's me!"

Transmitting Travel Times

It is becoming more commonplace now for fire fighters to abbreviate time and distance. Most people have probably used the term ETA (estimated time of arrival) at some point in their life. The question is, has it been used appropriately? Though travel time abbreviations are often specific to each agency, a few terms that are becoming more frequent but have very specific meanings are outlined in **TABLE 10-1**. As described later, most people misuse ETA and ETE (estimated time en route) and do not understand their meanings. Though you may not use these terms daily, when someone else does, it is important to understand what he or she means. Not understanding these terms can confuse a dispatch agency.

Aircraft Communications

Communicating with aircraft can be very challenging for both the aviation resource and the radio user on the ground. To make communicating easier, the best practice is to always speak to the aircraft from the pilot's perspective. The front of the aircraft is always referenced as the "twelve o'clock" position, and the rear of the aircraft is considered the "six o'clock" position. The pilot's right side of the aircraft is the "three o'clock" position, and the left side is the "nine o'clock" position. As the aircraft approaches your scene and you are trying to direct it in, your communication should sound something like this: "Helicopter zero echo hotel [0EH], I have a visual on you, and I am at your two o'clock. Please place the cargo on the orange marker panel that is directly in front of me." If you forget your references, you can always look at the cover of the IRPG to help you with your relation to aircraft **FIGURE 10-12**.

FIGURE 10-12 The cover of the IRPG displays aircraft references.

Other Communication Tips

Here are a few additional tips for communicating over a radio:

- Answer radio calls promptly. Have a pencil and notepad ready for messages.
- When communicating numbers, to prevent confusion, do not use the slang *oh* in place of zero. Using the word *zero* for the number "0" is preferable because it is specific and concise.
- When you are transmitting GPS coordinates, it is a good practice to say *decimal* instead of *dot*. Using the word *decimal* is very specific and assists with clear communication.
- If you end up in a situation where two people are trying to communicate with each other and you are in the middle and can hear both the sender and receiver, consider utilizing yourself as a human repeater.
- When you are on an incident that is using multiple frequencies, it is important when communicating to a receiver that you identify what frequency or channel you are talking on. If you forget to state what frequency you are on, it

TABLE 10-1 Common Abbreviations Used to Transmit Travel Times

Abbreviation	Meaning	Usage	Example
ETA	Estimated time of arrival	To identify a specific time when you think you will be arriving	"ETA seventeen twenty-two"
ETE	Estimated time en route	To describe how long in hours and minutes your travel time will be	"ETE forty-five minutes"
ETD	Estimated time of departure	To describe the approximate time you will be leaving a specific location	"ETD from drop point 232 will be eighteen forty-two"

SKILL DRILL 10-1
Using a Radio

Courtesy of the National Interagency Fire Center.

1. Listen to determine whether the channel is clear of any other traffic. Depress the PTT button, and wait at least 2 seconds before speaking. This delay enables the system to capture the channel without cutting off the first part of the message. Some systems sound a distinctive tone when the channel is ready.

2. Know what you are going to say before you start talking. Speak clearly and keep the message brief and to the point.

3. Speak across the microphone at a 45-degree angle, and hold the microphone 1 to 4 in (2.5 to 10 cm) from your mouth. Avoid wind blowing into the microphone when speaking. Never speak on the radio if you have something in your mouth.

4. Release the PPT button only after you have finished speaking.

can be difficult for a receiver when he or she is scanning multiple frequencies and channels.

To use a radio, follow the steps in **SKILL DRILL 10-1**.

LISTEN UP!

Be courteous. Swearing and obscene language spoken over the radio is prohibited.

Radio Care

Protection and Maintenance

To keep your radio in good working condition, perform the following protection and maintenance techniques:

- Protect your radio from water, dust, and retardant as best you can. If possible, keep a portable radio in a protective case. After each shift, wipe your radio with a dry cloth.
- Make sure that the accessory port is covered with a protective cap when no accessory is being used. There are copper contacts that can become oxidized and inoperable if these ports are not protected.
- If any functionality or clarity issues are noted, the radio must be turned in to a technician for repair.

If you find that your radio is not working as it should, consider the following troubleshooting tips before sending the radio for repair:

- Radio traffic overload
- Poor reception or transmission at your current location
- Low battery
- Loose antenna
- Operating on the wrong channel

Handling

One very bad habit some fire fighters have is using the antenna as a common holding feature or as a means to pull the radio from its case. While it may seem easier to manipulate a portable radio using the antenna, it is also the quickest way to damage it.

Portable radio antennae are made of coiled copper wire. The wires are designed to cover certain frequency bandwidths. Every time stress or a bend is forced upon the antenna, the copper wire develops internal stress cracks and diminishes the efficacy of the antenna.

High-gain and telescoping antennae are often used to boost signals in areas where coverage or range is poor. For these antennae to be effective, they need to be cared for and protected. When handling a portable radio, remember to use the body of the radio and not the antenna. Communications are a vital part of the wildland fire environment. The last thing you want to do is reduce your potential to communicate due to damaging your equipment.

After-Action REVIEW

IN SUMMARY

- There are three types of fire service radios:
  - Portable, or handheld, radio
  - Mobile radio
  - Base station radio
- Radios work by broadcasting electronic signals on certain frequencies, which are designated in units of megahertz (MHz).
- A frequency is the number of cycles per second of a radio signal; only those radios tuned to that specific frequency can hear the message.
- Radio cloning is a programming process that allows for the copying of frequency information from one radio and sending that copy to another to avoid manual programming.
- A radio channel is an assigned frequency or frequencies used to carry voice and data communications. There are several types of radio channels:
  - Simplex
  - Half duplex
  - Full duplex
  - Trunked (700/800 MHz)
- Radio traffic is field and geographic specific. There are three types of traffic:
  - Routine
  - Priority
  - Emergency
- When operating a radio, remember the following:
  - Use the sequence push, pause, talk, pause, release to prevent your messages from being cut off.
  - Transmit complete, accurate, and concise messaging by thinking about what you are going to say before you say it.
  - Keep messages to 15 to 20 seconds maximum.
  - Use plain language without coded messages.
  - Speak to aircraft from the pilot's perspective.
  - Use the word *zero* for the number "0"—not the slang *oh*.
  - Use the word *decimal* instead of *dot*.
- There are two common ways to communicate on a radio when talking to another user. They are generally "Hey you, it's me" and "It's me, hey you!"

- To keep your radio in good working condition, perform the following protection and maintenance techniques:
  - Protect your radio from water, dust, and retardant as best you can. After each shift, wipe your radio with a dry cloth.
  - Make sure that there is a protective cap on the accessory port when no accessory is being used.
  - If any functionality or clarity issues are noted, the radio must be turned in to a technician for repair.
- When handling a portable radio, use the body of the radio and not the antenna.

KEY TERMS

Base station radio A stationary radio permanently mounted in a building.

Channel An assigned frequency or frequencies used to carry voice and data communications.

Cloning Process that allows for the copying of frequency information from one radio and sending that copy to another radio using a set of cables.

Emergency traffic Urgent radio message that takes priority over all other radio communications.

Frequency The number of cycles per second of a radio signal.

Full duplex channels Radio systems that allow both parties to send information simultaneously.

Half duplex channels Radio systems that use one frequency to transmit and receive all messages; transmissions can occur in either direction but not simultaneously.

LUNAR Location, unit, name, assignment, and resources; acronym used to help remember the information needed when reporting a mayday, or emergency, message.

Mayday A verbal declaration indicating that a fire fighter is lost, missing, injured, or requires immediate assistance.

Mobile radio A two-way radio that is permanently mounted in a fire apparatus.

Portable radio A battery-operated, two-way, handheld transceiver.

Priority traffic Radio message that contains critical information.

Repeater A special base station radio that receives messages and signals on one frequency and then automatically transmits them on a second frequency.

Routine traffic Normal radio communication required in response to an incident.

Simplex channels Push-to-talk, release-to-listen radio systems that use one frequency to transmit and receive all messages; transmissions can occur in either direction but not simultaneously.

Trunked channels Radio systems that group users to available radio channels as needed through a computerized shared bank of frequencies; also called 700/800 MHz channels.

REFERENCES

Department of Natural Resources (DNR). "2019 Wildland Fire Portable Radio Recommendation." Accessed October 6, 2019. https://www.dnr.wa.gov/publications/rp_fire_portable_radio_recommendation2019.pdf?cekbiy.

National Wildfire Coordinating Group (NWCG). "RT-130, Wildland Fire Safety Training Annual Refresher (WFSTAR), N/A." Modified July 26, 2019. https://www.nwcg.gov/publications/training-courses/rt-130.

National Wildfire Coordinating Group (NWCG). "S-130, Fire Fighter Training, 2008." Modified August 21, 2019. https://www.nwcg.gov/publications/training-courses/s-130.

Wildland Fire Fighter in Action

You and your crew are assigned to patrol a section of line that a handcrew is prepping for a burnout operation. Currently, the fire is burning up on a ridge 200 ft (61 m) above the road. The road is narrow, cut into a hillside, and smoky and the wind is picking up. It is about 1600. This is your third shift on the fire, and the previous two days the fire was pushing back hard on your and the surrounding crews' efforts, which has made for many tired fire fighters. As your engine rounds a corner, you are stopped by a tree that has blown down. You and your crew prepare to remove the tree from the road so travel is opened up again. Just as the sawyer approaches the tree, a boulder 3 ft (0.9 m) in diameter rolls out of the canopy, strikes the fire fighter operating the chainsaw, and knocks the engine captain off the side of the road on the downhill side.

1. What is the first thing you should do?

- **A.** Call your division supervisor and notify him or her of an incident within an incident.
- **B.** Call the handcrew captain on the radio for assistance.
- **C.** Pick up the chainsaw and cut the road open so others can get to you.
- **D.** Get the first aid kit.

2. When transmitting the IWI and MIR, it will be necessary to say "break" if transmissions are longer than:

- **A.** 15 seconds.
- **B.** 20 seconds.
- **C.** 25 seconds.
- **D.** 30 seconds.

3. During your radio transmission for help, your radio starts to chirp, and the division supervisor states that you are cutting out. This might mean that:

- **A.** your radio is out of power.
- **B.** you are not pushing the PTT button hard enough.
- **C.** your scan switch is in the on position.
- **D.** someone else is talking on the radio.

4. Based on your location, it will make the most sense to fly out the injured crew. When directing the helicopter into your location you will want to direct them from:

- **A.** the pilot's point of view.
- **B.** your point of view.
- **C.** the uphill point of view.
- **D.** the injured fire fighter's point of view.

Access Navigate for flashcards to test your key term knowledge.

CHAPTER 11

Wildland Fire Fighter II

Orienteering and Global Positioning Systems

KNOWLEDGE OBJECTIVES

After studying this chapter, you will be able to:

- Describe the cardinal directions. (p. 234)
- Describe azimuth, or bearing. (p. 234)
- Describe the relationship between true north, magnetic north, and declination (pp. 234–235)
- Use a compass and a topographic map. (**NFPA 1051: 5.1.1, 5.1.2**, pp. 237–239)
- Describe how global positioning systems work. (p. 240)
- Describe latitude and longitude. (pp. 240–241)
- Identify map features:
  - Contour lines (pp. 242–244)
  - Map symbols and color coding (pp. 244–248)
  - Legal description (pp. 244–248)
  - Universal Transverse Mercator Grid (pp. 248–251)

SKILLS OBJECTIVES

After studying this chapter, you will be able to perform the following skills:

- Adjust a compass for declination. (pp. 235–237)
- Calculate personal walking pace in a 100–ft (30.5-m) interval. (pp. 240–241)
- Calculate personal walking pace in a 1-chain (20.1-m) interval. (pp. 240–241)
- Use a coordinate ruler to find your position on a topographic map. (pp. 251–257)

Additional Standards

- **NWCG PMS 475 (NFES 2865)**, *Basic Land Navigation*
- **NWCG S-130 (NFES 2731)**, *Firefighter Training*
- **NWCG S-190 (NFES 2902)**, *Introduction to Wildland Fire Behavior*

You Are the Wildland Fire Fighter

You are the senior fire fighter on a Type 3 engine module. Your crew consists of seven people and an engine with a chase truck. You and your crew were given a severity assignment that is located outside of your normal response area. On the first day, after the briefing by the local agency, your assignment is to find smoke from a lightning storm that passed through the night before. Your captain (engine boss) is in the chase truck and on an adjacent ridge looking at smoke way up on a hillside. You are new to the area, and you need to describe this location to the dispatch center.

1. Your home area is Colorado, and you just arrived in California. What would your declination adjustment be on your compass, and where would you go to find this information?
2. Because your crew boss is on an adjacent ridge, you suggest that you both shoot bearings to try to triangulate the smoke.
3. In which GPS format will you deliver the fire coordinates to dispatch?
4. If you do not have GPS, how would you find your location using a coordinate ruler?
5. If you do not have a GPS device or paper map, what programs would allow you to identify the needed data for GPS locations?

Access Navigate for more practice activities.

Introduction

This chapter discusses the use of a compass and the global positioning system (GPS) in wildland fire operations. It also covers compass reading, topographic map reading, the use of a GPS with a topographic map, and how to calculate personal walking pace using a map. It would be helpful to have a compass present while reviewing this chapter.

Compass Reading

The compass is small, portable, and fairly inexpensive. Compasses are used to orient yourself to magnetic north (discussed later) and to help you move in a straight line.

Simple compasses have a magnetized needle and are surrounded by a 0- to 360-degree azimuth, or bearing ring. A bearing, or **azimuth**, is a horizontal angle of a point measured clockwise from true north (discussed later). The measurement of these angles is based on the concept of Earth as a sphere. These angles are translated onto a compass. A **back azimuth**, or back bearing, is the opposite of an azimuth (an angle 180 degrees opposite of azimuth) and is useful when retracing your steps.

A more advanced compass has a magnetized needle surrounded by a rotating dial and an azimuth ring marked in 0- to 360-degree graduations, often every 2 degrees. The ring can be adjusted for declination, which is discussed later.

Cardinal Directions

There are eight cardinal directions on a compass:

- North
- Northeast
- East
- Southeast
- South
- Southwest
- West
- Northwest

Declination

A map is a graphic representation of Earth's surface, and all topographic maps are oriented to true north. **True north** (sometimes called true bearing) is north according to Earth's axis, or the top of the globe or map. On your compass rose (the compass figure used to display direction), you will see a star. The index line above the compass rose indicates true north **FIGURE 11-1**.

Magnetic north (where the compass needle actually points) is the magnetic north pole. Magnetic north changes over time and with location. The magnetic North Pole is currently located in the Baffin Island region of Canada.

The difference between magnetic north and true north is called **declination** (sometimes called magnetic declination or magnetic variation) **FIGURE 11-2**.

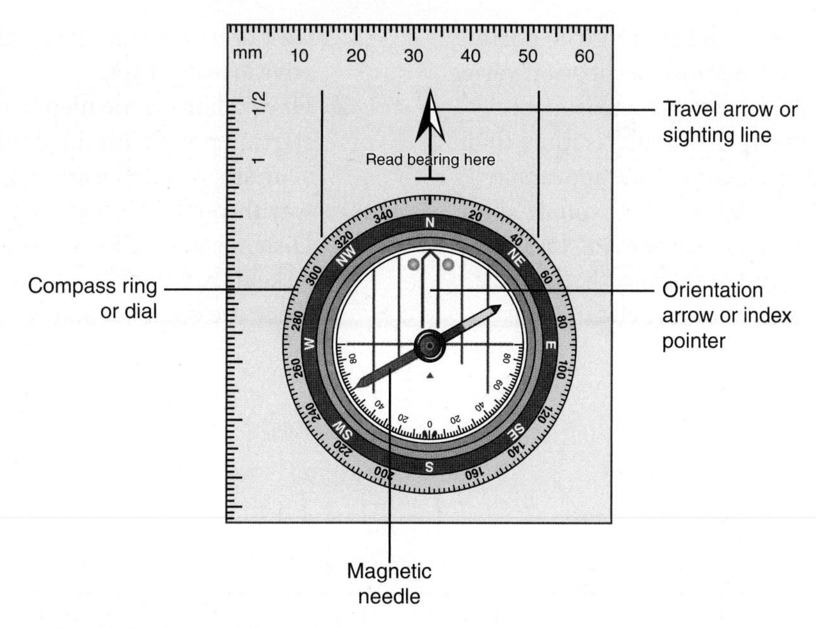

FIGURE 11-1 Parts of a compass.

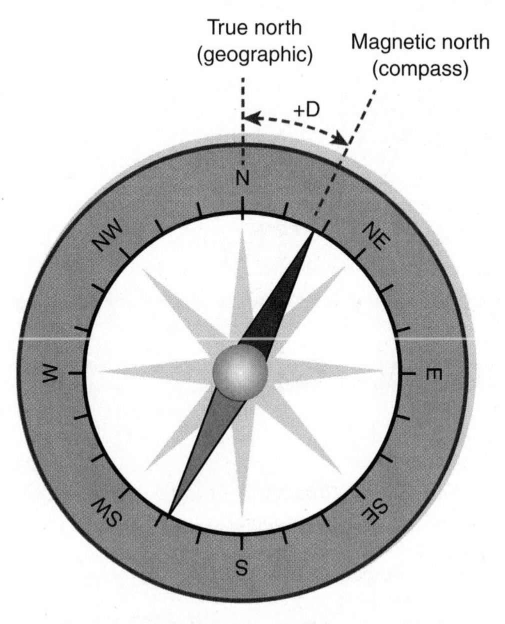

FIGURE 11-2 Declination is the difference between magnetic north and true north.

You need to take this declination into account when using your compass to find a location. This concept is very important because all maps are oriented to true north, not magnetic north. Depending on where you are in the country, you will need to adjust for the declination. If you have a topographic map, this variation can be found in the map legend on the map (map legend is discussed later in this chapter). Because the compass points with local magnetic fields, declination value will be needed to obtain true north.

There are many types of maps: highway maps, aeronautical charts, hazard maps, local geographic information system (GIS) maps, and military maps, to name a few. The most popular maps used by fire agencies are the **United States Geological Survey (USGS) topographic maps.** These maps show many features that are useful for the fire fighter, including roads, buildings, lakes, boundaries, and vegetation types. This chapter references use on USGS maps.

Directions for how to use your compass on a USGS map while accounting for declination will be discussed next.

LISTEN UP!

Declination can change over time, so check the revision date on your map before trusting your map's legend, or reference the National Oceanic and Atmospheric Administration website for the most current data.

Adjusting a Compass for Declination: Method 1

These directions assume that your orienting arrow lines up with the north indicator on your compass dial, meaning the compass has *not* been adjusted for declination.

1. Obtain the local magnetic declination for the area represented on your map. At the bottom of every USGS map is a diagram that displays the difference and direction between true north (represented as a star), grid north (abbreviated as GN), and magnetic north (abbreviated as MN). Magnetic declination is the number of

degrees and the direction between true north and magnetic north. Because declination varies over time, it is advisable to get a reasonably current figure. If your USGS map is more than 15 years old (the declination date appears in the diagram), for the information you need for your specific area, reference **FIGURE 11-3**. If magnetic north is east of true north, the local declination is positive. If magnetic north is west of true north, the local declination is negative **FIGURE 11-4**.

2. Draw a line on the map that connects your starting point with the destination. This is your azimuth, or map bearing. Extend the line all the way through the map border (the neat line).
3. Distance yourself from any nearby metal, such as keys, belt buckles, desks, cars, and fences. Place the compass on the map so the needle's

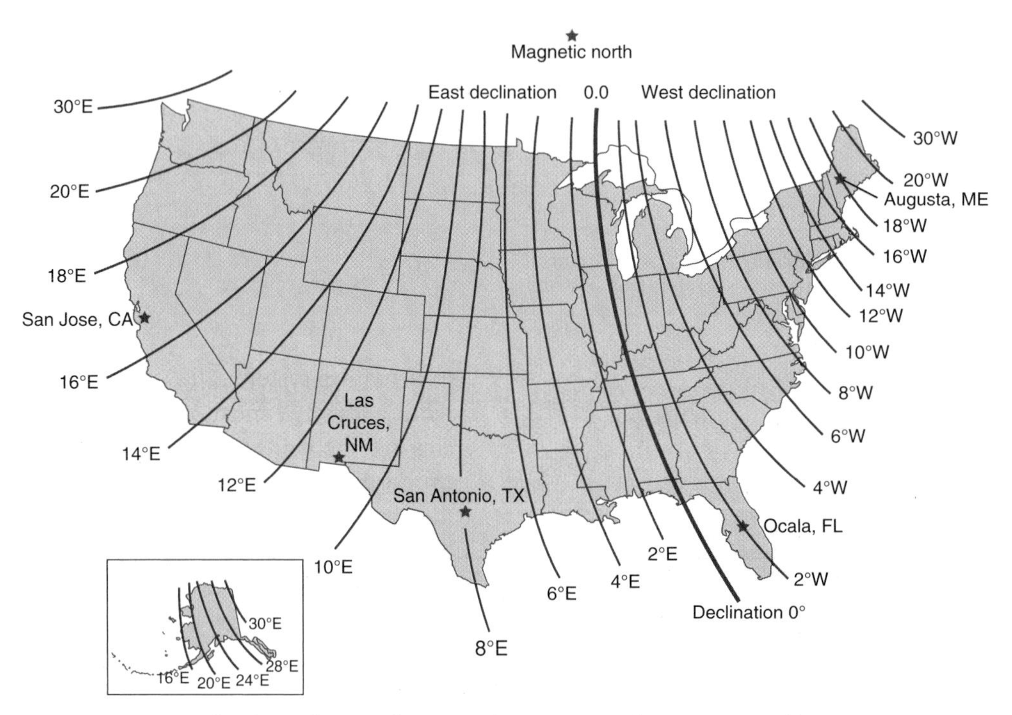

FIGURE 11-3 Declination reference for maps that are more than 15 years old.

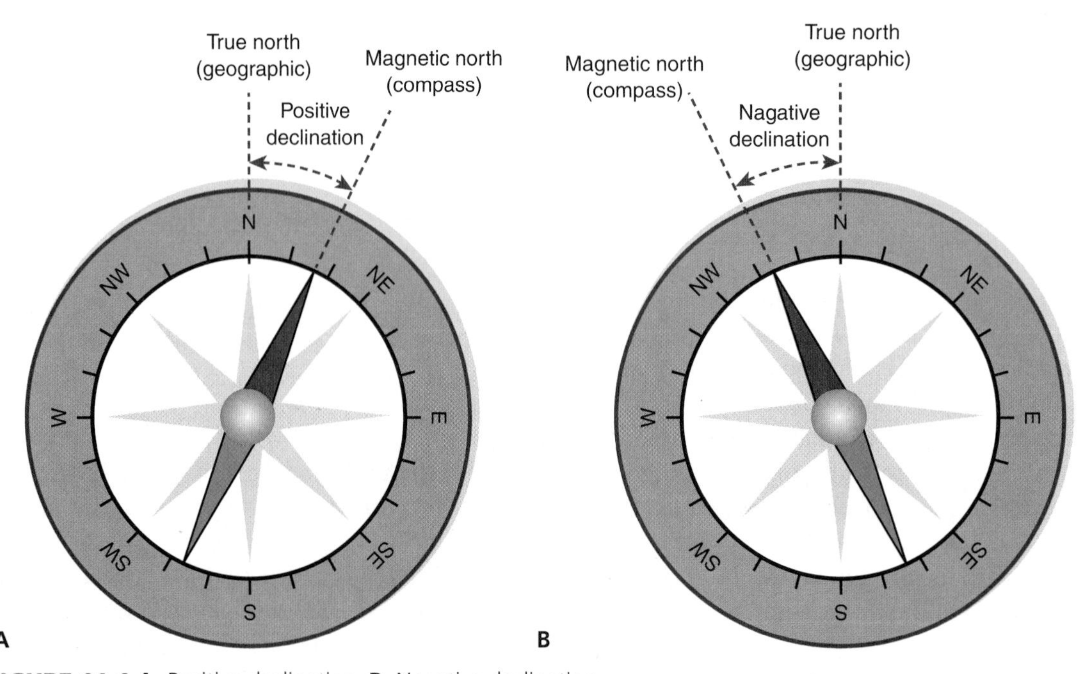

FIGURE 11-4 A. Positive declination. **B.** Negative declination.

pivot point is directly over the intersection of your map bearing and neat line.

4. Rotate the dial until the compass ring's north aligns with map north. Read your map bearing from the compass dial. Make sure the bearing agrees with your direction of travel; for example, if you intend to travel due east, the bearing is 90 degrees, not 270 degrees.
5. Do this step mentally; don't turn the compass dial. If the local declination is positive, then subtract the declination amount from the bearing you just derived. If the local declination is negative, then add the declination amount to the bearing you just derived.
6. Turn the compass dial until the number you calculated in step 5 lines up with the index line.
7. Lift the compass off the map, and, with the direction of the travel arrow pointing directly away from you, rotate your body and the compass all in one motion until the red magnetic needle overlays the orienting arrow.
8. Sight a landmark along this bearing, and proceed to it. Repeat this step until you reach your destination.

Adjusting a Compass for Declination: Method 2

These instructions describe how to navigate from a known location on a topographic map to another known location on the same map. These directions assume that your orienting arrow lines up with the north indicator on your compass dial, meaning the compass has *not* been adjusted for declination.

1. Obtain the local magnetic declination for the area represented on your map. At the bottom of every USGS map is a diagram that displays the difference and direction between true north (represented as a star), grid north (GN), and magnetic north (MN). Magnetic declination is the number of degrees and direction between true north and magnetic north. Because declination varies over time, it is advisable to get a reasonably current figure. If your USGS map is more than 15 years old (the declination date appears in the diagram), reference Figure 9-3. If magnetic north is east of true north, the local declination is positive. If magnetic north is west of true north, the local declination is negative.
2. Draw a line on the map that connects your starting point with the destination. This is your azimuth, or map bearing.
3. Distance yourself from any nearby metal, such as keys, belt buckles, desks, cars, and fences.
4. Place the compass on the map so the baseplate is parallel to the line you drew. Make sure the direction of the travel arrow points to your destination.
5. Rotate the dial until compass-ring north agrees with map north. Do not move the compass as you rotate the dial.
6. Remove the compass from the map, and, with the direction of the travel arrow pointing directly away from you, rotate your body and the compass all in one motion until the red magnetic needle overlays the orienting arrow.
7. If local declination is positive, then subtract the declination amount (turn the dial clockwise). If local declination is negative, then add the declination amount (turn the dial counterclockwise).
8. Again, with the direction of the travel arrow pointing directly away from you, rotate your body and compass all in one motion until the red magnetic needle overlays the orienting arrow. Sight a landmark along this direction of travel, and proceed to it. Repeat this step until you reach your destination.

Using a Compass with Adjustable Declination

Here are the directions for using a compass with adjustable declination:

A compass with adjustable declination allows you to rotate the orienting arrow independently of the compass dial. If you have such a compass, you can calculate your map bearing without adding or subtracting the amount of local magnetic declination. To calibrate your compass in this fashion, rotate the inner liquid capsule (or turn the screw with the key) until the orienting arrow deviates from the compass ring's north indicator by the amount and direction of the local magnetic declination. For example, if local declination is 10 degrees east of true north, rotate the inner liquid capsule (or turn the screw with the key) until the orienting arrow points to 10 degrees east. If using method 1, you do not need to add or subtract as indicated in step 5; just make sure that compass-ring north (not the orienting arrow) agrees with map north as directed in step 4. If using method 2, skip steps 7 and 8; again, be sure compass-ring north (not the orienting arrow) agrees with map north as directed in step 5.

Getting Your Map and Compass to Face the Same Direction

Place the compass on the map so the baseplate is parallel to the north-to-south map neat line (the map border).

Rotate the dial until compass-ring north agrees with map north. Add or subtract the amount needed to adjust for local magnetic declination (subtract if local declination is positive and add if local declination is negative); if your adjustable declination compass is already calibrated for local declination, you don't need to add or subtract; just make sure compass-ring north (not the orienting arrow) agrees with map north. Holding the map and compass steady (the baseplate should still be on the north-to-south map neat line), rotate both the map and compass in one motion until the red magnetic needle overlays the orienting arrow. Again, make sure there is no interference from metal when you perform this (such as rebar in concrete). Your map and compass are now oriented to true north. Compare the physical features around you with your map to help derive your location on the map.

Determining Percentage and Angle of Slope

Percentage of slope is determined by dividing the amount of elevation change by the amount of horizontal distance covered (sometimes referred to as "the rise divided by the run") and then multiplying the result by 100. The "run" assumes you're traveling on an idealized flat surface; it does not account for the actual distance traveled once elevation change is factored in.

Example 1. Let's assume your climb gains 1000 ft (304.8 m) in altitude (the rise) and the horizontal distance as measured on the map is 2000 ft (609.6 m) (the run):

1000 divided by 2000 equals 0.5.

Multiply 0.5 by 100 to derive the percentage of the slope: 50 percent.

Example 2. If we assume your climb gains 500 ft (152.4 m) in altitude (the rise) and the horizontal distance as measured on the map is 3000 ft (914.4 m) (the run):

500 divided by 3000 equals 0.166.

Multiply 0.166 by 100 to derive the percentage of the slope: 16.6 percent.

Example 3. If your climb gains 700 ft (213.4 m) in altitude (the rise) and the horizontal distance as measured on the map is 500 ft (152.4) (the run):

700 divided by 500 equals 1.4.

Multiply 1.4 by 100 to derive the percentage of the slope: 140 percent.

The angle of slope represents the angle that's formed between the run (remember, it's an idealized flat surface that ignores elevation change) and your climb's angular deviation from that idealized flat surface. To calculate this, you divide the rise by the run, and then obtain the inverse tangent of the result.

Example 4. Let's assume your climb gains 1000 ft (304.8 m) in altitude (the rise) and the horizontal distance as measured on the map is 2000 ft (609.6 m) (the run):

1000 divided by 2000 equals 0.5.

Press the INV button on your calculator (sometimes called second function).

Press the TAN button on your calculator.

Your angle of slope is 26.5 degrees.

Example 5. If your climb gains 1000 ft (304.8 m) in altitude (the rise) and the horizontal distance as measured on the map is 1000 ft (304.8 m) (the run):

1000 divided by 1000 equals 1.

Press the INV button on your calculator (sometimes called second function).

Press the TAN button on your calculator.

Your angle of slope is 45 degrees.

What Direction Am I Facing?

Let's say you know the aspect but not the number of degrees and you want to know what direction you are facing.

Using a compass without adjustable declination, make sure the direction of the travel arrow is pointing directly away from you. Now, rotate the compass dial until the red magnetic needle overlays the orienting arrow. Observe the reading at the index line. If local magnetic declination is positive, then add the necessary amount. If local declination is negative, then subtract the necessary amount. The number at the index line after adding or subtracting is the true direction you are facing.

If using a compass with adjustable declination, make sure the direction of the travel arrow is pointing directly away from you. If you haven't done so already, adjust the declination so that the orienting arrow deviates from the compass ring's north indicator by the amount and direction of the local magnetic declination. Turn the compass dial until the red magnetic needle overlays the orienting arrow. The number at the index line is the true direction you are facing.

How Do I Point Myself toward a Specific Direction?

In this case, you know the number of degrees but don't know in what aspect to face (direction to face). Let's assume you wish to face true north (0 degrees).

On a compass without adjustable declination, turn the compass dial until 0 is at the index line. If local declination is positive, subtract this amount from 0 on the compass ring (turn the dial clockwise). If local declination is negative, add this amount to

0 on the compass ring (turn the dial counterclockwise). With the direction of the travel arrow pointing directly away from you, rotate your body and compass together in one motion until the red magnetic needle overlays the orienting arrow. You are now facing true north.

On a compass with adjustable declination, turn the compass dial to 0. If you haven't done so already, adjust the declination so the orienting arrow deviates from the compass ring's north indicator by the amount and direction of local magnetic declination. With the direction of the travel arrow pointing directly away from you, rotate your body and the compass together in one motion until the red magnetic needle overlays the orienting arrow. You are now facing true north.

Why Am I Adding for One Situation But Subtracting for Another?

Let's assume the local magnetic declination is 10 degrees east of true north (a positive declination). Therefore, the needle always points to 10 degrees. Let's also assume that, although you don't know which direction you're facing, it coincidentally happens to be true north. If you rotate the dial until the red magnetic needle overlays the orienting arrow (with an implied question of which direction you are facing), the reading at the index line will be 350 degrees; you must add 10 to the 350 degrees to determine the true direction you are facing.

Conversely, if you dial 0 on the compass (to determine which direction you should face to be pointed toward true north) and then rotate your body and compass together in one motion until the red magnetic needle overlays the orienting arrow, the direction you'll face is 10 degrees, even though 0 is dialed at the index line. You must subtract 10 degrees by turning the dial clockwise to a reading of 350 degrees, and then rotate your body and compass again until the red magnetic needle overlays the orienting arrow.

If you're in an area where magnetic declination is positive, here's an easy way to remember whether to add or subtract:

What direction **am** I facing? A = add

What direction **should** I face? S = subtract

Determining Personal Walking Pace

A **pace** is two natural walking steps measured from the heel of one foot to the heel of the same foot after two steps **FIGURE 11-5**. Pace can be used to keep track of distance. Pace count will differ among people. Your personal pace will differ based on topography and weather (hills and strong winds) and the gear you are carrying.

You should determine your personal walking pace in both feet and chains. To determine your personal walking pace in feet, follow these steps:

1. Measure out a course to walk in a straight line for 100 ft (30.5 m).
2. Count each double step as you walk the distance in a straight line and with a natural pace.
3. Calculate your pace by dividing the course length by the number of double steps it took you. For example, a 100-ft (30.5-m) course walked in 20 paces (100 [30.5] divided by 20) would be a 5-ft (1.5-m) walking pace.

A **chain** is a unit of measure in land survey equaling 66 ft (20.1 m); 80 chains is equal to 1 mile (1.6 km). To

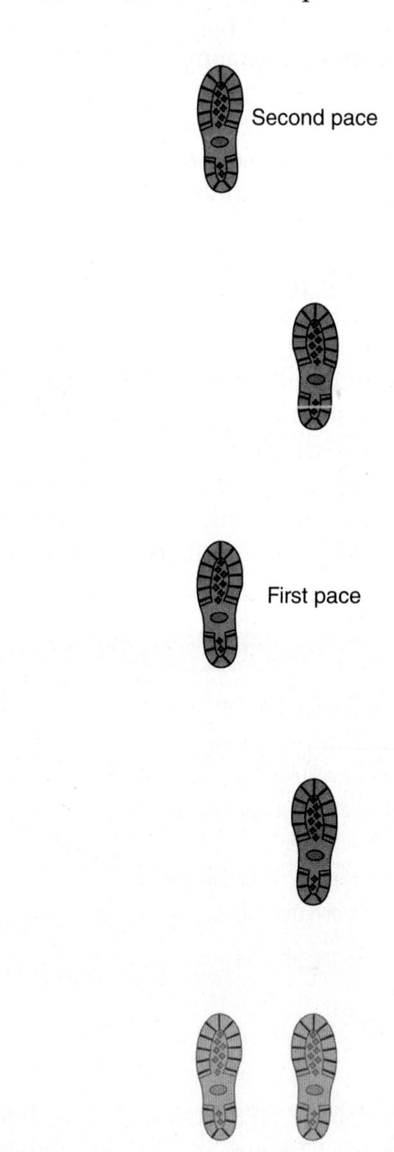

FIGURE 11-5 Walking pace.

determine your personal pace count in chains, follow these steps:

1. Measure out a course to walk in a straight line for 1 chain (66 ft [20.1 m]).
2. Count each double step as you walk the distance in a straight line and with a natural pace.
3. Calculate your pace by dividing the course length by the number of double steps it took you. For example, a 1-chain (66-ft [20.1-m]) course walked in 20 paces (66 [20.1] divided by 20) would be a 3.3-chain walking pace.

It is a good idea to walk the course several times to determine your personal walking pace average. You can determine your personal walking pace uphill and downhill by following the steps listed here and measuring out a walking course uphill or downhill. It is important to vary the slope to obtain an average.

It is important to walk in a straight line. Constantly check your location and topography because it is very easy to get off track. Additionally, in the real world, you may come across an obstacle that will require you to stop, determine your location, move around the obstacle, and determine your location again.

Global Positioning Systems

NAVSTAR GPS is an acronym that stands for Navigation Systems with Time and Ranging Global Positioning System. The **global positioning system (GPS)** is a system of navigational satellites operated by the U.S. Department of Defense and available for civilian use. It can track objects anywhere in the world with an accuracy of approximately 40 to 300 ft (12.2 to 91.4 m).

DID YOU KNOW?

Stanford University Professor Bradford Parkinson, now retired from the U.S. Air Force, is considered to be the father of the GPS, which was declared fully operational in January 1994. Initially, the military commanders scoffed at such a system, and it was viewed as a "Star Wars" type program. It wasn't until after one of the battles in the Gulf War that the military commanders realized how important specific locations were and that this was indeed the future of navigation.

Function

The space segment of the GPS involves 24 satellites that orbit Earth every 12 hours at a height of about 11,000 miles (17,702.8 km) above Earth's surface. At times, there may be more than 24 operational satellites because new ones are launched when older satellites need replacing. These satellites transmit microwave carrier signals back to Earth, which enable the user on the ground, with a receiver, to plot his or her location and obtain the current time, date, and elevation. It takes three satellites over a receiver to obtain a horizontal location on Earth's surface. Adding a fourth satellite enables you to determine your altitude as well.

The heart of the GPS master control system is located at Schriever Air Force Base in Colorado, where the satellites are constantly monitored. Corrections are made from the control facility to the satellite clock, and orbital data are also uploaded.

The user segment of the system is made up of GPS receivers that are hand carried or installed in aircraft, aboard ships, or on ground vehicles. Users can expect an accuracy of 0.06 mile (100 m) horizontally, 0.1 mile (156 m) vertically, and 340 nanoseconds of time.

LISTEN UP!

GPS can quickly provide you with six important pieces of information: latitude, longitude, altitude, speed, direction, and distance to destination.

Topographic Maps

Topographic maps, also called quadrangle, or quad, maps, are a representation of Earth's surface. To be able to locate ourselves on Earth's surface, there is a geographic coordinate system with positional values that was developed by projecting reference lines onto maps that represent Earth's surface. These lines are called parallels of latitude and meridians of longitude. Here, we will discuss two of these systems: latitude and longitude and the Universal Transverse Mercator (UTM) grid system.

Latitude and Longitude

A network of reference lines run around Earth. The horizontal lines that run east and west around Earth are called parallels; the vertical lines that run north and south are called meridians. These lines converge at the North and South Poles. With these reference lines in place, we have a coordinate system that can be used to locate any position on Earth's surface **FIGURE 11-6**.

Latitude is a point north or south of the equator, the equator being 0 degrees latitude. To find a location's latitude, you simply count the values of the lines either north or south of the equator. On a topographic map, the degrees, minutes, and seconds of latitude can be found in the right and left margins **FIGURE 11-7**. These horizontal values represent latitude.

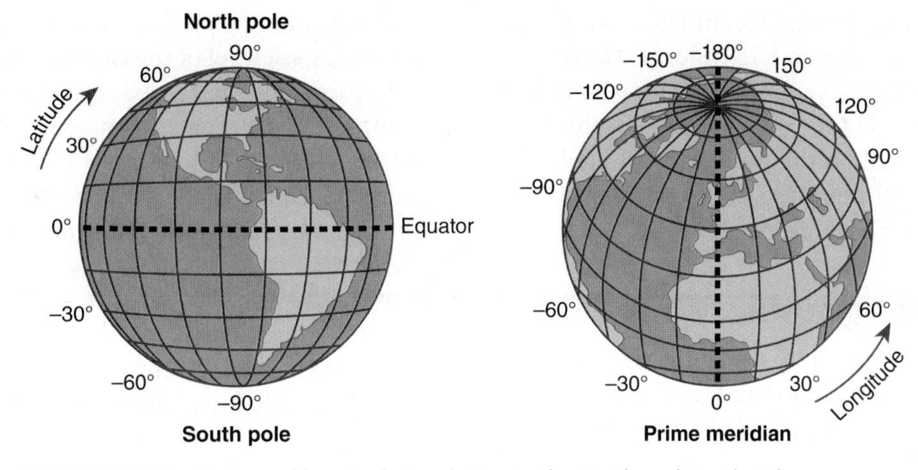

FIGURE 11-6 Latitude and longitude in relation to the North and South Poles.

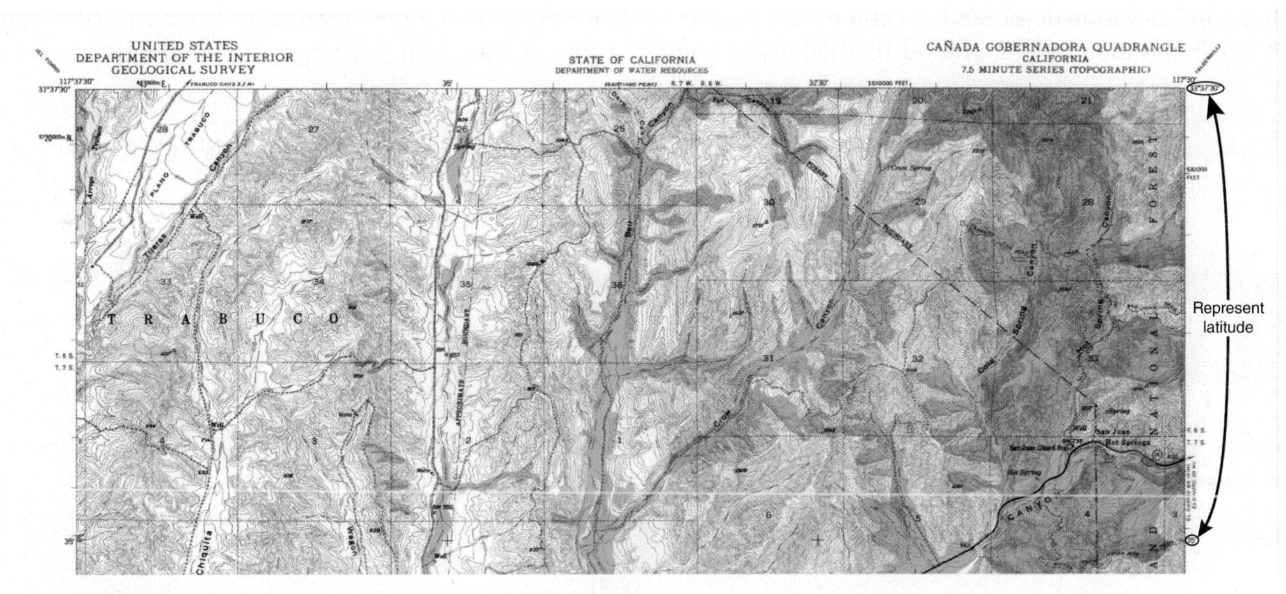

FIGURE 11-7 The horizontal values on a map represent latitude.

Courtesy of the U.S. Geological Survey.

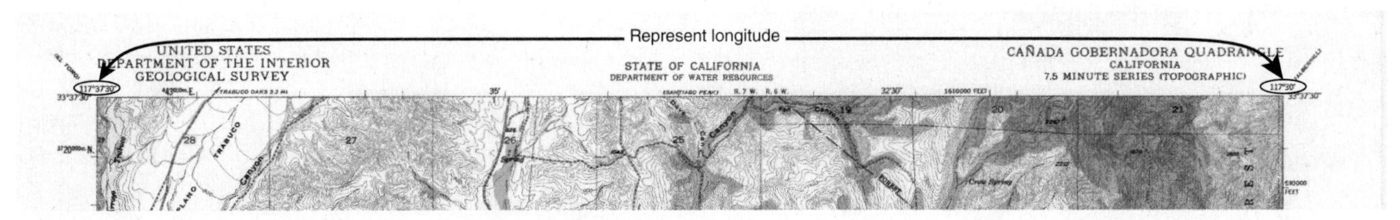

FIGURE 11-8 The vertical values on a map represent longitude.

Courtesy of the U.S. Geological Survey.

Longitude is a point on Earth's surface that is east or west of what we call the prime meridian. The prime meridian, defined as 0 degrees longitude, runs through Greenwich, England. To find a longitudinal point on Earth's surface, simply count the values of the lines either east or west of the prime meridian. On a topographic map, these values can be found in the top and bottom margins **FIGURE 11-8**. With this system of reference lines in place, you can now see how easy it is to locate a position on the map.

Before learning how to locate your position on a topographic map using longitude and latitude, two things must be defined: the smaller unit measurements used with latitude and longitude and what a topographic map is, along with a description of its features.

Each degree of latitude and longitude is broken down into 60 minutes, and each minute is broken down into 60 seconds. It must be emphasized that these measurements have nothing to do with time—they are angular measurements. One minute of latitude is equal to 1.15 miles (1.9 km). One minute of longitude at the equator is also equal to 1.15 miles (1.9 km). One second is equal to 33.82 yards (30.9 m) for both latitude and longitude. These figures change as the lines of longitude converge on the poles. A GPS handheld receiver is capable of reporting degrees, minutes, and seconds if it is set on that feature.

Types of Topographic Maps

Topographic maps show both the horizontal and vertical positions of Earth's features. There are two types of topographic maps: contour relief maps and shaded relief maps. The one that is used more commonly on wildland fires is the contour relief map. A contour relief map is a common method of representing the shape and elevation of the land. A contour line is a line of equal elevation on the ground that delineates the same elevation above sea level. It is referenced from the average sea level of the closest ocean.

A common map used on wildland fires is a 7.5-minute quadrangle map with a 1:24,000 scale **FIGURE 11-9**. The 7.5 refers to minutes in latitude or longitude. The 1:24,000 ratio is the map's representative fraction. One inch on the map equals 24,000 in. (60,960 cm), or 2000 ft (609.6 m), on the ground. The representative fraction on a map is typically found in the bottom margin area and is part of the map's legend. The **legend** is a key accompanying a map that shows information needed to interpret that map **FIGURE 11-10**.

Contour Lines

A topographic map is a representation of Earth's surface with mountains, manmade features, water features, and so forth. It is represented on a flat piece of

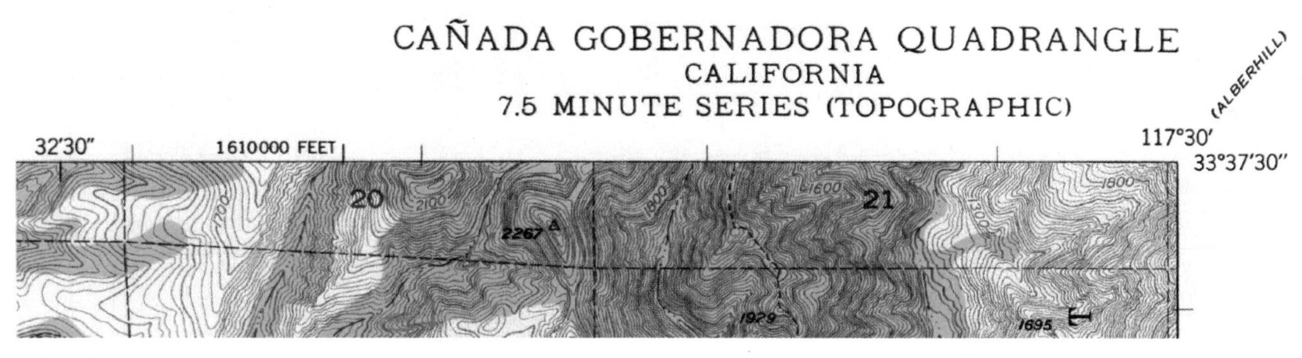

FIGURE 11-9 A 7.5-minute map.
Courtesy of the United States Geological Survey.

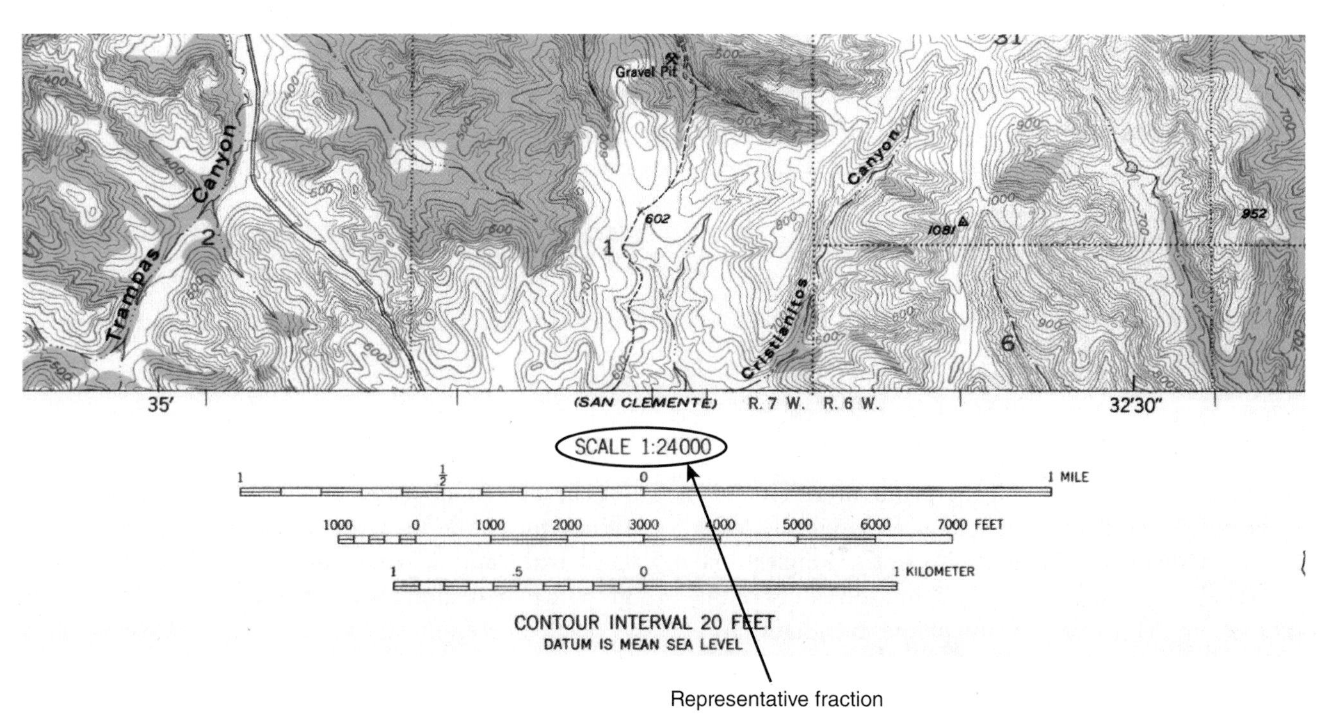

FIGURE 11-10 Map legend.
Courtesy of the U.S. Geological Survey.

paper; thus, there needs to be a way to show these terrain features. This is done by the use of contour lines **FIGURE 11-11**. Again, a contour line is a line of equal elevation on the ground that delineates the same elevation above sea level. Any point on that contour line is the same elevation. To find the distances between contour lines, you can look in the map's legend for the contour interval **FIGURE 11-12**.

There are several types of contour lines. To make it easier to read a topographic map, every fourth or fifth line is printed darker. These lines are called **index contours.** At some point along an index contour, you can find the elevation **FIGURE 11-13**. The lighter contour lines between the index contours are called **intermediate contours FIGURE 11-14**.

If the contour interval is unknown, then find two index contours adjacent to each other. Read the elevations and subtract one from the other. Then, divide the difference by 4 if the number of spaces between index contours is four; divide by 5 if the number of spaces is five. This will give the unknown contour interval. You may be required to figure the contour interval on the map displayed in an incident action plan.

A depression contour line represents areas either manmade or natural that are lower than the immediate surrounding terrain **FIGURE 11-15**. A good example would be a natural depression **FIGURE 11-16**. Depressions also contain no outlets for water drainage. Note the profile in **FIGURE 11-17**.

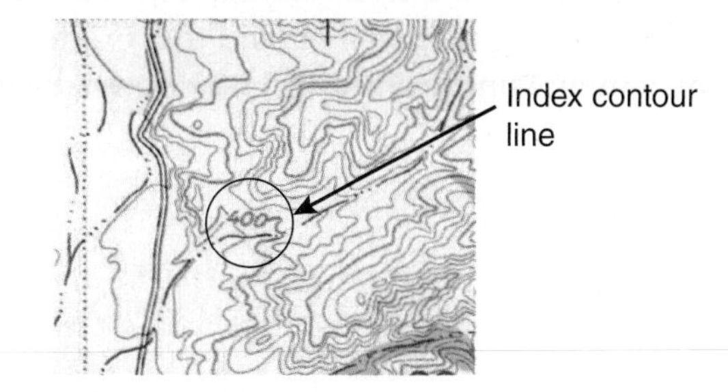

FIGURE 11-13 Index contour lines.
Courtesy of the U.S. Geological Survey.

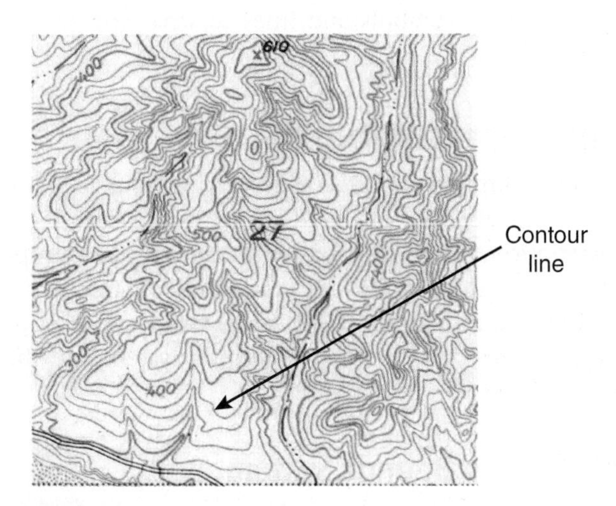

FIGURE 11-11 Contour lines.
Courtesy of the U.S. Geological Survey.

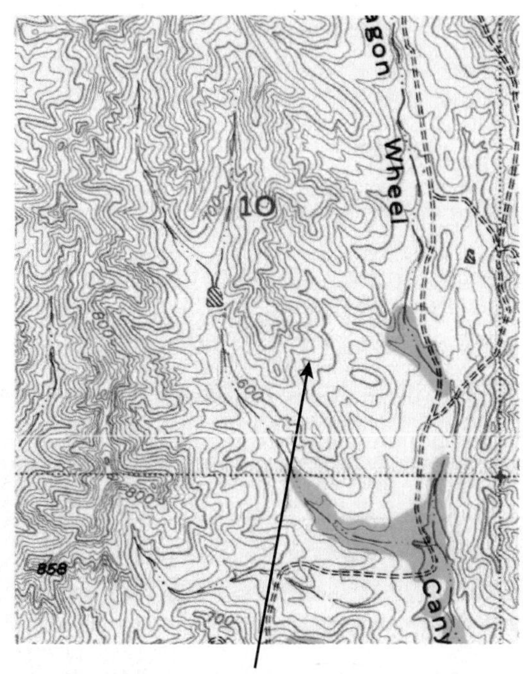

FIGURE 11-14 Intermediate contour lines.
Courtesy of the U.S. Geological Survey.

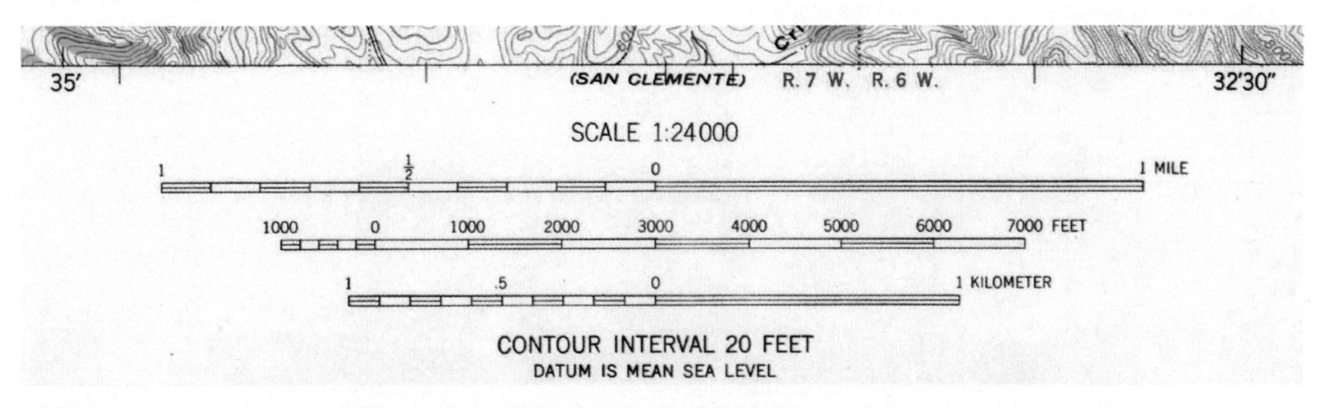

FIGURE 11-12 The contour interval is found in the lower margin of the map.
Courtesy of the U.S. Geological Survey.

The closer the contour lines are together, the steeper the slope will be **FIGURE 11-18**. Other terrain features examined are a hill, a ridge, a saddle, a valley, a spur, a draw, and a depression.

A hill is a land mass that usually tends to be round and falls off in all directions. A hilltop's crest is usually lower than the surrounding mountains **FIGURE 11-19**.

A ridge is a long elevation of land that is often located on a mountainside **FIGURE 11-20**. It usually has three downward slopes and one slope that projects upward.

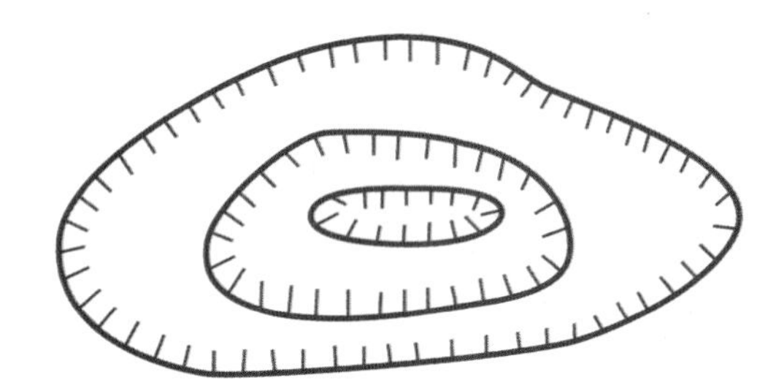

FIGURE 11-15 Depression contour lines.

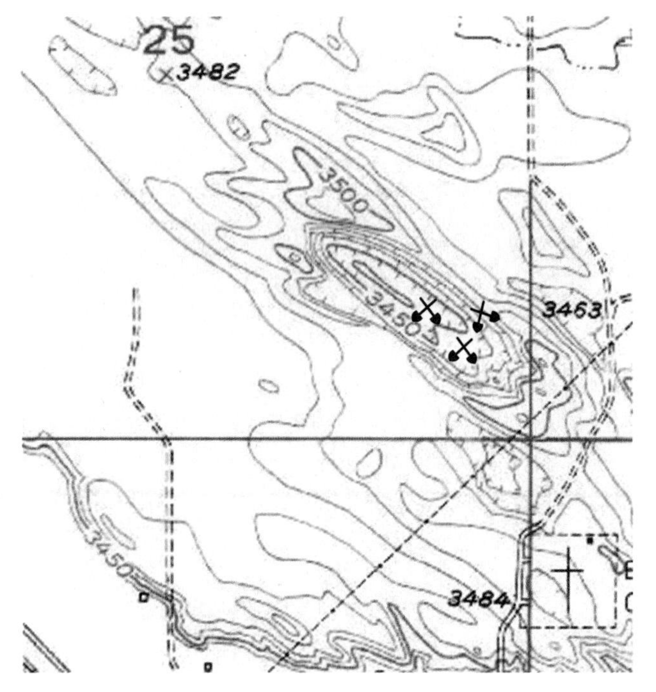

FIGURE 11-16 Gravel pits are represented on maps with a symbol showing two crossed shovels with the spades pointing downward.
Courtesy of the U.S. Geological Survey.

A saddle is a low topographic point between two hills or summits **FIGURE 11-21**. A saddle can be very deep or shallow; it can also be wide or narrow. Saddles have a channeling effect on a wildland fire (see Chapter 4, *Basic Wildland Fire Behavior*). Expect greater rates of spread through a saddle.

A valley is an area of lowland between two hills or mountains **FIGURE 11-22**. Sometimes, the valley floor has a stream or river in it. The sides of the valley floor slope upward, toward the adjacent hills or mountains.

Spurs are small ridges off the main mountain ridge **FIGURE 11-23**.

Map Symbols and Color Coding

Topographic map legends also contain details that describe the map symbols **FIGURE 11-24**. Topographic maps are printed in colors, which help in interpreting the map symbols more easily. Different colors mean different things. A list of colors commonly found on a topographic map are outlined in **TABLE 11-1**. Because not all maps are the same, look for the legend, and compare symbols and lines to what you see on the map.

Legal Descriptions

A very common format for describing your location or when receiving an incident location is the **legal description**, or the explanation of the land and property. Legal descriptions are markings on maps that divide land into sections. There are two legal description systems: the metes and bounds system and the Public Land Survey System (PLSS).

Metes and Bounds System

The metes and bounds system is a method of tracking boundaries with landmarks, lines, and words. The metes and bounds system usually includes lengthy descriptions in surveyor terminology. Firefighting agencies rarely use this system.

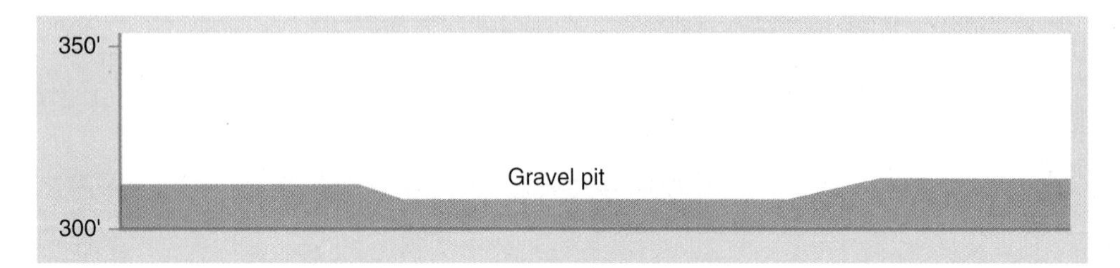

FIGURE 11-17 Gravel pit profile.

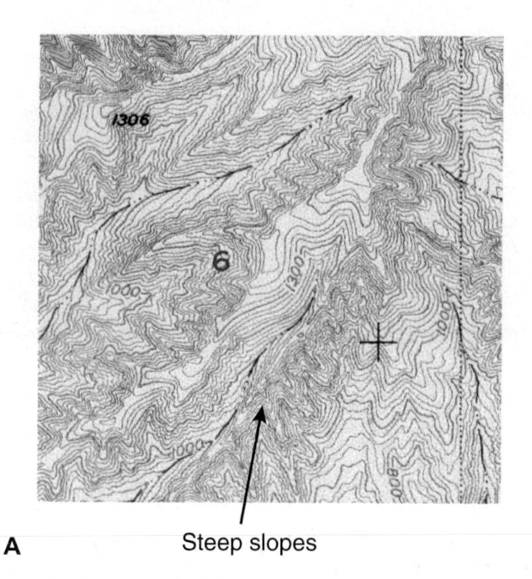

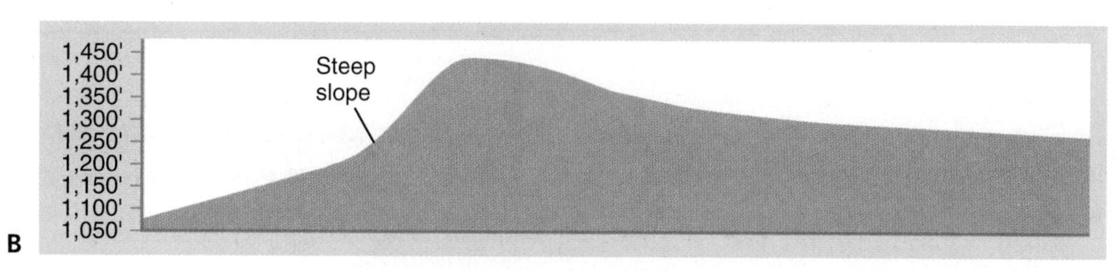

FIGURE 11-18 Steep slopes. **A.** Map view. **B.** Profile view.

A: © U.S. Geological Survey.

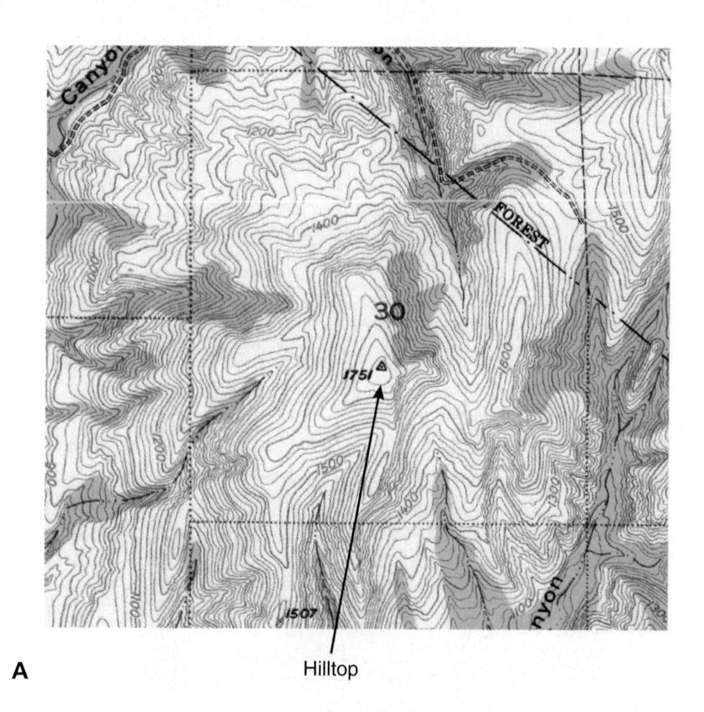

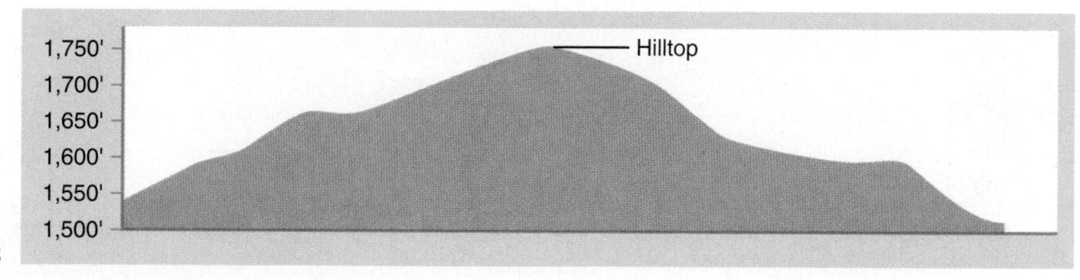

FIGURE 11-19 Hilltop. **A.** Map view. **B.** Profile view.

A: Courtesy of the U.S. Geological Survey.

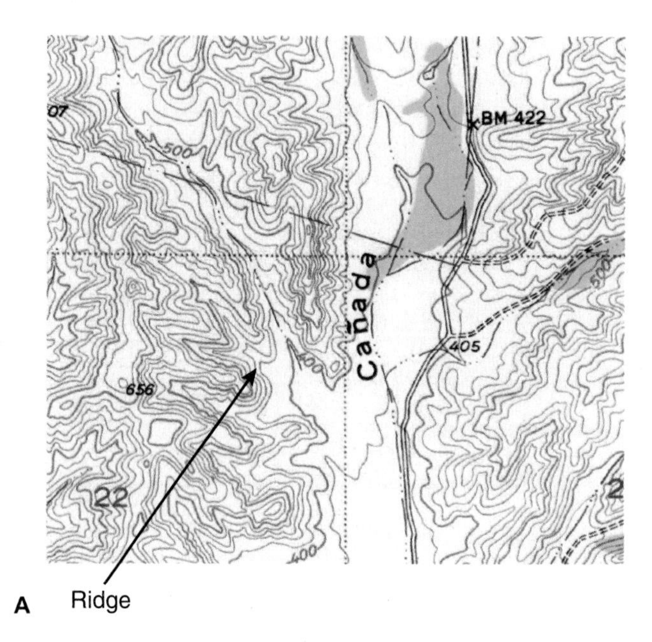

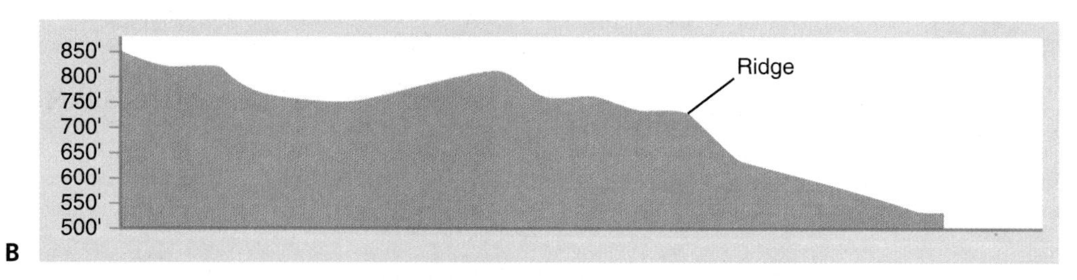

FIGURE 11-20 Ridge. **A.** Map view. **B.** Profile view.
A: © U.S. Geological Survey.

Saddle
1101
Water
Water
A

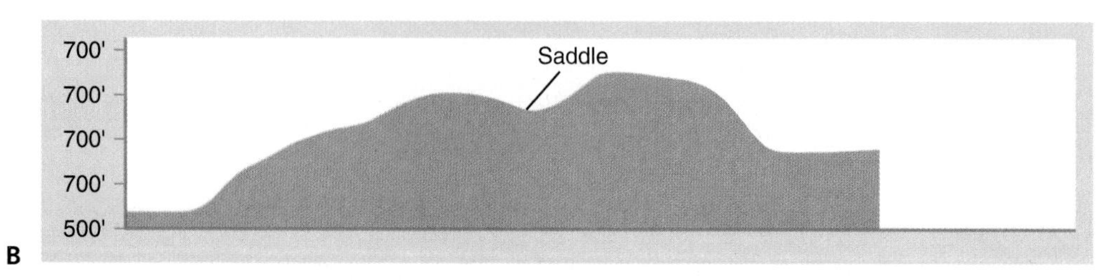

FIGURE 11-21 Saddle. **A.** Map view. **B.** Profile view.
A: © U.S. Geological Society.

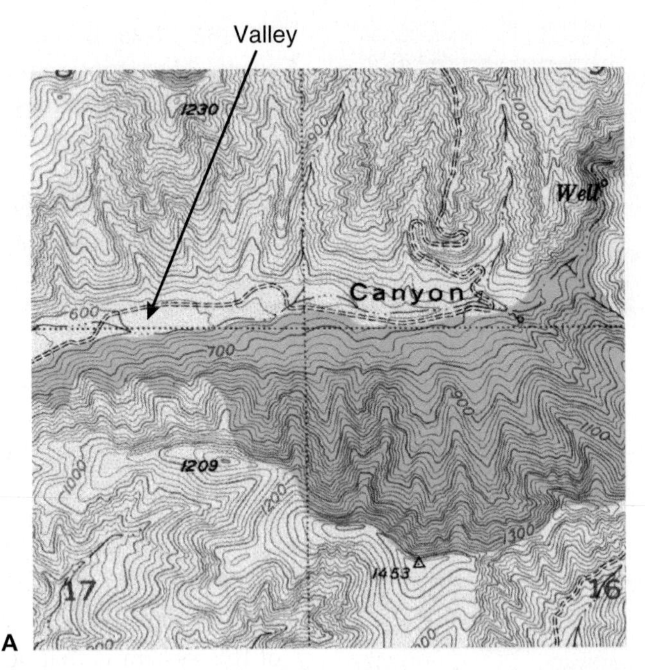

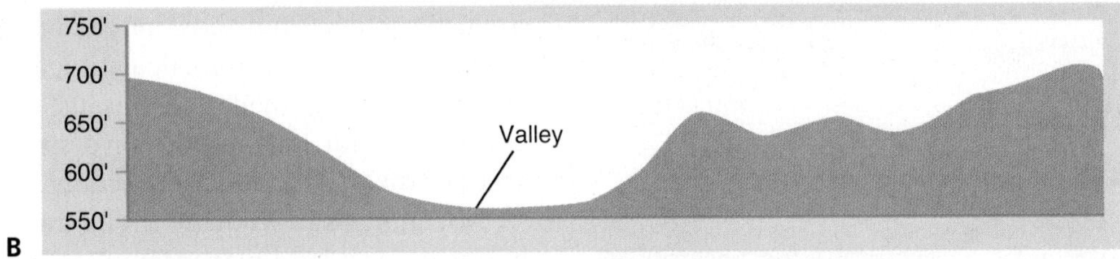

FIGURE 11-22 Valley. **A.** Map view. **B.** Profile view.

A: Courtesy of the U.S. Geological Society.

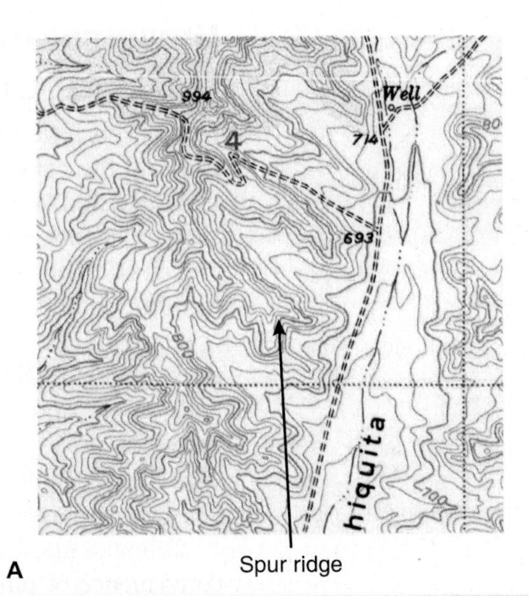

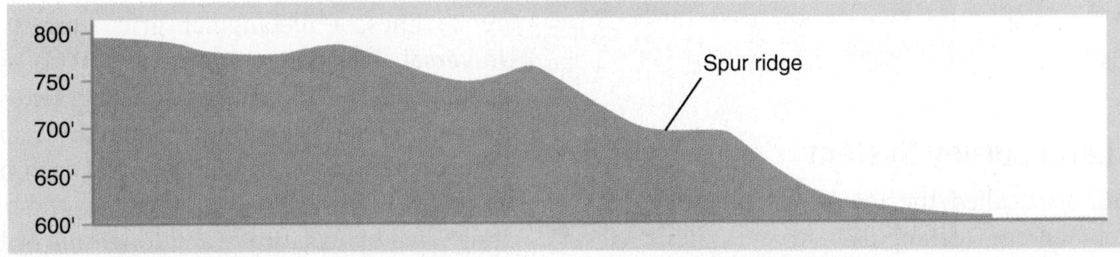

FIGURE 11-23 Spur ridge. **A.** Map view. **B.** Profile view.

A: Courtesy of the U.S. Geological Society.

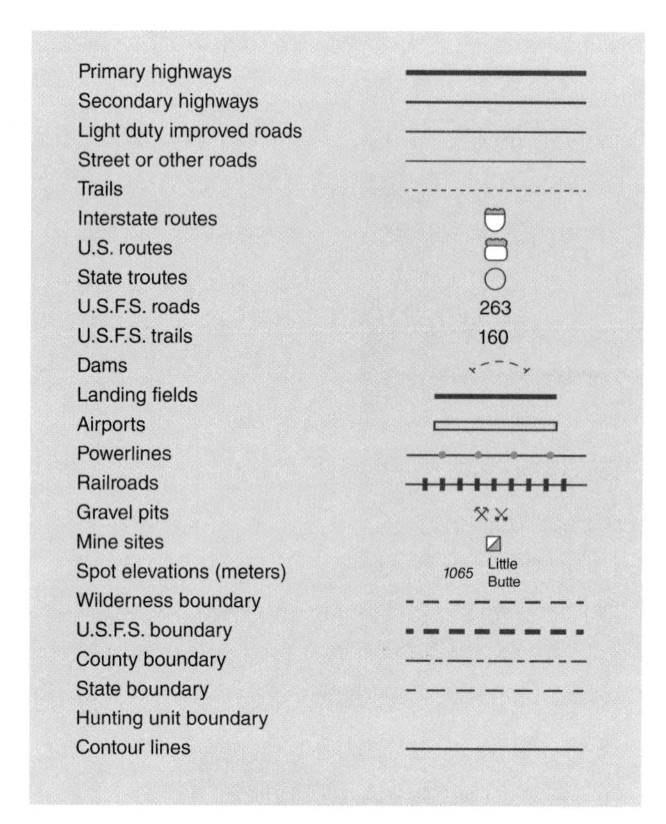

FIGURE 11-24 Map legend common details.

TABLE 11-1 Common Map Colors

Color	Description
Black	Most manmade or cultural features, such as roads and structures
Blue	Water features, such as lakes, rivers, and swamp areas
White	Little or no vegetation
Green	Vegetation features, such as woods, vineyards, and orchards
Brown	All relief features, contours, cuts, and fills
Red	Main roads, built-up areas, boundaries, and special features
Other	Special purposes: Their key is found in the legend of the map.

Public Land Survey System

The PLSS, also called the township and range survey system or the rectangular survey system, is a method of dividing and tracking public land locations and boundaries. The PLSS uses a grid system **FIGURE 11-25**. It helps to provide a representation of space. Most topographic maps up to the year 2006 include PLSS gridlines. Topographic maps published after 2007 usually do not include any PLSS data.

Though the PLSS does not include some small areas or the entire east coast, the system is still used by many, including surveyors, land management agencies, and laypersons. In wildland firefighting, it is used to estimate the size of a fire.

One of the reasons this system is used is that it is an easy way to reference distances by using simple math. One section is equal to 1 mi^2 (2.6 km^2). In that square mile, the area can be divided into 640 acres (259 hectares). If you were to divide that 1-mile (1.6-km) area into chains, you would have 80 chains. If 1 chain is 66 ft (20.1 m), 80 chains times 66 ft (20.1 m) equals 5280 ft (1609.3 m). This is significantly beneficial when trying to determine distance or size of a fire. Another benefit of this system is that you can use these simple measurements to determine how far you may need to go.

Because the math for calculating acres and chains is fluid, detailing smaller areas is just as simple to calculate. It is not uncommon for a size-up to include sections divided into quarter sections.

Not all maps in wildland firefighting use the PLSS; however, most land management agencies use this as their preferred method of location for initial dispatch.

Universal Transverse Mercator Grid

Mapmakers have always had to introduce distortion errors into their maps because of the mathematical treatment required to portray a curved surface on a flat map surface **FIGURE 11-26**. To best illustrate this, using a grapefruit, draw sixty 6 degree–wide vertical lines (zones) from top to bottom. Now, with a knife, cut off the top and bottom rind of the grapefruit (areas to be excluded from the map). Next, cut all the lines that are drawn, and peel the rind from the grapefruit. Each section represents a UTM zone once it is placed on a flat surface. This best illustrates the term **map projection**, which is a system used to project Earth onto a flat map surface.

To make maps simpler to use and to avoid the inconvenience of pinpointing locations on curved reference lines, a rectangular grid surface called the **Universal Transverse Mercator (UTM)** system was designed. The grid system consists of two sets of parallel lines that are superimposed on the map, each uniformly spread and perpendicular to each other. With the UTM grid system, any point on the map can be designated by its latitude and longitude or by the grid coordinates. A reference in one system can be converted into a reference in another system.

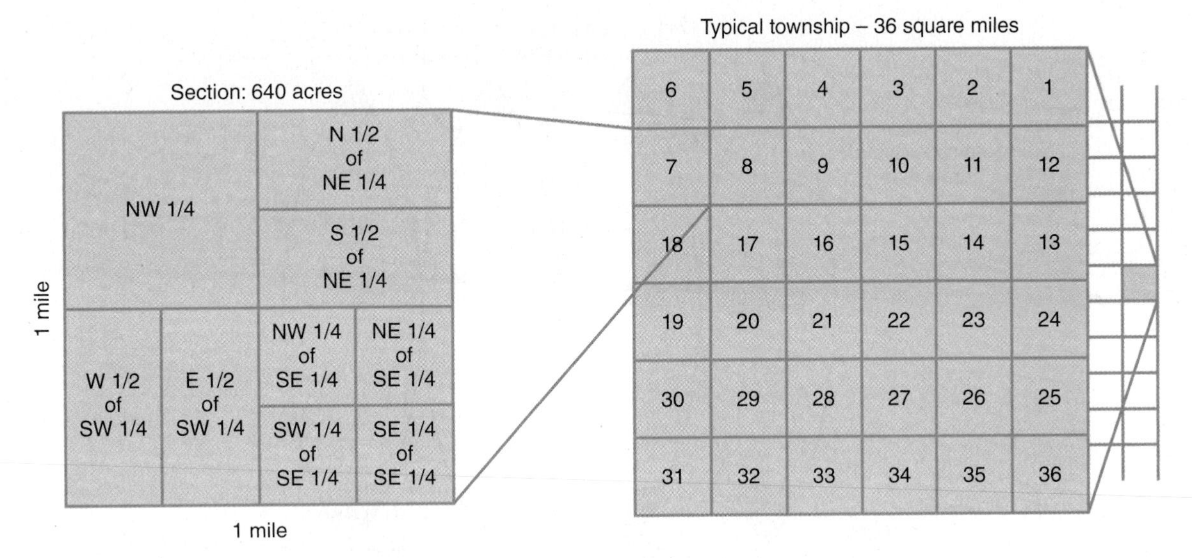

FIGURE 11-25 Public Land Survey System for legal description.

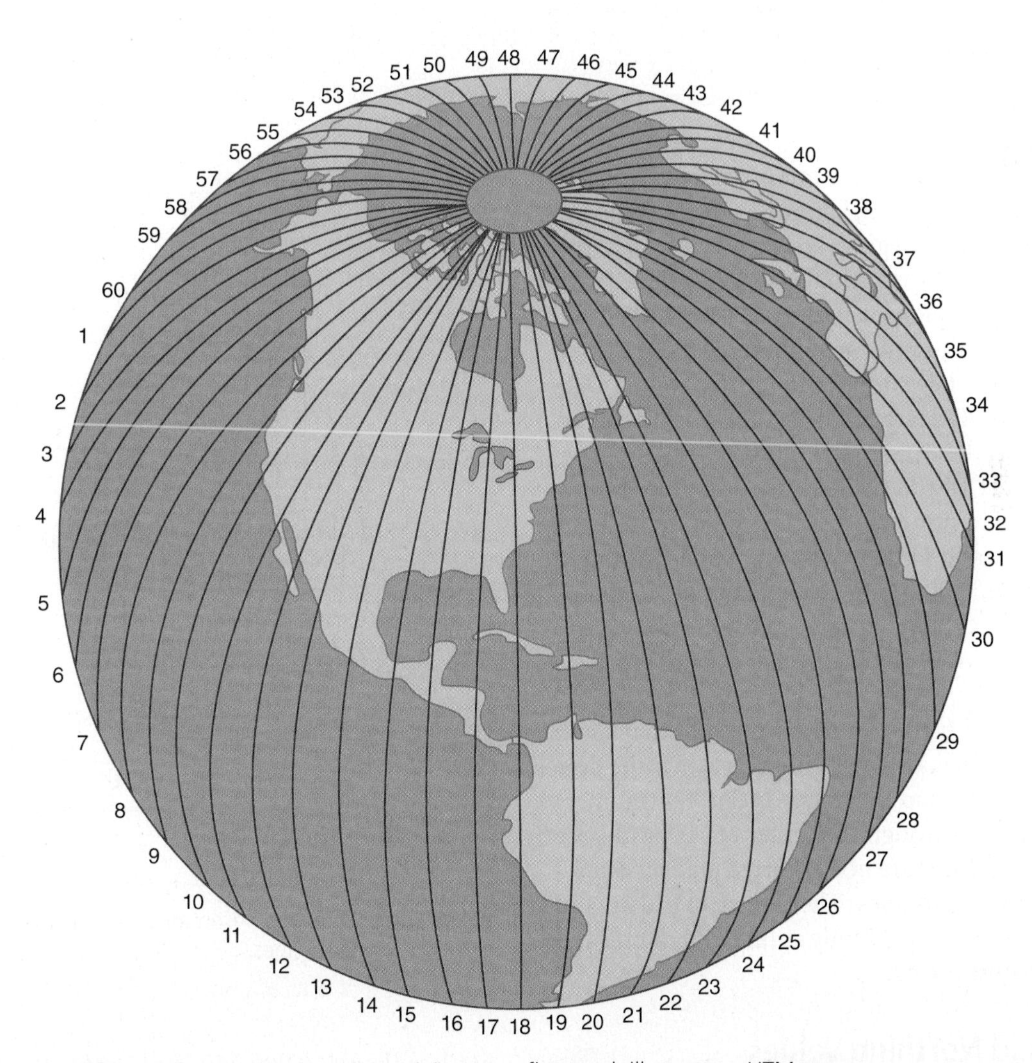

FIGURE 11-26 Each section of the globe, once flattened, illustrates a UTM zone.

UTM Zones

UTM is a special grid system consisting of 60 north–south zones that are only 6 degrees of longitude wide. Each zone is so narrow that, when it is peeled from the

LISTEN UP!

UTM is defined in degrees. Latitude and longitude are defined in degrees, minutes, and seconds.

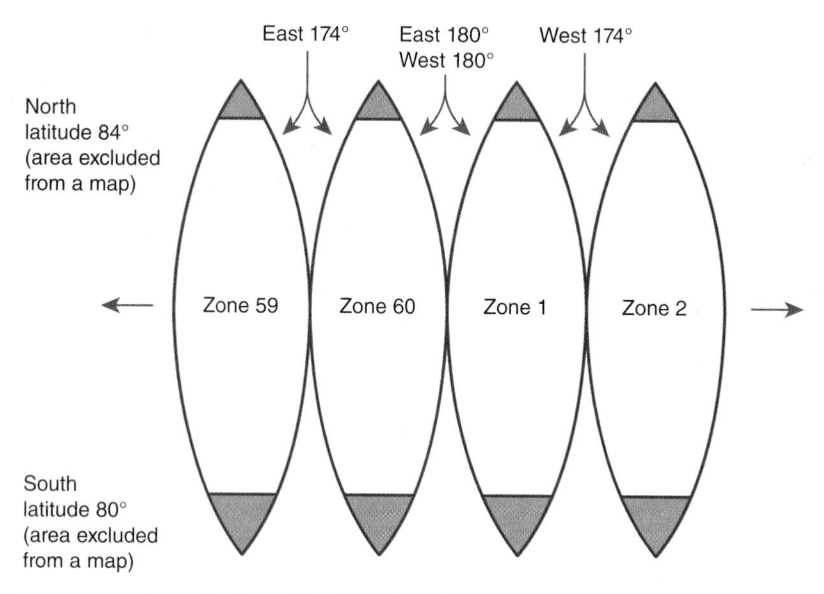

FIGURE 11-27 Each UTM grid zone is 6 degrees wide. Areas above north latitude 84 degrees and below south latitude 80 degrees are excluded from maps.

globe and flattened, it results in only minimal distortion. The areas above north latitude 84 degrees and below south latitude 80 degrees are excluded from maps due to the minimal distortion that takes place when that portion of the globe is flattened **FIGURE 11-27**.

All 60 UTM grid zones are numbered from west to east (left to right) beginning at the International Date Line, which is 180 degrees longitude. Zone 1 is a vertical area 6 degrees wide running between the meridians located at 180 and 174 degrees west longitude.

The zones are numbered progressively in an eastward direction and end at zone 60, which is located between 174 and 180 degrees east longitude. The conterminous United States is covered by 10 zones, from zone 10 on the West Coast through zone 19 in New England.

In each of the UTM zones, all the meridians and parallels experience a slight distortion when the segment is flattened (map projection). The central meridian, which runs through the center of each 6-degree zone, and the equator are not distorted and act as perpendicular UTM grid lines. The central meridian and equator also serve concurrently as lines of latitude and longitude **FIGURE 11-28**.

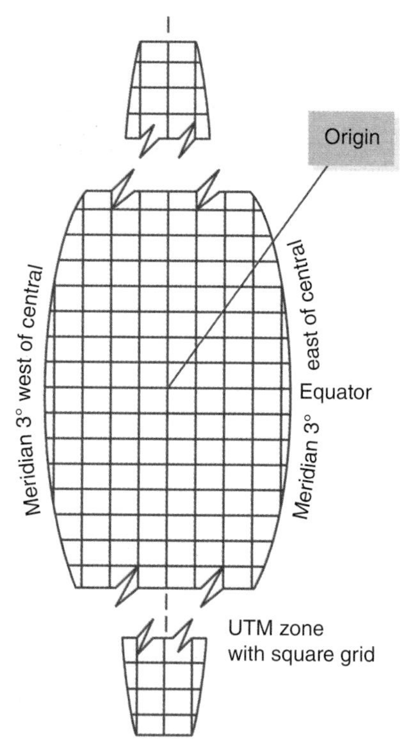

FIGURE 11-28 Flat representation from Earth's surface. Note the reference lines.

Easting and Northing Values

Each central meridian that runs through a 6-degree grid zone is assigned an easting value of 500,000 meters east (ME). **Easting value** is the distance in meters of the position east of the zone line. This helps you determine whether you are east or west of the central meridian. Grid values to the west of the central meridian are less than 500,000 ME, and grid values east of the central meridian are more than 500,000 ME.

The equator serves as the grid line for northing values. **Northing value** is the distance in meters of the position north of the equator, measured along a zone line. It is assigned a value of either 0000000 MN (meters north) or 10,000,000 MN, depending on whether you are in the northern or southern hemisphere. When your location is north of the equator, the true value assigned is 0000000 MN. Grid lines will increase in value as you move north of the equator. The value assigned for the southern hemisphere is 10,000,000 MN. However, remember that it is still referenced to the equator. Grid lines decrease in value as you move southward from the equator.

UTM coordinates, then, consist of an easting (in meters), a northing (in meters), and a zone number. To express the coordinates, read right and up.

Finding UTM Coordinates on a Map

To find UTM coordinates on a topographic map (7.5-minute series, scale 1:24,000), look at the margins of the map. Blue tick marks will be found that denote 0.62-mile (1000-m) grid lines. The marks on the top and bottom are called eastings; they are used to locate an east–west location. The blue tick marks on the left and right margins of the map are called northings; they are used to locate a north–south location.

Many of the newer topographic maps are printed with the UTM grids already in place. When using an older topographic map, use a large straightedge to draw lines at each 0.62-mile (1000-m) blue tick mark, from top to bottom on the map and from the left margin to the right margin. This preparation will create a UTM grid.

Here is an example of an easting coordinate 265,000 ME: The distance between 365,000 ME and 366,000 ME is 0.62 miles (1000 m). If you are traveling anywhere on the same topographic map, then you should be concerned with only the two large numerals. They are an abbreviation of grid lines commonly printed 0.62 miles (1000 m) apart. The last three numbers stand for meters. A northing value would look like this: 464,000 MN. Again, the distance between 464,000 MN and 465,000 MN is 0.62 miles (1000 m).

If easting numbers are increasing, you are headed in an easterly direction. If northing numbers are increasing, you are headed in a northerly direction.

The advantage in using the UTM grid system is that you will always be working with grid lines that are 0.62 miles (1000 m) apart. This makes it easy to determine your location on the map.

Using a Coordinate Ruler

A coordinate ruler will help you quickly locate your position on a topographic map. The one demonstrated in this text is called the Topo Tool Coordinate Ruler. It is made for use on a 7.5-minute topographic map with a representative fraction of 1:24,000.

Use

The coordinate ruler is a set of scales printed on clear plastic. These scales are used like a ruler to measure in different coordinate systems. The scales are sized to be used with the USGS system of 7.5-minute, 1:24,000 topographic maps that we use in the continental United States. The scales can be used to measure distance in feet and to determine slope. The PLSS quartered section is divided into 10-acre (4-hectare) squares, which can be used to estimate acreage.

Coverage

The UTM, feet, and PLSS scales are useable on any 1:24,000 map. The latitude and longitude scale is useable over the continental United States.

Conversions

Dealing with units like minutes and seconds can be difficult. The first three graphic scales are used to help with minute-second addition when finding or plotting coordinates **FIGURE 11-29**. The left side of each scale shows the ruler's numbers. The map coordinates are on the right. Note that the top and bottom numbers correspond to the minutes and seconds of the border lines (the tenths of minutes are left off). Suppose you locate a point with the ruler. The scale reads 1′43″. The right-side reference line is 17′30″. Use the scale starting with 7′30″. Go up the left side of the scale to the ruler reading 1′43″. The right side reads 9′13″ here. Because the tenths digit is 1, the point is at 19′13″.

Decimal Minutes

GPS receivers often give locations with latitude and longitude in degrees and decimal minutes. The graphic scale converts decimal minutes to and from seconds. When calling out coordinates and communicating them, it is vital that you remember to use the correct punctuation in the proper places:

Degrees: °

Minutes: ′

Seconds: ″

Suppose a GPS displayed a latitude of 35° 14.67″. Finding 0.67 minutes on the right side of the map corresponds to about 40 seconds on the left side. Thus, the coordinate could be expressed as 35° 14′40″.

It is very important to remember to use the proper format when communicating coordinates. The three generally accepted formats are shown in **TABLE 11-2**. Table 11-2 also shows a fourth format

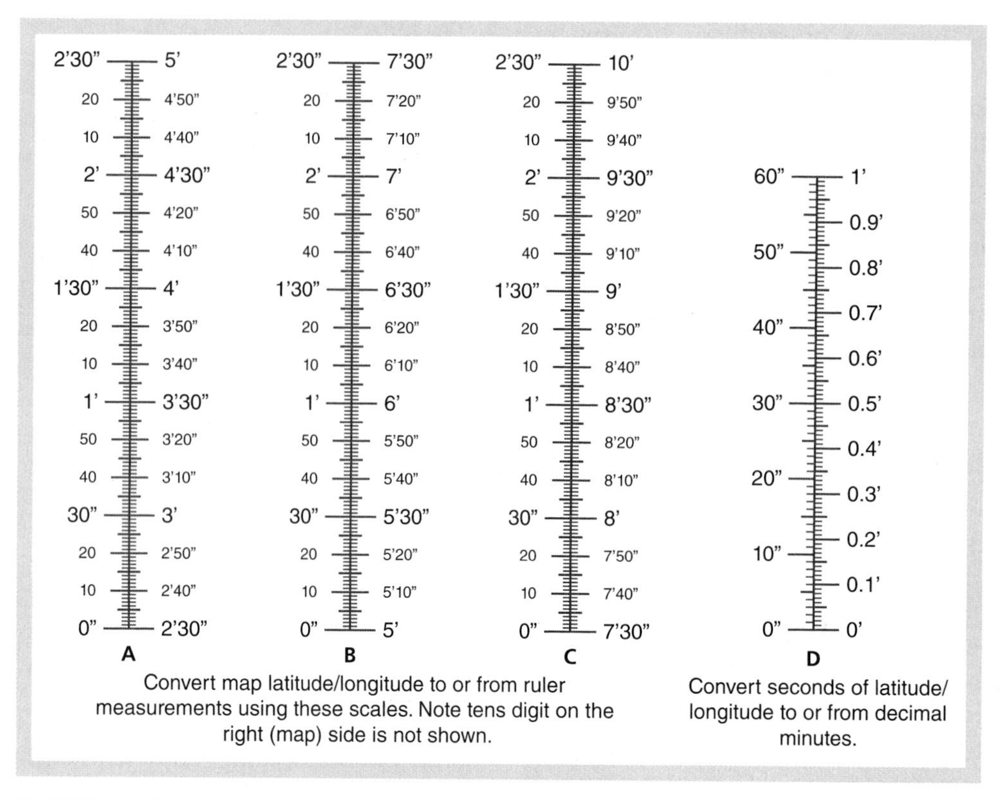

FIGURE 11-29 Conversion scales. Use the scales shown in **A**, **B**, and **C** to convert map latitude and longitude to or from ruler measurements. Use the scale shown in **D** to convert seconds of latitude and longitude to or from decimal minutes.

TABLE 11-2 Coordinate Communication Formats

Type	Format	Examples
Decimal degrees	DDD.DDDD°	48.3612°N 114.0812°W
Degrees and decimal minutes (DDM)	DDD° MM.MMM′	48° 36.12′N 114° 08.12′W
Degrees, minutes, and seconds (DMS)	DDD° MM′ SS.S″	48° 36′ 12″N 114° 08′ 12″W
Temporary flight restriction (TFR) requests	DDDDDDD	483612N 1140812W

used for requesting temporary flight restrictions (TFRs) in aircraft. This new format is an exception to the three generally accepted formats. The new format uses no punctuation and mirrors the decimal degrees format.

Depending on with whom you are interfacing, knowing the four formats will aid you in preventing confusion when trying to find a location, which means the nuances and incorrect use of punctuation will take you to three different locations. The most uniform format will be **degrees decimal minutes (DDM)**. Communication with aviation assets in wildland predominantly use the DDM format.

If you are in a situation when you may need to convert one format to another, here is an easy conversion formula to use. GPS being used outside of the wildland environment in some cases will be given in the degrees minutes seconds (DMS) format. To convert to DDM, the easiest way will be to divide the seconds (″) by 60.

Example 1. 48° 20′ 30″ → 30″ / 60 = 0.5′
→ **DDM** is 48° 20.5′

To convert degrees decimal minutes (DDM) to **degrees minutes seconds (DMS)**, you will need to multiply by 60.

Example 2. 48° 20.5′ → 0.5′ × 60 = 30″
→ **DMS** is 48° 20′30″

Orienteering etiquette is very important. Here are a few etiquette reminders:

- Use only one decimal point when writing latitude or longitude.

- Do not use any decimal points when writing latitude or longitude in the DMS format.
- Remember that there cannot be more than 60 seconds in the DMS format.
- For the sake of clarity, insert a "0" in front of any single-digit minutes. Many GPS units and map programs require two digits.
- Do *not* mix formats.

Additional Information

The coordinate ruler is silkscreen printed on clear plastic. The material is durable, but it gets superficial scratches. These scratches are not noticeable when the ruler is placed on a map. The printing on the ruler is quite durable as well, but it can be damaged. Use care to avoid unnecessary abrasion.

The ruler's scales (except the section) can be used in a pinch on 1:25,000 maps by using the sloping scale technique for both north–south and east–west coordinates.

Finding Latitude or Longitude

The coordinate ruler is sized to fit the common USGS 7.5-minute, 1:24,000 topographic maps as used in the continental United States.

Latitude and Longitude on 7.5-Minute Maps

A 7.5-minute topographic map is 7.5 minutes high and 7.5 minutes wide. The lower right corner has the smallest values of degrees, minutes, and seconds. There are latitude numbers along both sides of the map, showing the angular distance from the equator, and longitude numbers along the top and bottom margins, showing the angular distance west of the prime meridian.

There are two small lines or tick marks showing the location of additional latitude or longitude lines along each margin of the map. These intermediate marks are labeled with minutes and seconds only. Also, there are small crosses inside the map where these lines would intersect. Using a sharp pencil, draw lines connecting equal values of latitude and longitude. The long edges of the ruler enable you to draw these lines by using the crosses. This divides the map into nine rectangles, each 2.5 minutes (2′30″) on all sides **FIGURE 11-30**. These lines, along with the map margins, are used as reference lines of known latitude or longitude.

Now, imagine a series of horizontal and vertical lines forming a grid, sort of like a piece of window screen, laid on the map. Spaced 1 second apart, there would be 150 horizontal latitude lines and 150 vertical longitude lines in each 2.5-minute rectangle. If these lines were visible, the latitude of a point could be determined by counting the number of lines from a labeled reference line up to the point. Adding the number of lines (seconds of latitude) to the value of the reference latitude line would give the latitude of the point.

Finding Latitude

Latitude on a topographic map is found in just this way. The latitude and longitude scale is used just like a ruler to measure from a known latitude line to a point. Each ruler division shows the location of an imaginary latitude line. Measure from the lower labeled latitude (horizontal) line of the rectangle containing the point. Read the minutes and seconds from the ruler, and add this to the value of the reference line. This is the latitude of the point.

Latitude-Finding Tips. Paper maps vary somewhat in size with age and humidity. Rectangles typically do not measure exactly 2′30″ high. For demanding applications, the scale can be shifted slightly to minimize this inaccuracy when finding (or plotting) latitude. Align the lower line of the ruler on the lower rectangle border line if the point is in the lower quarter of the rectangle, and align the upper line of the ruler with the upper rectangle border line if the point is in the upper quarter. Center the ruler over the rectangle borders if the point is in the middle. Add the rule reading to the lower reference line **FIGURE 11-31**.

Adding the ruler's minutes and seconds of latitude to the latitude of the lower border line of the rectangle is treated like time addition. Using a piece of paper to add the coordinates is a good idea because it is easy to make a mistake with mental addition.

Finding Longitude

The longitude width of a 2.5-minute rectangle is not a constant distance because the longitude lines converge toward the poles. To compensate for this, the ruler uses a technique called sloping scale. If you are trained in drafting, you are probably familiar with this technique.

The latitude and longitude scale is too long to fit the width of a 2′30″ rectangle. However, if the scale is tilted, it can be adjusted so that the two end lines of the scale lie directly over the side borders of the rectangle. The location of each imaginary longitude line is then at the edge of the rule.

To find the longitude of a point, align the edge of the ruler just over the point while the ends of the scale

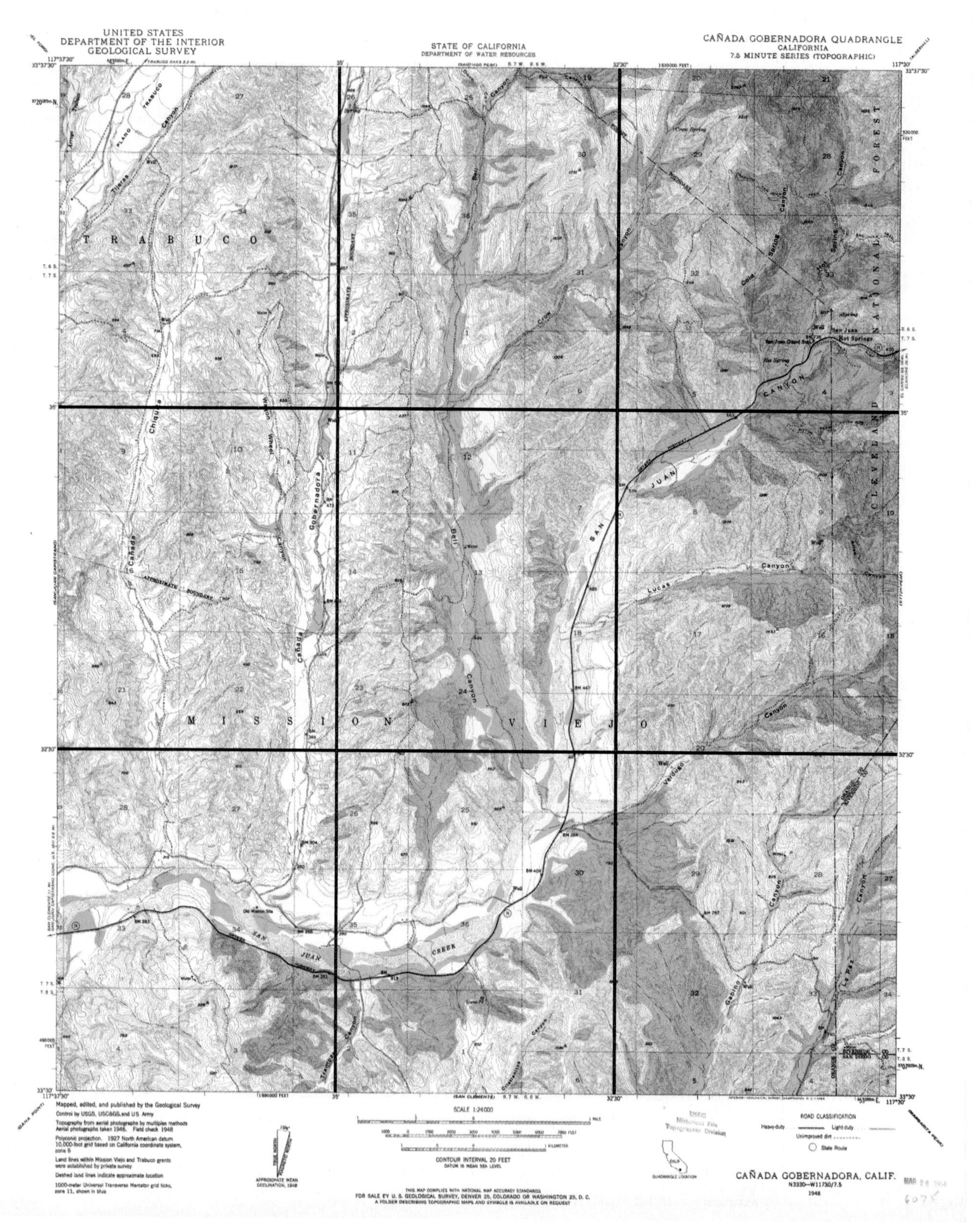

FIGURE 11-30 The grid lines on the map have been drawn. The interior crosses can be found where the arrow points to the comment "where both lines come together."

Courtesy of the U.S. Geological Survey.

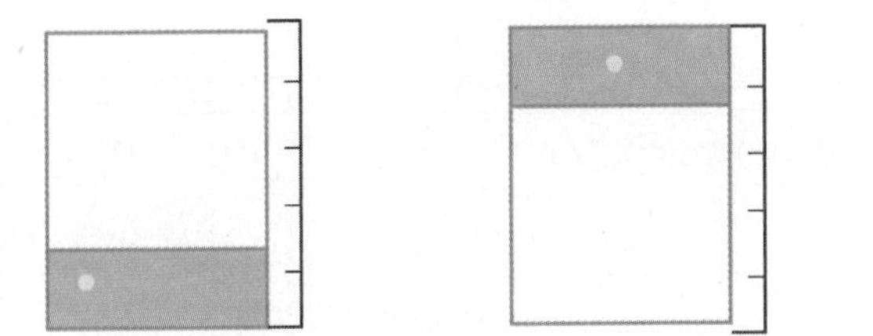

FIGURE 11-31 Shifting the scales to minimize latitude inaccuracies. Ruler scale is shown to the right of the rectangle for clarity.

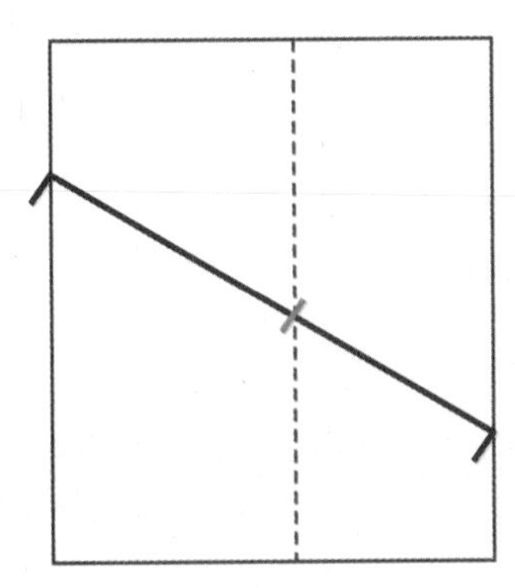

FIGURE 11-32 Tilt the scale to fit just within the rectangle borders. The longitude line of the point is shown as a dashed line.

are precisely aligned with the vertical border lines. Correctly aligning the scale is a little time consuming, but it is quite accurate. With some practice, it becomes easier. When finding longitude, remember that the longitude lines are all vertical; they do not follow the slope of the ruler graduations. The longitude lines lie just under the ruler marks at the very edge of the rule. The rule numbers must increase to the left. Add the ruler reading at the point on the right-side reference line **FIGURE 11-32**.

Longitude-Finding Tips. Use care in adding the rule reading to the value of the longitude reference line. Remember that angular degrees, minutes, and seconds are added just like time's hours, minutes, and seconds. Use the graphic scales mentioned previously to aid your arithmetic. Note that, because the borders of the rectangle are always multiples of 2′30″, the digit for seconds will be the same. Using the previous example in the Conversions section, the ruler reads 1′43″. The right-side reference line is 17′30″. Thus, it is not necessary to count each small graduation on the graphic aid—you know the coordinate will end with a 3. Just glance at the aid to see that 1′40″ corresponds to _9′10″ (the underline is for the missing tenths-of-minutes digit). So the ruler reading of 1′43″ added to the border line gives the point's minutes and seconds of longitude: 19′13″.

Because the ruler must be tilted or sloped to just fit the rectangle borders, it is helpful to extend the vertical reference lines out past the map proper into the margins. This allows the tilted ruler to still find the longitude of points near the upper and lower borders. However, small triangular-shaped areas at the top margin of the map may not be accessible to the sloped rule. If the rule starts going off the paper because the point is too far up, use the ruler just like a draftsman's triangle to project a small line down into the accessible region of the rectangle. Align the top line of the ruler with the top margin line of the map, and slide the edge of the scale up to the point. Now, mark the longitude of the desired point along the edge of the ruler farther down in the rectangle, where it can be measured.

Plotting

Plotting coordinates is the reverse of finding coordinates and is actually a little easier. Determine longitude first. By inspection of the coordinates and the reference lines, decide which rectangle contains the point. Then, align the ends of the ruler over the vertical rectangle borders (slope the ruler). Measure left the required number of minutes and seconds, and make a small vertical mark to indicate the position of the desired longitude line. Then, use the ruler like a draftsman's triangle to measure the required number of minutes and seconds up at a right angle from the horizontal latitude reference line. By aligning the ruler's lower line over the lower reference line and carefully sliding the ruler sideways to the longitude mark made previously, the edge of the ruler will lie directly along the longitude line of the point. Again, measure up the required number of minutes and seconds of latitude, and mark the location of the point. Erase the first longitude reference mark **FIGURE 11-33**. The latitude/longitude scale in this example is being used to find the longitude of an eastern Kentucky oil well. The ruler is used to connect the tick marks to the cross to form a rectangle with border lines of known longitude/latitude.

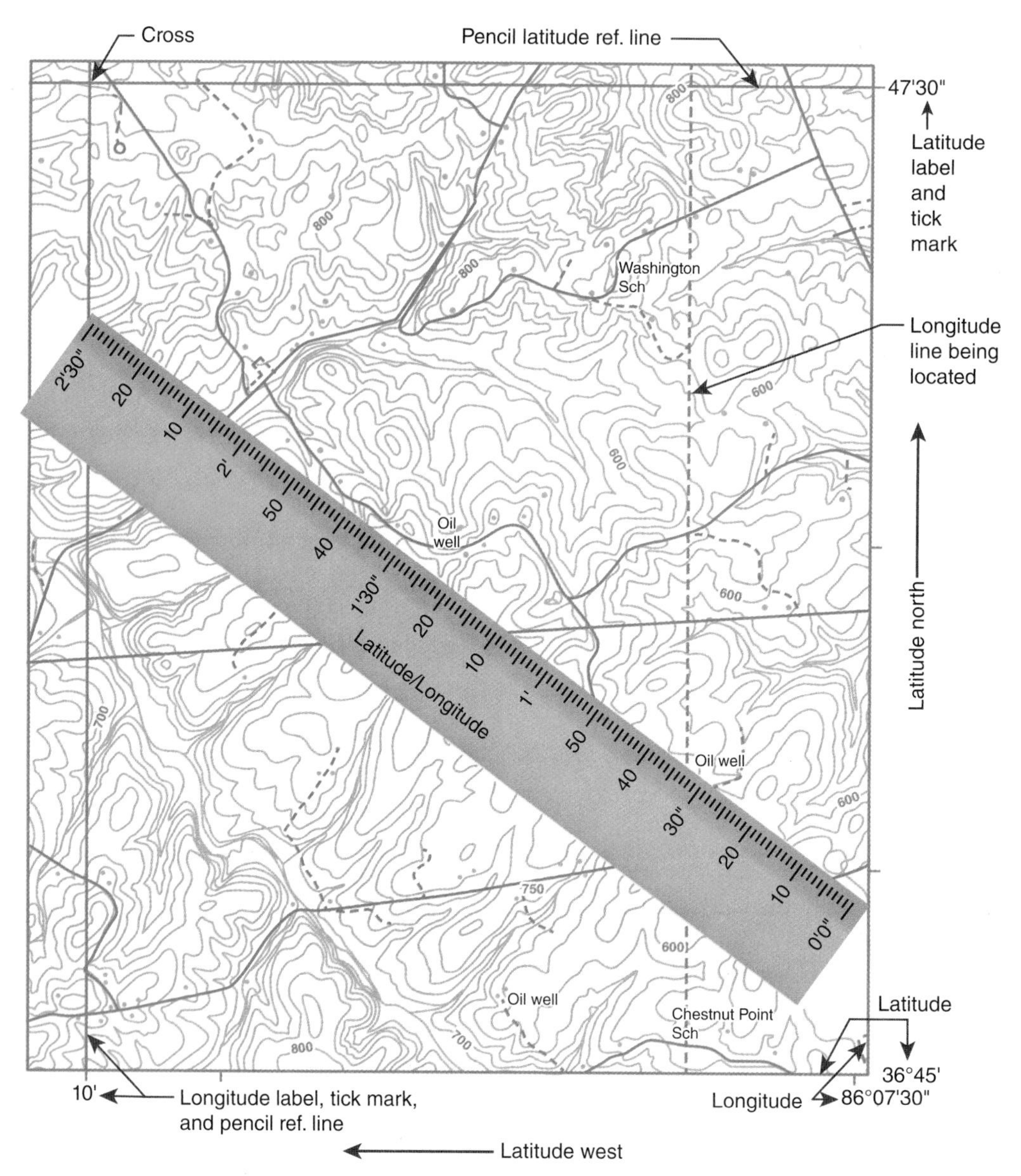

FIGURE 11-33 How to use the coordinate ruler to find longitude.
Courtesy of the U.S. Geological Survey.

Note the position of the coordinate ruler's upper and lower lines. This indicates how to tilt the ruler to fit a 2′30″ longitude distance. Add the 35″ on the scale to the value of the right-side reference line, 86°07′30″W, to get the longitude of the well, 86°08′05″W. The coordinates of this oil well are 36°45′47″N, 86°08′05″W.

Using the Coordinate Ruler to Find a UTM Grid

Finding UTM. The UTM system is very simple to use. The coordinates increase to the north and to the east (up and to the right). Measurements within a 0.62-mile (1000-m) grid square are taken from the lower left corner and are always additive. To find a UTM coordinate, measure from the west (left) border of the grid square to the point. This is the easting. Then, measure from the south (lower) border of the grid square to the point. This is the northing. The zone number is listed in the map margin.

Plotting UTM. Plot UTM coordinates on a map by first locating (by inspection) the 0.62-mile (1000-m) grid square containing the coordinate. Using the UTM scale, measure the proper number of meters from the lower left corner such that the sum of the left grid line and the ruler UTM scale equals the easting coordinate. Make a reference mark on the lower border, and then measure the proper number of meters up from this mark. Mark the point.

Advantages of UTM. One advantage of UTM coordinates is that, due to the small size of the squares,

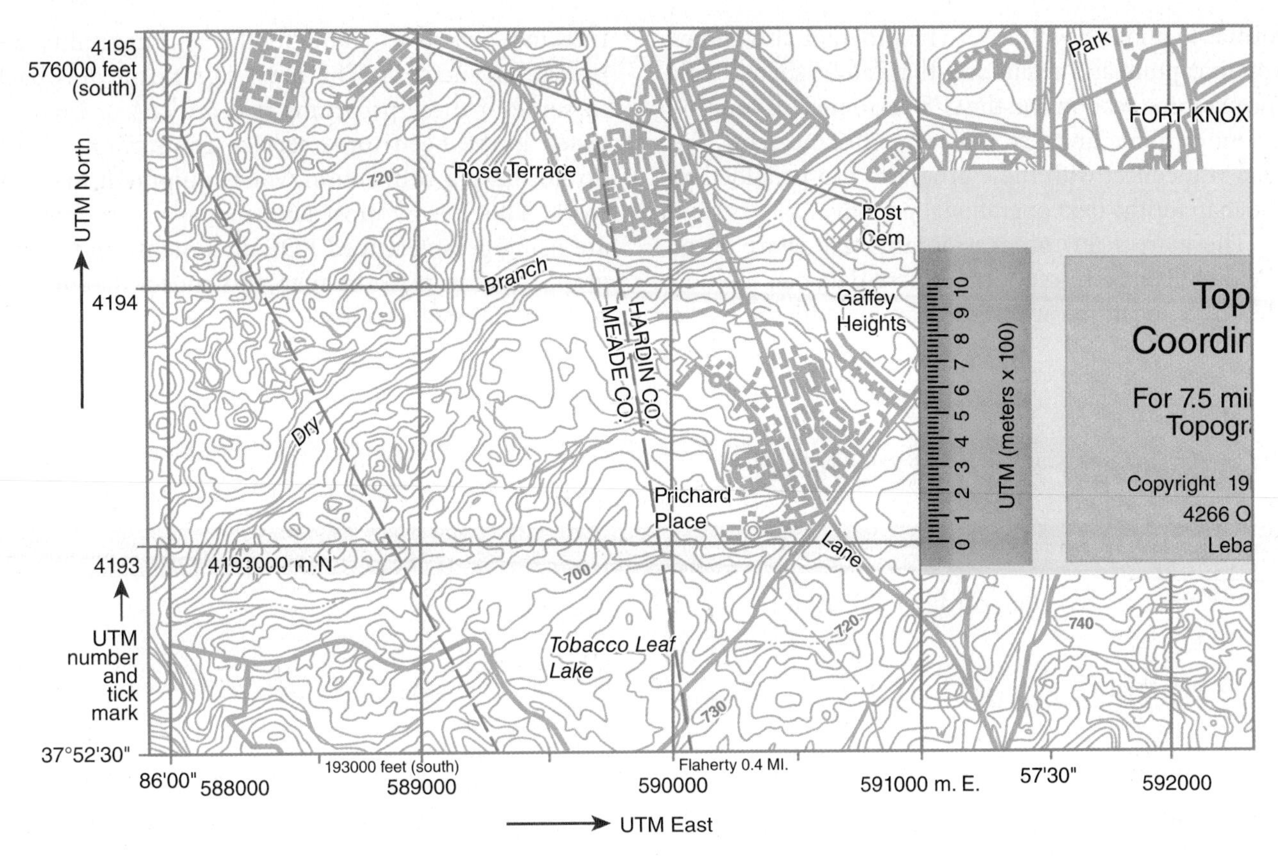

FIGURE 11-34 How to use the coordinate ruler to find UTM coordinates.

Courtesy of the U.S. Geological Survey.

folding the map does not affect coordinate accuracy as much as it does with other systems **FIGURE 11-34**.

As an example of UTM coordinate determination, suppose we want to find the UTM coordinates of the Fort Knox Bullion Depository. The UTM grid is not preprinted, but a note in the lower-left margin tells us that the 0.62-mile (1000-m) grid ticks are shown in blue (on the original topographic map, not this copy). The zone number is 6.

Examining the map borders, we note the UTM numbers and construct our grid by connecting matching tick marks on the opposite borders of the map. To locate one point, only four lines need to be drawn around the point. For convenience, expand a couple of the grid numbers to their full UTM coordinates. Remember, the two larger principal digits are in ten thousands and thousands of meters. Number any grid line that does not have a printed value. Note that the coordinates should increase from left to right and moving up.

The depository lies directly on the 591,000 ME line, so this is its east coordinate. Use the coordinate ruler's UTM scale to measure from the lower border line of the square containing the point up to the point itself. The north UTM coordinate is 4,193,140 MN, found by adding the scale reading of 459.3 ft (140 m) to the lower grid line number of 4,193,000 MN. The full UTM coordinates of the Fort Knox Bullion Depository are 591,000 ME; 4,193,140 MN; Zone 16. Remember, the order of the coordinates is "read right, up."

Newer Mapping Systems

Technology is evolving quickly. In the past four or five years, several newer mapping programs have been utilized to collect data and identify locations using smartphones and tablets.

These programs are beginning to provide enhanced benefits for real-time mapping, data collection, and information sharing for responding resources and triage during interface fire assignments. Most tablets and cell phones are GPS enabled and are just as accurate as the stand-alone GPS devices.

This GPS tracking capability is generally launched by taking real-world coordinates determined by a handheld device, such as a smartphone or tablet, and projecting that information onto a digital map (think Google Maps) in real time. When this occurs, the handheld device or tablet assumes the identity of a live-tracking GPS device. One of the most used programs is called Avenza Maps. One of the nice features of these types of programs is that they allow the user to switch from one format to another with ease. The program also allows you to capture and retain data that can be shared with the planning section and the GIS personnel on your incident or home unit. On most large fire incidents, the data you collect with these programs can be added to the map for the next operational period.

There are a few other programs that share all of the data live, or in real time. These data can include the GPS location, specific location, and real-time tracking, which can add to the accountability aspect of an incident. The most significant drawback to some of these programs that are cell or Internet based is the need for a service connection. If you do not have a connection, your equipment may not work. This is why most land management agencies are using georeferenced PDF documents and a program that can read the map and collect data for future operational periods.

After-Action REVIEW

IN SUMMARY

- A compass has eight cardinal directions:
  - North
  - Northeast
  - East
  - Southeast
  - South
  - Southwest
  - West
  - Northwest
- Declination is the difference between magnetic north and true north, and can change over time.
- Percent of slope is determined by dividing the change in elevation (the rise) by the horizontal distance covered (the run), then multiplying by 100.
- The angle of slope represents the angle formed between the run and your climb's angular deviation. To calculate, divide the rise by the run and obtain the inverse tangent of the result.
- To orient yourself in a specific direction, adjust the compass dial until the index line is in the aspect you wish to face, then rotate your body and compass in one motion until the red magnetic needle overlays the orienting arrow.
- There are two legal description systems: The Metes and Bounds System and the Public Land Survey System (PLSS).
  - The metes and bounds system is a method of tracking boundaries with landmarks, lines, and words.
  - The PLSS is a method of tracking boundaries with a compass and grid system.
- The Global Positioning System (GPS) is a system of navigational satellites that can track objects anywhere in the world with an accuracy of 40-300 feet.
- Topographic maps represent the Earth's surface with contour lines overlaid onto a Universal Transverse Mercator (UTM) grid. The contour lines represent certain elevations at regular intervals, with every fifth line (the index contour). The closer together contour lines are, the steeper the slope is.
  - Relief maps have physical elevation on the map, allowing you to see and feel the change in elevation relative to other areas of the map.
- The Universal Transverse Mercator (UTM) grid is a rectangular grid system consisting of two sets of parallel lines superimposed on a map, uniformly spread and perpendicular to each other. The UTM grid is used to represent the curvature of the Earth on a flat surface. Each section of the grid represents a 6° longitudinal slice of the Earth.

- Points within a 6° UTM grid zone assigned easting and northing values which help determine how far away the point is from the equator (northing value) or the central meridian of the zone (easting value).
  - The central meridian of each 6° grid zone is assigned an easting value of 500,000 ME (meters east). Points west of the central meridian have easting values less than 500,000 ME, while points east of the central meridian have easting values greater than 500,000 ME.
  - The equator is assigned different a different northing value depending on if the point on the map is located in the northern or southern hemisphere. When the point is north of the equator, the equator is assigned a value of 0 MN (meters north). When the point is south of the equator, the equator is assigned a value of 10,000,000 MN.
    - Points located on the southern hemisphere have a northing value of 10,000,000 MN minus its distance from the equator.
    - Points located on the northern hemisphere have a northing value equal to its distance from the equator.
    - Northing values always increase in value moving from south to north, whether in the north or south hemisphere.
- The coordinate ruler is a set of scales printed on clear plastic, used as a ruler to measure in different coordinate systems. They can be used to measure distance (in feet) and to determine slope.
- The latitude and longitude coordinates of a point on a map can be communicated in different formats. The most common format is degrees decimal minutes (DDM), which displays a point's map coordinates in degrees and minutes in decimals. Other formats are degrees, minutes, and seconds (DMS), decimal degrees, and temporary flight restriction (TFR).
  - Decimal Degrees format: DDD.DDDD°
  - DDM format: DDD° MM.MMM′
  - DMS format: DDD° MM′ SS.S″
  - TFR format: DDDDDDD

KEY TERMS

Azimuth The horizontal angle of a point measured clockwise from true north.

Back azimuth An angle 180 degrees opposite of azimuth; used when retracing steps.

Chain A unit of measure in land survey equaling 66 ft (20.1 m); 80 chains is equal to 1 mile (1.6 km).

Declination The difference between magnetic north and true north. Also called magnetic declination or magnetic variation.

Degrees decimal minutes (DDM) A format for communicating the latitude and longitude values of a point, displayed as DDD° MM.MMM′.

Degrees minutes seconds (DMS) A format for communicating the latitude and longitude values of a point, displayed as DDD° MM′SS.S″.

Easting value The distance, in meters, of a position east of a 6° UTM zone line. Each central meridian that runs through these grid zones is assigned an easting value of 500,000 ME. Positions west of the central meridian have easting values less than 500,000 ME, and positions east of the central meridian have easting values greater than 500,000 ME.

Global Positioning System (GPS) A system of navigational satellites that can track objects anywhere in the world with an accuracy of approximately 40 to 300 ft (12.2 to 91.4 m).

Index contour Every fourth or fifth line on a topographic map, printed darker to make it easier to read the map. Elevation is listed at some point along an index contour.

Intermediate contour The lighter contour lines between index contours.

Latitude A point north or south of the equator, measured in degrees, minutes, and seconds. The equator is 0° latitude.

Legal description The description of the land and property, commonly used to describe your location or when receiving an incident location.

Legend A key accompanying a map that shows information needed to interpret that map.

Longitude A point east or west of the equator, measured in degrees, minutes, and seconds. The prime meridian is 0° longitude.

Magnetic north The magnetic north pole. Magnetic north changes over time and with location.

Map projection A system used to project the Earth onto a flat map surface.

Northing value The distance, in meters, of a position relative to the equator. If you are in the northern hemisphere, the equator has a northing value of 0 m, and the northing value of a point located in the northern hemisphere is equal to its distance from the equator. If you are in the southern hemisphere, the equator has a northing value of 10,000,000 MN, and the northing value of a point located in the southern hemisphere is equal to 10,000,000 minus its distance from the equator.

Pace Two natural walking steps measured from the heel of one foot to the heel of the same foot after two steps. Pace can be used to keep track of distance. Pace count will differ among people and by terrain.

True north North according to Earth's axis.

United States Geological Survey (USGS) topographic maps Detailed topographic maps produced by the United States Geological Survey

Universal Transverse Mercator (UTM) A rectangular grid surface designed to make maps simpler to use and avoid the inconvenience of pinpointing locations on curved reference lines.

REFERENCES

Kjellstrom, Bjorn. *Be an Expert with Map and Compass: The Complete Orienteering Handbook*. 3rd ed. Hoboken, New Jersey: John Wiley & Sons, Inc., 2010.

National Wildfire Coordinating Group (NWCG). "Basic Land Navigation." PMS 475. NFES 2865. May 2016. https://www.nwcg.gov/sites/default/files/publications/pms475.pdf.

National Wildfire Coordinating Group (NWCG). "S-130, Firefighter Training, 2008." Modified August 21, 2019. https://www.nwcg.gov/publications/training-courses/s-130.

National Wildfire Coordinating Group (NWCG). "S-190, Introduction to Wildland Fire Behavior, 2019." Modified October 1, 2019. https://www.nwcg.gov/publications/training-courses/s-190/test/course-materials.

National Wildfire Coordinating Group (NWCG). S-341

United States Geological Survey (USGS). "Finding Your Way with Map and Compass." March 2001. https://pubs.usgs.gov/fs/2001/0035/report.pdf.

Wildland Fire Fighter in Action

You are assigned as a lookout on the Joy fire. The fire is an emerging incident with a lot of action, and the majority of the fire is midslope, which is causing the crews to construct underslung lines (cup trenches). The fire is about 120 acres (48.6 hectares), with group-tree torching and isolated instances of spotting throughout the day. In addition to taking weather, your job is to keep an eye out for any new starts and provide feedback for the crew and supervisors on previously established decision points for utilizing escape routes to get off the fireline. At about 1530, you notice a new column of smoke that seems to be a spot fire that will affect the escape route for everyone because it can't be seen by the crews.

1. How will you describe where the spot is using the township, range, and section description?

A. Using quarter sections
B. Using a map ruler
C. Using a GPS
D. Using the UTM grid

2. The fire is about 40 chains (804.7 m) from your location. How far is that in miles?

A. 2640 ft (804.7 m)
B. 5280 ft (1609.3 m)
C. 1320 ft (402.3 m)
D. 160 ft (48.8 m)

3. What is one of the most commonly used maps on a wildland fire?

A. 1:24,000
B. 1:50,000
C. 1:60,000
D. 1:30,000

4. GPS involves how many satellites?

A. 24
B. 20
C. 18
D. 16

5. What is the Universal Transverse Mercator grid system?

A. A 60-zone system consisting of 6-degree widths
B. A 6-zone system consisting of 60-degree widths
C. A 40-zone system consisting of 4-degree widths
D. A 4-zone system consisting of 40-degree widths

6. What does the term *7.5-minute topographic map* mean?

A. The latitude and longitude lines cover 7 minutes and 30 seconds.
B. The UTM lines cover 7 minutes and 30 seconds.
C. The GPS lines cover 7 minutes and 30 seconds.
D. The azimuth lines cover 7 minutes and 30 seconds.

Access Navigate for flashcards to test your key term knowledge.

CHAPTER 12

Wildland Fire Fighter II

Water Supplies and Operations

KNOWLEDGE OBJECTIVES

After studying this chapter, you will be able to:

- Identify common water sources. (pp. 264–268)
- Identify common wildland fire hoses. (p. 268)
- Describe different nozzle fittings and their uses. (p. 269)
- Identify common hose appliances and accessories. (pp. 269–272)
- Understand pump and water delivery system capabilities. (**NFPA 1051: 5.5.5**, p. 269)
- Describe how pumps operate. (N**FPA 1051: 5.5.5**, p. 274)
- Describe how drafting works. (**NFPA 1051: 5.5.5**, p. 274)
- Describe portable pump maintenance and preventive procedures. (**NFPA 1051: 5.3.2**, p. 274)
- Describe the responsibilities of the pump operator. (pp. 276–277)
- Describe portable pump setup. (**NFPA 1051: 5.5.1, 5.5.5**, pp. 277–278)
- Describe the differences between a simple and progressive hoselay. (pp. 278–280)
- Describe the different types of hose packs. (pp. 280–281)
- Describe wet mop-up operations. (**NFPA 1051: 5.5.1**, pp. 282–285)
- Understand basic water hydraulics. (**NFPA 1051: 5.5.5**, pp. 285–286)

SKILLS OBJECTIVES

After studying this chapter, you will be able to perform the following skills:

- Restrict water flow in a hose line with a hose clamp. (p. 269)
- Uncouple a hose with spanner wrenches. (p. 273)
- Operate a backpack pump. (p. 275)
- Unroll a hose. (p. 278)
- Construct a simple hoselay. (pp. 278–279)
- Construct a progressive hoselay. (pp. 279–280)
- Butterfly a hose. (pp. 282–283)
- Perform a doughnut hose roll. (p. 284)

Additional Standards

- **NFPA 1963**, *Standard for Fire Hose Connections*
- **NWCG S-130 (NFES 2731)**, *Firefighter Training*
- **NWCG S-211**, *Portable Pumps and Water Use*

You are the Wildland Fire Fighter

Your engine has been assigned to provide structure protection in a wildland/urban interface area. Your engine boss/captain has been assigned as the initial attack incident commander. As the senior fire fighter on the crew, it will be your responsibility to secure a water source, provide water to the rest of your crew, establish an anchor point, and begin a flanking attack. Water supplies are limited due to an overuse of water from hydrants, and water tenders are in short supply. You have noticed that a stream runs next to the house, and now you need to make some decisions.

1. What factors would you consider before coming up with a solution to your water shortage problem?
2. What are the pros and cons of your choices?
3. Would you have to change your tactics due to the water shortage?
4. How would the available equipment affect your decision? What alternatives could assist you in mitigating equipment needs?

Access Navigate for more practice activities.

Introduction

Fire fighters must understand the fundamentals of how to get water, where to find water, and how to deliver water in a way that will successfully stop the fire by removing the heat component of the fire triangle. This chapter will cover the basics of common water sources, the tools and accessories used to obtain water from these sources, hoselay techniques, mop-up operations, and basic hydraulics. What you will learn in this chapter will be your foundation to build on as you develop your skills and knowledge with water use on fire. Every fire situation you encounter will be different, and you will need to draw on these principles and creative thinking to accomplish your water supply needs for specific situations.

Water Sources

Water is a valuable commodity at a wildland fire, so use it wisely. Make every drop count. Depending on the size of the tank in your engine or the stream you are pulling from, using your water wisely will enable you to be effective in suppressing fire.

Upon arrival at the scene of a wildland fire, it is necessary to find a water source. Water is necessary to supply engine companies, handcrews, tenders, and water-dropping helicopters in use. Once water sources are identified, make them known to your team so incident resources can be supplied. This will include talking to the person responsible for making your fire maps and sharing the GPS coordinates to accurately identify the location of the water source.

Several types of water sources are discussed next.

Streams, Lakes, Ponds, and Irrigation Canals

A natural stream, irrigation canal, lake, or pond can be used to supply the engine or water tender with water. The area must be evaluated for ease of access. If the engine can get close enough, these natural water sources can be used. The engine's intake hose (hard suction hose) can be placed directly into the stream or canal to draft water, provided the water is deep enough and flowing sufficiently.

If the engine cannot get close enough to the water source, then another tool—a portable pump, floating pump, gravity sock, or ejector (discussed later in this chapter)—must be used to access the water supply.

Swimming Pools

Swimming pools can provide large supplies of emergency water. Access is not always easy for fire apparatus, though, because pools are usually found in residential backyards. Consider using a floating pump or a portable pump in cases where there are access restrictions.

LISTEN UP!

When working in a rural area, be very aware of and on the lookout for septic tanks and leach fields. As you size up the scene, try to identify where the tank is located in relation to the swimming pool. Overlooking this information could result in your fire engine falling into the septic tank and rendering your engine out of service.

Cisterns

A cistern is an underground water storage reservoir made of reinforced concrete **FIGURE 12-1**. Cisterns are found in some rural areas of the country and can provide water for firefighting purposes. The fire engine can get water from the cistern by using the hose connection found on the tank. At that point, the engine must either draft from the cistern or, if the tank outlet connection is above the engine, use gravity flow to fill the engine. In some cases, it is necessary to remove a cover and drop the intake hose into the cistern to draft from it.

Improvised Water Sources

It is common for wildland fire fighters to create many improvised water sources, called pump chances, over the course of a fire season **FIGURE 12-2**. In some remote locations and depending on the time of year, creeks and streams might not have the same water flow as in the early fire season or the spring. If water flow is sufficient but lacking depth that prevents use of gravity socks, floating pumps, or portable pumps (discussed later in this chapter), you may need to create your water source. This involves creating a dam to collect water and make a pool deep enough to prevent pump cavitation.

Most engines or crews carry large bags or Visqueen. In some cases, crews get creative with trash bags or natural materials available near the river. These materials are used in conjunction with logs, boulders,

FIGURE 12-2 Improvised water source. This example blocked the stream water with a trash bag and secured it with rocks.
Courtesy of Ryan Andrade.

FIGURE 12-1 A cistern.

rocks, and mud to create a dam that is sufficient for a pump to remove water. There is no right way of accomplishing this other than to make it happen.

However, if you do establish an improvised water source, it is very important to notify the resource adviser or operations section chief. It is also important to mark the area with a sign (and a name, if appropriate) and make sure that the improvised water source is added to the fire map for others to see.

Fire Hydrants

Few fire hydrants are found in wildland areas; they become more prevalent in the wildland/urban interface area.

Fire hydrants come in various sizes and shapes **FIGURE 12-3**. The threads on the hydrant may differ from jurisdiction to jurisdiction. It is important to have an assortment of adapters on the engine, especially if you are responding outside of your normal area. Some jurisdictions use 4-in. fittings and some use 5-in. fittings.

On a large wildland/urban interface fire, do not become heavily committed to hydrant use. Have an alternative water source. Studies have shown that, on most large wildland interface fires where numerous structures were being lost, the water system ultimately failed due to too great a demand from fire flow requirements.

In rural areas, dry hydrants may be available. A **dry hydrant**, also called a drafting hydrant or suction pipe, is simply a pipe that is threaded on one end from which water can be pumped by the fire engine **FIGURE 12-4**. The pipe is usually placed in a static water source, such as lakes or streams that run year-round. The pipe may also be used in conjunction with underground storage tanks or cisterns. A dry hydrant is not part of the domestic water supply but can be a reliable source of water from which to fill water tenders.

Before drafting water from a dry hydrant, flow water into the static water source to clear debris that may be against the strainer. Dry hydrants are not always maintained annually. Not pushing water through the strainer could reduce the available water needed when drafting occurs.

Water Tenders

Tank trucks that hold a large quantity of water, called water tenders or tankers, are widely used on wildland fires to transport water and water supplies to the

FIGURE 12-3 Fire hydrant.
Courtesy of Jeff Pricher.

A

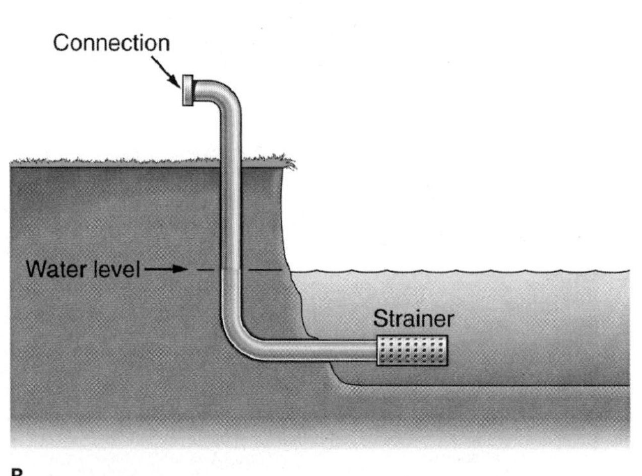

B

FIGURE 12-4 Dry hydrant. **A.** Aboveground view. **B.** Belowground view.
A: Courtesy of Jeff Pricher.

scene. They are usually equipped with a pump that enables them to fill from a drafting site and to offload water when needed. Many tenders are equipped with a dump valve for rapidly unloading water into a portable tank. Water tenders are discussed in more detail in Chapter 8, *Ground and Air Equipment*.

Portable Water Tanks

Portable water tanks (Port-A-Tanks) are containers used as reservoirs to store water that engines can use to refill. Large-capacity open tanks can be used to fill helicopters that have snorkels. Water tenders or helicopters in remote areas can be used to shuttle water to these portable tanks. If a stream or pond is close to the area where the portable tank is set up, then it can be supplied by a floating or portable pump.

Portable water tanks come in open and closed varieties. They come in many styles, the most common being a pillow, or pumpkin, shape or a self-supporting pyramidal design **FIGURE 12-5** and **FIGURE 12-6**. Collapsible tanks consist of a tubular metal frame with a synthetic or canvas duck liner **FIGURE 12-7**. Each type is foldable, which allows for ease of storage. Once folded, they are easily transported. Tank sizes vary from those normally employed for staged-type use to large 600- to 3000-gallon (2271- to 11,356-L) varieties.

FIGURE 12-5 Pumpkin-shaped portable water tank.

FIGURE 12-6 Pyramidal-shaped portable water tank.
Courtesy of Jeff Pricher.

FIGURE 12-7 Collapsible metal frame portable water tank.
Courtesy of SEI Industries LTD.

In a remote area or aircraft filling operation, it is not uncommon to observe several tanks connected to each other to supply water.

Hoses

A **hose** is a device used to deliver water from a water source to a fire. Wildland hoses normally come in 100-ft (30.5-m) lengths. There are typically two types of hoses used on wildland fires: attack hoses, which are used for fire suppression, and supply hoses, which deliver water from a water source for the attack. Techniques for rolling and unrolling a hose are discussed in the Hoselays section later in the chapter.

Forestry Fire Hose

A **forestry fire hose**, also called a forestry hose or soft-suction hose, is a lightweight, single-jacket, collapsible attack hose **FIGURE 12-8**. These hoses come in ¾-in. (19-mm), 1-in. (25.4-mm), or 1½-in. (38-mm) diameters. This small diameter provides for maneuverability through brush and trees. Forestry fire hoses are usually made out of an inner waterproof liner surrounded by an outer layer of high-strength woven fibers, such as nylon or, in some cases, rubber. Sometimes, these hoses are extended for hundreds of feet.

Hard Suction Hose

A **hard suction hose** is a noncollapsible hose used to supply water to the intake side of the fire pump (fire pumps are discussed later in this chapter) **FIGURE 12-9**. It drafts water from a source that is lower than the pump. The vacuum pressure created by a pump is needed to suck water out of a low location into the pump and out to the hose or into a tank.

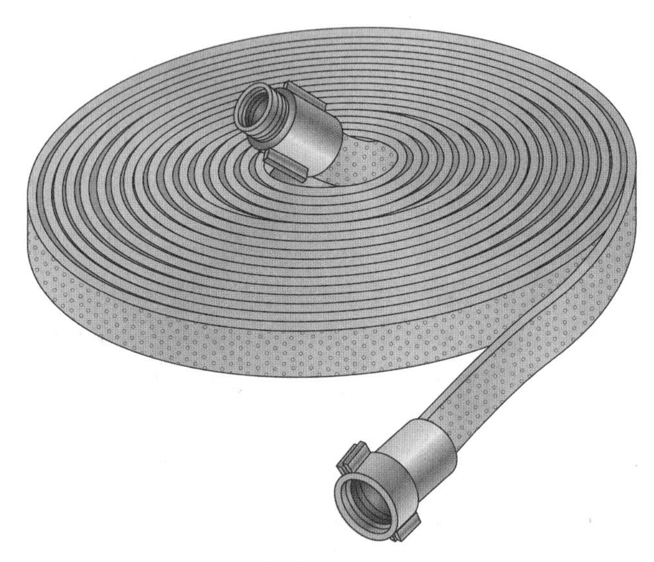

FIGURE 12-8 Forestry fire hose.

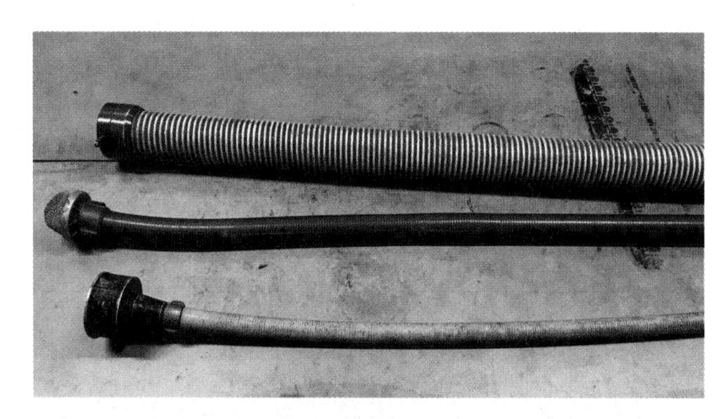

FIGURE 12-9 Hard suction hoses.
Courtesy of Jeff Pricher.

That pressure is so significant that rigid materials must be used to keep the hose from collapsing on itself. Typical soft suction, or soft, hoses are not able to withstand the pressure caused by a pump that sucks water. In this instance, the soft hose will collapse, and no water will be able to get to the pump. A hose that is delivering water under pressure to a pump (from a fire hydrant or another fire engine or pump) will not collapse unless the volume of water being pulled in is more than is being delivered by the pressurized water source.

Hard suction hose diameters are based on the capacity of the pump but can be as large as 6 in. (152.4 mm). These hoses are typically made of rubber or plastic.

Hose Appliances

There are many types of hose appliances. They include but are not limited to hose clamps, nozzles, gravity socks, and other connectors and accessories.

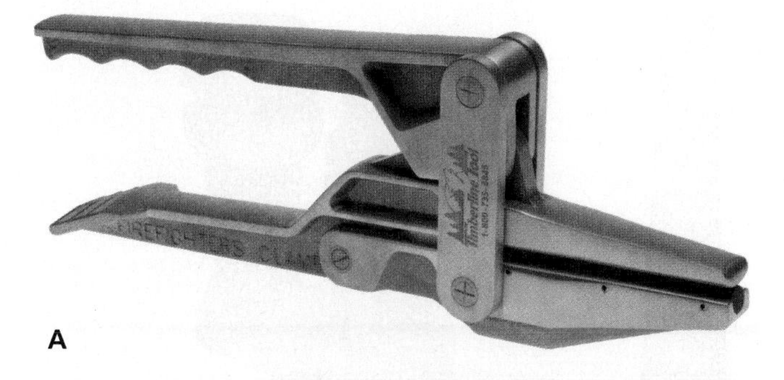

A

B

FIGURE 12-10 Hose clamp.
Courtesy of Timberline Tool.

Hose Clamp

A **hose clamp** is a crimping device used to temporarily stop the flow of water in a hose **FIGURE 12-10**. It allows you to extend your water supply without having to shut off the water source. It can also be used to stop water flow in a hose line if the hose ruptures or needs to be connected to a different appliance.

To restrict water flow with a hose clamp, turn off the nozzle or gated wye (if applicable). Place the hose clamp about 18 inches (45.72 cm) from the coupling, then open the nozzle to relieve pressure on the hose. Hose clamps can be used in both simple and progressive hoselays (discussed later in this chapter).

Nozzles

A **nozzle** is the spout connection to the end of a hose that water flows through. Nozzles are used to move water from the source or hose to the fire. There are several types of nozzles to choose from. The most common include fog stream, straight stream, and twin tip. **TABLE 12-1** and **FIGURE 12-11** outline common nozzles used on wildland fires. Information on how to use and maintain nozzle pressure and flow during hoselay is discussed later in this chapter.

TABLE 12-1 Common Wildland Nozzles and Usage

Nozzle/Fitting Type	Water Pattern	Usage	Gallons per Minute
Fog stream nozzle	Fog; fine water droplets	Fire attack across a large area or during close-up suppression, mop-up	10–95
Straight stream nozzle	Straight stream	Fire attack when fire is too hot to get close and is confined to a small area, mop-up	Varies
Twin tip "forester" nozzle	Fog and straight stream	Mop-up	Fog: 3–6 Straight: Varies
Air-aspirating foam nozzle	Foam	Fire attack, pretreatment	Varies

Gravity Sock

Used to take advantage of flowing water that is above the fire, a **gravity sock** is placed in a stream and anchored securely **FIGURE 12-12**. The large opening is placed in the stream, and the tail end has a hose connected to it. Gravity feeds the water into the sock and to the hose.

Connectors and Accessories

There are several types of hose connectors and accessories. **TABLE 12-2** and **FIGURE 12-13** outline the

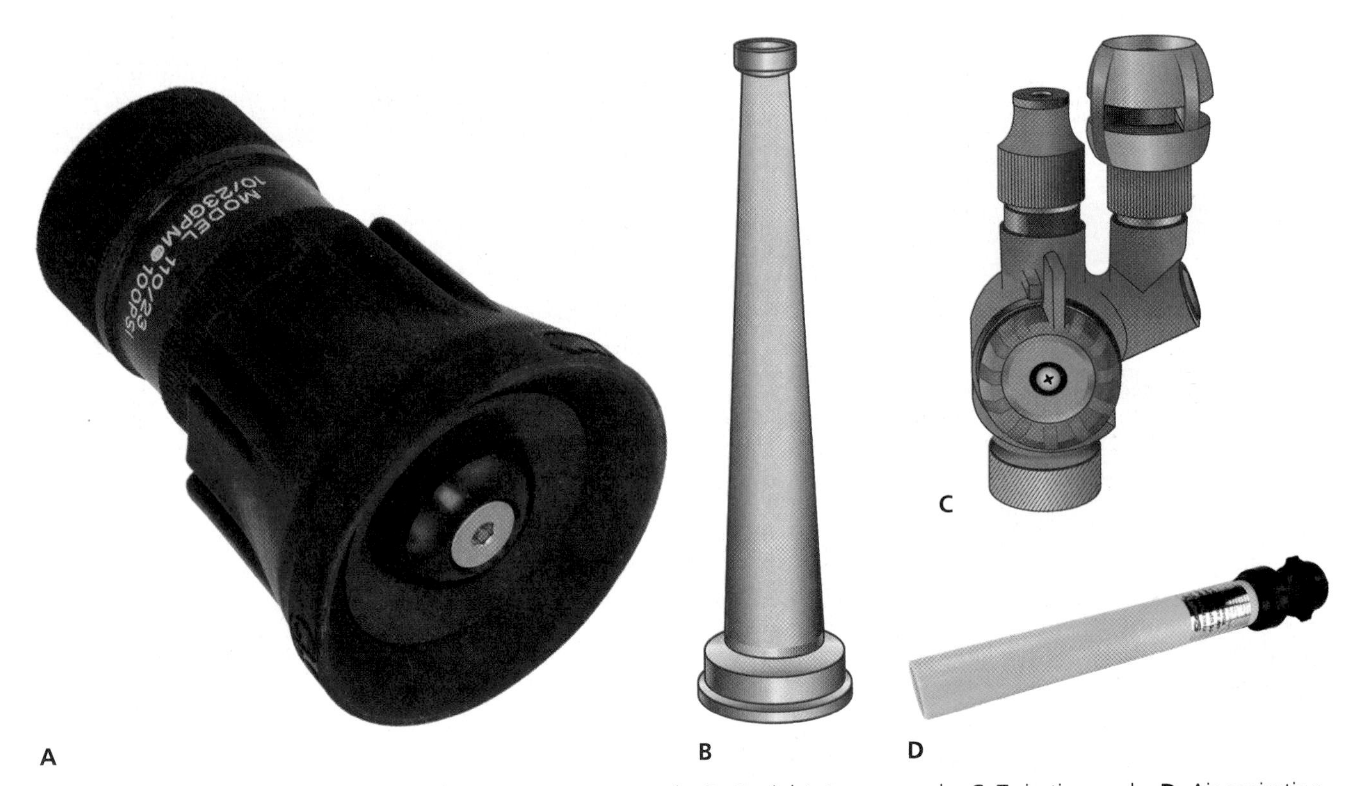

FIGURE 12-11 Common wildland nozzles. **A.** Fog stream nozzle. **B.** Straight stream nozzle. **C.** Twin tip nozzle. **D.** Air-aspirating foam nozzle.

A: Courtesy of C and S Supply Inc; **D:** Scotty Air Aspirating Foam Nozzle 57 Lt. 15 GPM, National Fire Fighter Corp. Retrieved from http://www.nationalfirefighter.com/store/p/3175-Scotty-Air-Aspirating-Foam-Nozzle-57-Lt-15-GPM.aspx.

FIGURE 12-12 A. Gravity sock. **B.** Gravity sock in use.

A: Courtesy of Cascade Fire Equipment; **B:** Courtesy of Chris Harper.

TABLE 12-2 Hose Accessories

Accessory	Description/Use
Adapter	Connects hoses with different thread types
Cap/plug	Provides thread protection and closes off hose pieces when not in use
Check and bleeder valve	Combination valve that keeps water from flowing back into the pump when pump is stopped; relieves pressure when pump is restarted
In-line shutoff valve	Enables water shutoff while the pump is still running
Pressure relief valve	Adjustable valve placed between the pump and the discharge hose to prevent overheating and release excess pressure on the pump when the nozzle is shut off
Reducer/increaser	Used to attach a smaller-diameter hose to a larger-diameter hose
Siamese connection/adapter	Combines two hose lines into one; can be improvised if needed by connecting two double female adapter and one double male adapter to a gated wye
Tee	Brings a hose off a main line; most do not have a shutoff valve
Washer/gasket	Rubber ring used with female accessory ends to provide a tighter seal
Wye	Divides a single hose line into two lines

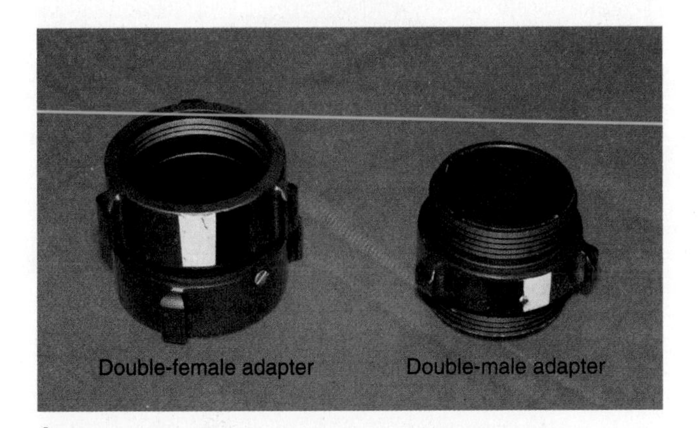

A

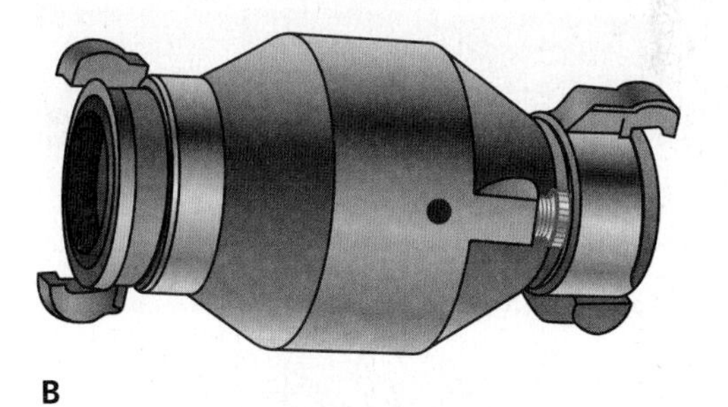

B

C

D

FIGURE 12-13 Hose accessories. **A.** Adapter. **B.** Check and bleeder valve. **C.** Reducer/increaser. **D.** Siamese connection.

E

F

G

H

FIGURE 12-13 *(Continued)* **E.** Improvised Siamese connection. **F.** Tee. **G.** Washer/gasket. **H.** Wye.

E: Courtesy of Jeff Pricher; **F:** Forestry Plain Hose T with Cap and Chain MODEL # T1.5 NST, C and S Supply Inc. Retrieved from https://www.cssupplyinc.com/forestry-plain-hose-t-with-cap-and-chain-model-t1-5-nst/; **G:** © Jones & Bartlett Learning. Photographed by Glen E. Ellman; **H:** 1" Gated Wye, Wildland Warehouse. Retrieved from https://www.wildlandwarehouse.com/shop/1-gated-wye/.

common ones. Some of these accessories require spanner wrenches **FIGURE 12-14**. To tighten and loosen them see **SKILL DRILL 12-1**.

LISTEN UP!

It is important to consider what tools and equipment your neighboring departments have. If you are in an area where a neighbor has different fittings, make sure you have a large supply of the appropriate parts so that joint suppression efforts are seamless.

FIGURE 12-14 Spanner wrenches.

© Jones & Bartlett Learning. Photographed by Glen E. Ellman.

SKILL DRILL 12-1
Uncoupling a Hose with Spanner Wrenches

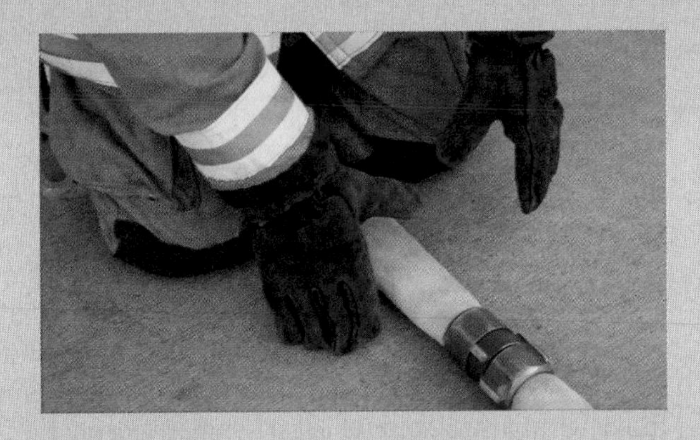

1 With the connection on the ground, straddle the connection above the female coupling.

2 Place one spanner wrench on the swivel of the female coupling, with the handle of the wrench to the left.

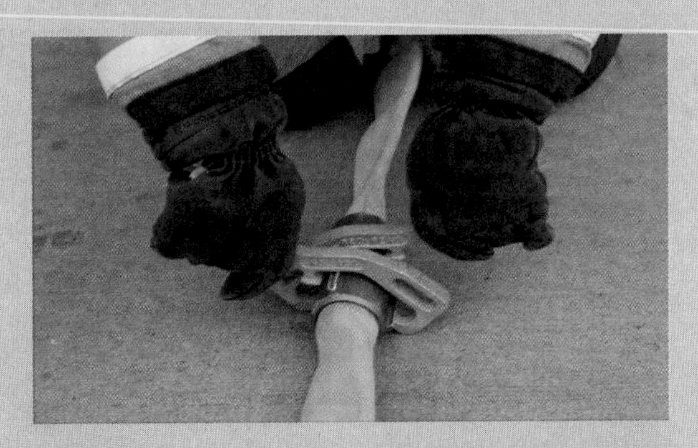

3 Place the second spanner wrench on the male coupling, with the handle of the wrench to the right.

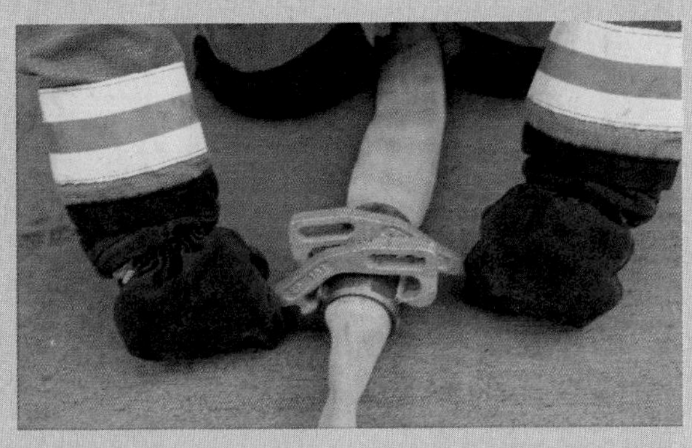

4 Push both spanner wrench handles down toward the ground, loosening the connection.

LISTEN UP!

It is important to always check the female side of the coupling to make sure a gasket is present and in good condition. If the gasket is missing or in poor condition, substantial leaking could result.

Portable Pumps

A **portable pump** is a small, usually gasoline-driven pump that can be carried to a water source and used to draft water **FIGURE 12-15**. **Drafting** is the process of drawing water from static water sources into a pump. Drafting is different than obtaining water from a fire hydrant because the water coming out of a fire hydrant is pressurized. Because drafting is done from a static water source, the pump acts as a vacuum to decrease atmospheric pressure. The atmospheric pressure outside of the hose in combination with the vacuum inside the hose uses gravity to pull the water into the pump. The pump then discharges the water under pressure. Portable pumps can deliver as much as 500 GPM (1892.7 L/min) and can operate on dry land next to a water source or physically in or on the water.

Portable pumps should be operated and maintained according to the pump manufacturer's instructions. A professional pump mechanic may be needed to perform significant repairs.

There are several types of pumps, including backpack, eductor, floating, and high-pressure portable pumps.

LISTEN UP!

Many factors will challenge drafting operations, such as temperature and altitude. It is important to know the capabilities and limitations of the pumps you work with.

FIGURE 12-15 Portable pumps can be used to draw water from natural water sources, such as lakes.

Courtesy of Jeff Pricher.

LISTEN UP!

Protect your ears! Most wildland fire fighters are exposed to sound levels beyond the daily recommended maximum. Whenever you are working around a pump and will be in the immediate vicinity of the motorized equipment for a significant period of time, it is essential that you wear hearing protection. It is not uncommon for pumps to run with an average decibel rating of 80–90, which can cause hearing loss.

Backpack Pump

A **backpack pump**, also called a backpack fire extinguisher or bladder bag, is a portable pump and tank fitted with straps so that it can be carried on a fire fighter's back **FIGURE 12-16**. It can be used in many situations, but it is particularly useful during wet mop-up operations (discussed later in this chapter).

Backpack pumps are designed to be refilled easily in the field, such as from a lake or stream, through a large opening at the top. A filter keeps dirt, stones, and other contaminants from entering the tank. When completely filled, a 5-gallon (18.9-L) backpack pump weighs approximately 45 pounds (20.4 kg).

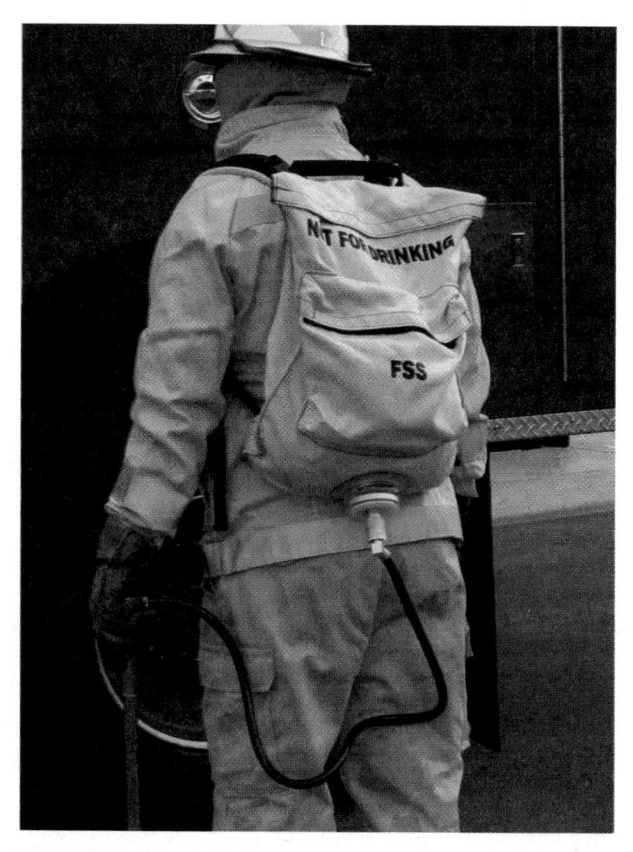

FIGURE 12-16 Backpack pump.

SKILL DRILL 12-2
Operating a Manual Backpack Pump

1. Fill the backpack with water from a water source. Seal the backpack closed.
2. With sturdy footing, lift the backpack carefully.
3. Move the backpack behind you, and put your arms through the carrying straps. Adjust the straps to fit your body.
4. To discharge, hold the pump in both hands. One hand should be stationary at the end of the nozzle while the other moves the piston/trombone in and out.
5. Direct water streams in a swinging motion parallel to the fire perimeter and at the base of the flame.
6. To produce a fog or straight stream, change the nozzle tip, or place a finger over the nozzle tip.

Most backpack pumps are operated via manual hand pumps. To operate a backpack pump, refer to **SKILL DRILL 12-2**.

If a backpack pump is not working properly, do the following:

- Check that there is water in the tank.
- Check the bottom outlet, nozzle tip, and inside the hose for blockage.
- Check that the ball in the check valve is not stuck in the open or closed position.

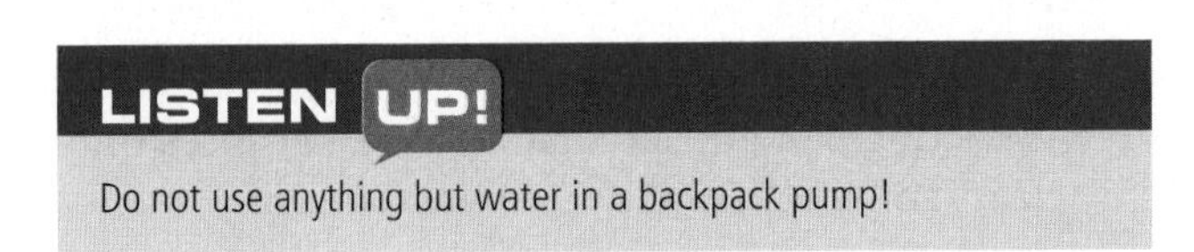

Do not use anything but water in a backpack pump!

Fire Eductor Pump

One tool that is becoming more common in wildland firefighting is the fire eductor. The **fire eductor pump**, also called an ejector or jet siphon, is a portable pump used to draft water from static sources **FIGURE 12-17**. This tool uses the Venturi effect from water discharged from the pump and tank to assist with lifting water and moving it greater distances than with hard suction hoses. The biggest benefit is that you will have usable and sustained water supply and flow. Drawbacks to this tool include needing enough hose supply to be successful and how much vertical lift is required to get from the water source to the pump. In some instances, a combination of soft and rigid hose is required for significant vertical obstacles. Training and consultation from the manufacturer will aid in your success in using this type of equipment.

FIGURE 12-17 Fire eductor pump.
Courtesy of Jeff Pricher.

Floating Pumps

Floating pumps are portable pumps that float on the water's surface **FIGURE 12-18**. A floating pump continuously draws water and feeds it to the hose. Some engines carry floating pumps. They work extremely well on interface fires where there are many

FIGURE 12-18 Floating pump.

residential swimming pools or in wildland areas with available lakes or streams. Oftentimes, floating pumps are used when water is not deep enough to allow for other types of drafting. These types of pumps have intake ports and strainers that go below the surface of the water and do not draw air into the pump.

High-Pressure Portable Pumps

High-pressure portable pumps are used to pump water from pools, lakes, ponds, streams, and other water sources. They can be used to supply firefighting hose lines and to fill portable tanks, engines, and water tenders from water sources that are inaccessible to fire vehicles.

The most common high-pressure portable pumps are the Mark-3 and the Wick 375 **FIGURE 12-19**. These pumps can be finicky. However, they have an impressive water-delivering capability. It is not uncommon for three to four high-pressure pumps to be able to push water 20 to 70 ft (6.1 to 21.3 m) in elevation over several miles of hose.

One of the nice features of these pumps is that the pump heads are interchangeable between the brands and can be swapped out in the field if one breaks. This is also the case for most light-brush engines.

Portable Pump Operator

The portable pump operator maintains and works the portable pumps. He or she maintains a consistent supply of water and pressure to the nozzle operators. The pump operator must remain alert at all times.

Tools and Equipment

In addition to standard personal protective equipment, the pump operator must have a warm coat, hard hat, gloves, earplugs, and established radio communications. The pump operator must also have a pump kit. The kit contains the following:

- Pump
- Fuel can and fuel
- Fuel supply line
- Suction hose
- Foot valve/strainer
- Tools
- Spare gaskets

FIGURE 12-19 Mark-3 high-pressure portable pump.

The pump operator's job is to ensure that all items in the pump kit are compatible and the proper sizes.

Responsibilities

The pump operator is responsible for the following:

- Selecting a pump site that is safe and causes the least amount of environmental damage
- Ensuring gas and oil are not spilled when refueling the pump.
- Determining or building a pump base with logs, sand, etc.
- Maintaining consistent water supply and pressure to nozzle operators
- Communicating, by radio or with hand signals, with the nozzle operators as needed **FIGURE 12-20**

In wildland firefighting, radios are almost always used over hand signals. Most of the time, brush and trees obstruct the view for both the pump operator and fire fighters. Still, it is important to know these hand signals in case of an emergency, such as radio failure.

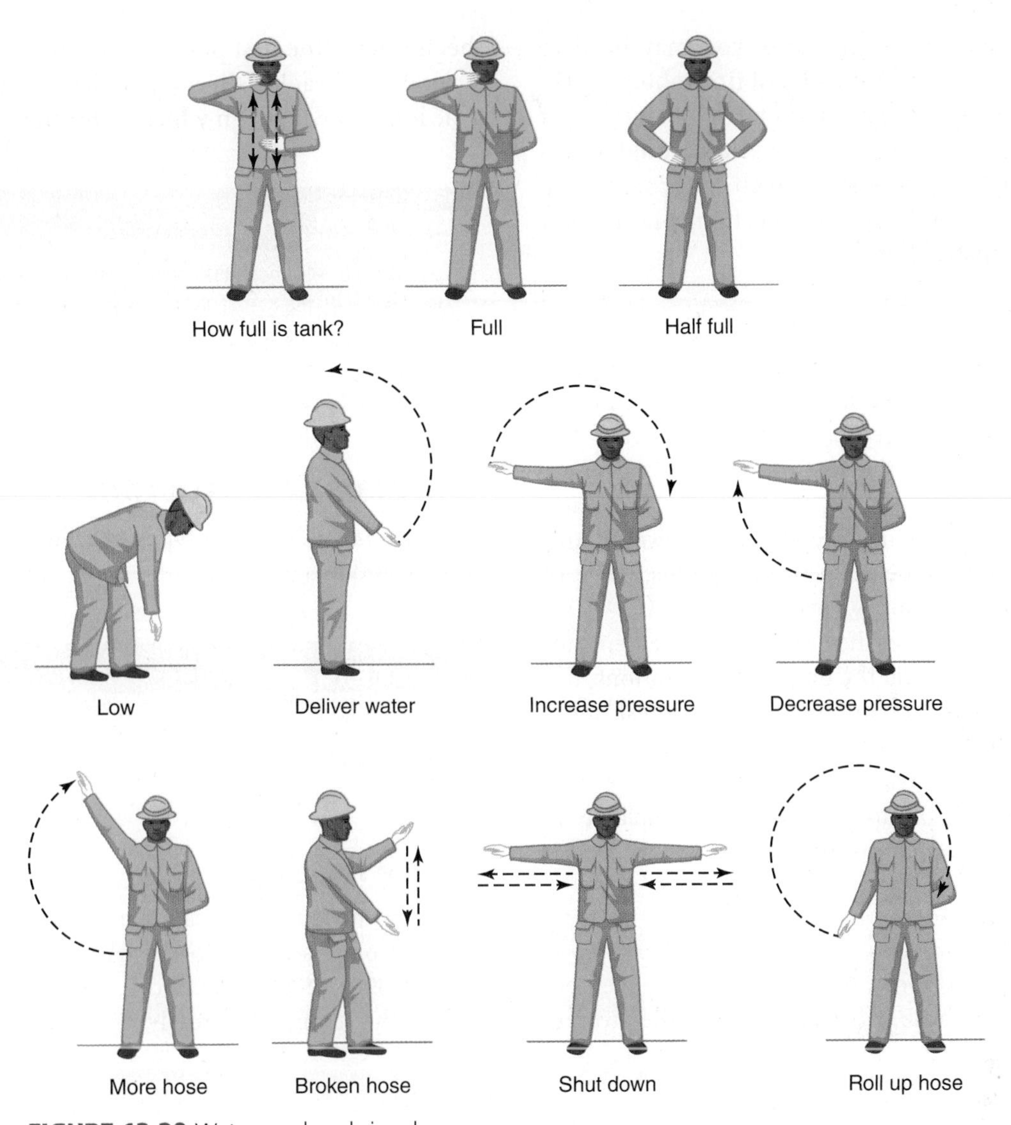

FIGURE 12-20 Water use hand signals.

LISTEN UP!

With the changes to the fuel makeup as required by the Environmental Protection Agency, it is essential to make sure you use "nonethanol" fuel in portable pumps. Prolonged use of ethanol-based fuel will damage your equipment. Not all fuel dispensers carry ethanol-free fuel, so ask before you fill up your fuel can.

Portable Pump Setup

There are three portable pump flow setups: staged, parallel, and series. It is important to understand the basic dynamics of these setups for longer hoselays and hoselays that deal with increased or decreased elevations. Depending on the fire setting, you may be required to use all three concepts or variations of the three. As the pump setups become more complicated, allow enough time to set up and plan for this operation.

Staged Setup

Staged pump configurations are pumps that are set up to include a single pump or two pumps that are not connected to each other. This concept is used when the available equipment creates a situation where one pump pulls water from a water source and pushes it to a portable tank that is used for distribution to an engine or another pump. Each pump operates independently, and the flow of the system is dependent on the individual pump ratings. It is important to have equally powerful pumps operating for this concept to work.

Parallel Setup

Parallel pump configurations are utilized to increase the volume of water for the intended use. Effective pumping will require two portable pumps pumping into one line. The main hose line is called the trunk line. Depending on the scenario, you may need to push a lot of water and have a lot of fire fighters working off of the trunk line. Firefighting operations or mop-up assignments with long hoselays would necessitate needing a lot of water, which is accomplished with two pumps. A Siamese connection is needed for this type of construction.

Series Setup

The series setup is a dual-pump effort that is used when increased pressure is needed for a specific use on a fire. In a series setup, the discharge from one pump is hooked into the suction of a second pump. The closer the pumps are to each other, the greater the chance for doubling the pressure (however, this does not always happen). In some settings, three or four pumps can be spread over a 3000-ft (914.4-m) hoselay in order to meet the pressure requirements for fire flow. Depending on the pump configurations, it is important to place the pump with the greatest pump capacity closest to the water source.

This operation takes planning to ensure that enough qualified pump operators are available to oversee the various pumps in the system.

Hoselays

A **hoselay**, or two words, hose lay, is the arrangement of connected fire hose and accessories on the ground, beginning at the first pumping location and ending at the point of water delivery. Hoselay essentially creates one long hose through which water travels.

LISTEN UP!

To properly unroll a hose, first, remove any storage ties, such as string or bands. Then, carefully unroll the hose so that the female end points toward the water source and the male end points toward the fire.

Hoselay Construction

The two most common types of wildland hoselay are simple and progressive.

LISTEN UP!

Always wear personal protective equipment when constructing a hoselay, regardless of your task.

Simple Hoselay

A **simple hoselay**, also called a single hoselay, can be thought of as a pump-to-discharge, or A-to-B, hoselay **FIGURE 12-21**. It includes hoses, fittings, and a nozzle that connects directly to a discharge port on a

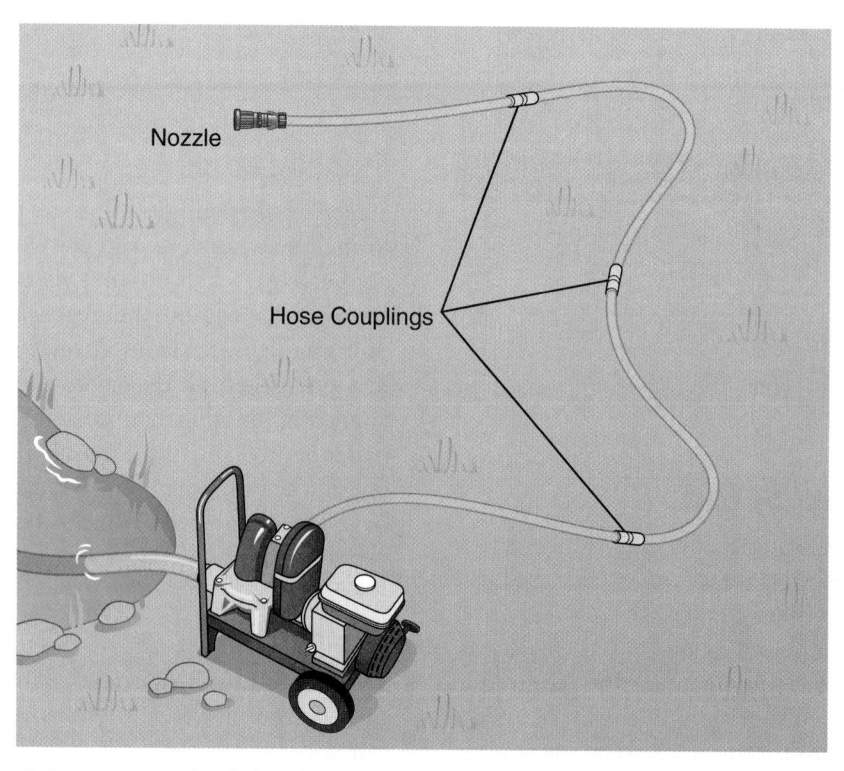

FIGURE 12-21 Simple hoselay construction.

pump. Once the hose is on the ground, water can flow through it. The simple hoselay connects hose in a line. It is the most basic form of water delivery.

There are several advantages to using a simple hoselay:

- High water pressures and high volumes of water are available for long distances and large fires.
- There are no valves to be accidentally shut off.

Despite the simple nature of water delivery with this system, there are a few disadvantages:

- Once water has been pushed to the nozzle, if the hoselay needs to be extended, the water usually needs to be shut off.
- Depending on the length of the hoselay, if a spot fire or a flare-up were to occur behind the nozzle, there would fire between you and the water source. This could compromise your escape route to your safety zone.

LISTEN UP!

Make sure to lay the hose so that it will not be damaged by fire, terrain, equipment, or apparatus. In some instances, you will need to be creative by digging under railroad tracks or digging a trench in a road that is regularly traveled. The trench will prevent damage to the hose.

- During mop-up, simple hoselays will require you to drag a lot of hose. This will decrease your overall energy as a result of having to work hard to move the hose.

Progressive Hoselay

A **progressive hoselay** is designed to quickly and continuously move water from the anchor point to the fire without having to shut the water off at the pump. Progressive hoselays are often used on fast-running fires and during mop-up. They allow for lateral hose lines to be attached to wyes branching from the main trunk hose line as more personnel and hose arrives at the scene **FIGURE 12-22**. Progressive hoselays typically include shutoff valves (wyes or tees) and hose clamps. The construction of a progressive hoselay allows for fire attack at several points.

LISTEN UP!

Most in-line tees do not have a shutoff valve and require the hose to be clamped with a hose clamp.

Progressive hoselays should always start from an anchor point. When deploying hose, effort should be made to deploy the hose in the green and far enough

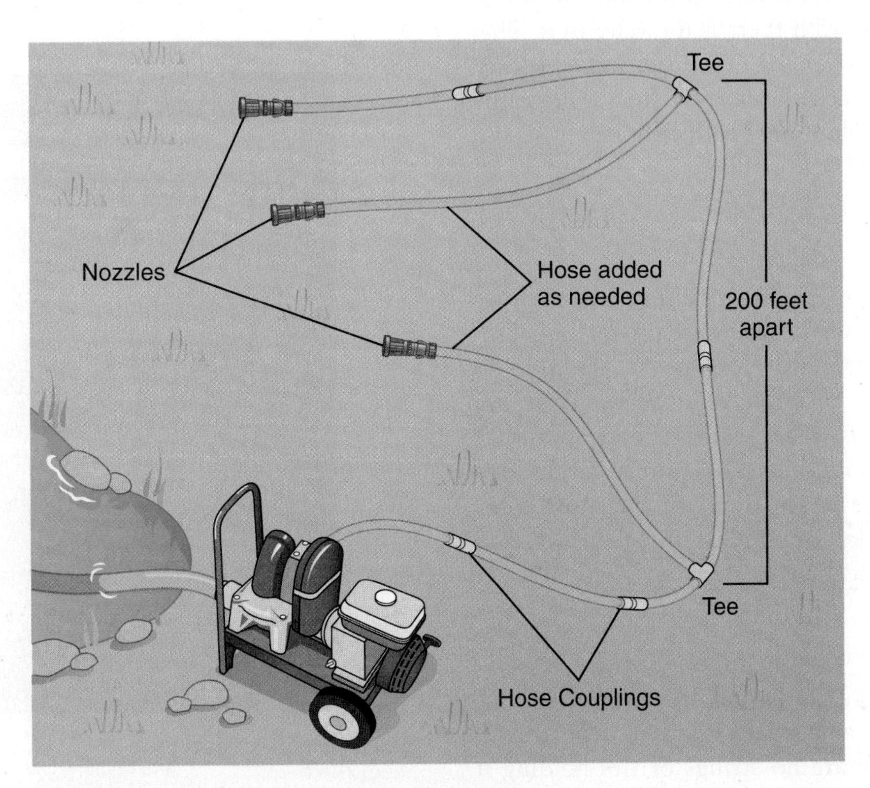

FIGURE 12-22 Progressive hoselay construction.

away to prevent the hose from being damaged by embers, flame, or heat.

Extend the hoselay as needed. This can be accomplished without turning off the pump, by stopping the water with a hose clamp. Once the new line has been added, the clamp is removed, and water starts flowing again.

Hoselay is extended with wyes or tees so that additional lateral hoses can be added as needed, when more personnel arrive.

There are several advantages to progressive hoselay, including:

- Lateral hose line allows for fire attack at several points
- The main trunk hose line can be turned off without the need of clamps
- Less water is used with a 1-in. (25.4-mm) nozzle

Disadvantages of progressive hoselay include the following:

- Very gear intensive
- Requires a lot of training
- Increased friction loss (this may be a factor on long hoselays or if elevation plays a role on your hose line)

It is important for you to become very familiar with the type of progressive hoselay that your agency uses. There are many types of hoselay construction that are used nationwide. While the setups may vary, all progressive hoselays follow the same principle: the nozzle does not move forward if there is no water in it. That is, dry hose is not advanced in a progressive hoselay. About 10 percent slack should be left in a hose line to allow for hose replacement with a shorter hose and easier movement of the hose line.

Hose Packs

A **hose pack** is a backpack containing a hose, hose accessories, and other hoselay equipment in a predetermined arrangement. Hose packs are used to quickly perform a progressive hoselay. They are usually used in areas where water tenders and other large vehicles cannot reach. There are several kinds of progressive hose packs. Some of the most common packs include the Travis, the Gansner, the Gnass, and the Pondosa.

Travis Packs

Travis packs utilize a configuration of 200 ft (61 m) of hose packed into a backpack for rapid deployment. Once packed, there are no strings or ties holding it together. Travis packs are great for minimizing bulk in limited storage compartments and being easy to pack **FIGURE 12-23**.

Gansner and Gnass Packs

Gansner packs are less gear intensive than Travis packs and utilize the hose as the backpack **FIGURE 12-24**. Gansners generally utilize 100 ft (30.5-m) of 1.5-in. (38.1-mm) hose for trunk line and 100 ft (30.5-m) of 1-in. (25.4-mm) lateral hose. The appliances that make this a Gansner are the 1.5-in. (3.8-cm) gated wye with one side of the wye fitted with a 1-in. (2.54-cm)

FIGURE 12-23 Travis pack.
Courtesy of Jeff Pricher.

FIGURE 12-24 Gnass pack.
Courtesy of Mike Gorsuch.

reducer. Building the Gansner takes time, and, to build it effortlessly, you need a flat surface. Depending on the setup, a hose clamp may be required to add additional sections. One of the benefits of this type of deployment is that, when the hose is charged, it forms a coil that allows for forward deployment with little if any entanglement of the hose.

Gnass packs are very similar to Gansner packs in that they also utilize the hose as a backpack. The central difference is that a Gnass pack utilizes in-line tees instead of gated wyes.

There are many ways to set up Gansner and Gnass packs. If you are using parachute cord, tuck all of the slipknot ends into the folds of the hose to prevent snagging and accidental deployment.

Pondosa Pack

The **Pondosa pack** is a backpack of sorts made out of straps with clasps so that a fire fighter can carry two or three 100-ft (30.5-m) doughnut-rolled hoses on their backs **FIGURE 12-25**. It allows for easy hose deployment during progressive hoselays.

Doughnut-rolled hose is folded in half and then rolled up so that both couplings end together and outside of the roll **FIGURE 12-26**. The hose is then stored on the Pondosa backpack.

When deploying a doughnut-rolled hose, the hose is rolled out *away* from the fire, charged, and then dragged forward. Be careful when rolling out the hose. Depending on the fuel type, it may snag on brush or bushes.

Performing a doughnut roll on a hose that fits into the Pondosa pack is discussed later in this chapter.

A

B

FIGURE 12-25 Pondosa pack. **A.** This Pondosa pack can carry two rolls of 100-ft (30.5-m) hose. **B.** Pondosa pack in use.

A: "C" Series Pondosa Pack, The Pondosa Pack. Retrieved from http://pondosapack.com/styles; **B:** The Pondosa Pack Hosepack. Retrieved from http://pondosapack.com/.

Retrieving Hoselays

While it is important to know how to deploy hose and set it up for water delivery, it is equally important to know how to break down sections of hose and safely store it for future use after suppression has been completed.

Before hose can be removed and retrieved, all of the water must be drained from it. This often works best when the hoses are uncoupled or detached and a fire fighter lifts the hose up to shoulder height and, starting from the downhill side, walks the entire length of the 100-ft (30.5-m) section. Once the fire fighter has walked the 100 ft (30.5 m) of hose, he or she can start to retrieve the hose from the uphill side. Any residual water will drain out with gravity.

If a hose has been identified as damaged, tie a knot in the hose. If you have flagging tape, tie it near the knot. After use, all hose should be inspected and tested

FIGURE 12-26 Doughnut-rolled hose.

Courtesy of Jeff Pricher.

to agency directives and NFPA 1962, *Standard for the Care, Use, Inspection, Service Testing, and Replacement of Fire Hose, Couplings, Nozzles, and Fire Hose Appliances*.

There are several techniques for retrieving and packaging hoses. The most common are the butterfly, or figure-eight, method and the doughnut roll.

Butterflying the Hose

The most common technique for retrieving and packaging hose is called butterflying the hose (also called the figure-eight method). Butterflying the hose utilizes no tools other than the hands and arms of a fire fighter. This method winds up the hose in a way that it can be looped over a tool so that several hundred feet can be carried at a time during retrieval. Some departments choose to store butterflied hoses in groups so that, in a pinch, the hose can be opened up and stretched out without tangles. To butterfly a hose, follow the steps in **SKILL DRILL 12-3**.

Doughnut Rolling the Hose

Another method used for retrieving and packaging hose is called the doughnut roll, or the single-doughnut roll. The doughnut roll is used when the hose will be put into use directly from the rolled state. With this arrangement and for easy access, both couplings are on the outside of the roll. The hose can be connected and extended by one fire fighter; it unrolls as it is extended. Doughnut-rolled hose can be carried in a Pondosa pack or stored as is. To perform a single-doughnut roll, follow the steps in **SKILL DRILL 12-4**.

Wet Mop-Up Operations

Wet mop-up uses water or water and soil together to extinguish burning materials. The mop-up method will vary depending on the fuel type.

In planning for wet mop-up, an assessment will need to be conducted to determine how much hose will be needed, the water supply, the number of fire fighters available, and the types of water-delivery appliances that will be used. All of these components will have to be calculated in conjunction with the slope and size of the fire. Depending on the terrain and the required depth, garden hose, or ¾-in. (19.1-mm) hose, could be included in the configuration. Once a fire is contained and controlled, mop-up operations begin.

SKILL DRILL 12-3
Butterflying a Hose

1 Hold either end of the hose in one hand so that the remainder of the hose is still on the ground and the last bit of hose from the ground is vertical to the hand.

2 Stretch out your arms so that your body resembles the letter *T*.

SKILL DRILL 12-3 (Continued)
Butterflying a Hose

3 With your arms stretched out, slowly twist your body in an alternating, side-to-side motion while wrapping the hose around each arm. Make sure that there is a cross in the middle every time you switch from one side of the body to the other.

4 Continue wrapping the hose until all of it is wrapped around your arms.

5 Once all of the hose is wrapped, pull one arm out so that the hose is hanging from one arm.

6 Tightly wrap the trailing end of the hose around the top of the crisscrossed section close to your arm. When it can no longer be wrapped, connect the two coupling ends. The hose is now ready to be carried or redeployed.

Courtesy of Jeff Pricher.

SKILL DRILL 12-4
Performing a Doughnut Roll

1 Lay the hose flat and in a straight line.

2 Locate the midpoint of the hose.

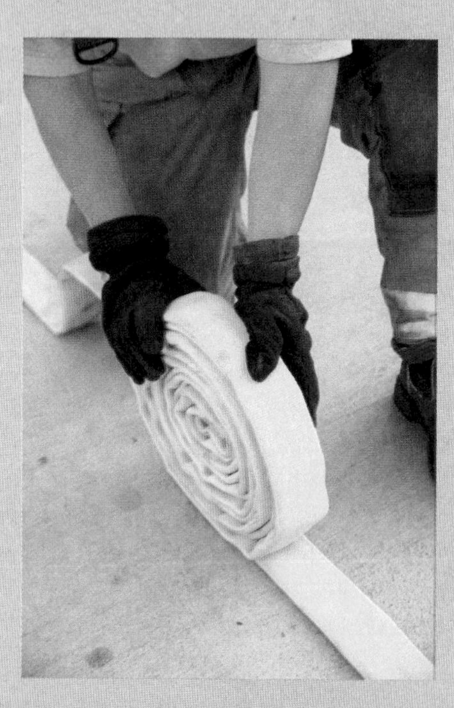

3 From the midpoint, move 5 ft (1.5 m) toward the male coupling end. From this point, start rolling the hose toward the female coupling.

4 If possible, at the end of the roll, wrap the excess hose of the female end over the male coupling to protect the threads on the male coupling.

Wet Mop-Up Techniques

Wet mop-up primarily consists of exposing burning materials using hand tools and then extinguishing the burning materials with water or water and soil. There are several techniques used to perform wet mop-up:

- Use a fine spray to apply water. This conserves water and is the most effective wet mop-up method.
- Apply water from the control line, and work your way toward the inside of the burned area.
- Apply water from the outside of a hot area, and work your way inward to the center.
- Use a straight stream to reach and wet a target.
- Spray dirt, stir, and spray again. Repeat as necessary.
- Apply foam.

When using water for mop-up, conservation of water will allow you to be the most effective. If you have a limited water supply, try not to waste water by using a straight stream to cool and turn the dirt. While you will expend more energy spraying a little water and then digging a bit, you will actually complete mop-up operations faster.

LISTEN UP!

When using a straight stream to reach a tree target, begin at the base of the tree and work your way up. For up-close mop-up, hand tools are essential. Haybales or similar fuels must be torn apart, sprayed, and spread out.

Wet Mop-Up Radius

Most agencies require the initial mop-up standard to be a 50-ft (15.2-m) radius around the fire. Once a 50-ft (15.2-m) depth all around the fire has been established, that depth will be increased as needed. The increase will be determined by agency policy and the fuel type. It is common to have to reach a depth of 200 ft (61 m), but it is rare to go beyond 300 ft (91.4 m). The deeper the mop-up operations are extended, the less safe the environment is for the fire fighter. The exposure to snag hazards increases exponentially. If you do have to extend your mop-up a considerable distance off the fire edge, ask to have a faller clear the area of snags. On large fires, it is very common to have falling teams ahead of mop-up operations to clear hazard trees for safety. If you identify a hazard tree, flag off the area, and stay clear until the hazard can be mitigated.

Safety Considerations

Common hazards associated with wet mop-up include dirt, rock, and ash particles reflecting and blowing back at the fire fighter when spraying hot rocks; white ash or steam from spraying a hot stump or rocks; spraying others if not careful with equipment; slippery footing; and eye irritation or damage due to lack of eye protection.

Basic Hydraulics

Hydraulics is the science of water in motion and at rest. Hydraulics explains how and why water is able to get from a water source to the end of the nozzle.

Fire fighters can save much time and reduce guesswork by planning ahead and considering basic hydraulics. Hydraulics allows you to know beforehand whether a hoselay will work. Hydraulic calculations allow you to save time and prevent fire-fighter injury and equipment damage.

Pump-Discharge Pressure

Many factors can decrease pump pressure: friction (water running against the inside of the hose line), appliances (such as wyes), and elevation.

Pump-discharge pressure (PDP) is the total pressure needed to overcome all friction, appliances, and elevation loss or gain while maintaining adequate nozzle pressure to deliver effective fire streams. This pressure is applied at the pump and can be increased or decreased by changing the throttle control on the pump. Each hose and nozzle combination has an optimal delivery pressure.

PDP calculations and formulas become more complicated as the complexity of fire attack operations increases, but all calculations must start with the basic PDP formula:

Pump-discharge pressure (PDP) =
nozzle pressure (NP) + friction loss (FL)

$$PDP = NP + FL$$

Calculating the nozzle pressure and friction loss are discussed next.

Nozzle Pressure

Nozzle pressure (NP) is the pressure that a nozzle is designed to operate at most efficiently. NP is measured in pounds per square inch (psi). The type of nozzle determines the nozzle pressure. It is important for you to know the pressure differences for the nozzles you are using. These differences will affect the PDP formula. Twin tip nozzles typically have an NP of 50 psi (344.7 kPa), and fog nozzles typically have an NP of 100 psi (689.5 kPa).

Friction Loss

After determining nozzle pressure, you need to determine friction loss (FL). **Friction loss** is the pressure lost from turbulence as water passes through hoses, fittings, adapters, and appliances.

Friction loss is calculated using the following formula:

$$FL = C \times Q^2 \times L$$

Where FL = friction loss in pounds per square inch;

C = the coefficient, a numerical measure that is a constant and specific to each hose type and diameter **TABLE 12-3**;

Q = the quantity of water flowing (in gallons per minute [GPM] or liters per minute [L/min]) divided by 100; and

L = the length of the hose in feet or meters divided by 100.

A friction-loss coefficient (C) is determined by the size of the hose and the roughness of the interior of the hose. As water flows through a hose, it creates friction. The smaller the hose, the higher the friction loss and the higher the coefficient. Generally accepted coefficient values can be found in fire publications. These values are outlined in **TABLE 12-3** and can be used in the friction-loss equation.

TABLE 12-3 Hose Friction-Loss Coefficients

Hose Diameter in Inches (Millimeters)	Coefficient
¾ (19.1)	1100
1 (25.4)	150
1 ½ (38.1)	24
1 ¾ (44.5)	15.5
2 (50.8)	8
2 ½ (63.5)	2

To determine friction loss in a hoselay, follow these steps:

For this example, we will be using a 2½-in. (63.5-mm) hose that is 200 ft (61 m) long and flows 300 GPM (1135.6 L/min).

1. Write the formula: $FL = C \times Q^2 \times L$
2. C: Select the coefficient (C) from the table that matches the hose diameter in the given scenario. The coefficient for the 2 ½-in. (63.5-mm) hose is 2. Therefore, C = 2.
3. Q: Determine the quantity (Q) of water through the hose line (300 GPM), and divide that amount by 100 to find Q: 300/100 = 3; therefore, Q = 3. Next, square that number: $3^2 = 9$; therefore, $Q^2 = 9$.
4. L: Determine the length of the hose you are calculating, and then divide by 100 to find L. The length of this hose is 200 ft (61 m), so 200/100 = 2; therefore, L = 2.
5. Multiply the results from steps 2, 3, and 4 to determine FL:

$$FL = C \times Q^2 \times L$$
$$FL = 2 \times 9 \times 2$$
$$FL = 36 \text{ psi } (248.2 \text{ kPa})$$

Calculating Pump-Discharge Pressure

As previously stated, PDP = NP + FL. To calculate PDP, find the nozzle pressure and friction loss; then plug those numbers into the equation.

Let's say that you are using a fog nozzle (100 psi, as noted earlier) with the hose mentioned in the Friction Loss section (2-in. [50.8-mm] hose that is 200 ft [61 m] long and flows 300 GPM [1135.6 L/min]). We determined that the FL for this hose was 36 psi (248.2 kPa). Plug these numbers into the PDP equation:

$$PDP = NP + FL$$
$$PDP = 100 + 36$$
$$PDP = 136$$

As mentioned earlier, PDP calculations become more complicated as the complexity of fire attack operations increases. You may need to calculate for head pressure (HP or H) or for additional appliances (A), such as wyes or tees. That formula looks like this: $PDP = NP + FL + A \pm HP$.

To obtain a better understanding of basic hydraulics and more complicated PDP calculations, it is recommended that you take the NWCG course S-211, Portable Pumps and Water Use.

After-Action REVIEW

IN SUMMARY

- Water is a valuable commodity at a wildland fire, so every drop counts. Find water sources after your arrival at the scene of a wildland fire and communicate their locations to your team.
  - Natural streams, irrigation canals, ponds, and lakes can be used to supply water.
- Before using a dry hydrant, clear any debris that may be against the strainer.
- Some portable water sources can be brought to the wildland fire scene, such as water tenders and portable water tanks.
- There are two types of hoses used on wildland fires: attack hoses and supply hoses.
  - Attack hoses are used for fire suppression, supply hoses deliver water from a source for the attack.
  - Forestry fire hoses are collapsible attack hoses with a small diameter to provide maneuverability through brush and trees.
  - Hard suction hoses are noncollapsible hoses used to supply water to the intake of a pump.
- There are several different types of nozzles to choose from, including fog stream nozzles, which produce fine water droplets; straight stream nozzles, which produce a straight stream of water; and twin tip nozzles, which produce both fine water droplets and a straight stream of water.
- Drafting is the process of drawing water from static water sources into a pump. This is different from obtaining water from a fire hydrant because the water coming out of a fire hydrant is pressurized.
- Fire eductor pumps use the venturi effect from water discharged from the pump and tank to assist with lifting water and moving it over greater distances than with hard suction hoses.
- In addition to standard PPE, pump operators must have a warm coat, hard hat, gloves, and ear plugs.
- There are three different portable pump flow configurations: parallel, series, and staged.
  - Parallel pump configurations increase the volume of water for the intended use by using two portable pumps pumping into one line. The main hose line is called the trunk line.
  - Series pump configurations are used when an increase of pressure is needed for a specific use on a fire. The discharge from one pump is hooked into the suction of a second pump.
  - Staged pump configurations use a single pump or two pumps that are not connected to each other. Each pump operates independently.
- Simple hoselays are the most basic form of water delivery. They connect hose in a line and include hoses, fittings, and a nozzle that connect directly to the discharge port of a pump.
- Progressive hoselays are designed to include shutoff valves (wyes or tees) that allow the hoselays to attack fire at several points.
  - Progressive hoselays can be extended without having to turn off the pump.
- Hose packs allow fire fighters to easily transport hose, hose accessories, and equipment to various locations. Common types of hose packs include:
  - Travis packs
  - Gansner packs
  - Gnass packs
  - Pondosa packs
- It is important to know how to break down sections of hose after fire suppression operations are completed and safely store it for future use.
  - The butterfly method winds up the hose in such a way that allows it to be looped over a tool so several hundred feet can be carried at once.
  - The doughnut hose roll is used when the hose will be put to use directly from the rolled state.
- Wet mop-up primarily consists of exposing burning materials using hand tools, then extinguishing them with water or water and soil.
  - Most agencies require the initial mop-up to be a 50-ft (15.2-m) radius around the fire. This radius will be increased as needed.

- Hydraulics is the science of water in motion and at rest.
- Many factors can decrease water pressure: friction, appliance use (wyes), and elevation.
  - Pump-discharge pressure (PDP) can be calculated with the following formula: PDP = NP + FL.

KEY TERMS

Backpack pump A portable pump and tank fitted with straps so it can be carried on a fire fighter's back.

Drafting The process of drawing water from a static water source into a pump.

Dry hydrant A pipe that is placed in a static water source and threaded on one end from which water can be pumped by the fire engine.

Fire eductor pump A portable pump used to draft water from static sources; also called an ejector or jet siphon.

Floating pumps Portable pumps that can float on the water's surface.

Forestry fire hose A lightweight, collapsible attack hose; also called a forestry hose.

Friction loss The pressure lost from turbulence as water passes through hoses and hose accessories.

Gansner packs Hose packs that utilize the hose for the backpack and use a 1.5-inch gated wye with one side fitted with a 1-inch reducer.

Gnass packs Hose packs that utilize the hose for the backpack and use inline tees rather than gated wyes.

Gravity sock A conical device, typically made of canvas, that has a large opening that is anchored in a stream, and a smaller opening with a threaded hose connection. Gravity feeds water into the sock and to the hose.

Hard suction hose A noncollapsible hose used to supply water to the intake side of the fire pump.

High-pressure portable pumps Pumps that supply pressurized water from water sources to firefighting hose lines.

Hose A device used to deliver water from a water source to a fire.

Hoselay The arrangement of connected fire hose and accessories on the ground, beginning at the first pumping location and ending at the point of water delivery.

Hose clamp A crimping device used to temporarily stop the flow of water in a hose.

Hose pack A backpack containing a hose, hose accessories, and other hoselay equipment.

Hydraulics The science of water in motion and at rest.

Nozzle The spout connection to the end of a hose that water flows through.

Nozzle pressure (NP) The pressure that a nozzle is designed to operate at most efficiently.

Pondosa pack A crude backpack made of straps with clasps that allow a fire fighter to carry two to three 100-foot doughnut-rolled hoses.

Portable pump A small pump that can be carried to a water source and used to draft water.

Portable water tanks Containers used as water reservoirs that engines can use to refill.

Progressive hoselay Hose connections designed to quickly and continuously move water from the anchor point to the fire. Additional hose lines can be attached without having to shut water off at the pump.

Pump-discharge pressure (PDP) The total pressure needed to overcome all friction, appliance, and elevation loss or gain while maintaining adequate nozzle pressure to deliver effective fire streams; calculated with the following equation: PDP = NP + FL.

Simple hoselay The connection of hoses from a pump to the point-of-discharge.

Travis packs Hose packs that utilize a configuration of 200 feet of hose for rapid deployment.

Wet mop-up The use of water or water and soil together to extinguish burning materials.

REFERENCES

Broyles, George, Corey R. Butler, and Chucri A. Kardous. "Noise Exposure among Federal Wildland Fire Fighters." *Journal of the Acoustical Society of America* 141, EL177 (2017). https://asa.scitation.org/doi/10.1121/1.4976041.

Fire Apparatus Manufacturers' Association (FAMA). "Fire Pump Selection Guide." Modified March 25, 2016. https://www.fama.org/wp-content/uploads/2016/03/1458909818_56f5327ab0545.pdf.

Haston, David V., and Dan W. McKenzie. "Friction Loss in Wildland Hose Lays." United States Department of Agriculture Forest Service. December 2006. https://www.fs.usda.gov/Internet/FSE_DOCUMENTS/stelprdb5239062.pdf.

National Fire Protection Association (NFPA). "NFPA 1963, Standard for Fire Hose Connections." 2019 ed. https://www.nfpa.org/codes-and-standards/all-codes-and-standards/list-of-codes-and-standards/detail?code=1963.

National Wildfire Coordinating Group (NWCG). "S-130, Firefighter Training, 2008." NFES 2731. https://www.nwcg.gov/publications/training-courses/s-130.

National Wildfire Coordinating Group (NWCG). "S-211, Portable Pumps and Water Use, 2012." https://www.nwcg.gov/publications/training-courses/S-211.

National Wildfire Coordinating Group (NWCG). *Water Handling Equipment Guide*. October 2003. PMS 447-1. NFES 1275. https://www.fs.fed.us/t-d/pubs/pdf/03511206.pdf.

Wieder, Michael A. *Fire Service Hydraulics and Water Supply*. 3rd ed. Stillwater, OK: Fire Protection Publications, Oklahoma State University, 2019.

Wildland Fire Fighter in Action

You are part of a structure protection group assigned to protect a subdivision. As you receive your morning briefing from your task force leader, you are given a challenging and unique assignment. Your engine crew has been assigned to establish four water sources for the group. Each of the four areas will require you to use all of your knowledge with respect to water delivery and sourcing. The terrain you are working in has areas in which you will need to lift water, control water going downhill, establish a water source, and use a static water source.

1. You have a large pond close to one of the homes; however, you cannot get close enough to draft from it because of the location of a fence and a septic system. Which of the following would be the best choice to pump water from the pond?
   - **A.** Gravity sock
   - **B.** Portable pump
   - **C.** Eductor pump
   - **D.** Floating pump
2. One of the homes has a swimming pool next to the road. Unfortunately, the pool is 20 ft (61 m) below the level of the road. Which one of these devices could be used to pump the water?
   - **A.** Backpack pump
   - **B.** Eductor pump
   - **C.** Floating pump
   - **D.** Gravity sock
3. Based on the distance from the other water sources, the only available option you have is to create an improvised water source. What items will be most useful to setting up your improvised water source?
   - **A.** Garbage bags, rocks, logs, and a fuel spill container
   - **B.** Rocks, backpack, and mud
   - **C.** Flagging, MRE containers, fire shelter, and traffic cones
   - **D.** Cardboard box, mud, rocks, and tree branches
4. The water source setup will be for a handcrew that is working the fire edge near the subdivision. The crew is working about 250 ft (76.2 m) below the road you are on, and next to the road is a natural spring that has a significant flow. Which device would be best to use in this situation?
   - **A.** Eductor pump
   - **B.** Gravity sock
   - **C.** Floating pump
   - **D.** Portable pump

Access Navigate for flashcards to test your key term knowledge.

CHAPTER 13

Wildland Fire Fighter II

Class A Foam and Fire-Blocking Gels

KNOWLEDGE OBJECTIVES

After studying this chapter, you will be able to:

- Explain what Class A foam is and how it can be used to extinguish fires. (pp. 292–293)
- Identify the types of Class A foam. (p. 294)
- Identify the differences between Class A foam solution types. (p. 294)
- Describe how foam is generated. (pp. 294–295)
- Describe how compressed air foam systems work. (pp. 295–297)
- Describe how aspirating air-foam nozzles work. (pp. 297–298)
- Explain the tactical applications of Class A foam. (pp. 299–301)
- Explain what a fire-blocking gel is and how it can be used to extinguish fires. (p. 301)
- Explain the differences between Class A foam and fire-blocking gel. (p. 301)
- Explain the tactical applications of fire-blocking gel. (pp. 301–302)
- Identify backpack gel applicators. (pp. 301–302)

SKILLS OBJECTIVES

This chapter contains no skills objectives for Wildland Fire Fighter II candidates.

Additional Standards

- **NFPA 11**, *Standard for Low-, Medium-, and High-Expansion Foam*
- **NFPA 1145**, *Guide for the Use of Class A Foams in Fire Fighting*
- **NFPA 1150**, *Standard on Foam Chemicals for Fires in Class A Fuels*
- **NWCG PMS 446-1 (NFES 2246)**, *Foam vs. Fire: Class A Foam for Wildland Fires*
- **NWCG S-130 (NFES 2731)**, *Firefighter Training*

You are the Wildland Fire Fighter

You are the company officer assigned to provide structure protection to a group of homes. The fuels close to the home are primarily shrubs. You do not have the defensible space around the home that you need. It is windy, and the relative humidity levels are low. Some of the engines are equipped with fire gel, and several have a compressed air foam system (CAFS) on them.

1. Between gel and CAFS, which will last longer if you pretreat one of the homes in the area?
2. You are prepping for a wet-line burnout around one of the structures. Which product will be more effective and efficient for your operation?
3. One of the homes you are going to protect has gutters full of pine needles that you cannot reach with a ladder. However, a hose stream will reach. Which product would you recommend on this home?
4. Can Class A foam and Class B foam be mixed in the same tank?

Access Navigate for more practice activities.

Introduction

Class A foam and fire-blocking gels are discussed in this chapter. Reliable Class A foam has been around since the mid-1980s and has proven itself to be a very effective tool for wildland fire suppression. Fire-blocking gel technology has been around since 1998. One of the first uses of this technology was the Slave Lake Fire in Alberta, Canada, where it was used to protect log decks at a lumber mill. The product worked, saving $60 million in property. It was then used in the 1998 Florida fires. The discussion in this chapter focuses on the application aspects of both of these suppression aids.

Class A Foam

Class A foam is an aerated solution intended for use on Class A, or woody, fuels as a firefighting aid. Class A foam is a hydrocarbon surface active agent called a surfactant **FIGURE 13-1**. A **surfactant** is a wetting agent that reduces the surface tension of water and allows it to better penetrate a fuel. It is intended for use on Class A fuels. Class A foam is formulated to make water a more effective fire suppressant. When added to water, the surfactants reduce the surface tension of the water, which allows the water to better penetrate the fuel. The spreading capabilities of water are also enhanced by water draining from the foam. Foam reduces combustion by cooling, moistening, and excluding oxygen.

FIGURE 13-1 Class A foam concentrate container.
Courtesy of Jeff Pricher.

Extinguishing with Class A Foam

Class A foam extinguishes a fire in four ways:

1. Absorbs heat (via the water held in the foam solution)

2. Isolates the fuel (via the insulating foam bubble structure)
3. Acts as a reflective barrier (the white color reflects radiant heat)
4. Interrupts the chemical chain reaction by reducing the surface tension in water to better penetrate organic materials, which lets the water penetrate deeper than the surface

Class A foams are designed to allow water to drain from the bubble structure in order to wet wildland fuels. The rate at which this happens is called the foam drain time. Fuels are wetted by the free liquid; therefore, a regulated released rate from foam to liquid is necessary.

Most Class A foam is biodegradable; therefore, potential environmental damage is greatly reduced when applied at the recommended application rates.

LISTEN UP!

Some Class A foams are not biodegradable and can have a detrimental impact on water quality and wildlife. Be sure to know what kind your agency uses.

Foam Mixing and Expansion Ratios

Costs are relatively low for Class A foam because of the low mixing ratio. Class A foam should be used at the mixing ratios outlined in **TABLE 13-1** or to the manufacturers' recommendations.

Class A foam is documented as being three to five times more effective in fire suppression and mop-up operations and reduces the chances of fire rekindle when compared to using plain water. This results in conservation of water supplies.

Foam can be applied using a wide range of expansion rates, depending on the amount of air that is mixed into the stream and the size of the bubbles that are produced. **Expansion** is the resultant increase in volume of a solution as air is introduced into it. It is a ratio of the volume of the foam in its aerated state to the original volume of the nonaerated foam solution. For example, a 10:1 expanded foam can enhance 200 gallons (757.1 L) of water to increase its effectiveness to the extent that it can potentially have the benefit of 2000 gallons (7570.8 L) of water. The expansion ratio is divided into three classes, related to how much foam is generated. Low-expansion foams are wetter and used for direct fireline applications, whereas high-expansion foams have more air, are drier, and stick better to vertical surfaces. They are generally used for structure protection. The three classes are outlined in **TABLE 13-2**. **TABLE 13-3** shows the four types of Class A foams that are produced.

TABLE 13-1 Class A Foam: Mixing Ratios

Ratio	Situation	Example
0.2%	Mop-up and overhaul operations	5 gallons (18.9 L) of class A foam at 0.2% will treat 2500 gallons (9463.5 L) of water
0.5%	Direct attack applications on a fire	5 gallons (18.9 L) of class A foam at 0.5% will treat 1000 gallons (3785.4 L) of water
1%	Exposure protection on homes	5 gallons (18.9 L) of class A foam at 1% will treat 500 gallons (1892.7 L) of water

Source: National Wildfire Coordinating Group, *Wildland Fire Incident Management Field Guide*, https://www.nwcg.gov/sites/default/files/publications/pms210.pdf.

TABLE 13-2 Class A Foam Expansion Ratios

Foam Type	Expansion Ratio
Low-expansion foam	Up to 20 times (1:1 to 20:1)
Medium-expansion foam	Up to 200 times (21:1 to 200:1)
High-expansion foam	Up to 1000 times (201:1 to 1000:1)

LISTEN UP!

Class A foam is documented as being three to five times more effective in fire suppression and mop-up operations and reduces the chances of fire rekindle compared to using plain water only. Low-expansion foams are wetter and used for direct fireline applications. High-expansion foams have more air, are drier, and stick better to vertical surfaces.

TABLE 13-3 Types of Class A Foams

Expansion Ratio	Foam Type	Drain Time
1:1	Foam Solution ■ Mostly water ■ Clear to milky in color ■ Immediately runs off vertical surfaces ■ Lacks bubble structure	Rapid
↑	Wet Foam ■ Watery ■ Lacks body ■ Very runny on vertical surfaces ■ Large to small bubbles ■ Fast drain time	↑
↓	Fluid Foam ■ Flows easily ■ Consistency similar to watery shaving cream ■ Flows readily from vertical surfaces ■ Medium to small bubbles ■ Moderate drain time	↓
20:1	Dry Foam ■ Mostly air ■ Looks like shaving cream or whipped cream ■ Clings to vertical surfaces ■ Medium to small bubbles ■ Slow drain time	Slow

Source: National Wildfire Coordinating Group, "Foam vs. Fire: Class A Foam for Wildland Fires," https://www.fs.fed.us/t-d/pubs/pdf/hi_res/93511208hi.pdf.

Class B Foam

Class B foam is not used for wildland firefighting; instead, it is used for flammable liquid fires and other applications. It is capable of a strong filming action but incapable of the efficient wetting of Class A fuels.

Another consideration with Class B foam is that certain types of Class B foam can cause cancer and other serious health problems. Some states have legislation that prohibits the storage and use of certain types of Class B foams. Some Class B foams also contain a "forever" chemical, meaning that they are not able to be broken down naturally by the environment.

If your apparatus has the ability to carry and dispense both types of foam, make sure you are using the appropriate type of foam.

LISTEN UP!

Different brands of foam should only be mixed if the manufacturer indicates that it is OK to do so. Never mix different classes of foam. This will cause gelatin-like substances to form, and they will clog or damage the foam system.

How Foam Is Generated—The Hardware

Class A foam can be batch mixed in an engine's water tank or injected into the discharge side of the pump tank through a proportioner. Compressed air can also be added to the foam after it has gone through the proportioner to make a compressed air foam.

Low-energy systems use only the energy of the water pump to educt air into the Class A foam solution. A good example is an aspirating foam nozzle that introduces air into the foam solution.

A high-energy system is a foam-generating device that combines the energy of air to the energy already produced by the water pump in the low-energy system. A good example of this is a compressed air foam system (CAFS), which is discussed later in this chapter.

Batch Mixing

Batch mixing is the process of manually adding a concentrated chemical solution, such as liquid foam, into in a tank, such as an engine's tank or a backpack pump. Batch mixing can be done but is not a preferred method of application. The advantage to batch mixing is that no hardware is required so the costs are minimal. However, the disadvantages outweigh the cost savings. One important step to remember is to mix the foam into the tank water. Do not put the foam in an empty tank because it will cause excessive foaming. Some of the disadvantages of batch mixing include:

- Possible corrosion to the tank
- Foam forming in the tank
- Excessive foam concentrate use
- Cleaning of the lubricants in the pumping system

- Excess foam in the tanks, causing the pump to cavitate
- Water refill difficulties as a result of foam bubbles in the water tank
- Needing to thoroughly clean out the tank and pump at the end of the assignment

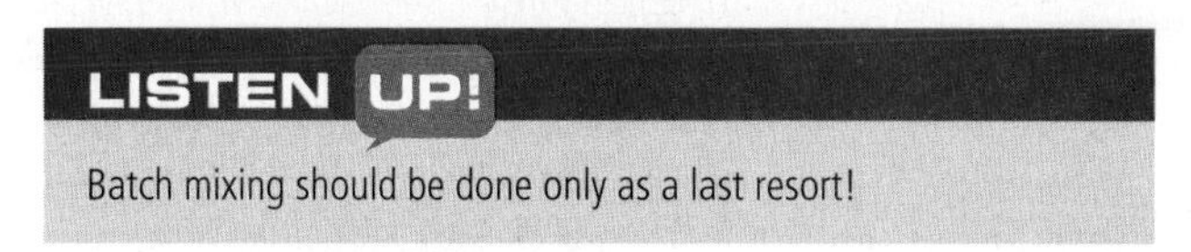

Proportioning Systems

A **proportioner** adds a predetermined amount of foam concentrate to the water supply to form a finished foam solution. Foam concentrate is generally added to the water supply on the discharge side of the pump. As the water is being pushed out of the pump, foam is injected into the water stream at the desired proportion, which creates the foam.

There are two types of proportioning systems: manually regulated systems and automatic regulating systems. Automatic regulating systems that inject foam concentrate directly into the discharge side of the water pump produce the most desirable results.

A foam concentrate proportioner should ideally have the following characteristics:

- Be proportional over the entire operating pressure and flow capacities of the water pump.
- Be unaffected by changes in engine pressure; changes in hose length and size; changes in water flow; and changes in nozzle adjustment, size, or elevation.
- Be suitable for use with CAFS.
- Be able to inject foam concentrate directly into the discharge side of the pump's water stream. It should be proportioned correctly to achieve the right ratio of foam concentrate to water. The foam solution should then flow directly through the engine's piping system into the hose lines. There should be no chance of the foam solution circulating back into the tank, pump, or engine piping system.
- Be highly reliable, simple in design, and easy to repair.
- Be able to proportion either Class A or Class B foams with proportions to 0.1 percent or less.
- Have an indicator showing how much foam concentrate is left in the system for use.
- Have enough foam concentrate to treat a full tank of water on the engine.
- Not cause a water pressure loss as water flows through the proportioning unit.
- Be able to operate without any external power source. Should the proportioner require the use of the engine's 12-volt electrical system to power it, then it should draw less than 30 amperes.
- Have corrosion-resistant storage tanks and be compatible with foam.
- Be easy to routinely flush.
- Have no galvanized pipe components.

Manually regulated proportioning systems require manual adjustment to the mix ratio when there is a change in pressure or flow through the proportioner. There are six types of manually regulated proportioners:

- Suction-side proportioning systems
- In-line eductor proportioning systems
- Bypass eductor proportioning systems
- Around-the-pump proportioning systems
- Balanced-pressure bladder tank proportioner
- Direct injection, manually regulated proportioning systems

Automatic regulating proportioner systems automatically adjust the flow of foam concentrate into the water stream, which automatically maintains the proper ratio of concentrate to water. There are three types of automatic regulating systems:

- Balanced pressure, Venturi proportioning systems:
  - Pump systems
  - Pressure tank systems
- Water-motor meter proportioning systems
- Direct injection, automatic regulating proportioning systems

With an automatic regulating system, once the mix ratio is set and the proportioner is operated, there should be no further need to adjust the proportioning unit.

Compressed Air Foam Systems

A **compressed air foam system (CAFS)** is used to deliver foam onto a fire during suppression. CAFSs are generally made up of an air compressor, a water pump, and foam solution. In CAFS, compressed air or gas is injected into the water stream after it has passed through the proportioner **FIGURE 13-2**. It is a high-energy foam system because pneumatic power (compressed air) is added to the water stream.

After combining the foam concentrate and air, foam solution flows through the hose line, where it is agitated by the inside of the hose lining. This action is called **scrubbing**. Scrubbing produces tiny uniform bubbles within the foam solution. It takes approximately 100 ft (30.5 m) of 1 1/2-in. (38.1-mm) hose to effectively scrub the foam solution. Scrubbing can also be done with the use of a device called a motionless, or static, mixer **FIGURE 13-3**. These types of mixers have no moving parts and utilize engineered-inserted flow disruptors to generate the desired foam effect.

It is important in CAFS for the static pressure of the air and foam concentrate to be the same. This ensures that there is no friction loss. It is also important to remember that, when there is not enough foam solution to mix with the air, inadequate mixing in the CAFS occurs. This situation is called **slug flow** because pockets of water (or slugs) and air move through

DID YOU KNOW?

The compressed air foam system (CAFS) was developed by Mark Cummins while he was working for the Texas Forest Service. He was issued a patent in 1982 for the system. In 1972, the Texas Forest Service used pine soap as the foaming agent, from which came the phrase "Texas snow job." In 1988, the CAFS received national attention during the Yellowstone National Park wildfires, during which it was successfully used to protect the four-story Old Faithful Inn.

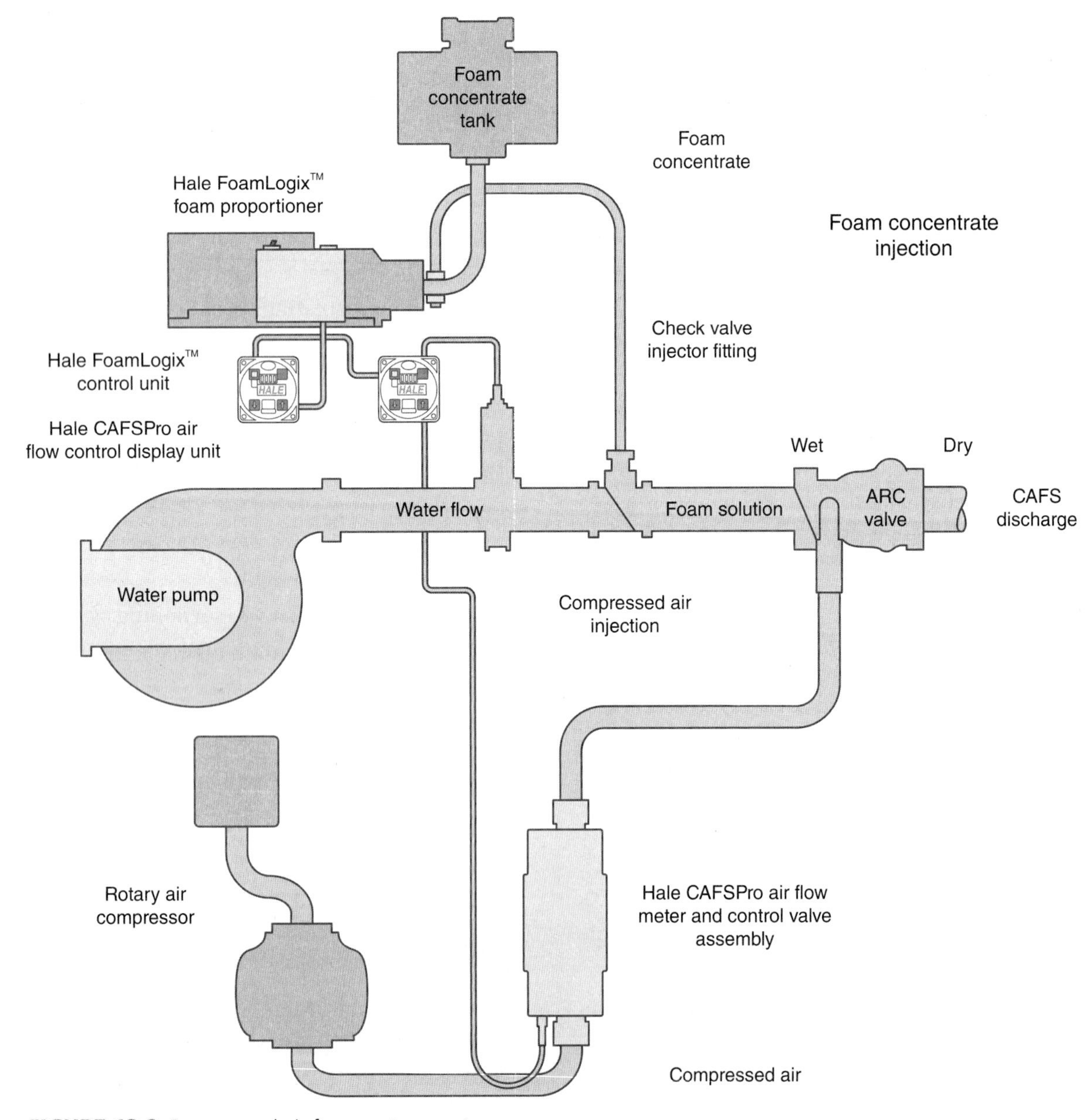

FIGURE 13-2 Compressed air foam system parts.

FIGURE 13-3 This is a cutaway example of a static mixer used to assist in creating different foam effects. There are no moving parts in this mixer, which reduces the potential for mechanical failure.

the nozzle, causing a hammer effect as these plugs pass through the hose and exit the nozzle. To remedy this problem, return to the use of regular water until you can correct the ratio of foam solution to air.

Configurations and Components of a CAFS

There are three CAFS configurations found on engines:

- The fire apparatus drives a pump, called a **centrifugal pump**, which expels water by force through the ports of a circular impeller rotating at high speed, and an air compressor.
- A single auxiliary engine mechanically drives both the centrifugal pump and the air compressor.
- The fire apparatus engine drives a centrifugal pump and an air compressor via a mechanical drive system.

A CAFS usually has the following components:

- Centrifugal pump
- Air compressor
- Foam concentrate proportioner system
- Drive or power system
- Control and instrument systems

Advantages and Disadvantages of a CAFS

There are several advantages to a CAFS:

- Foam can be pumped great distances in hose lines, as far as 5000 ft (1524 m).
- It can produce less head pressure when pumping up a slope.
- It creates a foam that is much drier than Class A foam, which reduces property damage, especially to structural property.
- It creates very light hoselays—half the hose is filled with air. A 1 1/2-in. (38.1-mm) hose line will carry 50 gallons (189.3 L) per minute of foam solution through 50-ft (15.2-m) sections of hose and 50 ft^3 (1.4 m^3) per minute of air (300 to 350 gallons [1135.6 to 1324.9 L] of atmospheric pressure air).
- Less foam concentrate is used.
- Foam with smaller bubbles is more uniform and lasts longer.
- It reduces air and water pollution. With less burning time, smoke and other contaminants have less time to disperse into the air. With less water use, less contaminated water runs off into rivers and streams.

Some of the drawbacks of a CAFS are as follows:

- It is very expensive up front.
- It requires additional training.
- The complexity of the system affects its reliability.
- It requires additional routine service and maintenance.

Aspirating Foam Nozzles

Proportioners add foam concentrate to the water supply to form a finished foam solution. To produce foam, air has to be introduced. In a low-energy system, this is accomplished by an aspirating air foam nozzle **FIGURE 13-4**. This special nozzle draws air into the water stream, usually by Venturi action, to create and mix the solution. The expansion chamber in the nozzle strengthens the bubbles before they are discharged. Aspirating nozzles are the least expensive and most reliable way to produce foam.

Generally, aspirating nozzles that have long reaches produce only wet, frothy foam. The foam expansion ratio will be less than 10:1.

Safety Practices Using Class A Foam

Class A foam is like a strong household detergent. You must wear your personal protective equipment, including safety goggles, when working with Class A foam because foam solution irritates the eyes.

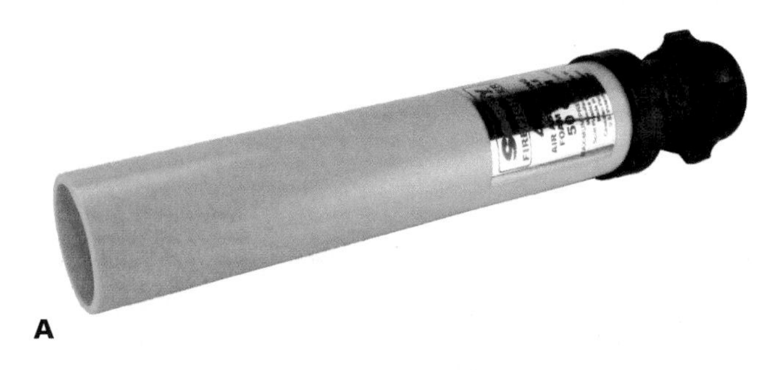

A

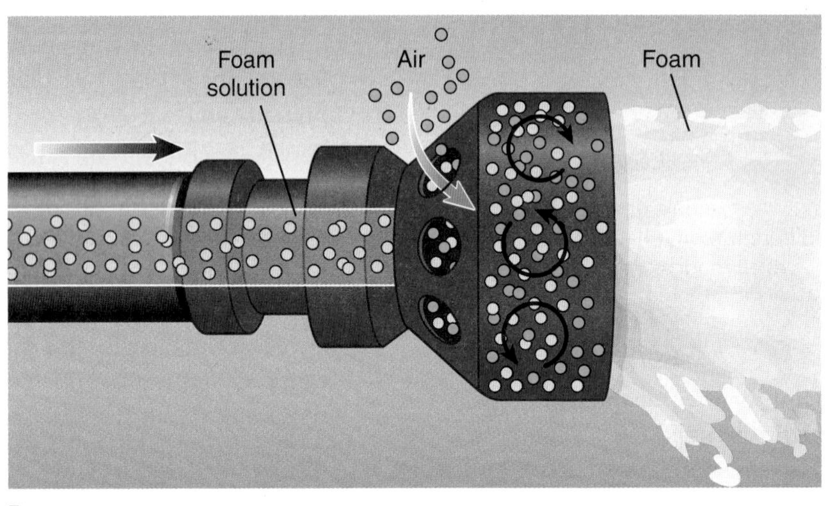

B

FIGURE 13-4 A. Aspirating air foam nozzle. **B.** Aspirating air foam nozzle function.

You should also wear waterproof gloves and, if possible, disposable overalls. Clothing or footwear that becomes wet with foam should be changed and washed before reuse. Foam can be very damaging to leather (i.e., your boots and gloves). Decontamination with clean water after exposure or contact with foam is best practice. Consider retreating footwear with boot grease or using the manufacturer's recommended treatment after repeated contact with foam.

Any equipment exposed to foam should be cleaned as soon as possible. Electronic equipment, such as smartphones, tablets, weather-reading gadgets, and communication equipment, should always be kept away from foam. If electronics come in contact with foam, clean immediately. The chemical properties of foam are very persistent in penetrating seals.

All foam containers should be labeled to alert personnel that they do not contain water and that the contents must not be used for drinking purposes. If foam solution is ingested, seek medical attention as soon as possible, and refer to the safety data sheet that came with the foam container. Large oral doses could produce narcosis.

Class A foam concentrate spilled on the ground can be slippery, so watch your footing if a spill occurs. Also, when using Class A foam on logs or downed trees, as you maneuver over them after applying the foam mixture, slips and falls can be increased if you are not careful.

Class A foam carries fewer environmental risks when used at the manufacturer's recommended mixing ratios, which are usually up to only 1 percent foam solution.

Significant foam concentrate leaks near a waterway or storm drain may need to be diked to prevent the product from mixing with the water and causing nuisance foaming and problems with municipal water treatment operations. Keep foam solutions out of ponds, lakes, and other water sources. They can cause harm to fish, plants, and other aquatic wildlife. It is a good practice to keep concentrates at least 100 ft (30.5 m) away from waterways.

Do not mix Class A and Class B foams together. They are not compatible, and combining these foams will cause the solutions to gel, which can result in damage to the equipment. Mixing them can also cause significant downtime for the required cleaning of the tanks and proportioners. This can also be true for foams of the same class made by different vendors. Even though they may be called the same thing, different vendors use different chemical properties that may not be compatible with a competing manufacturer's product.

LISTEN UP!

If you are working with aerial foam delivery, do not apply the foam directly into a waterway.

LISTEN UP!

Do not mix Class A and Class B foams together. Doing so can damage equipment.

Tactical Foam Application

Class A foam has been implemented and its tactical applications thoroughly tested in fire departments. Class A foam is three to five times more effective than water. One Southern California fire department performed an initial attack on a small grass fire with two Type 3 engines. One engine took the right flank of the fire with Class A foam, and the other used plain water in its attack. The one using plain water ran out of water, whereas the engine equipped with a Class A foam proportioner was able to complete the attack on its flank of the fire. In addition, the engine using Class A foam was able to extinguish the head of the fire and eventually complete the knockdown of the fire on the other engine's flank of the fire.

There are five tactical applications for Class A foam: direct attack applications on the fire's edge, indirect applications, mop-up operations, barrier protection, and structure protection.

Direct Attack Applications

Consult the manufacturer's recommendations for the mixing ratio of foam concentrate to water.

Foam solution (the mixture of water and foam concentrate) should be applied to the base of the flames. Also, while working the fire's edge, some of the foam solution can be directed toward unburned fuels.

Where there are hotter fuels that will take longer to extinguish, leave a foam blanket. A **foam blanket** is a layer of foam that forms an insulating and reflective barrier to heat, which will help smother the fuel and continue to wet it. Foam blankets can be used in fuel protection, fire attack, and mop-up operations (discussed later).

For foam blankets, the engine operator may have to increase the concentrate ratio amount. Creating a foam blanket requires a 1 percent concentrate level and an aspirating nozzle. A dual-function nozzle on a direct attack hose line works as a regular nozzle when using water solely as a wetting agent. If foam is desired, a slide can be moved forward to expose small teeth that cause the nozzle to then work as an aspirating nozzle. A wet foam solution will then be produced, provided the concentrate level is adjusted.

Mobile pumping from engines works well with Class A foam. The key is not to move too rapidly so you can completely extinguish the fire.

What makes Class A foam so effective is its ability to wet and cool fuels long after its initial application. This allows the fire fighter to move to a new area while the foam continues to work where it was initially applied.

Indirect Attack Applications

Class A foam can be used as a wet line from which to start a firing operation **FIGURE 13-5**. A 40-GPM (gallon per minute) (2.5-L/s) medium-expansion nozzle can be used for this purpose. It will produce a fluid foam solution that is wet enough to penetrate into the ground and surface fuels. This is important because foams that are too dry tend to hang on surface fuels and allow the fire to creep under the foam line.

The wet line is put down on the ground by an engine crew mobile pumping Class A foam through a medium-expansion nozzle. The foam line is generally 2.5 times as wide as the flame lengths. Be sure to coat all sides of the fuel, and make sure that the foam has penetrated the ground and surface fuels. The igniters involved in the firing operation follow along after the foam line has been constructed and fire from this line. In areas where it is not suitable to mobile pump, hose lines can be extended for this purpose. Also, consider ladder fuels because it may be necessary to supply foam to low-hanging limbs, brush, tree trunks, or canopy to help provide an insulting barrier.

FIGURE 13-5 Type 4 engine putting down a wet line from which a firing operation will be started.

Courtesy of Jeff Pricher.

LISTEN UP!

Wet lining is the preferred option in sensitive areas where soil disturbance is undesirable.

Mop-Up Operations

Class A foam works well during mop-up operations to penetrate fuels and put out deep-seated fires. Wet or fluid foam can be applied to heavier fuels. The blanket smothers the fire, holds water in solution longer so it releases slower, and helps hold down residual smoke levels **FIGURE 13-6**. Where there are pockets of deep-seated fire, such as in duff or litter, cover the area with a foam blanket. This is part of the standard mop-up process to reach buried fuels that hold heat. Class A foam also works well for deep-seated fires in log decks (stacked logs ready for the lumber mill) and stumps.

In mop-up operations, start mopping up along the edge of the burn and work inward, concentrating on hot spots. As a general rule, mop up 100 ft (30.5 m) in from the edge of the burn. At times of expected increase in wind activity or if close to an interface area, a 100 percent mop-up of the area can be done. Follow your departmental policy, and closely examine the fireline situation that exists.

Residual pockets of deep-seated fire can be seen by steam releasing through the foam blanket. Give these areas more attention.

Consider the use of medium-expansion nozzles where there are wide areas of smoldering surface heat. They will get the job done quickly and provide a foam blanket to smother fire, eliminate smoke, and cause a slow release of water into the fuel.

Mop-up operations without use of foam are discussed in the *Strategy and Tactics* chapter.

Barrier Protection

Class A foam works well for barrier protection because it insulates the fuel from the fire. In addition, it is white and reflects radiant heat. It can be applied to brush and grass, timber stands, low-hanging limbs to interrupt spread of fire to crown fuels, and log decks.

The length of time the barrier lasts depends on the fuel loading, moisture content, wind factor, relative humidity, and air temperature. Fluid foams provide effective blanketing and wetting properties. Compressed air foams remain longer than foam produced by aspiration nozzles. The compressed air foams produce a strong blanket and are very slow to wet the fuel.

Foam should be applied just before ignition because it is a short-term treatment option. Always evaluate the durability of the foam you are using. Wet foam provides a wet blanket that allows for rapid wetting. The opposite extreme is dry foam, such as compressed air foam, which insulates well because of the strong foam blanket produced and its very slow drain time.

The width of the foam line will depend on the fuel model present and the fire behavior indicators observed. Always remember to apply foam to all sides of the fuel whenever possible.

Structure Protection

Fluid foams and dry foams are best used in structure protection. They adhere well to vertical surfaces and the undersides of deck and roof projections on buildings **FIGURE 13-7**.

FIGURE 13-6 Foam blanketing.

FIGURE 13-7 Fluid and dry foams adhere well to vertical surfaces.

Again, apply Class A foam just before the arrival of the flaming front. In hot weather, foam produced by aspirating nozzles generally lasts about 30 minutes. Compressed air foam will last approximately 1 hour in hot weather.

Apply foam to all roof surfaces, outside walls, deck and roof eave projections, and supporting columns of the structure. Liquified petroleum gas tanks will also benefit from the application of Class A foam. In addition, survey any outbuildings, and, if time permits, use foam to protect them.

Fire-Blocking Gel

Fire-blocking gel, sometimes called fire gel or fire-retardant gel, is an absorbent and hydrating polymer mixed with water that aids in firefighting efforts by clinging to vertical surfaces and absorbing heat. The polymer comes in liquid or powder form. It is added to water to thicken it. Once mixed, the water is suspended in the gel, which creates a heat barrier.

One of the main differences between Class A foams and fire-blocking gels is that the bubbles in gel solutions are water filled **FIGURE 13-8**. Class A foam has air-filled bubbles that burst at a much lower temperature when heat is applied. The water-filled bubbles found in gels stack on each other and provide an excellent barrier against radiant heat. Fire-blocking gel also has an extremely long evaporation time, which allows it to be applied long before the arrival of the flaming front. As one layer of gel starts to burn off, more water-filled bubbles slide down to cover the exposed area, attacking two sides of the fire triangle.

FIGURE 13-8 Barricade fire-blocking gel.
Courtesy of Barricade International.

Better atomization of the water stream takes place as a result of the addition of the gel to the hose stream. This exposes greater wetting-surface areas, which in turn provides for a greater cooling effect. Thinner gel solutions are therefore used when greater extinguishment properties are desired. More gel concentrate is added when a thicker gel is desired for its insulating value. Gels as thick as 1 in. (2.5 cm) can be applied to exposed vertical surfaces.

Gel Applications

Fire gel can be applied in one of three ways. The first is to batch mix in the engine water tank and then run it through the pump. Check with the gel company before batch mixing it in the water tank and running the gel through the pump because centrifugal pumps are designed to pump water and not a water–gel mixture. The second way gel is applied is at the end of the hose stream with a backpack application **FIGURE 13-9**. The friction loss in a gel-filled hose line is many times that of water; therefore, backpack applications seem to work well. The third application method is an eductor/proportioner mounted on an engine or offloaded unit placed on the ground.

Gel can also be used in both fixed-wing aircraft and helicopters, where the mixed solution can be pumped into the holding tank for aerial delivery. If the helicopter is using external buckets, instead of a fixed tank, the solution is then pumped into the bucket carried by the aircraft. Aviation uses are normally in the 0.8 to 1.7 percent viscosity ranges. Gel for helicopter operations at 0.8 to 1.3 percent weighs about 8.35 pounds (3.8 kg) per gallon.

Fire-blocking gels can be applied several hours before the flaming front arrives, and they will still have good insulating values. Fire gel clings to glass, metal, roof overhangs, wood or painted siding, vegetation, rubber, and other flammable solids. Gel will provide protection for several hours even under severe atmospheric conditions.

Fire fighters can coat a structure with gel solution and then retreat to a safe zone as a high-intensity fire moves through the area. After the flaming front passes, engine companies can be redeployed to put out any residual fire left in the area. There have been tests on both exposed vegetative materials and the sides of structures where a 3500°F (1927°C) propane torch was applied to the gel solution for 10 minutes or longer. The plant materials did not burst into flames, and the side of the structure did not burn. Certainly, these temperatures are much hotter than those produced in a wildland fire.

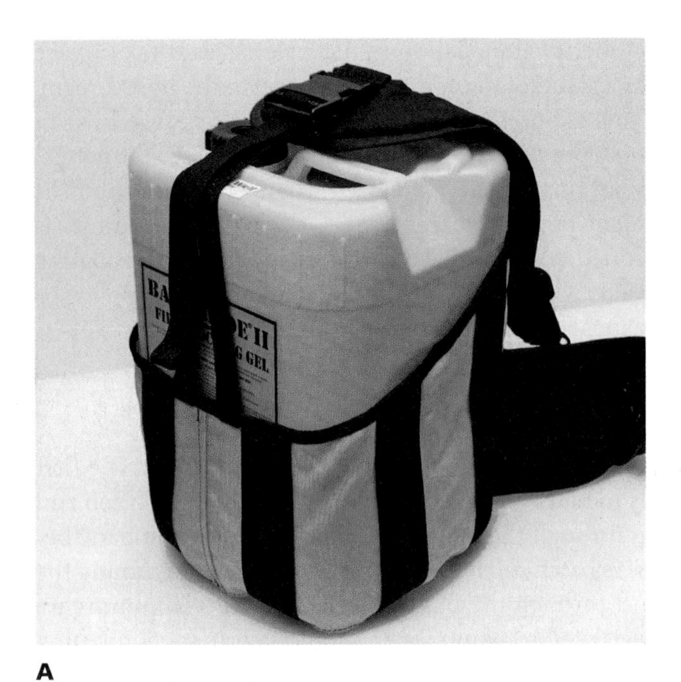

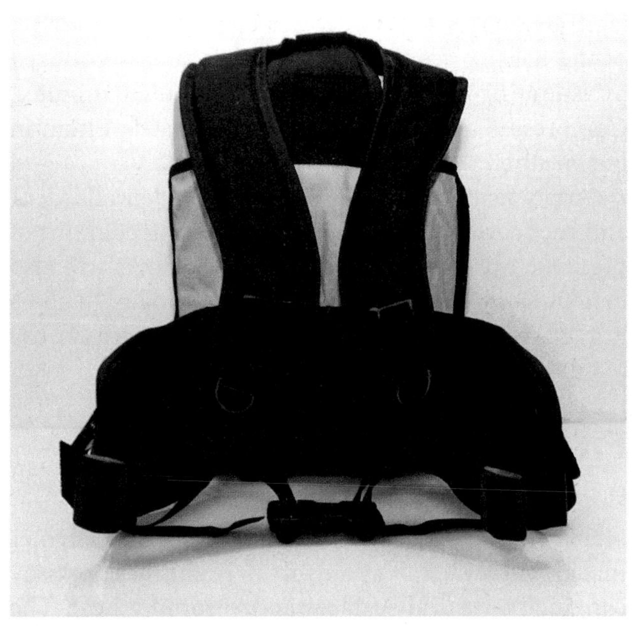

A B

FIGURE 13-9 Backpack gel applicator.
A. Courtesy of Barricade International; B. Courtesy of Barricade International.

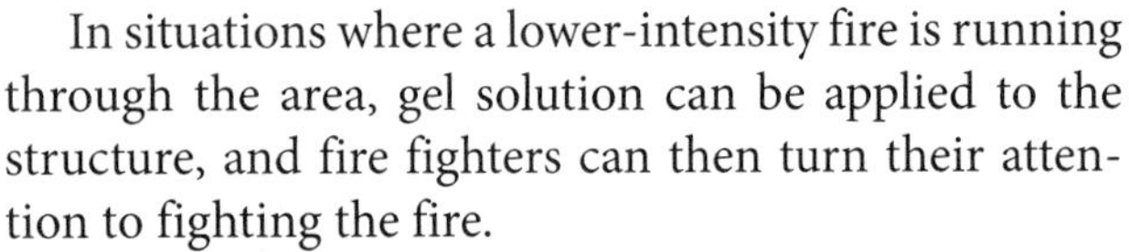

In situations where a lower-intensity fire is running through the area, gel solution can be applied to the structure, and fire fighters can then turn their attention to fighting the fire.

Fire-blocking gels can also be used to provide barrier protection, as a wet line in firing operations, and in mop-up operations.

Fire gel can also be used in aviation operations on both fixed-wing aircraft and helicopters. A USDA-approved food-grade fugitive dye can be introduced into the mixed gel to provide superior visibility for pilots to locate previous gel drops.

DID YOU KNOW?

Private companies and corporations are now using fire gel. They include utility companies, private logging companies, oil refineries, fuel tank farms, and chemical plants, all of which deal with fire as a major hazard on a daily basis. These companies have taken steps to train employees in the techniques of controlling and extinguishing fires with the use of fire gel.

The Southern California Edison Company, an electric service provider, has used fire gel to protect power poles during wildfires. They are a very large utility company and have had great success with gel.

After-Action REVIEW

IN SUMMARY

- Class A foam is a surfactant that is intended for use on Class A fuels and is formulated to make water a more effective fire suppressant.
- Class A foam extinguishes a fire in four ways:
  - Absorbs heat
  - Isolates the fuel
  - Acts as a reflective barrier
  - Interrupts the chemical chain reaction
- Most Class A foam is biodegradable; therefore, potential environmental damage is greatly reduced when it is used at the recommended application rates.
- Costs are relatively low for Class A foam because of the low mixing ratio.

- Class A foam is documented as being three to five times more effective than water in fire suppression and mop-up operations and reduces the chances of fire rekindle compared to using plain water.
- Expansion is the resultant increase in volume of a solution as air is introduced into it. It is a ratio of the volume of the foam in its aerated state to the original volume of the nonaerated foam solution.
- Class B foam is not used for wildland firefighting. If your apparatus has the ability to carry and dispense both types of foam, make sure you are using them appropriately.
- Class A foam can be batch mixed in an engine's water tank or injected into the discharge side of the pump tank through a proportioner. Air can also be added to the foam after it has passed through the proportioner to make a compressed air foam.
- Proportioners add foam concentrate to the water supply to form a finished foam solution. There are two types of proportioning systems:
  - Manually regulated proportioner systems
  - Automatic regulating proportioner systems
- A compressed air foam system (CAFS) is used to deliver foam onto a fire during suppression. CAFSs are generally made up of an air compressor, a water pump, and foam solution.
- Proportioners add foam concentrate to the water supply to form a finished foam solution. To produce foam, air has to be introduced. In a low-energy system, this is accomplished by an aspirating air foam nozzle.
- You must wear your personal protective equipment, including safety goggles, when working with Class A foam because foam solution irritates the eyes.
- Do not mix Class A and Class B foams together. They are not compatible, and combining these foams will cause the solutions to gel, which can result in damage to the equipment.
- There are five tactical applications for Class A foam: direct attack applications on the fire edge, indirect applications, mop-up operations, barrier protection, and structure protection.
- One of the main differences between Class A foams and fire-blocking gels is that the bubbles in gel solutions are water filled.
- Fire gel can be applied in one of three ways:
  - Batch mix in the engine water tank and then run it through the pump
  - Backpack and hose stream application
  - Eductor/proportioner mounted on an engine or offloaded unit placed on the ground for use
- Gel can also be used in both fixed-wing aircraft and helicopters.
- Fire-blocking gels can be applied several hours before the flaming front arrives, and they will still have good insulating values.
- Fire gel clings to glass, metal, roof overhangs, wood or painted siding, vegetation, rubber, and other flammable solids.
- Gel will provide protection for several hours, even under severe atmospheric conditions.

KEY TERMS

Batch mixing Manually adding and mixing a concentrated chemical into a solution in a tank or container.

Centrifugal pump Pump that expels water by centrifugal force through the ports of a circular impeller rotating at high speed.

Class A foam An aerated solution intended for use on woody (Class A) fuels as a firefighting aid by reducing the surface tension of water and allowing it to better penetrate fuels.

Compressed air foam system (CAFS) A high-energy system used to deliver foam onto a fire; made up of an air compressor, a water pump, and foam solution.

Expansion The ratio of the volume of the foam in its aerated state to the original volume of the nonaerated foam solution.

Fire-blocking gel An absorbent and hydrating polymer mixed with water that aids in firefighting efforts by clinging to vertical surfaces and absorbing heat.

Foam blanket A layer of foam that forms an insulating, smothering, and reflective barrier to heat; used for fuel protection, fire attack, and mop-up operations.

Proportioner Device that adds a predetermined amount of foam concentrate to water to form a foam solution.

Scrubbing Agitating foam solution and air within a confined space (usually a hose) to produce tiny, uniform bubbles.

Slug flow Discharge of distinct pockets of water and air due to the insufficient mixing of foam concentrate, water, and air in a compressed system.

Surfactant Any wetting agent; a formulation that, when added to water in proper amounts, materially reduces the surface tension of the water and increases the penetration and spreading abilities of the water.

REFERENCES

Avsec, Robert. "A Case for CAFS: 5 Ways It Can Help You." FireRescue1. Published February 25, 2013. https://www.firerescue1.com/fire-products/apparatus-accessories/articles/1410826-A-case-for-CAFS-5-ways-it-can-help-you/.

Colletti, Dominic J. "Compressed-Air Foam Mechanics." *Fire Engineering* 3, no. 147 (1994). https://www.fireengineering.com/articles/print/volume-147/issue-3/features/compressed-air-foam-mechanics.html#gref.

National Fire Protection Association (NFPA). "NFPA 11: Standard for Low-, Medium-, and High-Expansion Foam." 2016 ed. https://www.nfpa.org/codes-and-standards/all-codes-and-standards/list-of-codes-and-standards/detail?code=11.

National Fire Protection Association (NFPA). "NFPA 1145: Guide for the Use of Class A Foams in Fire Fighting." 2017 ed. https://www.nfpa.org/codes-and-standards/all-codes-and-standards/list-of-codes-and-standards/detail?code=1145.

National Fire Protection Association (NFPA). "NFPA 1150: Standard on Foam Chemicals for Fire in Class A Fuels." 2017 ed. https://www.nfpa.org/codes-and-standards/all-codes-and-standards/list-of-codes-and-standards/detail?code=1150.

National Foam. "A Firefighter's Guide to Foam." Accessed October 7, 2019. http://www.foamtechnology.us/Firefighters.pdf.

National Wildfire Coordinating Group (NWCG). "Foam vs. Fire: Aerial Applications for Wildland Fires." PMS 446-3. NFES 1845. October 1995. https://www.fs.fed.us/t-d/pubs/pdf/hi_res/95511209hi.pdf.

National Wildfire Coordinating Group (NWCG). "Foam vs. Fire: Class A Foam for Wildland Fires." PMS 446-1. NFES 2246. October 1993. https://www.fs.fed.us/t-d/pubs/pdf/hi_res/93511208hi.pdf.

National Wildfire Coordinating Group (NWCG). "S-130, Firefighter Training, 2008." Modified August 21, 2019. https://www.nwcg.gov/publications/training-courses/s-130.

United States Fire Administration (USFA), Federal Emergency Management System (FEMA). "Compressed Air Foam for Structural Fire Fighting: A Field Test." January 1994. https://www.usfa.fema.gov/downloads/pdf/publications/tr-074.pdf.

United States Forest Service. "Interagency Policy for Aerial and Ground Delivery of Wildland Fire Chemicals Near Waterways and Other Avoidance Areas." Modified November 29, 2012. https://www.fs.fed.us/rm/fire/wfcs/Interagency_Policy_Aerial-Ground_%20Delivery_112912_508.pdf.

Wildland Fire Fighter in Action

After your return from a fire in which you successfully protected structures in its path, you were assigned to train other fire fighters in your battalion. You will need to answer and explain the following questions to the students:

1. What is a surfactant?
2. What is drain time as it applies to Class A foam?
3. Describe what medium-expansion foam is.
4. What is a high-energy foam system?
5. Can Class A and Class B foam solutions be mixed in a proportioner?
6. What does the term *slug flow* mean?
7. What is fire-blocking gel?
8. Do fire-blocking gels experience drain-time problems?

Access Navigate for flashcards to test your key term knowledge.

CHAPTER 14

Wildland Fire Fighter II

Handcrew Operations

KNOWLEDGE OBJECTIVES

After studying this chapter, you will be able to:

- Identify the different types of handcrews. (pp. 308–311)
- Describe the certifications and pre-operations required before chainsaw use. (p. 312)
- Identify the personal protective equipment required for chainsaw use. (pp. 312–313)
- Describe the blood bubble. (pp. 313–314)
- Identify the parts of a chainsaw. (pp. 314–317)
- Describe how to fuel a chainsaw. (**NFPA 1051: 5.1.2, 5.3.2, 5.3.3, 5.5.3, 5.5.4**, p. 317)
- Describe how to start a chainsaw on the ground. (**NFPA 1051: 5.1.2, 5.5.3, 5.5.4**, p. 318)
- Describe how to start a chainsaw between your legs. (**NFPA 1051: 5.1.2, 5.5.3, 5.5.4**, p. 318)
- Describe how to start a chainsaw over a log. (**NFPA 1051: 5.1.2, 5.5.3, 5.5.4**, p. 318)
- Describe how to stop a chainsaw. (**NFPA 1051: 5.1.2, 5.5.3, 5.5.4**, pp. 318–319)
- Describe how to maintain and store a chainsaw. (**NFPA 1051: 5.3.1, 5.3.2, 5.3.3**, p. 319)
- Identify the proper tool order in a handline. (pp. 321–322)
- Understand handcrew production rates. (pp. 322–323)

SKILLS OBJECTIVES

After studying this chapter, you will be able to perform the following skills:

- Fuel a chainsaw. (**NFPA 1051: 5.1.2, 5.3.2, 5.3.3, 5.5.3, 5.5.4**, p. 317)
- Maintain and store a chainsaw. (**NFPA 1051: 5.3.1, 5.3.2, 5.3.3**, p. 319)
- Operate as part of a handcrew. (pp. 319–323)

Additional Standards

- **NWCG S-212**, *Wildland Fire Chainsaws*

You Are the Wildland Fire Fighter

You are a squad boss on a Type 2 IA crew, and you arrive on a wildfire burning in grass and shrub. You have been with the crew for the past three wildfire seasons and are in your first role as a squad boss. On this assignment you are Charlie squad, and your job is to construct a hand line in this fuel type. You are following Alpha squad, which is made up of three saw teams. They are already working ahead by brushing the line and prepping for the diggers. You are following them on a hillside that has a 22 percent slope with patchy and continuous fuel continuity. Your squad and diggers will be the first crew creating line. Bravo squad will be chasing you because they were the lead on the last assignment. As the squad boss, you will need to make strategic and deliberate choices that will establish positive fire control on your line. You will need to determine the following:

1. What is the proper tool order and placement of hand tools?
2. What do you think your fireline production rate will be in this fuel type?
3. What will the spacing be between handcrew members?
4. How would you determine your fireline width?
5. How will you determine whether LCES/LACES is in place?

Access Navigate for more practice activities.

Introduction

A **handcrews** is a team of 18 to 20 fire fighters who use hand tools to actively suppress low flame–production fires. Handcrews are used to construct hand lines (firelines constructed with hand tools) directly along the fire's edge, where intensity levels allow their use **FIGURE 14-1**. Handcrews often operate on remote parts of the fireline that are inaccessible to motorized equipment and in areas that are protected by environmental regulations that prohibit the use of mechanized equipment. The removal of fuels by hand may be a preferred method when working close to structures.

Handcrew members can be an established team or a group of fire fighters, or they can be assembled from several engine or air crews to serve in the capacity of established crews. Regardless of their makeup, all fire fighters need to have a good understanding of the fundamentals of handcrew operations. Lastly, handcrews can be called up for various assignments based on their unique capabilities. For example, handcrews were used during Hurricane Katrina in Louisiana to assist in moving patients into military aircraft and injured fire fighters off the fireline in difficult terrain **FIGURE 14-2**.

Handcrew Deployment

Because of the ability of handcrews to operate in steep and remote terrain, they are often the first line of defense in wildland firefighting. Their primary deployment scenario is utilization of a direct attack method where they apply suppression tactics directly to the burning fuel. Handcrews also perform indirect attack fireline operations ahead of the fire, such as a line constructed some distance from the fire's edge. They then burn out the unburned fuels by intentionally setting a fire between the control line and the fire's edge. Handcrews use this tactic where fireline intensity levels are too great for a direct attack on the fireline. Backfires and burnout basics are discussed in more detail in Chapter 15, *Basic Firing Operations*. The fireline is usually started on a flank from an anchor point. This tactic may shorten the fireline by constructing the line across the unburned fingers and thus burning out the intervening fuels.

Handcrew Types

Handcrews are typed depending on their training and experience levels. Standards for the typing of crews are found in the *Wildland Fire Incident Management Field Guide*. There are three types of handcrews: Types 1, 2, and 3. Crews involved in aviation operations can be made up of Type 1 or 2 members.

Type 1 Crews

Crews are credentialed based on the experience and qualifications of the fire fighters. With this in mind, there are two kinds of Type 1 crews: interagency hotshot crews and firefighting crews that meet the Type 1 crew requirements.

A B

FIGURE 14-1 A. Before handcrew efforts. **B.** After handcrew efforts.
Courtesy of Jeff Pricher.

FIGURE 14-2 Handcrews were used during Hurricane Katrina to assist in moving patients and injured fire fighters.
Courtesy of Jeff Pricher.

An **Interagency hotshot crews (IHC)**, also just called a hotshot crew, is an intensively trained fire crew. They are the most experienced crews with the highest levels of training and preparedness to battle the most serious wildfires. The members receive more than 80 hours of training annually, and at least 80 percent of them must have at least one season of fire-suppression experience. Hotshot crews are very versatile and have the flexibility to adapt to several types of needs on the various levels of fire.

For a crew to be identified as an IHC, it must have at least seven permanent employees (superintendent, assistant superintendent, three squad bosses, and two senior fire fighters). Additional requirements include a minimum of 40 hours of training together as a crew and being identified as a national resource.

LISTEN UP!

In 2018, hotshots averaged over 100 days on fire assignments, and each crew traveled, on average, more than 50,000 mi (80467.2 km) during the fire season. If you think you have what it takes to be a hotshot, the 2018 fitness requirements include the following:

- Chin-ups: Between four and seven, based on body weight
- Push-ups: 25 in 60 seconds
- Sit-ups: 40 in 60 seconds
- 1.5-mi run: Less than 10 minutes and 35 seconds

DID YOU KNOW?

Hotshots are called hotshots because they were originally fire fighters assigned to the hottest part of the fire. The first forest known to have adopted this idea was the Cleveland National Forest in the late 1940s. Today, hotshots are the most intensively trained and versatile type of handcrew.

A **Type 1 crew** can be broken into squads for constructing firelines and performing complex firing operations. They have no fireline use restrictions and are fully equipped and mobile. Type 1 crews are considered the most experienced and physically fit and have the ability to adapt during a fire assignment. Based on the required depth of the crew members' qualifications and makeup, crews can be split up to fill different functions or have individuals pulled from the crew to fill various overhead (Incident Command System [ICS]) roles on a developing incident.

Type 2 Crews

Type 2 crews are not trained to the same level as Type 1 crews, but they can construct firelines and complete complex firing operations. There are two configurations of Type 2 handcrews: Type 2 initial attack (IA) crew and Type 2 crew.

In a **Type 2 IA crew**, 60 percent of the handcrew must have been a part of the crew for at least one season. The crew needs to have three agency-qualified sawyers and Incident Commander (IC) Type 4 or 5 qualified squad bosses. Type 2 IA crews are often sought after on emerging incidents due to the flexibility of the crew. They have the capability to be broken into squads for fireline construction and firing operations. Type 2 IA crews have more capabilities than Type 2 crews.

A **Type 2 crew** must have at least 20 percent of the crew with one season of experience in fire suppression. Type 2 crews construct firelines and complete firing operations as directed.

Type 2 IA crews and Type 2 crews are the backbone of firing operations. These crews are often tasked with the bulk of the fire assignments. It is important to understand that, while the assignments may not be as high profile as the hotshot assignments, Type 2 crews have the most important job on the fire in operations. Without these crews, fires would not be able to be controlled and contained safely and efficiently.

Other variations of Type 2 crews include military, Snake River Valley (SRV), Native American Crews (NAC), Emergency Firefighter Crews (EFF, Alaska) and private contract crews.

DID YOU KNOW?

Many states use inmate labor for fire operations. An **inmate crew** is composed of prison inmates and/or wards. The inmates are usually part of volunteer firefighting programs where they can earn time off of their sentences. Use these general guidelines if you are working with inmate fire crews:

- Inmate crews are not to be split up.
- In fire camp, separate sleeping areas are provided.
- Do not offer food, drinks, or cigarettes to inmates.
- Do not talk to or befriend inmates; maintain work-related contact only.
- Do not offer to take messages back to their friends or family.
- If an inmate is injured, make sure a correctional officer or sheriff accompanies that individual to the hospital.
- Keep male and female inmate crews separate.
- Check the rest and feeding requirements of inmate crews.
- Inmate crews must be accompanied by trained correctional officers accustomed to supervising inmates.
- The use of inmate crews is usually limited to the state in which they are based.
- Contact with inmates shall be made through the correctional officer in charge of the crew.
- Inmates are the responsibility of the correctional officer in charge.

Be sure to consider these requirements when planning inmate use on the fireline.

LISTEN UP!

Firefighting can involve many people from different cultural backgrounds. Areas of cultural differences can include:

- Food and diets
- Housing
- Clothing and dress
- Religion
- Language
- Behavior and customs

Professionalism in fire suppression operations requires respecting these differences.

Type 3 Crews

A **Type 3 crew** has limited to no experience. A good example of a Type 3 crew would be a crew used to help facilitate camp-area operations. These crews are also known as camp crews. While little experience is required for these crews, they can provide the first opportunity or stepping-stone to another crew type.

Air Crews

Air crews work with aircraft, such as helicopters and fixed-wing aircraft. Some paid air crews are given additional training that allows them to be deployed from helicopters or fixed-wing aircraft. There are several types of air crews, including but not limited to smokejumper, helitack, and rappel crews.

Smokejumpers

The U.S. Department of Agriculture (USDA) Forest Service and Bureau of Land Management have a smokejumper program. Smokejumpers are fire fighters who deploy from fixed-wing aircraft by parachute into remote areas for wildfire-suppression efforts. A **smokejumper** operates in a team of two or more.

Smokejumpers are considered to be among the most trained, physically fit, and experienced operators in the suppression environment. Fire fighters who are selected as candidates are required to participate in rookie training. Rookie training includes a grueling set of physical conditioning standards over a 5-week period. Returning jumpers receive a 2-week refresher course, which usually includes several hours of medical training. Many smokejumpers go on to earn an Emergency Medical Technician (EMT) certification.

LISTEN UP!

Smokejumpers are self-sufficient for 48 to 72 hours and need little logistical support.

All of these requirements are based on the need for a smokejumper to be self-sufficient, work very long hours, carry heavy backpacks full of gear, and maintain an operational tempo for up to 21 days or more depending on the assignment. It is not uncommon for smokejumpers to be moved from state to state, depending on the needs during a fire season. In the 2019 fire season, there were over 200 smokejumpers in Alaska.

Helitack Crews

Agencies that deploy handcrews from helicopters either land their crews on the ground before deployment or rappel them from the helicopter. These crews are called helitack crews and rappel crews. A **helitack crew** is a crew of fire fighters specially trained in the use of helicopters for fire suppression. Helitack crews perform duties similar to other initial attack handcrews. When a fire is reported in a remote location, a helicopter will be dispatched, carrying two or three fire fighters, to the newly reported fire. Helitack crew members can be deployed on small fires to construct a line around the fire. They can also be supported with water drops from the aircraft that brought them to the fire while they are involved in hand-line construction. On extended attack fires, they can be used to support water drops, cargo delivery, crew shuttling, or reconnaissance.

Rappel Crews

In remote areas where road access is limited and smokejumpers are not available or their use is not practical, a **rappel crews** is an effective suppression choice. Rappelling is a technique of deploying fire fighters from hovering helicopters in areas where the helicopter is unable to land. Rappelling involves fire fighters wearing a full-body harness while using a friction device to slide down a rope from the helicopter to the ground. A rappeller is the fire fighter who is rappelling.

Rappelling heights can range from 30 ft (9.1 m) in brush to 250 ft (76.2 m) in a timber model. Crew members carry 30 pounds (13.6 kg) of personal gear or more at all times and, depending on the assignment, up to 300 pounds (136.1 m) of fire gear or more, which may be lowered down to them from the helicopter. Rappellers are also in the category of the most trained, physically fit, and experienced operators in the suppression environment. Fire fighters who are selected as candidates are required to participate in rookie training. Rookie training includes a set of physical conditioning and standards. Rappellers are also required to be self-sufficient on fires for up to 14 days at a time. It is not uncommon for rappellers to be assigned to advanced duties on fires to clear helispots for landing helicopters. This allows for IHC, Type 1, Type 2A, and Type 2 crews to be ferried from a camp or a helibase to the fireline, which cuts down on the transportation or hiking time and increases efficiency. On some occasions, rappellers will join forces with IHCs or smokejumpers when extra fire fighters are needed. When they have completed their assignment, they usually hike out to a road and are picked up.

LISTEN UP!

A spotter is a smokejumper who selects the drop location and oversees all crew actions as fire fighters depart from an aircraft to the ground. The spotter does not leave the aircraft. From the aircraft, the spotter communicates with dispatch, does a fire size-up, selects the jump site, checks winds, selects the exit point for the jumpers, and determines the aircraft pattern for safe deployment of the jumpers.

Chainsaws

Handcrew tools consist of hand and power tools necessary to construct hand lines. Handcrews use a variety of tools for cutting and scraping. Common tools, such as the Pulaski and McLeod, are discussed in Chapter 7, *Hand Tools and Equipment*.

A power tool that is frequently used in heavier fuels is the chainsaw. A **chainsaw** is a portable power tool with handles and an engine on one end and a set of sharp, rotating teeth on the other end **FIGURE 14-3**. It can be used to rapidly cut a variety of fuels, from large trees to heavier brush types, in falling and clearing operations. Chainsaw operations are an integral component of hand-line construction and safety mitigation on all fire incidents.

LISTEN UP!

Chainsaws are an effective but dangerous power tool. When the correct saw is used in conjunction with the appropriate level operator or faller, exposure to the faller and crew members is significantly reduced. Fallers who cut beyond their training and skill level are at a significantly higher risk for injury. Between 2012 and 2018, accidents involving cutting operations or tree strikes were consistently one of the leading causes of injuries and fire-fighter fatalities.

Certifications

Chainsaws are relatively simple to use and very efficient. However, it is extremely important to have a baseline of knowledge before utilizing this tool. The information that is shared in this chapter is basic information. To become proficient and competent with this tool, consider taking a power-saw class, such as NWCG S-212, *Wildland Fire Chainsaws*. After completing the class, it is imperative that the trainee sawyer work with a certifier or mentor to acquire the necessary skills to become certified.

Chainsaw certification varies by state and agency. The national credentialing of fallers includes task performance evaluation and practical experience that covers three distinct levels. These levels were recently changed from A, B, C to 3, 2, and 1. The old A level is equivalent to the new 3 level, and the old C level is equivalent to the new 1 level.

Level 1/C is a professional chainsaw operator with the skills to cut any tree and teach new fallers to evaluate lower-level faller performances. Level 3/A is an entry-level chainsaw operator who still needs certification. This level operator is only able to cut brush and trees that propose no significant challenges.

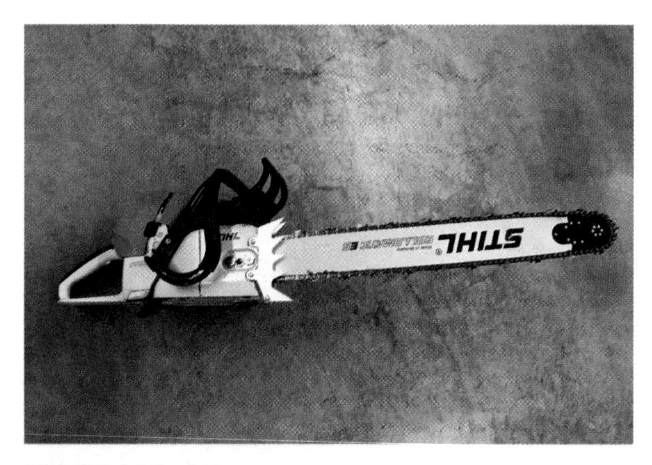

FIGURE 14-3 Chainsaw.
Courtesy of Jeff Pricher.

Pre-Operations

Fully analyze a situation before using a chainsaw. Just because you are certified to use a chainsaw does not necessarily mean you should use it. Any cutting operation should include a hazard analysis. Decision making always falls upon the chainsaw operator. Remember that it is always OK to turn down a cutting operation if you feel like you are underqualified or the assignment seems dangerous.

Falling or cutting operations generally require two people: the chainsaw operator, known as the **faller** (sometimes called the sawyer or cutter), and the swamper. The **swamper** is the fire fighter who assists the chainsaw operator. The swamper assists in clearing brush and small trees and serves as a lookout when a road or trail is in the vicinity of the falling operation. Coordination between the faller and the swamper is paramount. This includes predefined hand signals or other communication expectations. In many crew operations, it is not uncommon for the swamper and faller to swap positions to reduce fatigue and foster a learning environment for the newer cutters.

PPE during Chainsaw Operations

Any fire fighter operating as a swamper or faller should wear basic PPE, including a long-sleeve shirt, a helmet, gloves, and boots. Additional PPE during chainsaw operations includes chainsaw chaps, hearing protection, and eye protection.

Chainsaw Chaps

Chainsaw **chaps** are specialized pieces of PPE that cover the legs **FIGURE 14-4**. Chainsaw chaps are constructed of several layers of Kevlar under a protective nylon cover. If a chainsaw strike were to occur in the leg area, the chain (which produces a ripping effect) will cut through the nylon and grab on to the Kevlar layers. When this occurs, the Kevlar strands are pulled into the power head, which binds up the sprocket and chain drivers. This action can stop the chainsaw in several hundredths of a second.

FIGURE 14-4 Chainsaw chaps.
Courtesy of Jeff Pricher.

FIGURE 14-5 Mesh goggles.
Courtesy of Jeff Pricher.

Chainsaw chaps come in the following lengths: 28, 30, 32, 36, and 40 in. (71.1, 76.2, 81.3, 91.4, and 101.6 cm). Chainsaw chaps should overlap the firefighting boot by 2 in. (5.1 cm). This allows for coverage when the knee is bent.

Hearing Protection

Hearing protection for the faller and swamper is crucial. Hearing loss is a reality with continued chainsaw operation (and any power-tool operation) and no ear protection. To prevent hearing loss, anyone in the immediate vicinity of the power tool as well as the faller and swamper must wear earplugs or earmuffs with at least a 28-decibel rating.

Eye Protection

Eye protection is essential during falling and other power-tool operations. Eye protection should meet the American National Standards Institute (ANSI) standard. Mesh goggles or glasses are a favorite of fallers and swampers **FIGURE 14-5**. They are preferred because they do not fog like typical lensed eye protection. Unfortunately, like all tools, mesh goggles or glasses have a specific use. If the sawyer is cutting in a situation where very fine dust may be produced, such as in damaged wood, it will penetrate the mesh and enter the cutter's eyes. Mesh goggles or glasses should be used only in live timber or in situations where large wood chips will be created by the saw.

Not all eye protection is created equal. Sunglasses are not an acceptable form of eye protection, especially if the sunglasses do not provide wraparound protection. Several injuries have been catalogued where a wood chip or other foreign debris bounced off the cheek, off the lens, and into the eye. Foreign debris in the eye is very painful and, in some instances, can cause permanent damage.

Safety

The Blood Bubble

When the faller is running the saw, it is crucial to make sure you are clear of the blood bubble. The **blood bubble** is the radius around the faller where the tip of the chainsaw can reach **FIGURE 14-6**. It is not uncommon for the faller to be working in a swivel motion, so anyone and anything inside the blood bubble can potentially be injured by the chainsaw.

FIGURE 14-6 The blood bubble is the radius around the faller operating a chainsaw where the tip of the chainsaw can reach.

LISTEN UP!

Good communications must be maintained between the saw team and adjoining personnel in order to avoid injuries. Visually account for adjoining personnel before starting the operation and post lookouts as necessary.

Medical Plans

Fallers and swampers should always have a safety and medical plan in place prior to chainsaw operations. Most fallers and swampers carry compression bandages and/or tourniquets as part of their safety equipment. Refer to Chapter 3, *Safety on Wildland Fires*, for information on medical incident reports and incident within an incident processes.

Chainsaw Anatomy

A chainsaw has three components: the chain, the bar, and the power head. All of these components need to be clean and well maintained, and the chain sharp, to prevent accidents or injury.

LISTEN UP!

When falling, it is not uncommon for dead branches to break off as the tree falls. Always look up and do not work under overhead hazards. Always have an escape route before falling hazard trees.

Chain

The chain is integral to being able to perform at the needed level of operation. Chains must be sharp to cut through the wood fiber as fast as possible to reduce the faller's time at the stump and limit the exposure of being hit by widow-makers or other weakened parts of a tree.

The chain used for wildland operations is vastly different from the chain commonly used in structure-fire operations. Do not use structure chain in the wildland environment. Structure chain is not as sharp as the wood chain and can cause damage to the power head by overheating the bar and internal components.

The parts of the wildland chain include a left-hand cutter; a right-hand cutter; and a driver, or drive link. The driver is the part that makes contact with the sprocket on the power head **FIGURE 14-7**.

The cutters have two distinct parts, the depth gauge (raker) and the cutting tooth. The depth gauge determines how much wood is chipped away **FIGURE 14-8**. If a depth gauge is too tall, less wood will be cut. If a depth gauge is too short, too much wood may be cut. In either event, efficiency is lost. There are special tools that are used to determine the height of the depth gauge in relation to the cutter.

Any sharpening of the saw chain must be completed by a trained fire fighter or professionally sharpened. There are indicators on the chain that determine when a chain cannot be sharpened any longer. This is usually identified by an etched line on the top plate of the cutter.

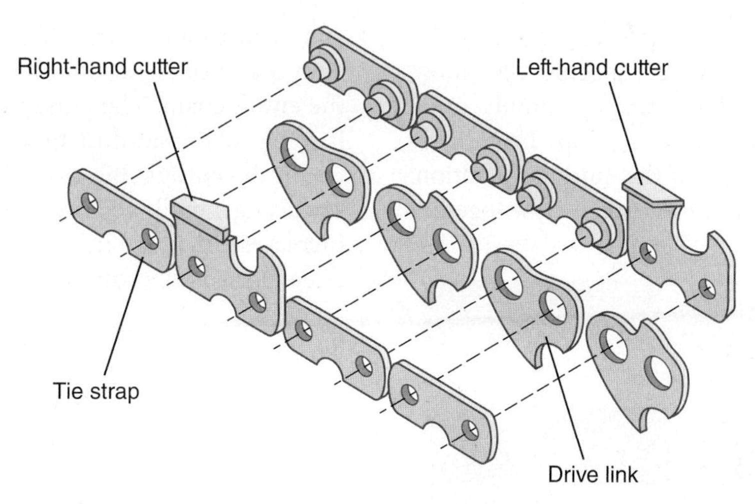

FIGURE 14-7 Chainsaw chain parts.

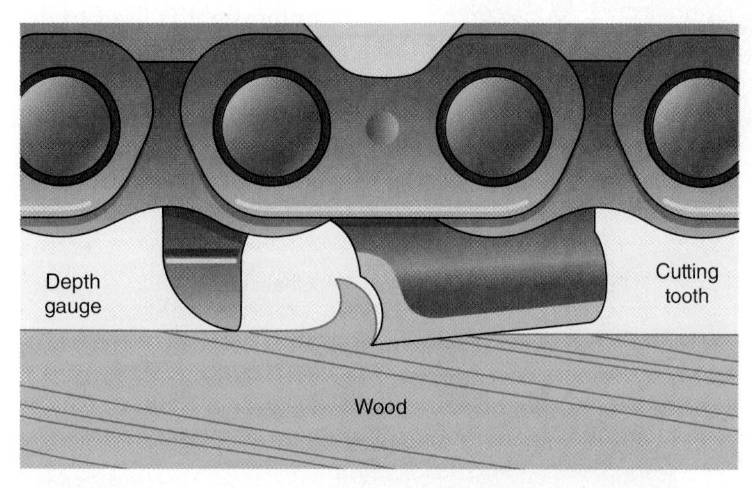

FIGURE 14-8 The chain's depth gauge determines how much wood is chipped away.

With the metal-to-metal contact, it is essential that the chain and bar be well lubricated. As the driver travels the length of the bar, it pulls bar oil from the oil reservoir through the oil hole located on the bar. As this oil is captured, it lubricates the chain in the bar groove as it makes its way around the bar. The chain will move at speeds of 50 to 55 mi/h (80.5 to 88.5 km/h).

Bar

The chainsaw bar guides the chain in the cyclical motion that is necessary for ripping and cutting through wood and other brush. There are many varieties of bars that can be used on a power head, including steel bars, lightweight bars, and solid nose. It is important that the bar matches the chain and the sprocket on the power head. A mismatched bar can be a recipe for disaster, causing malfunctioning and severe injury. Most bars have a stamp that includes the length, the number of cutters, and other important information **FIGURE 14-9**.

There are several types of chains, which are specific to the cutting operation. Chains can be categorized by

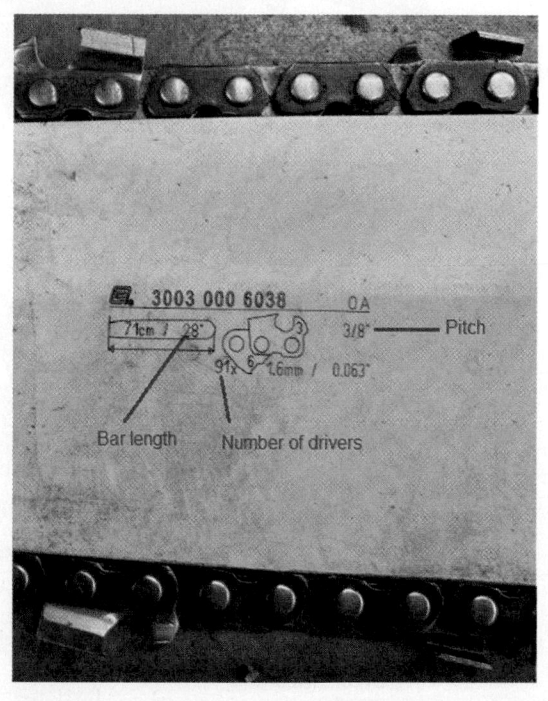

FIGURE 14-9 Chainsaw bar stamp information.
Courtesy of Jeff Pricher.

the type of cutter (chisel, chipper, and semi-chisel) and should be chosen for the specific operation. Chisel and semi-chisel chains are commonly used. Chains are either full skip or semi-skip. This difference is specific to the need of the cutting operation. The semi-skip chain alternates between close together and spaced-out cutters. The full-skip chain consistently leaves out one cutter in the sequence throughout the entire chain. The reason for the loss of the cutter is to allow for sawdust to accumulate in the gap between the cutters. Big trees that will generate a lot of sawdust generally require a full-skip chain. This chain increases efficiency and reduces time at the stump; reduced time at the stump creates safety.

FIGURE 14-10 Chainsaw power head.
Courtesy of Jeff Pricher.

Power Head

The final main component of the chainsaw is the power head. The power head is essentially everything short of the bar. It houses the motor, stores the fuel, and allows you to safely handle the chainsaw **FIGURE 14-10**. A black line on the top of the power head called the gunning sight is used to determine the angle, depth, and general direction a tree is expected to fall when cutting at a specific angle.

Understanding the parts of a chainsaw will help you to understand the performance of this tool as well as the areas that need to be maintained **FIGURE 14-11**. On the

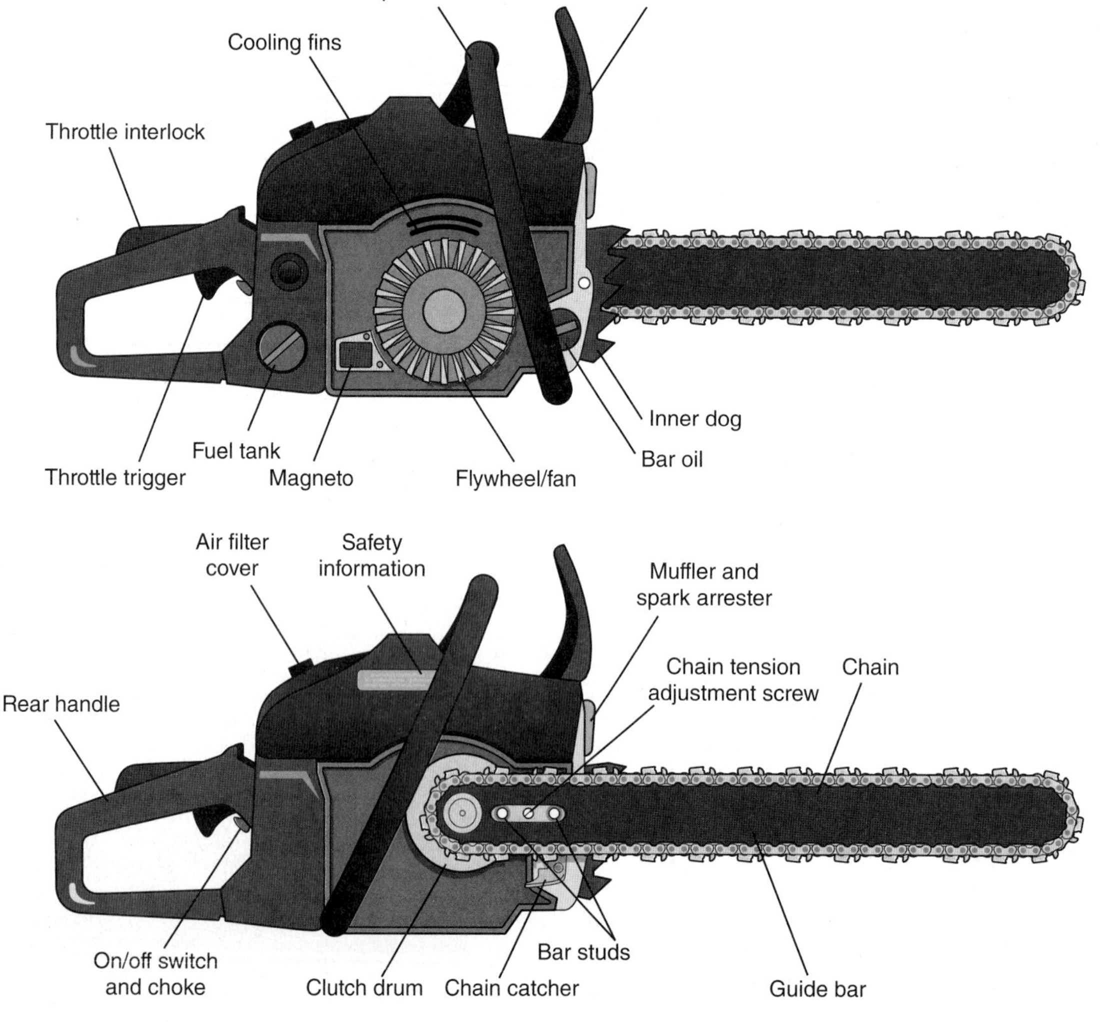

FIGURE 14-11 Chainsaw parts.

saw, adjustments can be made to the bar oil, the rpm, and the idle. If you are not trained on how to perform this maintenance, do *not* make any adjustments. Adjusting these components requires a tachometer or a very comprehensive understanding of the chainsaw settings.

Operations

Fueling

A chainsaw should be fueled on bare ground and away from an ignition source or at least 20 to 30 ft (6.2 to 9.1 m) away from the fireline **FIGURE 14-12**. Fuel vapors can ignite in areas within 20 ft (6.2 m). The recommended fuel mixture for most chainsaws is 50:1 non-ethanol gasoline to oil. Follow the manufacturer's recommendations for the saw you are using.

The gasoline that is used must not contain ethanol. Fuel that contains ethanol is prone to disintegrating the gaskets, rubber, and other plastic parts of the chainsaw. Ethanol-free fuel is not available everywhere, so ask before you fill your fuel cans.

To replenish the fluids in a chainsaw, follow these steps:

1. Wait for the saw to cool. Stage yourself and the saw from 20 to 30 ft (6.2 to 9.1 m) away from the fireline and any ignition sources.
2. Fill the bar oil reservoir. Forgetting to replenish the bar oil could lead to a catastrophic failure of the chain and bar interface.
3. Open the fuel container slowly as you look away, and cover the fuel port with a rag or your gloved hand. With the recent EPA requirements of fuel and fuel containers, "fuel geysering" has significantly increased. This phenomena occurs when the fuel vapors are not able to escape the container and pressure builds up. To combat this, it is important that you cover the fuel port and look away as you slowly open the container. If you fail to do this, it is very probable that a geyser could occur and spray fuel into your eyes.

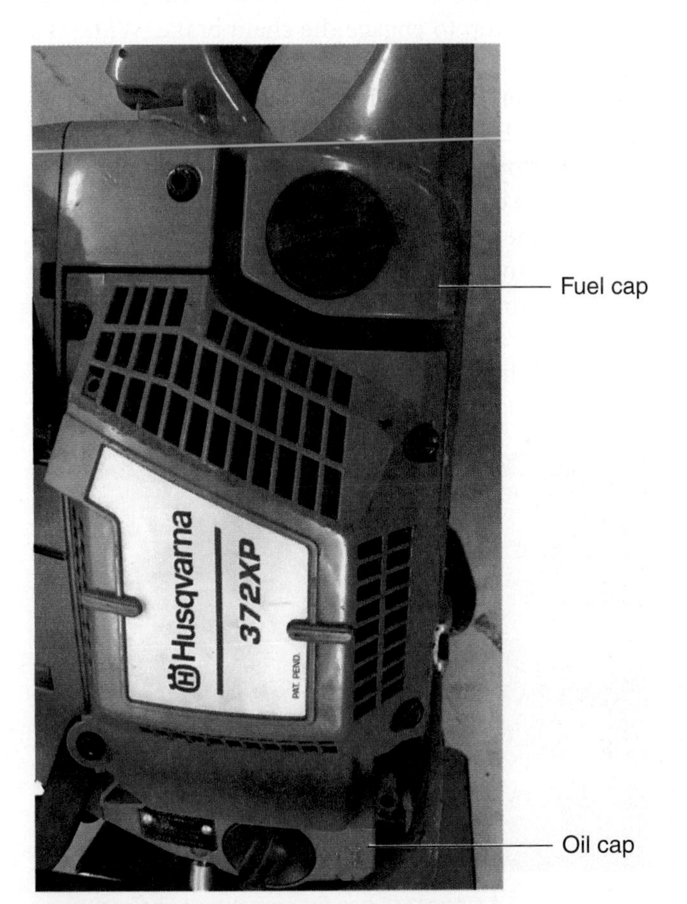

FIGURE 14-12 Chainsaw fuel and oil caps.
Courtesy of Jeff Pricher.

One of the more common chainsaw fuel containers in wildland firefighting is the split fuel container, known as a dolmar **FIGURE 14-13**. This fuel container can carry 1 gal (3.8 L) of bar oil and 1.5 gal (5.7 L) of mixed fuel.

In extended cutting operations away from a large fuel source, fuel bottles meant for camping can also be used. These bottles generally contain 1 L (0.3 gal) of fluid. Be sure to read the label before emptying to avoid putting bar oil in the fuel container or vice versa. As noted earlier, be sure to open all fuel containers slowly, place your hand or a rag over the top, and look away to prevent fuel from spraying into your eyes.

Starting a Chainsaw

There are three Occupational Safety and Health Administration (OSHA)-approved methods for starting a chainsaw, and they are identified in all manufacturers' documentation: on the ground, between your legs, and over a log.

FIGURE 14-13 Dolmar split fuel can.
Courtesy of Jeff Pricher.

The starting procedure for some saws may be different. Make sure you read the manufacturer's documentation and understand the recommended procedures.

To start a chainsaw on the ground, follow these instructions:

1. From the right side of the power head, remove the bar's protective cover.
2. Check that the chain brake is engaged.
3. Set the saw controller to the choke position.
4. Reaffirm that the chain brake has been engaged. If your saw has one, depress the decompression button to reduce the friction on your pull of the starter cord.
5. Before pulling the starter cord, yell out "Starting saw!" so that anyone in your blood bubble or working in the vicinity of the saw will be alerted to your operation.
6. Place your right foot on the right side of the power head (on the larger side of the plate), and pull up on the pull cord as hard as you can. Pull as many times as necessary until you hear the power head pop or the motor sound as though it is going to start. Immediately after hearing the pop, take the saw off of full choke, and place the saw in half choke.
7. Pull the starter cord again, and the saw should start within one to four pulls. If you are successful in starting the saw, the saw will immediately run in high idle, which is normal.
8. With all chainsaws, when the saw starts and is in high idle, immediately tap the throttle (barely pull on the trigger/throttle), and the saw will drop into a normal idle. It is important to do this because, if you do not tap the throttle and the saw remains in high idle, you can damage the power head.

To start a chainsaw between your legs, follow the same starting procedures regarding the choke setting and listening for the pop. The hand that is holding the handle needs to be stretched out straight in a stiff-arm position. The throttle handle should be pinched between your legs and the bar, facing away from your body. As with all of the starting procedures, make sure you yell out "Starting saw!" before pulling the pull cord. Most sawyers prefer this starting method because it reduces fatigue from having to lean over.

The third method follows the same starting guidelines regarding the choke setting and voice commands. The major difference is that the chain brake is not engaged as the saw is rolled over a log when the pull cord is pulled. This method is effective for logging operations but is the least preferred method for wildland fire fighters because, of the approved starting procedures, it provides the least control over the saw.

Operation Stance

When operating a chainsaw, check your stance. Your feet should be spread apart in a wide, balanced stance, with feet and legs out of the direct line of the bar **FIGURE 14-14**.

LISTEN UP!

Do not use the "drop start" method to start your chainsaw. This method involves dropping the chainsaw while hanging on to it and using gravity to help you pull the pull cord. When you use this method, you have no control over the chainsaw. Starting a chainsaw in this manner is an OSHA violation.

Stopping a Chainsaw

It is important to be familiar with the chainsaw you intend to use prior to using it. Each saw manufacturer has a different procedure for stopping chainsaws. In general, it can be accomplished in one of two ways: either toggle the switch from run to stop, or slide the switch from run to off. Wait until the chain comes to a complete stop to engage the chain brake. While the chain brake is designed to stop the chain, it is only intended for use for emergencies or when the saw kicks back. Engaging the chain brake before the chain has fully stopped can wear down the clutch plate and damage the equipment.

Once the saw is off and the chain brake has been engaged, store the saw with the choke in the closed position. Storing the saw with the choke open can lead to the saw constantly flowing fuel and keep the spark plug wet, which makes starting the saw more difficult.

FIGURE 14-14 Correct stance for operating a chainsaw.
Courtesy of Jeff Pricher.

Additionally, if the saw is stored with the choke open, there is a potential for fuel to leak out into a storage compartment.

Operating Safety Tips

When operating a chainsaw, follow these safety tips:

- Allow the motor to cool before fueling.
- Do not fuel the chainsaw within 20 ft (6.1 m) of an ignition source.
- Do not fuel the chainsaw on the tailgate of a truck.
- Before cutting, analyze overhead obstructions, widow-makers, and other loose debris. Maintain a safe working area that is clear of debris.
- Use a wedge to redistribute a tree's weight to the desired falling location and to prevent a tree from sitting back.
- Never cut above your head.
- Never cross your arms over the power head. If you end up reaching and inadvertently invert the saw, switch hands so that your arms are not crossed.
- Place your hand behind the chain brake as much as possible.
- Always grip the chainsaw with both hands, ensuring that thumbs and fingers encircle the handle.
- If your thumb is resting on the top of the handle and the chainsaw kicks back, you could lose control of the chainsaw.
- Always plan for kickback. In the event the chainsaw kicks back, you will be prepared.
- Engage the chain brake only with the hand that rests on the handle. Do not use the control hand (hand that operates the throttle) to engage or disengage the chain brake. This is a bad habit that could result in injury to the operator. Use the top hand.
- If you are walking more than 2 ft (0.61 m) with the chainsaw, the chain brake must be on.
- If you are walking more than 10 ft (3.0 m) with the chainsaw, the chainsaw should be turned off.
- If you are cleaning or repairing the chainsaw, it should be turned off.
- If you are tired, stop cutting. Fatigue is a common cause of accidents or injury.

Storage and Maintenance

It is important to properly secure the chainsaw for storage and transport. The chainsaw bar should always be covered when the saw is not in use or when transporting the saw to or from a cutting area. Some agencies purchase commercially available bar covers.

FIGURE 14-15 Bar cover made from out-of-service fire hose.
Courtesy of Jeff Pricher.

Others opt for the less expensive and practical bar covers that are made from old, out-of-service fire hose **FIGURE 14-15**. When walking with the chainsaw over your shoulder, make sure that the power head is facing up, toward the sky. This is important in case the faller falls. If the power head is facing down, toward the ground, the saw will fall very close to the body. When the power head is facing up, it will fall away from the body, lessening the potential for injury.

Chainsaws should be anchored to their storage location to prevent them from moving during transport.

It is important to note that the transport of chainsaws also involves the transport of flammable and combustible liquids. It is important to follow agency guidelines and regulations regarding the safe transport of these materials.

To keep a chainsaw in top operating condition, maintenance is required after each use. Consult the manufacturer's information for the appropriate maintenance of your specific saw. Having the right tools and spare parts is imperative for operating and maintaining a saw. It is recommended that you carry several spare parts with you in the field in the event you lose or break a part. An example of a fully stocked saw kit is shown in **FIGURE 14-16**.

Handcrew Arrival Procedures and Use of Hand Tools

A fire crew can arrive on the fireline in a **crew-carrying vehicle**, also called a CCV or a crummy **FIGURE 14-17**. The crew can also arrive in crew busses or pickups. In this configuration, they can easily be split into squads. The following is an example of a tool-up

FIGURE 14-16 Fully stocked chainsaw kit.
Courtesy of Jeff Pricher.

A

B

FIGURE 14-17 A. Crew bus. **B.** Crummies.
A: Courtesy of Jeff Pricher; **B:** Courtesy of Joe Lowe.

procedure. Crew members are seated in the vehicle according to the hand tool they will be using. When the crew arrives on the fireline, the swamper unloads from the vehicle first. His or her first duty is to place wheel chocks under the vehicle. After that is done, the crew boss unloads and instructs the swamper on which tools to issue. The safety person, usually the last crew member with a shovel in the handcrew lineup, unloads from the CCV. He or she prepares to issue canteens.

The crew boss then gives the order to unload the rest of the crew from the CCV. The crew boss then gives the tool order. This establishes the order in which hand tools are placed during a line-cutting operation. The tool order is dependent on the fuel type in which the hand line will be cut.

After the crew members receive the tool order, they line up—in full safety gear—at the rear corner of the driver's side of the vehicle. The swamper then issues the hand tools to the crew. After crew members receive the tools, they proceed to the right side of the CCV and receive a canteen. When all hand tools and canteens have been issued, the crew proceeds to the hook-line position. Crew members keep a 10-ft (3.0-m) distance between each other.

In general, the makeup of crews will fall into two sections. Depending on what region you work in or the preferences of the crew boss, these sections are the **cutting section** and the **scraping section**. The responsibilities of the cutting section on a fireline include sawing, chopping, and grubbing of woody fuels. Once the fuel has been cut, it is then removed from the fireline. The scraping section completes the hand line. It removes duff, grass, and smaller brush from the fireline. McLeods, shovels, and fire brooms are used in the scraping section. The crew leader makes the decision on the attack method to use and the tool order. Again, this decision depends on the type of fuels in which the crew will be working.

Hand-Line Construction Principles

A handline is constructed progressively, with members of the handcrew spaced 10 ft (3.0 m) apart. This section of the line is called "your dime." Each member of the handcrew takes a stroke with his or her hand tool and then moves one step forward. With each member of the crew removing some fuel, the line progresses forward more rapidly. Toward the end of the line, you should be able to see a clean line down to mineral soil. The last person in the line (the drag shovel) ensures that the line is clean and complete.

The fireline should be constructed no wider than necessary to stop the fire. Time and energy are best used to construct more hand line. The general guideline for fireline width is to make it 1.5 times the dominant fuel height. In timber, the objective is to construct a line that will stop burning surface fuels as well as the lower aerial fuels. The rule is to construct this line 20 to 30 ft (6.1 to 9.1 m) wide and then provide a scraped line between 3 and 4 ft (0.9 and 1.2 m) wide. Most firelines are unsuccessful at stopping an active crown fire.

Tool Order

The crew leader establishes the tool order, which is dictated by the fuel type the crew is operating in. The next sections describe the tool order for woody fuels, such as shrub.

Brush Hook and Chainsaw

The crew leader is near the front of the operation. He or she is followed by the lead hook or first saw. This crew member is the first person who operates a chainsaw or brush hook. He or she is the lead person on the crew and locates the line as instructed by the crew leader. The first hook and first saw cut the brush only wide enough to walk through. He or she usually disposes of the brush that has been cut.

Handcrew members who follow the lead hook or lead saw are spaced at least 10 ft (3.0 m) apart. This is done for safety reasons because people will be using hand tools. The brush-hook operators who follow the lead hook widen the line to the specifications set by the crew leader. Each brush-hook operator cuts some brush and leaves, leaving some to be cut by the next brush-hook operator.

LISTEN UP!

Handcrew members who follow the lead hook or lead saw are spaced at least 10 ft (3.0 m) apart to avoid accidental injuries.

Disposal of Cut Brush

After the brush has been cut, it must be disposed of properly. All previously burned or burning materials need to be scattered into the burned area. Unburned cut fuels are scattered in the unburned areas. Disposing of cut brush can take the entire crew or several crews.

The three methods used to disperse or dispose of brush follow:

1. Toss method. The brush is thrown and scattered from where it is cut. This works only if the surrounding brush is sparse or low in height.
2. Window method. If the surrounding brush is too tall and dense for the toss method, then you need to pull brush to an opening, called a *window*, and dispose of it. If a natural opening is not available, then you may need to create one.
3. Bank method. This method is used when the surrounding brush is too tall or dense for the toss or the window method. In this case, a double-width line is cut, and all the cut brush is compacted against the standing brush on the side away from the burned area.

Pulaskis

Pulaskis are normally placed behind the last brush hook or chainsaw operator. These grubbing tools are used to remove stumps and stobs (previously cut and pointed branches and stumps close to the soil) and to cut roots that cross the control line. They are also used to trench where it is necessary **FIGURE 14-18**. The mattock edge of the tool is used for this purpose.

FIGURE 14-18 A cup trench is used to catch rolling material on a slope that could otherwise start fire below the fireline. A high berm on the outermost downhill side of the trench helps the cup trench catch material. Also known as gutter trench.

Trenching is done on steep hills during handline construction. The purpose of trenching is to prevent any rolling firebrand materials from crossing the control line.

Pulaskis have an ax on the side opposite the mattock edge. Therefore, this tool is sometimes moved up in front of the lead brush hook to cut down heavier fuels. In harder soils, Pulaskis are used to loosen soil, especially if dirt is needed to be thrown on the fire.

McLeods

McLeods are scraping tools, and they generally follow the crew members carrying Pulaskis. The rake side of the tool is used to rake away litter, duff, and other loose materials.

The tool is used to break up rats' nests of woody debris and piles of brush at window areas.

Shovels

Shovel operators usually follow the McLeods. Shovels are used to finish the control line and remove fuels that were missed by the McLeod operators.

Additional shovel operations follow:

- Hot materials can be carried or thrown back into the burn.
- Trenches can be cleared out.
- Windows can be scattered with the use of a shovel.

Again, as with all of the tools mentioned, keep 10 ft (3.0 m) of space between workers.

Other Fuel Types

In lighter fuels, the hand-tool order changes to fit the fuel type. In this case, there may be two Pulaskis as the first hand tools and then a mix of scraping tools with a Pulaski in the middle of the crew.

Additional Fireline Construction Rules

To have a complete understanding of fireline construction, it is necessary to study the following additional rules:

- Clean the entire fireline down to mineral soil for all or part of the width, except in fuels such as bog, peat, or tundra.
- Cover stumps and logs just outside the fireline with dirt to protect them from radiant heat when the fire reaches the fireline.
- Cut low-hanging limbs from trees on either side of the fireline that could cause fire to spread across the line.
- The hotter and faster the fire burns, the wider the control line must be. Seven factors used to determine fireline width are
  1. Slope
  2. Fuel
  3. Weather and wind direction (Is the wind steady? Is it blowing toward or away from the control line?)
  4. The part of the fire where the control line will be constructed: head versus flank
  5. Possibility of cooling
  6. Size of the fire
  7. Fireline conditions and intensity levels
- Make the fireline no wider than necessary. Time can be better spent constructing more fireline and encircling the fire.
- Always watch for rolling hot materials that can start spot fires below the crew.

Handcrew Production

Handcrews are an important fireline tool. They should train constantly and stay in good physical condition. They should know how to use their tools, especially chainsaws. They are a team, and teamwork gets the job done.

The basic mission of a handcrew is to suppress the fire and construct hand lines by using a combination of hand and power tools. Handcrew operators should be supported tactically and logistically on the fireline if they are going to be used to their maximum potential. Also, remember to rest these crews because the physical demands placed on them are great. Handcrew production rates will decrease as the operational shift goes on.

Handcrew Production Rates

It is important for a fire manager to understand an estimate of handcrew production rates. A **handcrew production rate** is used during planning to estimate how long it will take to complete a section of line with the available handcrew resources before the fire arrives. It can also be used as a basis for ordering handcrews for the fire.

TABLE 14-1 can be referenced as a general guide for handcrew production. It gives the reader an idea of the capabilities of a Type 1 crew.

TABLE 14-1 Initial Attack Handcrew Production

Fire Behavior Fuel Model	Type	Conditions Used In	Construction Rate in Chains per Hour (Meters per Hour)
1	Short grass	Grass	4.0 (80.5)
		Tundra	1.0 (20.1)
2	Open timber	All	3.0 (60.4)
	Grass understory		
3	Tall grass	All	0.7 (14.1)
4	Chaparral	Chaparral	0.4 (8.0)
		High pocosin	0.7 (14.1)
5	Brush (2 ft [0.6 m])	All	0.7 (14.1)
6	Dormant brush, hardwood slash	Alaska black spruce	0.7 (14.1)
		All others	1.0 (20.1)
7	Southern rough	All	0.7 (14.1)
8	Closed timer litter	Conifers	2.0 (40.2)
		Hardwoods	10.0 (201.2)
9	Hardwood litter	Conifers	2.0 (40.2)
		Hardwoods	8.0 (160.9)
10	Timer (litter and understory)	All	1.0 (20.1)
11	Light logging slash	All	1.0 (20.1)
12	Medium logging slash	All	1.0 (20.1)
13	Heavy logging slash	All	0.4 (8.0)

Modified from: https://www.nifc.gov/nicc/logistics/references/Wildland%20Fire%20Incident%20Management%20Field%20Guide.pdf.

After-Action REVIEW

IN SUMMARY

- Handcrews are teams of 18 to 20 fire fighters who use hand tools to actively suppress low flame–production fires directly along the fire's edge. They can be an established team of fire fighters, or they can be assembled from several engine or air crews.
  - Handcrews have the ability to operate in steep and remote terrain, so they are often the first line of defense in wildland firefighting.

  - The primary deployment scenario of a handcrew is the utilization of a direct attack method, though they can also perform indirect attack fireline operations.
  - Handcrews are typed depending on their training and experience levels:
    - Interagency hotshot crews (IHCs)
    - Type 1 crews
    - Type 2 IA crews
    - Type 2 crews
    - Type 3 crews
- Air crews work with aircraft, such as helicopters and fixed-wing aircraft. There are several types: helitack, rappel, and smokejumper crews
  - Helitack crews are fire fighters specially trained in the use of helicopters for fire suppression.
  - Rappel crews are fire fighters who rappel from hovering helicopters in areas where the helicopter cannot land.
  - Smokejumper crews deploy from fixed-wing aircraft by parachute into remote areas for wildfire-suppression efforts.
- A chainsaw is a portable power tool used in heavier fuels. It has a sharp, rotating blade of teeth that can rapidly cut a variety of fuels, from large trees to heavy brush.
  - Chainsaw certification varies by state and agency, and cover three distinct levels: 3, 2, and 1, with 1 being the highest.
  - Chainsaw operations generally require two people: the chainsaw operator, or faller, and the assistant, or swamper.
  - Proper PPE, such as chainsaw chaps and eye protection, is required for falling operations.
  - Be sure to follow the proper procedures for starting, operating, stopping, and storing the chainsaw.
- A handline is constructed progressively, with handcrew members spaced about 10 ft (3 m) apart. Each member of the handcrew takes a stroke with his or her hand tool and then moves one step forward.
  - The fireline should be no wider than is necessary to stop the fire. In general, the guideline for fireline width is 1.5 times longer than the dominant fuel height.
  - After brush has been cut, it must be disposed of properly.
    - Previously burned or burning fuels need to be scattered into burned area
    - Unburned cut fuels are scattered in the unburned areas
- The crew leader establishes the tool order of a hand-line. Though hand tool order changes to fit the fuel type, tools are generally ordered as such:
  - Crew leader
  - Lead hook or first saw
  - Brush-hooks
  - Pulaskis
  - McLeods
  - Shovels
- Handcrew production rates are used during planning to estimate how long it will take to complete a section of fireline with available handcrew resources before the fire arrives.

KEY TERMS

Air crews Crews that work with aircraft, such as helicopters and fixed-wing aircraft.

Blood bubble The radius around the faller where the tip of the chainsaw can reach.

Chainsaw A portable tool, either gasoline-powered or electric, with handles and an engine on one end and a set of sharp, rotating teeth lining a blade on the other end.

Chaps Specialized pieces of PPE, constructed of several layers of Kevlar under a protective nylon cover, that cover the legs during chainsaw operations.

Crew-carrying vehicle (CCV) A vehicle specialized for carrying a fire crew and their equipment.

Cutting section A division of a fire crew responsible for the sawing, chopping, and grubbing of woody fuels.

Faller A chainsaw operator.

Handcrew A team of 18 to 20 fire fighters who use hand tools to actively suppress low flame-production fires by constructing hand lines.

Handcrew production rate A measure of how long it will take to complete a section of line with available handcrew resources before the fire arrives.

Helitack crew A crew of fire fighters specially trained in the use of helicopters for fire suppression.

Inmate crew A crew composed of prison inmates and/or wards; usually part of volunteer firefighting programs.

Interagency hotshot crew (IHC) A kind of Type 1 crew; the most experienced, highly and intensively trained type of fire crew that have the skills to battle the most serious wildfires.

Rappel crew A crew of fire fighters who rappel down from a hovering helicopter in areas where the helicopter is unable to land.

Scraping section A division of a fire crew responsible for the removal of duff, grass, and smaller brush from the fireline, and completing the hand line.

Smokejumper crew A team of two or more fire fighters who deploy from fixed-wing aircraft by parachute into remote areas for wildfire-suppression efforts.

Swamper A fire fighter who assists the chainsaw operator.

Type 1 crew A highly experienced, and physically fit crew that can be broken up into smaller squads for constructing firelines and performing complex firing operations.

Type 2 crew A crew that constructs firelines and completes firing operations, with 20 percent of its members having at least one season of experience in fire suppression.

Type 2 IA crew A crew with more experience and capability than Type 2 crews, with 60 percent of its members having at least one season of experience in fire suppression, qualified sawyers, and Incident Commander Type 4 or 5 qualified squad bosses.

Type 3 crew A crew with limited to no experience in fire suppression; also known as a camp crew.

REFERENCES

California Department of Forestry and Fire Protection. *Fire Crew Firefighting Handbook 4200*. Ione, CA, 2004.

National Interagency Fire Center. *Interagency Standards for Fire and Fire Aviation Operations* (Red Book). 2020. https://www.nifc.gov/policies/pol_ref_redbook.html.

National Wildfire Coordinating Group (NWCG). "S-211, Portable Pumps and Water Use, 2012." Modified October 23, 2019. https://www.nwcg.gov/publications/training-courses/s-211.

National Wildfire Coordinating Group (NWCG). "S-212, Wildland Fire Chainsaws, 2012." Modified October 23, 2019. https://www.nwcg.gov/publications/training-courses/s-212.

National Wildfire Coordinating Group (NWCG). "S-230, Crew Boss (Single Resource), 2012." Modified October 23, 2019. https://www.nwcg.gov/publications/training-courses/s-230.

National Wildfire Coordinating Group (NWCG). *Wildland Fire Incident Management Field Guide*. PMS-210. NFES 002943. January 2014. https://www.nwcg.gov/sites/default/files/publications/pms210.pdf.

Wildland Fire Fighter in Action

It is June 7, and you are the squad boss on a Type 2 IA crew that was just assigned to a division in an area where there are several structures. The fire you are on started on May 8 and has just escalated to a Type 1 fire. The division group supervisor wants you and another fire fighter to serve as lookouts for the division. At 1430, the division supervisor orders all crews to the safety zone. It is now 1500, and, as you start to make your way to the safety zone, you realize that your escape route has been cut off.

1. What would your trigger points be based on?
   - **A.** Time of day
   - **B.** Current and expected fire behavior
   - **C.** Proximity to the fire
   - **D.** All of the above
2. What would be your first concern as you realize you will need a plan B?
   - **A.** Communicating your situation to your supervisor
   - **B.** Retreating down the fireline
   - **C.** Pulling out your fire shelter for possible deployment
   - **D.** Starting to burn out a deployment zone
3. As you make your way to your secondary escape route, you come across two stone buildings. What would you do?
   - **A.** Look for a hose to wet down a deployment area.
   - **B.** Scrape an area down to mineral soil on the lee side of the building for a shelter deployment.
   - **C.** Use the building as a temporary refuge area (TRA).
   - **D.** Look for a garden hose to set up a defensible position for the impending flame front.
4. What other resources would be best suited to assist you in this situation?
   - **A.** A very large air tanker (VLAT)
   - **B.** A lookout from another location to advise you of the fire activity and location
   - **C.** A Type 1 helicopter with an 800-gal (3028.3-L) tank to pretreat the fuels before the flame front arrives
   - **D.** The IRPG

Access Navigate for flashcards to test your key term knowledge.

CHAPTER 15

Wildland Fire Fighter II

Basic Firing Operations

KNOWLEDGE OBJECTIVES

After studying this chapter, you will be able to:

- Describe the purpose of conducting firing operations. (**NFPA 1051: 5.5.1, 5.5.3**, p. 328)
- Understand firing-operation permissions. (pp. 328–329)
- Describe the term *backfire*. (pp. 329–330)
- Describe the term *burnout*. (pp. 330–331)
- Identify preplanning considerations for firing operations. (p. 331)
- Identify fuel, weather, and topography considerations for firing operations. (pp. 331–332)
- Describe general firing operations rules. (p. 332)
- Identify and describe common firing devices and their components:
  - Fusee (pp. 333–335)
  - Drip torch (pp. 334–337)
  - Flare launcher (pp. 338–339)
  - Other firing equipment and field-expedient firing methods (p. 339)
- Identify a go/no go checklist (pp. 340–341)
- Describe different firing techniques:
  - Line firing (pp. 340–342)
  - Strip firing (pp. 340–344)
  - Flank firing (pp. 342–344)
  - Ring firing (pp. 344–345)

SKILLS OBJECTIVES

After studying this chapter, you will be able to perform the following skills:

- Light a fusee. (pp. 334–335)
- Operate a drip torch. (pp. 336–337)
- Operate a flare launcher. (pp. 338–339)

Additional Standards

- **NWCG S-130 (NFES 2731)**, *Firefighter Training*
- **NWCG S-219**, *Firing Operations*

You Are the Wildland Fire Fighter

The fire you are on has just become a wind-driven fire on a day that has a fire weather watch in place. You are the engineer on a fire engine staffed with six. The engine boss is ahead of you and the rest of the crew scouting a potential place to make a stand around a subdivision with a flame front rapidly approaching. The crew members you are working with have 3 to 4 years of experience each and are prepared to make the stand. Over the radio, the engine boss calls for you to line out the crew to begin an immediate burnout from a wet-line in an effort to cut off the fire and protect all the structures in your area. Over the radio, she advises you that the hoselay will be approximately 600 ft (182.9 m) long and will require her to light a counterfire to ease the wildfire's push on the line as you bring fire with you.

1. How will you line out the crew?
2. How many firing devices will you need?
3. What does your anchor point look like, and will it accommodate your operation?
4. What type of hoselay will you be using or preparing for?
5. How will you ensure that LCES/LACES (lookouts, awareness, communication, escape routes, and safety zones) is in place?

Access Navigate for more practice activities.

Introduction

In wildland firefighting, we sometimes fight fire with a fire that we intentionally light **FIGURE 15-1**. In certain situations, when fires are too intense for direct attack with hose lines, firing operations may be necessary.

Firing operations always present some risk, especially if done incorrectly. Strategic values must never take precedence over safety concerns. As a wildland fire fighter, it is essential that you understand the concepts presented in this chapter and wildland fire behavior before getting involved in any firing operation.

FIGURE 15-1 Fire fighter burning out with a drip torch. Fuel modification can assist in reducing hazards.
Courtesy of Joe Lowe.

Remember that once fire is put on the ground, it cannot be taken back. There is an incredible responsibility that comes with using this fire-suppression technique. It is paramount that all personnel involved in firing operations have a solid understanding of the hazards associated with burning, where other crews are working, the different types of fire spread, environmental components (fuel loading, topography, and weather effects), and how fire behaves and its associated effects.

Firing-Operation Permissions

Some states have provisions in their government codes that address firing operations and who should perform them. For example, California addresses backfire operations in its Public Resource Code (Section 4426) and identifies who should perform this operation. Check with your state to see whether there is a government code that addresses firing operations.

In many firefighting organizations in this country, it is necessary to acquire approval, through the chain of command, from the division supervisor before proceeding with a backfire or burnout operation. The only exception would be an emergency situation to save a life. Some agencies permit the crew boss to initiate burnout operations. Before going out on the fireline, determine the local agency policy regarding firing operations.

Every firing operation should be run by one person who focuses his or her total attention on supervising the firing team. That person should be experienced and trained in such operations. He or she should be constantly aware of weather changes, problem areas developing on the line, others working in and around the area, and fireline intensities being created by the firing operation. He or she must always be thinking about safety and fire behavior.

Another consideration is the sociopolitical aspect related to firing operations. The media and the public are encouraged to see all defensive actions used to suppress wildfire, including firing operations. If things go wrong during these firing operations, the public's perception of fire fighters can change and it can become difficult for the public to accept firing operations as an efficient fire suppression tactic. For this reason, every firing operation should be run by an experienced, trained individual.

In completing a successful firing operation, specific elements need to be in place. The success of the burn demands a thorough plan that is communicated and understood, including all the communication, safety, and coordination aspects for everyone involved. It is important to make sure that you receive the appropriate information. When receiving a briefing for general fire assignments, most fire fighters reference the inside back cover of the *Incident Response Pocket Guide* (IRPG; Briefing Checklist) to make sure that all the important assignment components are covered. The same is true when burning. Some of the elements of a good burning operation are discussed next.

Backfire

Backfire is defined as a fire intentionally set by fire fighters along the inner edge of a control line in an attempt to burn an area of vegetation. By removing fuels through the application of intentionally applied fire, a buffer area of already burned fuel is created. This buffer area is expected to slow or stop the fire or change the direction of the fire's convection column **FIGURE 15-2**. Backfiring is an indirect tactic usually used only when other fire-control methods have been judged impractical. It is usually a preplanned event with sufficient resources available to hold the fireline during the firing operation and sufficient resources held in reserve to rapidly attack any spot fires and **slop-over**, which is a fire edge that crosses a control line or natural barrier intended to confine the fire.

When we talk about the backfire being influenced by the main fire, we mean that an indraft effect takes place as the main fire approaches. **Indraft** is the process of air being drawn into the larger, main fire. The main fire has a strong need for oxygen because it generally has a strong convective column. Thus, the fire being lit will be drawn toward the base of the approaching larger main fire (see Figure 15-2). This is called the indraft effect. On some large fires, backfiring can be used to change the direction and spread of the main fire.

When you set a backfire, you must allow enough time (based on fuel factors, topography, weather, and fire behavior) for the backfire to develop and move away from the control line. If the distance between the backfire and main fire is too great, then consider the use of a counterfire (discussed later). Give yourself enough time; the worst thing that can happen is that both fires meet next to the control line, jeopardizing that which you are trying to protect. Always remember to proceed at a rate that can hold the fire you are lighting. Never create more fire than you can control.

There is no firm set of rules as to when to light a backfire because timing is based on the fuels present, the topography, existing fire conditions, and the

FIGURE 15-2 Backfire.

> **LISTEN UP!**
>
> Never create more fire than you can control.

weather. These factors must be evaluated and wildland fire-prediction skills used to determine when to light the backfire, so it is important to use the most experienced fire crew member for this operation.

A **counterfire**, or auxiliary fire, is a fire set between the main fire and the backfire to hasten spread of the backfire and redirect the main wildfire. A counterfire is used when the distance is too great between the main fire and the backfire **FIGURE 15-3**. Sometimes, a counterfire requires the aid of firing devices, such as a flare launcher.

Counterfires are one of the most technically challenging firing operations and should *only* be initiated by those experienced with both firing operations and the specific fuel type they will be burning in. If a counterfire isn't implemented perfectly, it can sometimes lure the main fire towards the control line and create an even larger fire front.

Burnout

Burnout is defined as setting fire inside a control line to consume the fuel between the edge of the fire and the control line **FIGURE 15-4**. The control line is not complete until all fuels have been removed between the fire's edge and the control line. Large islands of unburned fuel within the fire area that are close to the control line need to be removed before the control line can be considered secure. Burnout is considered a direct attack method.

FIGURE 15-3 Counterfire.

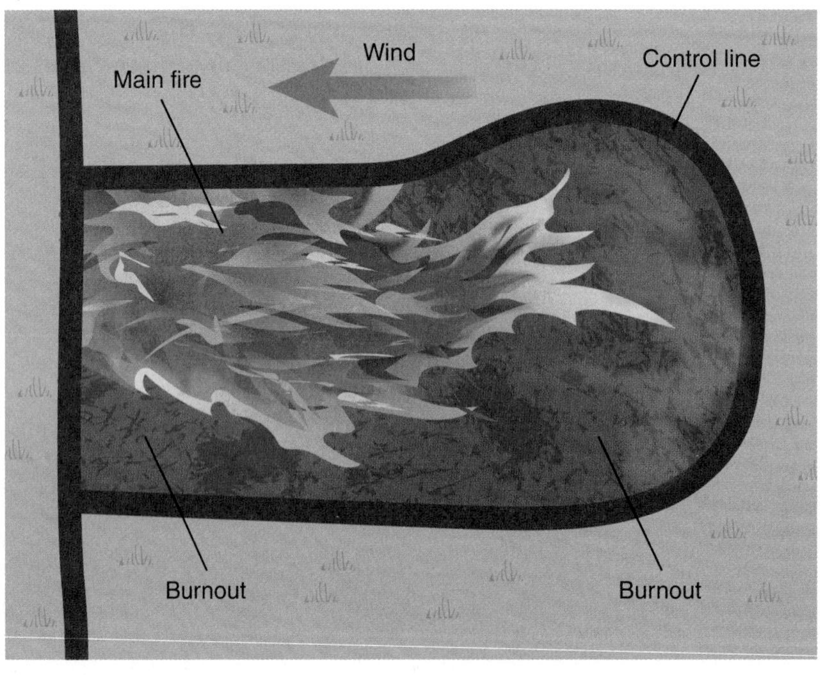

FIGURE 15-4 Burnout.

LISTEN UP!

Be aware of aircraft working in your area and make sure all are aware of your firing plan. This could prevent the aircraft from trying to put out the fire that you intentionally started.

Handcrews also burnout as they simultaneously cut line from fingertip to fingertip because this reduces the total amount of line that needs to be constructed. Fire is carried from finger to finger while the line is being constructed.

Burnout takes advantage of line-construction techniques by cutting across fingers and around concentrations of spot fires. It also takes advantage of natural barriers.

Firing-Operation Preplanning

Firing operations and prescribed burning should be organized and planned unless they are done as a result of an emergency. A complete briefing should be given to the crews involved in the firing operation and appropriate assignments made. Crews need to be assigned to work as firing forces, holding forces, mop-up forces, and reserve forces.

Before the operation is started, objectives need to be defined regarding what is to be accomplished. As a part of this process, time frames need to be determined for completion of the operation. Consideration needs to be given to the ignition sequence, when the firing operation is to commence, and what firing method will be used. A topographic map designating the location of the main fire should be shown to the crew that will be doing the firing operation.

You will also need to consider the amount and type of supplies needed to complete the operation. Planning also needs to be done to anticipate future needs of those doing the firing operation, and drop points for supplies and personnel need to be established.

Discussion regarding safety zones and escape routes needs to take place before any fire is put on the ground. All fire fighters must consider LCES/LACES at all times. Primary, alternate, and emergency plans should also be in place and included in briefings before firing operations begin.

The topographic map is used to show anticipated problem areas to the fire crews who will be doing the burning. A problem area can be defined as an area that could delay the operation, has the potential for slop-over, or could simply cause confusion in the operational area. Examples may be topographic features, such as saddles, sharp turns along a roadway, or narrow canyons. The area should be scouted and problem areas identified.

Before the firing operation begins, a decision should be made about the chain of command and what crews are in the division. Any fire-weather forecasts and fire-behavior predictions should be reviewed. This information can be found in an incident action plan. Communications on any wildland incident are a concern. The communications plan should be reviewed, and an alternate plan should be in place. A plan also needs to be in place for medivac procedures in case someone is hurt.

A small test fire should be set to evaluate fire behavior and controllability of your firing operation.

If these points are covered, the operation will go smoother, and the objectives will be met and accomplished safely.

Fuel, Topography, and Weather Considerations

Fuel, topography, and weather need to be evaluated before the firing operation begins. These factors are the key to the intensity levels and flame lengths that will be created. Never create more fire than can be controlled.

Light Fuels

Light fuels are easily ignited by the tools at hand and generally produce low-heat intensities on the fireline. They generally burn clean and spread rapidly. They can be held with a smaller control line than other fuel types. Of all the fuel types, light fuels (1-hour time-lag fuels) are the most susceptible to changes in relative humidity. Therefore, their moisture content can change rapidly. This fuel type has a greater tendency to spot than other fuel types.

Medium Fuels

Medium fuels produce greater heat intensity and flame lengths than light fuels. They can be ignited by the tools at hand; however, they normally require a drip torch. This type of fuel is generally less susceptible to spotting than light fuels. Medium fuels are more readily ignited at night than light fuels due to the slower effect of humidity on them. These fuels require increased control-line widths and an increased resource deployment to hold the line.

Heavy Fuels

Heavy fuel types produce great fireline intensities and are a problem for fire crews. They need large control lines and are very susceptible to embers blowing across the control line due to the fuel height. Heavy fuels often require ignition sources that are not readily available, such as a helitorch or a Terra Torch.

Topography

Topographic features present some of the greatest obstacles to successful firing. Saddles present special problems because they are areas of intense air movement. They have to be fired simultaneously downhill from the high points to the middle. Narrow canyons present special problems because the canyon walls are close together and the potential for spotting is great. Hooks and sharp bends are a cause for concern because they have the potential to expose the line to slop-over. Midslope roads call for extreme care because unburned fuel is either above or below the operation and the likelihood of spotting is higher. Some topographic features can work for you, such as natural barriers, whereas others are cause for concern. Evaluate these features before proceeding with the firing operation.

Weather

Winds need to be evaluated when planning the firing operation. Firing with the wind, meaning, the wind blows away from the control line, produces an increased fire intensity and a more rapid rate of spread. More heat builds, causing a strong convective column and perhaps more spotting. The flammable vegetation in the fire's path can be removed by the fire more quickly. It will take less time to complete the backfire/burnout operation. A smaller control line is needed to hold the firing operation.

When firing against the wind, it will take longer to complete the operation because the fire will spread at a slower rate. It will also be harder on crew members because the smoke and heat will be blown back toward them.

When the firing operation is done from the fire's flank, you should fire against the wind from a well-anchored point. The wind should be steady, but watch for wind shifts.

Other weather concerns to watch for include the following:

- Be aware of any thunderstorm activity developing in the area.
- Evaluate the smoke column. It will provide information on atmospheric stability and burning conditions.
- Monitor high-level winds because they may become surface winds later in the day.
- Monitor changes in humidity.
- Watch for swirling wind currents (eddies) created by topographic changes or features. These winds are also referred to as turbulence.

Firing Operations in Structural Areas

Special precautions must be taken with firing operations in or around structures. The firing operation may require additional resources assigned strictly to protect the structures. The citizens living there should have been evacuated before starting the operation because the firing operation may cause traffic congestion and reduced visibility.

Always consider the amount of defensible space around the structures, the construction types, where the houses are located in relation to the topography, the fuel loading, and proximity to the houses that you are trying to protect with the firing operation.

General Firing Operations Rules

Review the following general rules to ensure an effective and safe firing operation. Ultimately, LCES/LACES must be in place at all times.

- Every firing operation needs an anchor point, a safe place to start the operation from, and a predetermined termination point.
- All crew members need to be briefed on accessible escape routes and safety zones. Depending on the topography, time, and location, it may be deemed necessary to mark the escape route path beforehand with "escape route flagging" or other designated flagging. Ensure all crew members know to look for flagging.
- Generally, firing operations should start at higher elevations and work down so that the intensity level of the fire being lighted can be controlled. Firing operations that are started at the bottom of a hill can produce uncontrollable fire intensities because more fuel is available to burn and the slope affects the fire spread. There are two exceptions to this rule:
  - When the upper control lines are secure enough that there is no possibility of an escape or slop-over
  - When strong downslope winds are driving the fire and they are overcoming the effect of the slope

Other potential firing problems include the following:

- Areas with lots of dead fuel mixed with live fuels.
- Areas with heavy fuel buildup.
- Power lines.

- Aerial retardant drops. Coordinate these so they do not interfere with the firing operation.
- Any situation or environment that falls within the 10 standard fire orders and 18 watch outs.

LISTEN UP!

LCES/LACES must be in place at all times during firing operations.

LISTEN UP!

It is not uncommon to have an escape route flagged for 2 to 5 miles (3.2 to 8 km) on forest roads that have multiple junctions. If flagging is deemed necessary, make sure the junctions are well flagged so that crew members do not inadvertently take a wrong turn.

Firing Devices

Three commonly used firing devices are discussed in this section: the fusee, the drip torch, and the flare launcher. These devices may be used in a firing operation, so you should become familiar with the characteristics of each. Remember that you should always be wearing full personal protective equipment (PPE) and protective eyewear during firing operations.

Fusee

A **fusee** (sometimes called a backfire torch, a backfire fusee torch, or forest fire torch), is a colored flare designed to ignite backfires. Fusees are light and portable and produce a hot flame. The phosphorus material burns at 1400°F (760°C). Fusees can be broken and thrown and also be relit. Fusees are readily available on most fires, and most fire fighters carry several of them in their fire packs.

Note that fusees are very different from road flares. Road flares burn for 15 to 30 minutes and make a bright light. Fusees burn for less than 15 minutes and drip "slag." The slag is what starts the fire. The fumes from fusees can be very noxious and should not be inhaled. It is important to always try to orient yourself upwind of the smoke. To extinguish a fusee, place the tip into the dirt and snuff it out.

Parts of a Fusee

The burning/striking end of the fusee is stored with a protective paper and wax covering. There is a black cloth strip that is used to open the burning end, exposing the striker compound and uncovering the scratch strip **FIGURE 15-5**. Opposite the striking and burning end is a cardboard tube. The tube is connected to the fusee with wax **FIGURE 15-6**. The cardboard tube is used for connecting more than one fusee together and also extends the firing operation **FIGURE 15-7**. This allows the fire fighter not to have to bend over. When the first fusee burns to the end, it will ignite the next fusee so that a burning operation can continue without pauses.

FIGURE 15-5 Striking/burning end of a fusee.
Courtesy of Jeff Pricher.

FIGURE 15-6 The cardboard tube is used to connect two or more fusees together.
Courtesy of Jeff Pricher.

Fusee Storage

Fusees must be kept dry. Do not store them where they can rub together or be damaged. Whenever possible, transport them in their original containers. Note that the cardboard tube is easily damaged during storage, so it is important to protect the cardboard tube. To do

FIGURE 15-7 Connect several fusees together using their cardboard tubes.
Courtesy of Jeff Pricher.

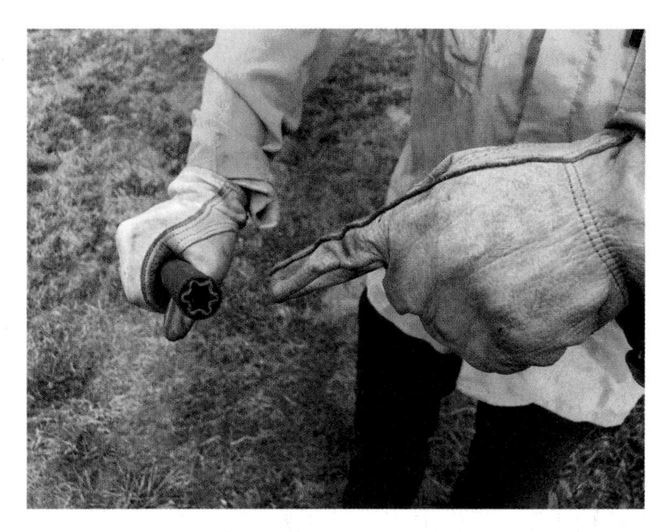

FIGURE 15-8 Store each fusee in a clean cardboard tube. This protects both the fusee and the tube.
Courtesy of Jeff Pricher.

this, remove the tube from the fusee. When the tube is free, scrape off any additional glue and wax that is attached to the fusee, or it will make it hard to push the tube back on the fusee. After everything has been removed, stow the cardboard tube back on the fusee so that none of the tube is hanging off the end of the fusee and is subject to crushing **FIGURE 15-8**. This will allow you to easily and efficiently assemble more than one fusee.

Lighting a Fusee

To light a fusee, put on full PPE and eye protection to protect yourself from fusee particles, fumes and slag, and high-intensity light. Then, follow the steps in **SKILL DRILL 15-1**.

Drip Torch

A **drip torch** is a canister that drips ignited fuel from a spout to a wick, which becomes saturated with fuel

SKILL DRILL 15-1
Lighting a Fusee

1 Pull on one side of the black cloth, which will expose the scratch strip and allow you to remove the protective cap.

2 To light the fusee, rub the scratch strip against the striker compound, away from your body.

SKILL DRILL 15-1 (Continued)
Lighting a Fusee

Courtesy of Jeff Pricher.

3. Once lit, hold the fusee away from your clothes and body. Stand on the upwind side of the flame. For large firing operations, connect additional fusees.

4. To extinguish, place the tip of the fusee into the dirt and snuff it out. Dispose in accordance with agency policy.

FIGURE 15-9 Drip torch.
© Brian Melley/AP/Shutterstock.

and burns continuously **FIGURE 15-9**. It is a very simple and effective tool that works on all fuel types. The contents of the canister are a mixture of one part gas to three parts of diesel fuel, and they last about 1 hour. This firing device is to be used with care because it can create a lot of fire very quickly.

LISTEN UP!

Do not exceed the gasoline mix ratio. Too much gasoline may cause the torch to burn too intensely or explode upon ignition, which can cause burn injuries.

Parts of a Drip Torch

The small brass knob on the top of the canister is the container's vent or breather valve. When the drip torch is being used, the valve has to be open, or the fuel will not flow out in an orderly fashion. The small brass nut with the chain attached is the discharge, or sealing, plug, which keeps the fuel in the container and prevents it from flowing out of the spout when the spout is stowed for transportation **FIGURE 15-10**. The loop in the spout is called a fuel trap, which stops the fuel from flashing back into the canister.

Drip-Torch Care and Storage

If you will be taking a break from firing or are preparing the drip torch for storage, make sure to put out the flame on the drip torch. Methods to extinguish the wick on the drip torch vary by agency. The two most common methods are to let the wick burn dry or to use a gloved hand to quickly grasp the wick to smother the fire (do not grasp the wick for more than 2 seconds). Whatever you do, do not attempt to blow out the flames. Doing so could cause you to inhale the flammable liquid or blow the flammable liquid onto someone else. Extinguish the wick by letting it burn dry. Once the flame is out, stand the drip torch upright, and let it cool completely. Next, remove the lock ring, reverse the spout and put it inside the tank, and

FIGURE 15-10 A. Discharge, or sealing, plug. **B.** Breather valve.
Courtesy of Jeff Pricher.

replace the lock ring. Then, replace the discharge plug, and close the breather valve. Clean and fill the drip torch before storing for a fire season.

Operating a Drip Torch

Hold the drip torch by the handle in an upright position until you are ready to use it. To operate a drip torch, put on full PPE and eye protection. Then, follow the steps in **SKILL DRILL 15-2**.

LISTEN UP!

When using the drip torch, make sure that the torch is kept away from your body, specifically your legs. Several burn injuries have occurred over the past few years as result of fire fighters accidentally pouring fuel on their legs.

SKILL DRILL 15-2
Operating a Drip Torch

1 Unscrew the discharge plug, and screw it onto the plug holder. Make sure the plug is tightly secured.

2 Remove the spout with the loop (fuel trap) from the canister. Make sure the loop and wick are pointing away from the drip-torch handle. When using this device, the fuel mix should saturate the wick. Screw down the brass lock ring to secure the spout in place.

SKILL DRILL 15-2 (Continued)
Operating a Drip Torch

3 Open the breather valve.

4 To light the drip torch, pour a small amount of the fuel on the ground, and light it with a lighter or fusee so that a flame is present. Then, bring the canister wick in contact with the small amount of fuel that is burning on the ground. The wick will ignite on the drip torch. The drip torch is now ready for use. Keep fuel flowing so the wick does not burnout. Burn from the top of a hill downward, and keep the torch away from you.

5 To extinguish the drip torch, place the canister in an upright position or close the breather valve and let the wick burn dry. Do not return the spout to inside the tank until you are sure that the wick and the spout are sufficiently cool.

Courtesy of Jeff Pricher.

Flare Launcher

A **flare launcher** (also called a flare gun or a Very pistol) is a handgun-shaped device that expels flares. Flare launchers are used in situations where a fire needs to be started at a distance from the control line (up to approximately 100 yards [548.6 m]). They are effective in areas with light to medium fuels.

Firequick is a brand of fire launcher commonly used for wildland firefighting **FIGURE 15-11**. It requires special field training for proper use.

The 1-in. (2.5-cm) flare system consists of a flare (also called a stubby), a launcher, and blank cartridges. The device comes with two flare size options that can be used with different cartridges. Make sure the cartridges are always matched to the respective flares.

The launching system was designed in conjunction with the U.S. Department of Agriculture (USDA) Forest Service. It is the only fire-starting flare system designed specifically for that agency. The system is certified by the Bureau of Alcohol, Tobacco, and Firearms (BATF) as an industrial tool and not as a firearm. It is often found in USDA Forest Service caches.

Preparing the Flare Launcher for Use

Wipe down the launcher with an absorbent cloth, removing any excess oil and dirt. Holding the launcher on its right side in the palm of your right hand, press on the cylinder release button, and remove the cylinder pin, allowing the cylinder to fall into your right hand. With the cylinder removed, load the cylinder with blank cartridges.

With the launcher in the palm of your right hand and the barrel pointing away from you, replace the cylinder in the launcher, and insert the cylinder pin while pressing the pin release button. Release the button, and ensure the pin is locked in place by attempting to remove the pin without pressing the release button.

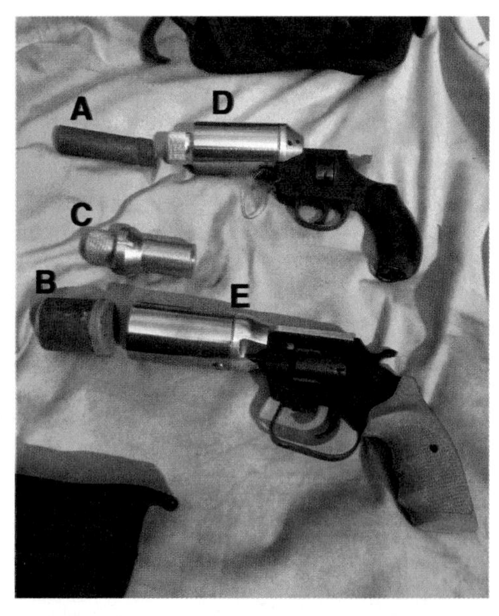

FIGURE 15-11 Firequick flare launchers and flares. **A.** Small flare. **B.** Large flare. **C.** Removable part used in both old and new Firequick devices that, when removed, allows for larger flares to be used. **D.** An older-style starter pistol called the Firequick II. This device has a removable cylinder (not shown). **E.** A current starter pistol called the Firequick III. This device does not have a removable cylinder.

Courtesy of Jeff Pricher.

Loading the Flare

Do not cock the hammer or place your finger inside the trigger guard while loading the flare. Using one hand to grasp the launcher by the grip, insert the capped end of the flare into the muzzle of the launcher. The flare will extend out of the muzzle about half an inch when fully inserted into the barrel. The flare will fit tightly and may need extra pressure to be fully inserted into the barrel.

Holding the launcher in one or both hands, point the launcher in the proper direction, and elevate the muzzle about 15 degrees or the angle for maximum range and ignition after the flare hits the ground. Cock the hammer, and pull the trigger to fire the launcher. The launcher can also be fired by just pulling the trigger in double-action mode; however, without the hammer being cocked, there may not be enough room inside the trigger guard for a gloved finger. Practice firing the flares to gain confidence in your ability to hit the target area.

Safety Precautions

Follow these safety precautions when using the flare launcher:

- Ensure you initiate the firing of the device from a position that allows for a clear shot at your target.
- Always keep the launcher pointed away from personnel and equipment.
- Always notify all personnel in the immediate area that you will be firing a flare launcher.
- Dispose of all spent cartridges.
- Only qualified personnel should use this device.
- Always wear ear and eye protection when using the launcher.
- The launcher is not classified as a firearm by the BATF; however, the launcher should be treated as a firearm, and general firearm safety rules should be followed.
- Always treat the launcher as if it were loaded.

- Always check the cylinder for cartridges when you first handle the launcher.
- Never transport the launcher while it is loaded with cartridges.
- Never point the launcher at anything you do not want to burn.
- If you are planning to travel by air with a launcher, unload it and place it in a hard case in your checked baggage. Make sure you notify the ticket agent.

LISTEN UP!

The flare launcher was designed for emergency use, not for extended use. After a number of shots, its aluminum firing pin can fail.

Unloading the Launcher

Place the right side of the launcher in the palm of your hand. While pressing on the cylinder release button, remove the cylinder pin, allowing the cylinder to fall into your right hand. Using the cylinder pin as a punch, remove the spent cartridges from the cylinder, and replace the cylinder in the launcher.

Maintenance

After each firing session, spray the launcher with gun oil or light machine oil, and return the launcher to its storage compartment. Inspect the launcher periodically for paint buildup in the barrel, and use a lacquer thinner to clean it.

Troubleshooting

The most common malfunction for the launcher is the sticking of the cylinder when attempting to fire a flare. This indicates that the cylinder pin is not locked in place and is working loose. Keeping the launcher pointed in a safe direction, center the cylinder, and replace the pin, ensuring it is locked in place by trying to remove it without pressing on the release button. Should the launcher become inoperable for any reason, contact the manufacturer for recommended actions. Replacement cylinders are available if required. If you end up in a situation where a cartridge does not fire, present the firing device to your supervisor, or refer to your agency and/or manufacturer's procedures for removal and disposal.

Other Firing Equipment

There are many other types of firing devices, such as hand-thrown flares, the Terra Torch flamethrower, helitorch aerial ignition device, Premo MK III aerial ignition system, and the aerial flare ignition system **FIGURE 15-12**. These devices require special field training beyond the scope of this text. Field-expedient methods include dragging already burning fuels with a tool or lighting an oil-soaked rag with a match.

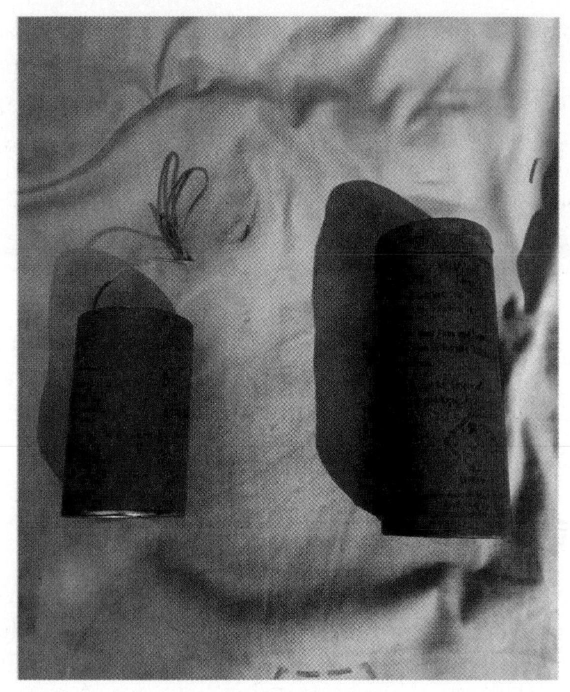

A

B

FIGURE 15-12 Other firing devices: **A.** Hand-thrown flares. **B.** Terra Torch flamethrower in use.

A: Courtesy of Jeff Pricher; **B:** Photo courtesy of John Pennell/Army photo.

Crew Assignments

Firing teams do the actual lighting and must focus their attention on the actual firing operation. They must also be aware of what is happening around them (LCES/LACES) and should be thinking about safety at all times.

Holding forces follow the firing personnel and protect them from injuries from the fire they are lighting.

They also prevent the fire of the firing operation from slopping over the control line, and they pick up spot fires. Holding forces also strengthen critical points on the control line. When a handcrew is used for holding, it is not uncommon for half of the crew to face the fire and the other half to face away from the fire to be on the lookout for potential spot fires.

Mop-up forces must remain behind to secure the line and do mop-up. These forces may start out as part of the holding forces. *Reserve forces* are held in reserve to be used when needed. They can act as a separate tactical unit to pick up slop-over, rapidly attack a spot fire, or provide structure protection when needed.

Firing Techniques

As a fire fighter who may be involved in firing operations, make sure that there is a firing plan and that the plan is known and understood by everyone. Ensure that all elements of the plan have been put in place before putting fire on the ground.

Patience is a large component of firing operations. All too often, when firing techniques are implemented, too much fire is used. Fire needs to have a bit of natural influence, meaning that it is OK if the fire that was just put down is not burning fast. Sometimes, it is difficult to wait for terrain and wind and fuel influences to take hold. Firing bosses and prescribed fire burn bosses have years of experience. This skill will take years to develop.

Remember, a good firing plan will have the following elements:

- Authorization
- A go/no-go checklist **FIGURE 15-13**
- Objectives
- Current and forecasted weather
- Reconnaissance of the area to be burned
- A briefing (that includes communications, equipment used, organization, contingencies, decision points, and areas of concern)
- Safety
- A test fire
- An ignition plan
- A holding plan

The last element that should occur is an after-action review of the operation.

There are several techniques used in firing operations. The three most commonly used techniques are strip firing, flank firing, and ring firing. More details on these techniques and additional techniques, including dot firing, chevron firing, and concentric firing, are discussed in more detail in an NWCG S-219, Firing Operations class.

Line Firing

The most common firing technique is called line firing or perimeter firing. **Line firing** involves one person burning along the inner edge of the control line to a termination point **FIGURE 15-14**. Line firing is extremely safe as you are standing on the inner edge of the control line or slightly in the unburned fuels and can easily step back on the control line.

If you are firing against the wind or effect of slope, a backing fire will result and it will burn slower with less intensity. Firing against wind and slope will take more time to complete the firing operation so plan accordingly. Also monitor closely for spot fires across your control line.

If you are firing with the wind or effect of slope you can expect a head fire to result. This will develop a more intense fire moving away from the control line. It will allow for shorter burnout times. It is extremely important to monitor fire conditions and changes in weather due to the intensity of the firing operation you just initiated.

Strip Firing

Strip firing is a firing technique that involves setting fire to one or more strips of fuel to create indrafts and reduce the intensity at the control line **FIGURE 15-15**.

Strip firing requires two or more fire fighters. The firing team members are located in the green at a distance that varies depending on several factors: fuels, topography, wind, and the need to regulate the fire intensity. More than one strip is often fired in a parallel alignment allowing strips to burn together. When using the strip firing technique, always evaluate what will be driving the fire you are lighting: wind or topography. Depending on the wind or slope, the number one lighter may not be in the lead position **FIGURE 15-16**. The safety zone is the control line or a planned area that has been identified.

Strip firing utilizes the 1-2-3/3-2-1 firing pattern. The 1-2-3/3-2-1 firing pattern is a technique that requires several igniters. Each igniter is assigned to a number (1, 2, or 3). The number 1 position is always located closest to the control line. The wind and terrain will dictate whether the number 1 or the number 3 position is leading. The igniters move parallel to the control line in a staggered fashion, igniting strips and letting them burn together to a termination point.

Project Name:______________________________

Unit Name:______________________________

PRESCRIBED FIRE GO/NO-GO CHECKLIST

(Prescribed Fire Plan, Element 2B)

Preliminary Questions	**Circle YES or NO**
A. Have conditions in or adjacent to the ignition unit changed, (for example: drought conditions or fuel loadings), which were not considered in the prescription development? If **NO** proceed with the Go/NO-GO Checklist below, if **YES** go to item B.	YES NO
B. Has the prescribed fire plan been reviewed and an amendment been approved; or has it been determined that no amendment is necessary? If **YES**, proceed with checklist below. If **NO**, **STOP: Implementation is not allowed. An amendment is needed.**	YES NO

GO/NO-GO Checklist	**Circle YES or NO**
Have ALL permits and clearances been obtained?	YES NO
Have ALL the required notifications been made?	YES NO
Have ALL the pre-burn considerations and preparation work identified in the prescribed fire plan been completed or addressed and checked?	YES NO
Have ALL required current and projected fire weather forecast been obtained and are they favorable?	YES NO
Are ALL prescription parameters met?	YES NO
Are ALL smoke management specifications met?	YES NO
Are ALL planned operations personnel and equipment on-site, available and operational?	YES NO
Has the availability of contingency resources applicable to today's implementation been checked and are they available?	YES NO
Have ALL personnel been briefed on the project objectives, their assignment, safety hazards, escape routes, and safety zones?	YES NO
If all the questions were answered "**YES**" proceed with a test fire. Document the current conditions, location and results. If any questions were answered "**NO**", DO NOT proceed with the test fire: Implementation is not allowed.	
After evaluating the test fire, in your judgment can the prescribed fire be carried out according to the prescribed fire plan and will it meet the planned objective? **Circle: YES or NO**	

Burn Boss Signature:______________________________ Date:______________

FIGURE 15-13 A go/no-go checklist aids in the decision-making process during firing operations.

Courtesy of National Wildfire Coordinating Group.

FIGURE 15-14 Line firing.
Courtesy of California Department of Forestry and Fire Protection.

FIGURE 15-15 Strip firing.
Courtesy of South Dakota Wildland Fire Academy.

FIGURE 15-16 During strip-firing operations, the escape route is a step onto the control line.

Each igniter should have an escape route that is not going to be affected by any of the other igniters. The igniters need to allow enough space so that their fires can't grow together and become too intense. All firing team members must stay in visual contact and maintain safe spacing. They must also move at the same speed. Study **FIGURE 15-17** through **FIGURE 15-20** for the correct placement of the firing team members in adverse winds, favorable winds, adverse slopes, and favorable slopes.

Flank Firing

Flank firing is a firing technique consisting of treating an area with lines of fire. Lighting is done into the wind or perpendicular to the slope, allowing the fire to spread at right angles into the influence of wind or slope **FIGURE 15-21.** Flank firing helps to modify wildfire intensity. It works best in light, continuous fuels. It is used to increase consumption of the fuels without creating a wall of fire. Flank firing also helps to modify

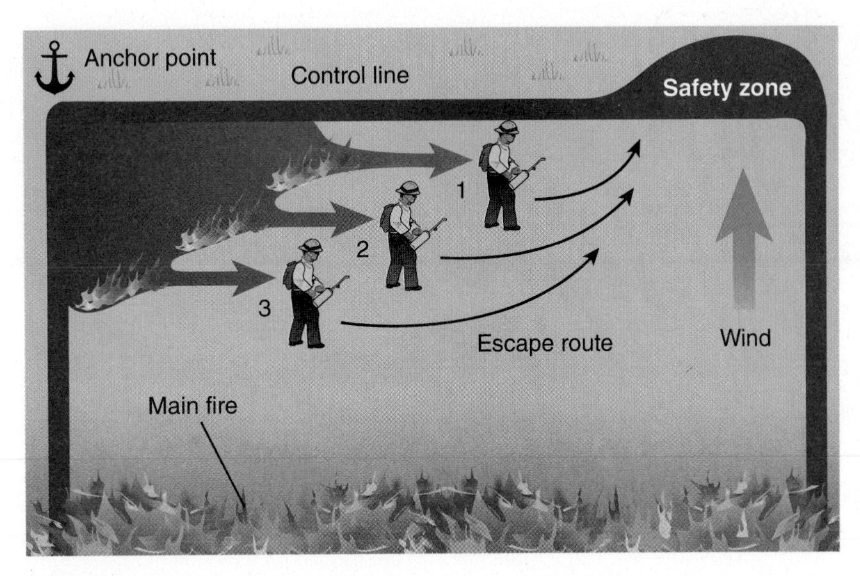

FIGURE 15-17 1-2-3 firing pattern in adverse winds.

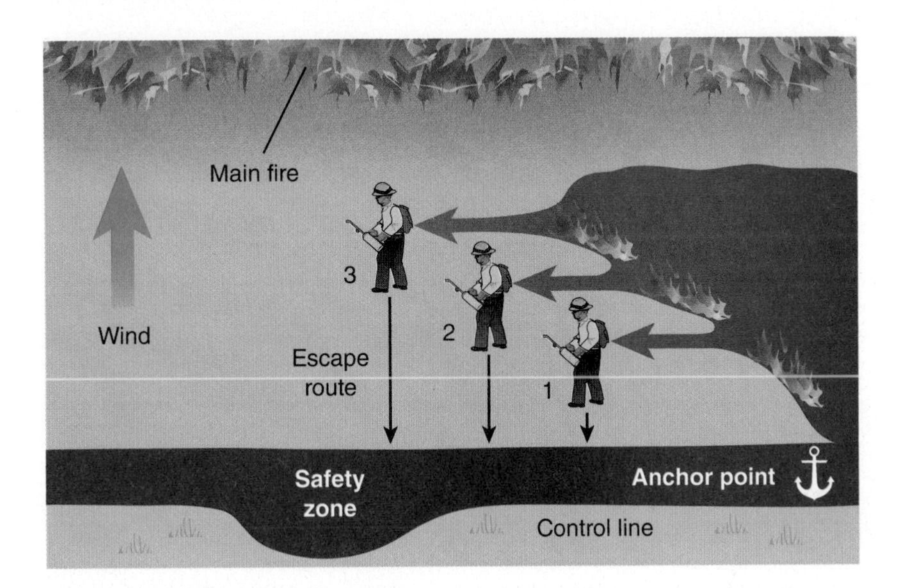

FIGURE 15-18 3-2-1 firing pattern in favorable winds.

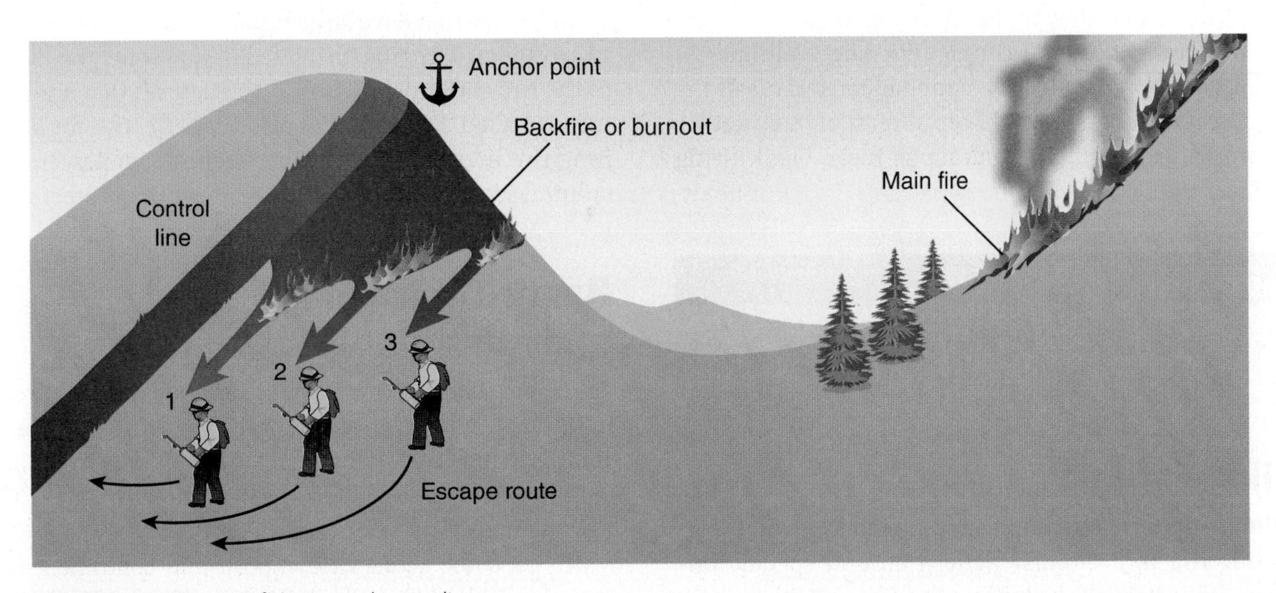

FIGURE 15-19 1-2-3 firing on adverse slopes.

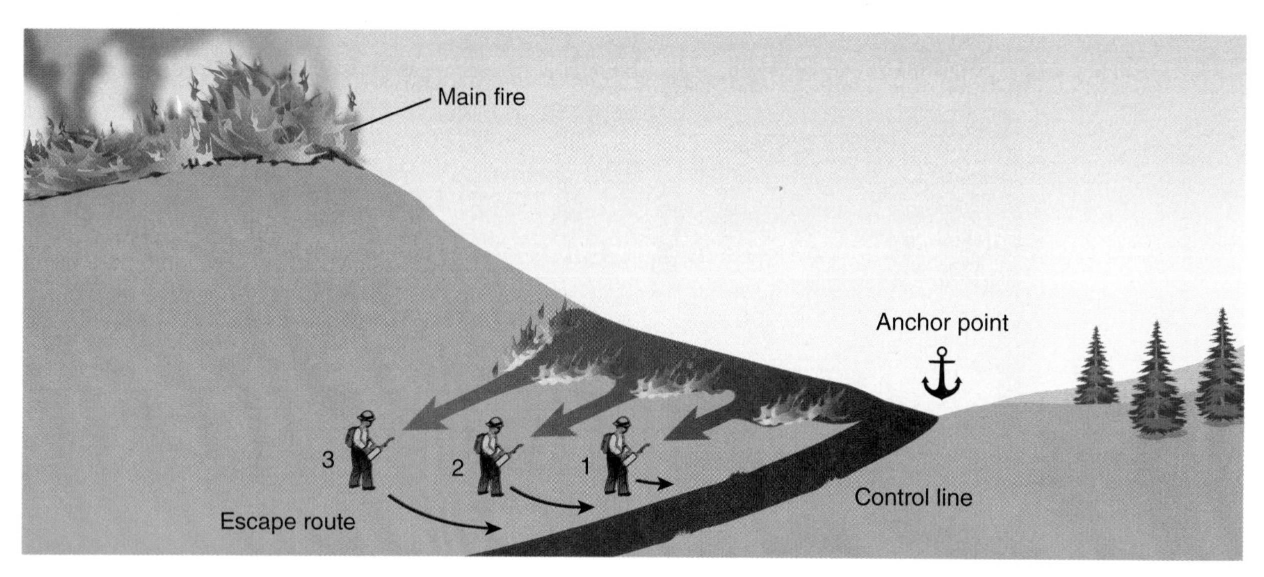

FIGURE 15-20 3-2-1 firing on favorable slopes.

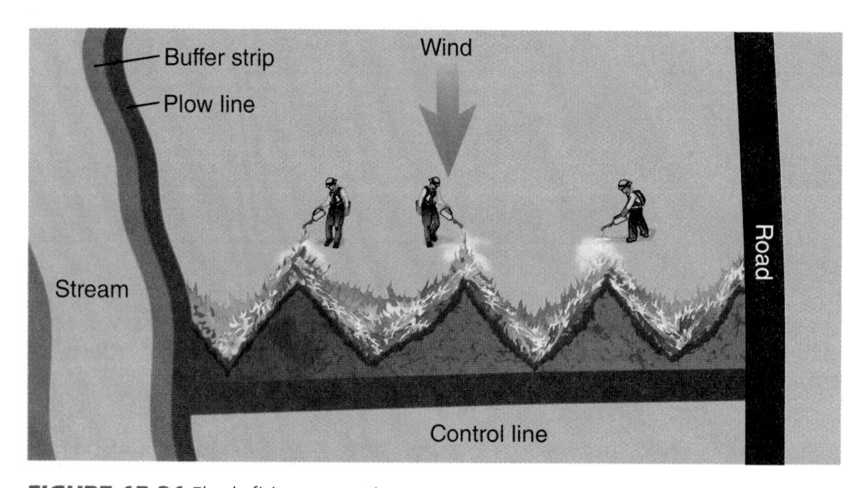

FIGURE 15-21 Flank-firing operations.

wildfire intensity. This method of firing can stand little variation in wind direction so monitor weather and wind direction closely. Flank firing requires expert crew coordination and timing. For safety, all lines of flank fire should be ignited simultaneously and all firing crew members should keep abreast of one another It works best in light, continuous fuels. Flank firing utilizes the 1-2-3/3-2-1 firing pattern.

LISTEN UP!

When flank firing, be aware of wind shifts. Inconsistent winds can turn the fire against you.

Ring Firing

Ring firing is a firing technique that involves igniting fire in a circular pattern around specific objects to establish a buffer line against the main wildfire **FIGURE 15-22**. This technique works well in structure-protection situations where there are pockets of houses that have large expanses of vegetation between them. The ring-firing technique splits the head of the fire and moves it around structures that are being protected. Ring firing is also very effective for limiting fire intensity and damage to power poles, trees, cultural sites, and other resources needing protection

Hands-On Training and Additional Skills

There are several other firing techniques not covered in this text, such as dot firing, chevron firing, and concentric firing. To further develop your skills in firing techniques, it is suggested that you take the NWCG S-219, Firing Operations class.

To get some actual field exposure in controlled situations, find a prescribed fire school or attend the

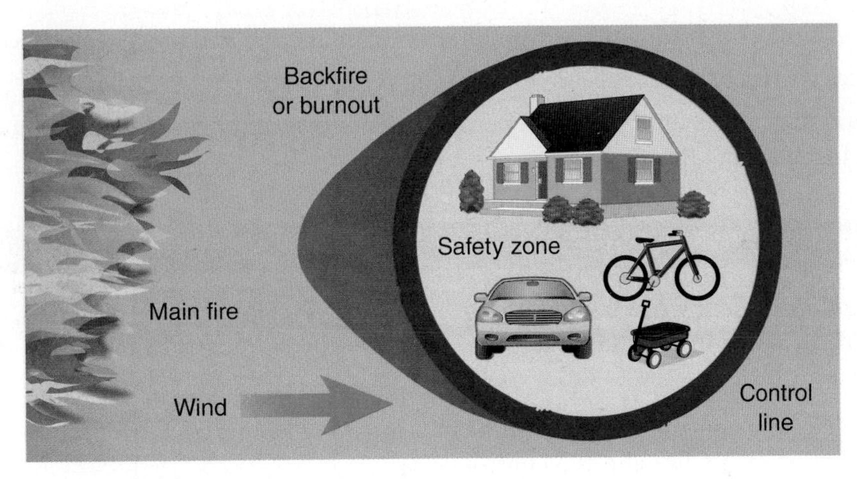

FIGURE 15-22 Ring firing.

hands-on training in the S-219 class. Another option would be to find out when a controlled burn is scheduled in your area. Contact the local agency representative who is in charge of the burn, and ask to observe the operation. Select a mentor at the burn site, and talk with him or her about the operation and techniques to be used during the burn.

Each year, the USDA Forest Service also conducts a fire school. This is an excellent way to learn firing operations in addition to other wildland skills. In the Southern California area, it is held annually at the Camp Pendleton Marine Base near Oceanside, California.

After-Action REVIEW

IN SUMMARY

- Oftentimes in wildland firefighting, fires are intentionally lit in order to burn the fuel that would be used by the main fire. By removing fuels through these intentional firing operations, a buffer area is created. This area is expected to slow or stop the fire and/or change the direction of the fire's convection column.
  - Some states have provisions that address firing operations and who should perform these operations.
  - Firing operations should be run by a single person whose entire focus should be on supervising the firing team.
- Backfiring is a tactic usually used only when other fire control methods are impractical. A backfire is a fire intentionally set along the inner edge of a control line in an attempt to burn a buffer area in the direction of the fire, stopping its progression.
  - Backfires must be given time to develop and move away from the control line.
- Control lines are not complete until all fuels have been removed between the fire's edge and the control line.
- Firing operations should be organized and planned, unless done in an emergency. Involved crews should be completely briefed and given appropriate assignments. All objectives need to be defined, as do time frames for completion of the operation.
  - Other concerns involve fuel, topography, weather forecasts, fire behavior predictions, and medivac procedures.
  - Firing operations require special precautions when they occur in or around structures. Additional resources may be necessary.
- Three commonly used firing devices are the fusee/backfire torch, the drip torch, and the flare launcher.
- There are different crew assignments responsible for different parts of firing operations.
  - Firing teams do the actual lighting of the fire and focus their attention on the actual firing operation.

- Holding forces protect the firing teams from injuries from the fire and prevent slop-over.
- Mop-up forces remain behind to secure the line and do mop-up.
- There are several techniques used in firing operations. Four commonly used techniques are line firing, strip firing, flank firing, and ring firing.

KEY TERMS

Backfire A fire intentionally set by fire fighters along the inner edge of a control line in an attempt to burn an area of vegetation.

Burnout Setting fire inside a control line to consume the fuel between the edge of the fire and the control line.

Counterfire A fire set between a main fire and a prescribed backfire to hasten spread of the backfire and redirect the main wildfire; also called auxiliary fire.

Drip torch A canister that drips ignited fuel from a spout to a wick, which becomes saturated with fuel and burns continuously. Also called an orchard torch.

Flank firing A firing technique that involves lighting a fire into the wind or perpendicular to the slope, allowing the fire to spread at right angles.

Flare launcher A handgun-shaped device that expulses flares.

Fusee A colored flare designed to ignite backfires; also called a backfire torch or forest fire torch.

Indraft The process of air being drawn into a larger, main fire from a backfire.

Line firing A firing technique that involves burning along the inner edge of the control line to a termination point.

Ring firing A firing technique that involves igniting fire in a circular pattern around specific objects to establish a buffer line against the main wildfire.

Slop-over A fire edge that crosses a control line or natural barrier.

Strip firing A firing technique that involves setting fire to one or more strips of fuel to create indrafts and reduce the intensity of the fire at the control line; also called line firing.

REFERENCES

California Department of Forestry and Fire Protection. "CDF Firing Operations: C-234, Intermediate Firing Methods." https://www.mendofb.org/wp-content/uploads/2014/06/CalFire-Firing-techniques.pdf.

International Association of Fire and Rescue Services. "Prevention & Control: Using Back Fire to Combat Wildfire." July 23, 2018. https://www.ctif.org/news/prevention-control-using-back-fire-combat-wildfire.

National Wildfire Coordinating Group (NWCG). *NWCG Standards for Ground Ignition Equipment.* PMS 443. February 2019. https://www.nwcg.gov/sites/default/files/publications/pms443.pdf.

National Wildfire Coordinating Group (NWCG). "S-130, Firefighter Training, 2008." https://www.nwcg.gov/publications/training-courses/s-130.

National Wildfire Coordinating Group (NWCG). "S-219, Firing Operations, 2014." https://www.nwcg.gov/publications/training-courses/s-219.

Wildland Fire Fighter in Action

You are assigned to Division X-ray on a fire that straddles a state border. The incident command post is over 2 hours away, and for the past 2 days you were told to spike out in a soccer field of a very small community. You are working for a division supervisor who is very new, and you and two other engine crews are ordered to start burning around a farm that contains three historical buildings. The main fire is 1 mile (1.6 km) away, and the fuels around the farm are 3-ft (0.9-m) tall grass with patchy sagebrush. One-quarter mile away is a ponderosa pine timber stand in moderately steep terrain. Communication is very inconsistent, and the plan seems as though it was put together very quickly. You ask for a spot weather forecast from the division supervisor and are told that it has been submitted to the National Weather Service. You advise the division supervisor that you are concerned that any fire could endanger the historical buildings.

1. What type of checklist would you think is most important at this point?
   - **A.** Go/no-go checklist
   - **B.** After-action review
   - **C.** Complexity analysis
   - **D.** Contingency
2. What is the essential component of starting a firing operation?
   - **A.** Weather briefing
   - **B.** Anchor point
   - **C.** Decision point
   - **D.** Working firing devices
3. In an effort to establish a controlled fire in light flash fuel, which of the following firing techniques would be the most dangerous to use?
   - **A.** Dot firing
   - **B.** Chevron firing
   - **C.** Strip firing
   - **D.** Ring firing
4. You are very concerned about the efficacy of this burn and the potential for the fire to get out of control. What element of the burn plan needs to be identified before the burn?
   - **A.** Test fire
   - **B.** Communications
   - **C.** Briefing
   - **D.** Contingency plan

Access Navigate for flashcards to test your key term knowledge.

Courtesy of Joe Lowe.

Appendix A

NFPA 1051 Correlation Guide

NFPA 1051, *Standard for Wildland Firefighting Personnel Professional Qualifications*, 2020 Edition (Chapters 4 and 5)

NFPA Job Performance Requirement	Corresponding Chapter(s)	Corresponding Page(s)
4.1 General. The Wildland Fire Fighter I shall meet the JPRs defined in Sections 4.1 through 4.5.	1, 2, 3, 4, 5	See below.
4.1.1 General Prerequisite Knowledge.		
Fireline safety, use, and limitations of personal protective equipment	3	N/A
AHJ policy on fire shelter use	3	N/A
Basic wildland fire behavior	4	N/A
Fire suppression techniques	5	N/A
Basic wildland fire tactics	5	N/A
The fire fighter's role within the AHJ incident management system	2	N/A
AHJ required first aid	1	N/A
NFES 1077, *Incident Response Pocket Guide (IRPG)*	1	N/A
4.1.2 General Prerequisite Skills.		
Basic verbal communications	3	N/A
Use of required personal protective equipment	3	N/A
4.2 Human Resource Management. No JPRs at this level.	N/A	N/A
4.3 Preparedness.	1, 2, 3, 4	See below.
4.3.1 Definition of Duty.		
Activities in advance of fire occurrence to ensure safe and effective suppression action	1, 2, 3, 4	N/A

(*continues*)

NFPA Job Performance Requirement	Corresponding Chapter(s)	Corresponding Page(s)
4.3.2 Maintain assigned personal protective equipment, given the standard equipment issue, so that the equipment is serviceable and available for use on the fireline and defects are recognized and reported to the supervisor.	3	See below.
4.3.2 (A) Requisite Knowledge.		
Maintenance of personal protective equipment, including inspection, the recognition of unserviceable items, and proper cleaning procedures.	3	56–66
4.3.2 (B) Requisite Skills.		
(No requisite skills required at this level.)	N/A	N/A
4.3.3 Maintain assigned suppression hand tools and equipment, given tools and equipment and AHJ maintenance specifications, so that assigned equipment is maintained and serviceable and defects are recognized and reported to the supervisor.	7	See below.
4.3.3 (A) Requisite Knowledge.		
Inspection of tools and assigned suppression equipment	7	165–166
The recognition of unserviceable items	7	162–163
Required maintenance techniques	7	161–167
4.3.3 (B) Requisite Skills.		
Sharpening and other maintenance techniques for assigned suppression equipment	7	161–167
Use of required maintenance equipment	7	161–167
4.3.4 Maintain personal gear kit, given a deployment and AHJ policies, so that mobilization response readiness meets AHJ requirements.	3	See below.
4.3.4 (A) Requisite Knowledge.		
Contents of a personal gear kit, given the type and duration of the incident, and AHJ policies	3	60
4.3.4 (A) Requisite Knowledge.		
(None specified.)	N/A	N/A
4.4 Mobilization. No JPRs at this level.	N/A	N/A
4.5 Suppression.	5, 6, 7, 8	See below.

NFPA Job Performance Requirement	Corresponding Chapter(s)	Corresponding Page(s)
4.5.1 Definition of Duty.		
All activities to confine and extinguish a wildland fire, beginning with dispatch.	5, 6, 7, 8	N/A
4.5.2 Assemble and prepare for response, given an assembly location, an assignment, incident location, mode of transportation, and the time requirements, so that arrival at the incident with the required personnel and equipment meets AHJ guidelines.	5, 8	See below.
4.5.2 (A) Requisite Knowledge.		
Equipment requirements	5	126–127
AHJ time standards	5	127–128
Special transportation considerations (weight limitations)	8	172
AHJ safety and operational procedures for various transportation modes	8	172
4.5.2 (B) Requisite Skills.		
(None specified.)	N/A	N/A
4.5.3 Recognize hazards and unsafe situations, given a wildland or wildland/urban interface fire and the standard safety policies and procedures of the AHJ, so that the hazard(s) and unsafe condition(s) are communicated to the supervisor and appropriate action is taken.	3, 4, 5, 6	See below.
4.5.3 (A) Requisite Knowledge.		
Basic wildland fire safety	3, 6	78, 144–148
Fire behavior	3, 4	48–49, 66–68, 115–116
Suppression methods	5	131–132
4.5.3 (B) Requisite Skills.		
(None specified.)	N/A	N/A
4.5.4 Construct a fireline, given a wildland fire, AHJ line construction standards, suppression tools, water or other suppression agents, and equipment, so that the fireline conforms to the construction standard.	5, 7	See below.
4.5.4 (A) Requisite Knowledge.		
Principles of fireline construction, techniques, and standards	5	133–135

(continues)

NFPA Job Performance Requirement	Corresponding Chapter(s)	Corresponding Page(s)
4.5.4 (B) Requisite Skills.		
Proper use of hand tools	7	160–161
Proper use of fire stream practices	5	136–137
Proper use of agent application	5	136–137
4.5.5 Secure the fireline, given a wildland fire and suppression tools, water or other suppression agents, and equipment, so that burning materials and unburned fuels that threaten the integrity of the fireline are located and abated.	5	See below.
4.5.5 (A) Requisite Knowledge.		
Line improvement techniques and safety considerations	5	133–138
4.5.5 (B) Requisite Skills.		
Use of basic ignition devices only under direct supervision	5	135
4.5.6 Describe the methods to reduce the threat of fire exposure to improved properties, given a wildland/urban interface fire, suppression tools, and equipment, so that improvements are protected.	4, 6	See below.
4.5.6 (A) Requisite Knowledge.		
Wildland fire behavior	4, 6	97–99, 148–150
Wildland fuel removal	4, 6	89, 146–150
Structure protection methods	6	148–150
Equipment and personnel capabilities	6	148–150
4.5.6 (B) Requisite Skills.		
The use of methods to protect improvements	6	148–150
4.5.7 Mop-up fire area, given a wildland fire, suppression tools, and water or other suppression agents and equipment, so that burning fuels that threaten escape are located and extinguished.	5	See below.
4.5.7 (A) Requisite Knowledge.		
Mop-up principles, techniques, and standards	5	136–137
4.5.7 (B) Requisite Skills.		
Use of basic tools and techniques to perform mop-up operations	5, 7	136–137, 156–159

NFPA Job Performance Requirement	Corresponding Chapter(s)	Corresponding Page(s)
4.5.8 Patrol the fire area, given a wildland fire, suppression tools, and equipment, so that a containment of the fire area is maintained.	5	See below.
4.5.8 (A) Requisite Knowledge.		
Patrol principles, techniques, and standards	5	137
4.5.8 (B) Requisite Skills.		
Observe, identify, and take action on potential threats	5	137
5.1 General. Prior to progressing to the Wildland Fire Fighter II level, the Wildland Fire Fighter I shall meet JPRs defined in Sections 5.1 through 5.5.	9, 10, 11, 12, 14	See below.
5.1.1 General Prerequisite Knowledge.		
The Wildland Fire Fighter II role within the incident management system	9	N/A
Basic map reading and compass use or other locating device	11	N/A
Radio procedures	10	N/A
Chainsaws	14	N/A
Pumps	12	N/A
Record keeping	9	N/A
5.1.2 General Prerequisite Skills.		
Orienteering	11	N/A
Operation of chainsaws	14	N/A
Operation of pumps	12	N/A
Record keeping	9	N/A
Radio use	10	N/A
5.2 Human Resource Management.		
5.2.1 Evaluate the readiness of assigned crew members, given a wildland fire, an assigned task, and AHJ equipment standards, so that crew members are equipped and supplied for suppression duties.	9	See below.
5.2.1 (A) Requisite Knowledge.		
AHJ standards and personnel inspection procedures	9	195–200

(*continues*)

NFPA Job Performance Requirement	Corresponding Chapter(s)	Corresponding Page(s)
5.2.1 (B) Requisite Skills.		
Inspect members' personal protective equipment, tools, and supplies	9	195–200
Inspect members' qualifications	9	200
Inspect members' physical fitness level	9	200–201
5.2.2 Brief assigned personnel, given an assignment, supporting information, and equipment requirements, so that the personnel are informed of specific tasks, standards, safety, operational, and special interest area considerations.	9	See below.
5.2.2 (A) Requisite Knowledge.		
Incident and task information necessary to carry out assignments	9	203–204
5.2.2 (B) Requisite Skills.		
Briefing skills	9	203–204
5.2.3 Lead wildland fire fighters in the performance of a task, given an assignment and performance standards, so that the task is completed within the standards in accordance with AHJ guidelines.	9	See below.
5.2.3 (A) Requisite Knowledge.		
Leadership techniques for small groups	9	202–203
5.2.3 (B) Requisite Skills.		
The application of requisite knowledge to lead wildland fire fighters to complete the task in a manner that is within the standard	9	203
5.3 Preparedness.	9	See below.
5.3.1 Definition of Duty.		
Responsibilities in advance of fire occurrence to ensure that tools, equipment, and supplies are fire ready	9	N/A
5.3.2 Maintain chainsaws and portable pumps as designated by the AHJ, given AHJ maintenance specifications, supplies, and tools, so that equipment is maintained and serviceable and defects are recognized and repaired.	12, 14	See below.
5.3.2 (A) Requisite Knowledge.		
Maintenance procedures for chainsaws as designated by the AHJ	14	319
Maintenance procedures for portable pumps as designated by the AHJ	12	274

NFPA Job Performance Requirement	Corresponding Chapter(s)	Corresponding Page(s)
5.3.2 (B) Requisite Skills.		
Preventive maintenance and repair of chainsaws as designated by the AHJ	14	319
Preventive maintenance and repair of portable pumps as designated by the AHJ	12	274
5.3.3 Inspect tools and equipment, given AHJ specifications, so that suitability of the tools and equipment for fire use is ensured.	9	See below.
5.3.3 (A) Requisite Knowledge.		
Tool and equipment inspection guidelines	9	195–200
5.3.3 (B) Requisite Skills.		
The ability to inspect tools and equipment according to guidelines	9	195–200
5.4 Mobilization. No JPRs at this level.	N/A	N/A
5.5 Suppression.	12, 13, 14, 15	See below.
5.5.1 Definition of Duty.		
All activities to contain and extinguish a wildland or wildland/urban interface fire, beginning with dispatch	9, 10, 11, 12, 13, 14, 15	N/A
5.5.2 Select fireline construction methods, given a wildland fire and line construction standards, so that the technique used is compatible with the conditions and meets the AHJ standards.	14	See below.
5.5.2 (A) Requisite Knowledge.		
Resource capabilities and limitations	14	323
Fireline construction methods	14	320–322
5.5.2 (B) Requisite Skills.		
(None specified.)	N/A	N/A
5.5.3 Reduce the risk of fire exposure to improved properties, given a wildland or wildland/urban interface fire and available tools and equipment, so that improvements are protected from fire.	9, 14, 15	See below.
5.5.3 (A) Requisite Knowledge.		
Knowledge of fire behavior in both wildland and improved properties	9	204–206
The effects of fuel modification to reduce the hazard	15	328–331
NWCG S-215, *Fire Operations in the Wildland/Urban Interface*	9	203

(*continues*)

NFPA Job Performance Requirement	Corresponding Chapter(s)	Corresponding Page(s)
5.5.3 (B) Requisite Skills.		
The use of tools and equipment to protect the improved property	14	321–322
5.5.4 Operate a chainsaw, given an assignment at a wildland fire and operational standards, so that the assignment is completed in a safe manner.	14	See below.
5.5.4 (A) Requisite Knowledge.		
AHJ standards for chainsaw operation and safety	14	311–314
Tool selection	14	319
Personal protective equipment used during saw use	14	312–313
5.5.4 (B) Requisite Skills.		
Site preparation	14	313–314
Handling and cutting techniques	14	317–318
Use of wedges	14	319
Use of saws	14	317–319
Equipment storage	14	319
Equipment transportation	14	319
5.5.5 Operate water delivery equipment, given an assignment at a wildland fire and operational standards, so that the proper equipment is selected, desired nozzle pressure is attained, and flow is maintained.	12	See below.
5.5.5 (A) Requisite Knowledge.		
Basic hydraulics	12	285–286
Pump and water delivery system capabilities	12	269
Operation of pumps	12	276–277
Basic drafting and associated equipment	12	274–277
5.5.5 (B) Requisite Skills.		
Placement, operation, and system setup	12	274–277
5.5.6 Operate a portable radio, given AHJ policies, so that communication is clear, concise, and accurate.	10	See below.

NFPA Job Performance Requirement	Corresponding Chapter(s)	Corresponding Page(s)
5.5.6 (A) Requisite Knowledge.		
AHJ operational standards for portable radios	10	225–228
5.5.6 (B) Requisite Skills.		
Operation of portable radios	10	228
5.5.7 Secure the area of suspected fire origin and associated evidence, given a wildland fire and AHJ procedures, so that all evidence or potential evidence is protected from damage or destruction and reported to a supervisor.	9	See below.
5.5.7 (A) Requisite Knowledge.		
Knowledge of types of evidence	9	208
Knowledge of the importance of site security and evidence preservation	9	208–209
5.5.7 (B) Requisite Skills.		
Evidence preservation techniques	9	208–209
Use of marking devices for site security	9	209
5.5.8 Serve as a lookout, given an assignment at a wildland fire as per AHJ procedures, so that fire fighters are updated or warned when conditions change.	9, 10	See below.
5.5.8 (A) Requisite Knowledge.		
Basic fire behavior and how to recognize hazardous situations	9	206
Communications methods	9, 10	206–207, 225–227
Communications equipment	9, 10	206–207, 220–221
Communications procedures	9, 10	206–207, 225–227
5.5.8 (B) Requisite Skills.		
The ability to accurately describe fire behavior and changes in fire behavior through verbal communication	9	206–207
The ability to accurately describe fire behavior and changes in fire behavior through hand signals	9	207
The ability to accurately describe fire behavior and changes in fire behavior through communication equipment	9, 10	206–207, 225–227

Appendix B

NWCG S-190 Correlation Guide

Courtesy of Joe Lowe.

NWCG S-190, *Introduction to Wildland Fire Behavior*, 2020 Edition

NWCG Job Performance Requirement	Corresponding Chapter(s)	Corresponding Page Number(s)
Unit 0: Course Introduction to Wildland Fire Behavior. (Course Objectives).	4	See below.
■ Describe the basic terminology used in wildland fire.	4	N/A
■ Identify and discuss the fire triangle.	4	N/A
■ Identify and discuss key characteristics of the primary wildland fire environment components - fuels, weather, and topography.	4	N/A
■ Identify critical fire weather factors that, combined with receptive fuels, may result in extreme fire behavior.	4	N/A
■ Recognize how alignment of fuels, weather, and topography can increase the potential for extreme fire behavior.	4	N/A
Unit 1: Basic Concepts of Wildland Fire.	4, 5	See below.
Describe the basic terminology used in wildland fire.	4, 5	N/A, 129
Describe the elements of the fire triangle.	4	88–89
Describe the methods of heat transfer.	4	90
Unit 2: Fuels.	4	See below.
Describe the term *fuels*.	4	88
Describe how fuel type and fuel characteristics affect fire behavior.	4	90–98

NWCG Job Performance Requirement	Corresponding Chapter(s)	Corresponding Page Number(s)
Unit 3: Temperature and Moisture Relationships.	4	See below.
Describe dry bulb temperature, wet bulb temperature, dew point, and relative humidity.	4	114
Describe how temperature and relative humidity can influence wildland fire behavior.	4	113–114
Determine relative humidity (RH) and dew point by using a psychrometric table and given inputs.	4	114
Unit 4: Topography.	4	See below.
Identify topographic features found in the wildland fire environment.	4	99–102
Describe the basic characteristics of topography and how they can affect wildland fire behavior.	4	99–102
Unit 5: Atmospheric Stability, Winds, and Clouds.	4	See below.
Describe atmospheric stability and discuss the effects on fire behavior.	4	111–112
Describe wind and its effects on fire behavior.	4	102–104
Explain cloud classifications and their impact on fire behavior.	4	106
Explain the similarities between smoke layers and clouds in relation to impact on fire behavior.	4	112
Unit 6: Critical Fire Weather.	4	See below.
Describe critical fire weather conditions.	4	104–112
Describe critical fire weather events that can impact fire behavior:	4	104, 106–111
(a) Cold fronts	4	104
(b) Thunderstorms	4	106–109
(c) Foehn winds	4	109–111
(d) Other local fire weather phenomena	4	N/A

(*continues*)

NWCG Job Performance Requirement	Corresponding Chapter(s)	Corresponding Page Number(s)
Unit 7: Alignment.	4	See below.
Describe how the primary wildland fire environment components—fuels, weather, and topography—are made more complex by interaction with each other.	4	115–116
Describe how alignment of these components greatly increases the potential for extreme fire behavior.	4	115–116

Courtesy of Joe Lowe.

Appendix C

NWCG S-130 Correlation Guide

NWCG S-130, *Firefighter Training*, 2008 Edition

NWCG Job Performance Requirement	Corresponding Chapter(s)	Corresponding Page Number(s)
Unit 0: Introduction.		
Unit 0A: Logistics and Overview (Course Objectives).	3, 5, 12, 14	See below.
Explain what the LCES (Lookouts, Communications, Escape Routes, and Safety Zones) system is and how it relates to the Standard Firefighting Orders.	3	N/A
Construct fireline to required standards using various methods.	5, 14	N/A
Strengthen, reinforce, and use holding actions on a fireline.	14	N/A
Extinguish the fire with or without the use of water.	5, 12	N/A
Complete assigned tasks in a safe and efficient manner.	5	N/A
Given an assignment in a wildfire environment, describe factors in that environment that could impact safety.	5	N/A
Unit 0B: Basic Terminology.	4, 5	See below.
Identify nine parts of a fire.	5	N/A
Define nine fire behavior terms.	5	N/A
Discuss five other useful firefighting terms.	4, 5	N/A
Unit 1: Firefighter Preparedness.	3, 9	See below.
Explain the importance of the proper use and maintenance of their assigned Personal Protective Equipment (PPE).	3	56
Develop a list of personal gear needed for an extended period away from their home station.	3	59

(*continues*)

NWCG Job Performance Requirement	Corresponding Chapter(s)	Corresponding Page Number(s)
Explain the firefighter's accountability for personal and agency property.	3	56
List the benefits of maintaining a high level of physical fitness and health.	3	79
Explain how eating well and staying hydrated can reduce firefighter fatigue.	3	79–80
Explain the importance of keeping personal gear and the assigned area in fire camp clean and organized.	3, 9	56, 195
Unit 2: Introduction to ICS.	2	See below.
Describe the firefighter's chain of command.	2	23
Define an incident and describe how the incident management structure is organized.	2	22–23
Describe the general responsibilities of each section in the Incident Command System (ICS):	2	
a) Command		24
b) Operations		27
c) Planning		30
d) Logistics		31–32
e) Finance and Administration		33
Name the two positions above the firefighter in the chain of command.	2	N/A
Unit 3: Resource Types.	14	See below.
Explain different types of crew organizations commonly used in initial attack and extended attack.	14	308–311
Explain the importance of respecting cultural differences in terms of food, standards of behavior, dress, and customs.	14	310
Unit 4: Risk Management.		
Unit 4A: Watch Out Situations and Standard Firefighting Orders.	3	See below.
Identify the common denominators on tragedy fires.	3	48–49
Given a scenario, identify the appropriate Watch Out Situations.	3	48–49

NWCG Job Performance Requirement	Corresponding Chapter(s)	Corresponding Page Number(s)
Apply the appropriate Standard Firefighting Orders to minimize the potential for serious injury or death.	3	46–48
Unit 4B: LCES.	3	See below.
Describe how LCES is related to the 10 Standard Firefighting Orders.	3	50
Define escape route, escape time, and safety zone.	3	49–50
Identify travel barriers that will affect escape time.	3	50
List the three types of safety zone categories and describe one example of each.	3	49
Describe a general guideline for determining safety zone size to avoid radiant heat injury.	3	49
Identify the limitations of utilizing the Incident Response Pocket Guide safety zone guidelines.	3	49
Describe the difference between deployment sites and safety zones.	3	49–50
Unit 4C: Fire Shelter.	3, 9	See below.
Explain the two most important functions of the fire shelter.	3	59
Discuss the inspection and care of the fire shelter.	3, 9	66, 200
Discuss the last resort survival options.	3	50, 66–69
Discuss entrapment and deployment site.	3	50, 62
Demonstrate the correct deployment procedures for the fire shelter in 25 seconds or less.	3	62–65
Unit 4D: Potential Hazards and Human Factors on the Fireline.	3, 9	See below.
Define safety.	3	46
Define Situation Awareness and describe why it is important.	3	51
State the five communication responsibilities.	3, 9	51, 201
Identify potential hazards in the fire environment.	3	52, 68–78
Define the Risk Management Process and describe why it is important.	9	203
Describe actions that foster teamwork.	9	203

(continues)

NWCG Job Performance Requirement	Corresponding Chapter(s)	Corresponding Page Number(s)
Unit 5: Transportation Safety.	8	See below.
Develop a list of three safety procedures to follow when traveling by each of the following:	8	172
(a) Vehicle	8	172
(b) Boat	8	172
(c) Helicopter	8	172
(d) Fixed-wing aircraft	8	172
(e) On foot	8	172
Unit 6: Hand Tools.	7	See below.
Given hand tools, personal protective equipment, and proper maintenance tools, check the condition of each item, perform field maintenance, and identify those needing replacement.	7	165–166
Demonstrate the proper sharpening techniques for commonly used tools.	7	161–164
Given a description of three fireline jobs and a choice of tools, state the tool that would be used for each job.	7	N/A
Demonstrate the proper methods of carrying and passing tools.	7	161
Demonstrate the proper spacing when using hand tools.	7	160–161
Demonstrate the proper placement, near a fireline, of one or more tools when not in use.	7	167
Given a swatter or gunnysack and personal protective equipment, check the condition of the fire swatter and perform field maintenance.	7	166
Unit 7: Firing Devices.	15	See below.
Describe two hazards to operators when using a fusee.	15	333–334
Demonstrate or simulate how to ignite, use, and extinguish a fusee.	15	334–335
Describe hazards to operators when using a drip torch.	15	335–336
State the proper fuel mixture for a drip torch.	15	335
Prepare a drip torch for use and ignite.	15	336–337
Demonstrate the safe use of the drip torch.	15	336–337
Extinguish a drip torch and prepare it for storage.	15	335–337

NWCG Job Performance Requirement	Corresponding Chapter(s)	Corresponding Page Number(s)
Describe two field-expedient methods for igniting wildland fuels.	15	340
Unit 8: Use of Water.	12	See below.
Given a backpack pump and a source of water, demonstrate how to properly operate and maintain the pump.	12	275
Correctly identify common hose components and accessories.	12	271–272
Describe the process of correctly unrolling a hose.	12	278
Use a hose clamp and/or field-expedient method to restrict water flow in a charged line.	12	269
Describe and demonstrate the two hose lay methods.	12	278–280
Correctly identify the water use hand signals.	12	277
Demonstrate the nozzle settings for straight stream and fog spray.	12	269
Describe a fire situation when the straight and fog spray nozzle water streams would be used.	12	269
Describe protective measures for hose and fittings when in use or being transported.	12	271, 283–284
Describe hazards to hose lays.	12	279–280
Retrieve deployed hose using two methods.	12	281–282
Identify and mark non-serviceable sections of hose and couplings.	12	281–282
Unit 9: Suppression.	3, 4, 5, 6, 8, 14	See below.
Describe three methods for breaking the fire triangle.	4, 5	89, 129
Describe three methods of attack on a fire.	5	131–133
List three suppression techniques and describe their uses.	5	133–134
Describe the blackline concept.	5	135
Describe four kinds of fire control line.	5	133–134
Name four threats/hazards to an existing control line when fire is burning inside the line.	5	135

(continues)

NWCG Job Performance Requirement	Corresponding Chapter(s)	Corresponding Page Number(s)
Describe the proper follow-up procedures for a dozer or tractor plow fireline.	8	177–179
Describe two kinds of coordinated crew techniques used for fireline construction, and, with at least four additional personnel, construct a fireline utilizing these techniques.	5	134–135
Describe safety procedures to follow when in an area where retardant/water drops are being made.	3	78
Describe five safety procedures to follow when working around engines, tractor plows, and dozers.	3, 8	76–78
Demonstrate the proper use of appropriate hand tools during fire suppression activities.	7, 14	160–161, 321–322
Demonstrate the construction of a cup trench on a steep slope.	14	321
Unit 10: Patrolling and Communication.		
Unit 10A: Patrolling.	5	See below.
Describe ways to communicate with designated personnel.	5	137
Describe a systematic method of locating spot fires.	5	137
Describe considerations when patrolling a fire.	5	137
Unit 10B: Radio Communication.	10	See below.
Describe frequencies and how they affect radio communications.	10	221–222
List four elements of proper radio use procedures.	10	225–228
Transmit a message clearly using proper procedure and language.	10	228
Describe three radio troubleshooting practices used to improve radio reception or transmission.	10	228
Describe precautions and care to protect the radio from damage.	10	228–229
Unit 11: Mop-Up and Securing the Fireline.	5, 12	See below.
Describe and demonstrate how to extinguish burning materials by chopping, scraping, and mixing them with soil and water.	5, 12	136–137, 285

NWCG Job Performance Requirement	Corresponding Chapter(s)	Corresponding Page Number(s)
Describe precautions to take when applying water to hot materials and demonstrate proper techniques for doing so.	12	285
Describe a systematic method of mop-up and give two reasons for using this method.	5	136
Describe how each of the four senses aid in detecting burning materials.	5	136
Discuss the importance of breaking up and dispersing machine piles and berms adjacent to the control line.	5	136
Demonstrate the technique of cold trailing on a simulated fire perimeter.	5	133
Discuss factors that determine the amount of additional work required for a water or retardant line.	12	N/A
Given a constructed control line, strengthen the line to facilitate holding by rearranging and fireproofing fuels adjacent to the line.	5	137
Unit 12: Fire Exercise.	3, 5, 7, 8, 9, 12	See below.
Demonstrate proper travel procedures en route to and from a fire.	8	172
Demonstrate proper use, handling, and maintenance of hand tools.	7	161–167
Construct progressive handline.	5	134–135
Construct a leap frog handline.	5	135
Construct simple and progressive hoselays.	12	278–280
Use escape routes to promptly retreat to a safety zone.	3, 5	49–50
Participate in an "after action review."	9	211–212
Unit 13: Hazardous Materials.	6	See below.
Develop a working definition of hazardous materials.	6	147
Explain the general guidelines when reacting to a possible hazardous materials emergency.	6	147
List and explain the six steps in the D.E.C.I.D.E. process.	6	147

(*continues*)

NWCG Job Performance Requirement	Corresponding Chapter(s)	Corresponding Page Number(s)
List and explain the clues for detecting the presence of hazardous materials.	6	147–148
Unit 14: Wildland/Urban Interface Safety.	6	See below.
Identify the wildland/urban interface watch out situations.	6	144–148
Identify personnel safety concerns in wildland/urban interface fires.	6	144–148
Unit 15: Pump Operations. (Optional).	12	See below.
Identify common types of pumps.	12	274–276
Identify the responsibilities and personal protective equipment of the portable pump operator.	12	276
Demonstrate the proper setup and operation of a portable pump.	12	277–278
Unit 16: Map Reading and Use of the Compass. (Optional).	11	See below.
Identify the symbols depicted on a locally used map with legend.	11	248
Name and describe the land survey system used locally.	11	248
Locate a legal description of a point on a map.	11	244–248
Identify major topographic features, both in the field and on a map.	11	240–248
Demonstrate the ability to read and shoot an azimuth (bearing) off a compass.	11	235–237
Explain the importance of knowing the proper declination of an area.	11	235
Determine one's walking pace on flat and sloped ground in one chain and 100 foot intervals.	11	239–240
Given a compass with proper declination and a local map, determine an azimuth and a distance and be able to navigate through wildland terrain between two points.	11	234–239
Unit 17: Wildland Fire Investigation. (Optional).	9	See below.
Describe items to watch for when traveling to, arriving at, and during initial attack that might show the origin and/or cause of the fire.	9	208–209
Given a simulated situation, record and report all information that will help in determining fire cause and origin.	9	209–210

NWCG Job Performance Requirement	Corresponding Chapter(s)	Corresponding Page Number(s)
Given a simulated fire situation, designate and protect the area of fire origin.	9	208–209
Unit 18: Cultural Resources. (Optional).	6	See below.
Define the phrase "Cultural Resources."	6	150
Describe the effects of fire and fire management activities on cultural resources.	6	151
Describe the steps to protect cultural resources during fire management activities.	6	151

Appendix D

NWCG L-180 Correlation Guide

Courtesy of Joe Lowe.

NWCG L-180, *Human Factors in the Wildland Fire Service*, 2014 Edition

NWCG Job Performance Requirement	Corresponding Chapter(s)	Corresponding Page Number(s)
Module 1: Working in the Wildland Fire Service.		
Introduction. This module introduces the learner to the concepts of human factors on the fireline and situation awareness. After completing this module, the learner will be able to identify the relationship between situation awareness, observation, communication, perception, and reality.	1, 3	See below.
Agree that firefighters have a responsibility to reduce errors.	1	9
Agree that firefighters have a responsibility to learn and improve their performance.	1	9
Agree that firefighters have a responsibility to be aware of their situation.	3	51
Describe the relationship between situation awareness, reality, and perception.	3	51
Describe the relationship between situation awareness, information, observations, and communication.	3	51–52
Module 2: Communication.		
Introduction. This module introduces one of the most important concepts in this course: communication. Effective communication is essential for good situation awareness, and it plays a part in all other human factors in the Wildland Fire Service.	9	See below.
Recall common sender-receiver communication errors.	9	201
Construct a direct statement from an indirect statement.	9	201–202
Identify existing standard communication procedures and opportunities to initiate standard communication procedures in the work environment.	9	201–202
Describe the Five Communications Responsibilities.	9	201
Agree that firefighters have a responsibility to communicate.	9	201

NWCG Job Performance Requirement	Corresponding Chapter(s)	Corresponding Page Number(s)
Module 3: Barriers to Situation Awareness.		
Introduction. After completing this module, learners will understand what physical and internal barriers impact situation awareness and how these barriers can be overcome.	3	See below.
Identify hazardous attitude barriers and their impacts on situation awareness.	3	52
Identify stress reaction barriers and their impacts on situation awareness.	3	81
Agree that firefighters have a responsibility to minimize barriers.	3	52
Module 4: Decision Making.		
Introduction. After completing this module, the learner will be able to describe the decision cycle, understand the importance of preplanning, and use the Risk Management Process as a decision tool to assist in preplanning.	9	See below.
Describe the process of preplanning and its role in decision making.	9	203
Describe the situation awareness self-check tool and its role in decision making.	9	203
Agree that firefighters have a responsibility to minimize risk.	9	203
Module 5: Team Cohesion.		
Introduction. After completing this module, the learner will be able to understand the relationship between teamwork and the human performance concepts presented in this course.	9	See below.
Describe the relationship between teamwork and the human performance concepts discussed in the previous lessons.	9	203
Agree that firefighters have a responsibility to work as a member of a team.	9	203
Module 6: End of Course Simulation.		
Introduction. At the completion of this module, the student will have practiced decision-making skills related to one or two core learning objectives by responding to variable conditions presented in a simulation.	N/A	N/A

Glossary

Courtesy of Joe Lowe.

10 standard fire orders A set of rules designed to reduce firefighter deaths and injuries, and increase firefighting efficiency.

18 watch outs An expansion of the 10 standard fire orders that identifies specific areas of concern in the wildland environment.

A

Aerial fuels Fuels that are greater than 6 ft (1.8 m) in height. Include trees, branches, foliage, and tall brush.

After-action review (AAR) A post-incident debrief that encourages self-evaluation.

Air crew A crew that works with aircraft, such as helicopters and fixed-wing aircraft.

Air operations branch Ground-based branch assignment for management of air assets used for the incident.

Alaska fire swatter A swatter with several strips of rubber on the end, rather than one thick, flat piece.

Anchor point An area close to where the fire started and from which vegetation has already been burned.

Apparatus A motor-driven vehicle designed for the purpose of fighting fires.

Arduous pack test A physical fitness test for wildland fire fighters; also called the work capacity test.

Area ignition Occurs when several individual fires throughout an area cause the main body of fire to produce a hot, fast-spreading fire condition. Also called simultaneous ignition.

Area of origin The location where a wildland fire begins.

Aspect The direction of a slope face in relation to a cardinal compass point.

Atmospheric stability The degree to which vertical motion in the atmosphere is enhanced or suppressed.

Axe A tool with a short handle and steel blade used for chopping.

Azimuth The horizontal angle of a point measured clockwise from true north.

B

Back azimuth An angle 180 degrees opposite of azimuth; used when retracing steps.

Backfire A fire intentionally set by fire fighters along the inner edge of a control line in an attempt to burn an area of vegetation.

Backpack pump A portable pump and tank fitted with straps so it can be carried on a fire fighter's back.

Base station radio A stationary radio permanently mounted in a building.

Batch mixing Manually adding and mixing a concentrated chemical into a solution in a tank or container.

Blackline method A method in which a buffer of preburned fuels is created between the fire's edge and the rest of the fire.

Blood bubble The radius around the faller where the tip of the chainsaw can reach.

Box canyons Canyons with three steep walls that create very strong upslope winds. Also known as dead-end canyons.

Branches Organizational level having functional or geographical responsibility for major aspects of incident operations; can only be used in the operations section or the logistics section.

Briefing A meeting designed to provide information and instruction on an incident.

Brush hook An axe with a curved blade head rather than a standard axe head.

Burning period The part of each 24-hour period in which fires spread most rapidly, typically, from 10:00 AM to sundown.

Burnout Setting fire inside a control line to consume the fuel between the edge of the fire and the control line.

C

Centrifugal pump Pump that expels water by centrifugal force through the ports of a circular impeller rotating at high speed.

Chain A unit of measure in land survey equaling 66 ft (20.1 m); 80 chains is equal to 1 mile (1.6 km).

Chain of command A series of management positions, in order of authority, through which decisions are made.

Chainsaw A portable tool, either gasoline-powered or electric, with handles and an engine on one end and a set of sharp, rotating teeth lining a blade on the other end.

Channel An assigned frequency or frequencies used to carry voice and data communications.

Chaps Specialized pieces of PPE, constructed of several layers of Kevlar under a protective nylon cover, that cover the legs during chainsaw operations.

Chimney Terrain feature that has a channeling effect on the convective energy of fire.

Class A foam An aerated solution intended for use on woody (Class A) fuels as a firefighting aid by reducing the surface tension of water and allowing it to better penetrate fuels.

Cloning Process that allows for the copying of frequency information from one radio and sending that copy to another radio using a set of cables.

Combination tool A sturdy and lightweight tool with many uses: grabbing, digging, smothering, scraping, cutting, and prying; also called a combi tool or Colby tool.

Command Has decision-making responsibility and overall responsibility for management of the incident.

Command staff Consists of the public information officer (PIO), the safety officer, and the liaison officer. These officers report directly to the IC.

Communications unit Distributes and maintains all forms of communications equipment on the incident and is responsible for developing an incident communications plan, included in the IAP, for the most effective use of the communications equipment.

Compactness Refers to the spacing between fuel particles, sometimes, called density.

Compensation and claims unit Responsible for ensuring that all forms required for a worker's compensation claim are filled out correctly and maintains records of injuries or illnesses incurred during the incident.

Compressed air foam system (CAFS) A high-energy system used to deliver foam onto a fire; made up of an air compressor, a water pump, and foam solution.

Control line A constructed or natural barrier used to control a fire.

Convergence Occurs when horizontal air currents merge together.

Cost unit Within the finance and administration section and responsible for tracking costs, analyzing cost data, making cost estimates, and recommending cost-saving measures.

Counterfire A fire set between a main fire and a prescribed backfire to hasten spread of the backfire and redirect the main wildfire; also called auxiliary fire.

Crew A team of two or more fire fighters.

Crew-carrying vehicle (CCV) A vehicle specialized for carrying a fire crew and their equipment.

Critical fire weather Weather that can quickly increase fire danger and cause extreme fire behavior, such as weather fronts and foehn winds.

Critical winds Winds that totally dominate the fire environment and easily override the upslope/downslope and upvalley/downvalley winds; include foehn winds, thunderstorm winds, glacier winds, and frontal winds.

Crown fires Fires that advance from the tops of trees or shrubs, more or less dependent on a surface fire.

Cultural resource Any object relating to past human life that can serve as a potential resource for understanding the past.

Cutting section A division of a fire crew responsible for the sawing, chopping, and grubbing of woody fuels.

D

Dead fuel moisture The amount of water contained in dead plant tissue.

Declination The difference between magnetic north and true north. Also called magnetic declination or magnetic variation.

Degrees decimal minutes (DDM) A format for communicating the latitude and longitude values of a point, displayed as DDD° MM.MMM'.

Degrees minutes seconds (DMS) A format for communicating the latitude and longitude values of a point, displayed as DDD° MM'SS.S".

Demobilization unit Responsible for developing an efficient incident demobilization plan to ensure that incident personnel and equipment are released in an orderly, safe, and cost-effective method after they are no longer needed.

Deployment site A last-resort location where a fire shelter must be deployed; used when access to escape routes and safety zones has been compromised.

Detonation line A control line created by using explosives.

Direct attack A method of fire attack in which fire fighters apply a treatment such as wetting or smothering to the fire or by physically separating the burning from unburned fuel.

Direct communication Language that is concise and factual and provides closure.

Divisions Separate an incident into geographic areas of operation. They are established to regain and maintain span of control when available resources exceed the span of control of the operations chief. Divisions are managed by the division supervisor or group supervisor.

Documentation unit Provides incident duplication services, maintains, stores, and packs incident files for legal, historical, and analytical purposes.

Downbursts Strong and, sometimes, damaging winds that spread out upon meeting the ground and produce erratic gusting winds.

Downdrafts Winds that are part of the normal flow pattern of a thunderstorm.

Dozer A tracked vehicle with a front-mounted blade used for moving fuels and exposing soil.

Drafting The process of drawing water from a static water source into a pump.

Drip torch A canister that drips ignited fuel from a spout to a wick, which becomes saturated with fuel and burns continuously. Also called an orchard torch.

Dry hydrant A pipe that is placed in a static water source and threaded on one end from which water can be pumped by the fire engine.

Dry line front A boundary separating moist and dry air masses. Also called a dew point line.

E

Easting value The distance, in meters, of a position east of a 6° UTM zone line. Each central meridian that runs through these grid zones is assigned an easting value of 500,000 ME. Positions west of the central meridian have easting values less than 500,000 ME, and positions east of the central meridian have easting values greater than 500,000 ME.

Eddy A circular-like flow of air or water drawing energy from a flow of much larger scale.

Emergency traffic Urgent radio message that takes priority over all other radio communications.

Escape route A preplanned route that fire fighters take to move to a safety zone.

Escape time The time it takes for crew members to make it to a safety zone.

Excavator A tracked vehicle with a rotating cab and a long, jointed neck with a bucket.

Expansion The ratio of the volume of the foam in its aerated state to the original volume of the nonaerated foam solution.

Extended attack (EA) A fire that exceeds the capabilities of the initial attack fire resources and requires additional resources.

F

Facilities unit Within the logistics section and provides the layout, activation, and management of all incident facilities.

Failure zone Any area that can be reached by any part of a falling tree.

Faller A chainsaw operator.

Feller buncher Mechanized logging equipment used for accelerated fireline construction by cutting down and removing trees. Also called harvesters.

Finance and administration section Responsible for all incident costs and financial considerations; includes time, procurement, compensation and claims, and cost of units.

Finger A long, narrow extension of a fire, caused by a change in weather, topography, or fuel.

Fire-blocking gel An absorbent and hydrating polymer mixed with water that aids in firefighting efforts by clinging to vertical surfaces and absorbing heat.

Fire Danger Pocket Cards Outline potential fire dangers in a given area based on historical fire and weather data.

Fire eductor pump A portable pump used to draft water from static sources; also called an ejector or jet siphon.

Fire investigator An individual trained to examine a scene and determine a point of ignition.

Fireline A barrier purposefully constructed by fire fighters to control a fire.

Fire rake A long-handled rake.

Fire shelter A reflective cloth tent that offers protection from fire, heat, smoke, and ember showers in an entrapment situation.

Fire swatter A long-handled tool with a thick, flat piece of rubber on the end used to smother flames.

Fire triangle Model for explaining the behavior of wildland fires. Consists of three elements: fuel, oxygen, and heat.

Fixed-wing air tanker A fixed-wing aircraft generally used to drop long-term fire retardant to slow the spread of fire.

Flame length The measurement from the average flame tip to the middle of the flaming zone at the base of the fire; measured on a slant when the flames are tilted due to effects of wind and slope.

Flaming front zone Where continuous flaming combustion is taking place.

Flank The edge between the head and heel of the fire that runs parallel to the direction of fire spread.

Flank firing A firing technique that involves lighting a fire into the wind or perpendicular to the slope, allowing the fire to spread at right angles.

Flanking attack A direct fire suppression attack that involves placing a suppression crew along the flank of a fire.

Flare launcher A handgun-shaped device that expulses flares.

Floating pumps Portable pumps that can float on the water's surface.

Foam blanket A layer of foam that forms an insulating, smothering, and reflective barrier to heat; used for fuel protection, fire attack, and mop-up operations.

Foehn winds Warm, dry, and strong general winds that flow down into the valleys when stable, high-pressure air is forced across and then down the lee slopes of a mountain range. Also known as Chinook, east, north, Santa Ana, and Wasatch winds.

Food unit Within the logistics section and responsible for providing meals for incident personnel.

Forestry fire hose A lightweight, collapsible attack hose; also called a forestry hose.

Frequency The number of cycles per second of a radio signal.

Friction loss The pressure lost from turbulence as water passes through hoses and hose accessories.

Front A boundary between two air masses of different properties.

Frontal lifting Occurs when a moving, cooler air mass pushes its way under a warmer air mass and forces the air up a slope.

Fuel continuity The degree of continuous distribution of fuel particles in a fuel bed.

Fuel moisture Refers to the amount of moisture present within a fuel source. It is an important factor in how easily a fuel ignites, burns, and spreads.

Fuels Materials that store energy and therefore are combustible.

Full duplex channels Radio systems that allow both parties to send information simultaneously.

Fusee A colored flare designed to ignite backfires; also called a backfire torch or forest fire torch.

G

Gansner packs Hose packs that utilize the hose for the backpack and use a 1.5-inch gated wye with one side fitted with a 1-inch reducer.

General staff Group of incident management personnel reporting to the IC, consisting of the operations section chief, planning section chief, logistics section chief, and financial and administration chief.

General winds Large-scale winds produced by pressure gradients associated with high- and low-pressure differences. Also known as gradient winds.

Global Positioning System (GPS) A system of navigational satellites that can track objects anywhere in the world with an accuracy of approximately 40 to 300 ft (12.2 to 91.4 m).

Gnass packs Hose packs that utilize the hose for the backpack and use inline tees rather than gated wyes.

Gravity sock A conical device, typically made of canvas, that has a large opening that is anchored in a stream, and a smaller opening with a threaded hose connection. Gravity feeds water into the sock and to the hose.

Ground equipment Heavy, mechanized equipment used to construct firelines.

Ground fuels Combustible material lying on or beneath the ground or subsurface litter. Ground fuels include root systems, deep duff, rotting buried logs, and other woody materials in various states of decomposition.

Ground support unit Within the logistics section and responsible for fueling, maintaining, and repairing vehicles and transporting personnel and supplies.

Groups Established to divide the incident into functional areas of operation.

H

Half duplex channels Radio systems that use one frequency to transmit and receive all messages; transmissions can occur in either direction but not simultaneously.

Handcrew A team of 12 to 20 fire fighters who use hand tools to actively suppress low flame-production fires by constructing hand lines.

Handcrew production rate A measure of how long it will take to complete a section of line with available handcrew resources before the fire arrives.

Hand line A fireline that is constructed using hand tools.

Hard suction hose A noncollapsible hose used to supply water to the intake side of the fire pump.

Hazardous attitude A negative human characteristic that can affect situational awareness, crew safety, and assignment success.

Hazardous material A substance capable of causing harm to the environment, people, and property.

Hazard tree A tree with a structural defect that can cause the entire tree or part of the tree to fall.

Head The most intense and rapidly spreading part of the fire.

Heat transfer The process by which heat is imparted from one body to another, through conduction, convection, and radiation.

Heel The area of a fire close to the area of origin, also called the rear of the fire.

Helicopter An aircraft with revolving head rotors.

Helispot A temporary location where supplies, equipment, or personnel are transported or picked up by helicopters.

Helitack crew A crew of fire fighters specially trained in the use of helicopters for fire suppression.

High-pressure portable pumps Pumps that supply pressurized water from water sources to firefighting hose lines.

Hoe A digging tool with a sharp, flat blade on one or both sides of the head.

Horizontal continuity Refers to the way that wildland fuels are distributed at various levels.

Hose A device used to deliver water from a water source to a fire.

Hose clamp A crimping device used to temporarily stop the flow of water in a hose.

Hoselay The arrangement of connected fire hose and accessories on the ground, beginning at the first pumping location and ending at the point of water delivery.

Hose pack A backpack containing a hose, hose accessories, and other hoselay equipment.

Hydraulics The science of water in motion and at rest.

I

Incident A natural or human-caused occurrence that requires emergency response to prevent or minimize loss of life or damage to property and natural resources.

Incident action plan (IAP) A document that contains objectives reflecting the overall incident strategy, as well as specific tactical actions and supporting information for the next operational period; the plan may be written or oral.

Incident commander (IC) The individual responsible for all incident activities, including the development of strategies and tactics and the ordering and release of resources.

Incident Command System (ICS) An adaptable emergency management system that allows users to adopt an integrated organizational structure equal to the complexity and demands of single or multiple incidents of any type or size.

Incident Response Pocket Guide (IRPG) A pocket-sized book that outlines standards and reference information for wildland fire incident response.

Incident within an incident (IWI) An accident or emergency that occurs on the scene of a fire.

Index contour Every fourth or fifth line on a topographic map, printed darker to make it easier to read the map. Elevation is listed at some point along an index contour.

Indirect attack An indirect fire attack that involves building a fire control line along a predetermined route, based on natural fuel breaks or favorable breaks in the topography, and then burning out the intervening fuel.

Indraft The process of air being drawn into a larger, main fire from a backfire.

Initial attack (IA) A fire that is controlled by the first dispatched resources, without significant reinforcements.

Inmate crew A crew composed of prison inmates and/or wards; usually part of volunteer firefighting programs.

Interagency hotshot crew (IHC) A kind of Type 1 crew; the most experienced, highly and intensively trained type of fire crew that have the skills to battle the most serious wildfires.

Intermediate contour The lighter contour lines between index contours.

Inversion A layer in the atmosphere that acts like a lid or cap over a fire.

Island An area of land that untouched by fire, but is surrounded by burned land.

L

Ladder fuels Refers to the vertical arrangement of fuels; allow a fire to climb from ground fuels to surface fuels to aerial fuels.

Latitude A point north or south of the equator, measured in degrees, minutes, and seconds. The equator is 0° latitude.

Leader's intent A clear, concise statement that outlines what individuals must know in order to be successful for a given assignment.

Leapfrog method A system to build a fireline in which each crew member is assigned a task at a specific section of line until the task is completed. After the task is completed, the crew member calls out and everyone ahead of them moves up the line.

Legal description The description of the land and property, commonly used to describe your location or when receiving an incident location.

Legend A key accompanying a map that shows information needed to interpret that map.

Liaison officer Member of the command staff responsible for coordinating with agency representatives from assisting and cooperating agencies.

Line firing A firing technique that involves burning along the inner edge of the control line to a termination point.

Live fuel moisture Ratio of the amount of water to the amount of dry plant material in living plants.

Live-to-dead fuel ratio The amount of dead fuel present in a fuel bed.

Local winds Result of local temperature differences and are most influenced by terrain factors.

Logistics section Orders resources (facilities, services, and supplies) and develops the transportation, communications, and medical plans.

Longitude A point east or west of the equator, measured in degrees, minutes, and seconds. The prime meridian is 0° longitude.

Lookout An individual designated to detect and report fires from a vantage point.

LUNAR Location, unit, name, assignment, and resources; acronym used to help remember the information needed when reporting a mayday, or emergency, message.

M

Machine line A control line constructed using a variety of heavy equipment and machinery.

Magnetic north The magnetic north pole. Magnetic north changes over time and with location.

Map projection A system used to project the Earth onto a flat map surface.

Marine inversions Result of cool, moist air from the ocean spreading over low-lying land. Common around large bodies of water.

Masticator A type of mechanized equipment used to grind brush and debris while driving through it. Also referred to as a mulcher.

Mayday A verbal declaration indicating that a fire fighter is lost, missing, injured, or requires immediate assistance.

McLeod A combination scraping and raking tool; one side of the head has a blade and the other side has a rake.

Mean sea level Average height of the sea for all stages of the tide over a 19-year period.

Medical incident report (MIR) A written document or verbal report that outlines the details of an accident.

Medical unit Provides and maintains first aid and emergency medical treatment stations for incident personnel.

Mobile pump The operation of a water pump while in motion.

Mobile radio A two-way radio that is permanently mounted in a fire apparatus.

Mop-up The action of extinguishing or removing burning material near control lines.

N

Narrow canyons Steep canyon walls that are close together, possibly creating problems for fire crews if fire from one canyon wall jumps to the opposite canyon wall.

National Fire Danger Rating System (NFDRS) Compares current and historic data related to weather, live and dead fuel moistures, and fuel types to rate the level of potential fire danger.

National Fire Protection Association (NFPA) Develops and maintains nationally recognized minimum consensus standards on many areas of fire safety and specific standards on hazardous materials.

National Wildfire Coordinating Group (NWCG) An organization made up of several wildfire management agencies that was designed to coordinate programs of the participating agencies to standardize trainings and policies, avoid duplicative efforts, and provide a means of constructively working together to establish efficient wildland fire response efforts.

Night inversion Results from air cooling as it comes in contact with Earth's surface. It is the most common type of inversion, found predominately in mountainous terrain and inland valleys. Also called a radiation inversion.

Northing value The distance, in meters, of a position relative to the equator. If you are in the northern hemisphere, the equator has a northing value of 0 m, and the northing value of a point located in the northern hemisphere is equal to its distance from the equator. If you are in the southern hemisphere, the equator has a northing value of 10,000,000 MN, and the northing value of a point located in the southern hemisphere is equal to 10,000,000 minus its distance from the equator.

Nozzle The spout connection to the end of a hose that water flows through.

Nozzle pressure (NP) The pressure that a nozzle is designed to operate at most efficiently.

O

Objectives Statements of what is to be accomplished.

One-lick method A progressive system used to build a fireline in which each crew member does one to several licks of work and then moves forward to make room for the person behind them.

Operational leadership Management technique that takes human factors and processes into consideration; includes improving systems and processes, taking charge, motivating fire fighters, demonstrating initiative, and directly communicating.

Operational period Period of time scheduled for execution of a given set of tactical actions, usually not exceeding 24 hours.

Operations section Responsible for all tactical operations at the incident.

Organic material Any natural matter, such as leaves, rotting logs, or duff.

Orographic lifting Occurs in mountainous areas where the heated moist air is forced up as a result of the presence of a slope and topographic features.

P

Pace Two natural walking steps measured from the heel of one foot to the heel of the same foot after two steps. Pace can be used to keep track of distance. Pace count will differ among people and by terrain.

Parallel attack A method of fire attack in which a fire control line is built parallel to the fire's edge.

Patrol vehicle Small pickup trucks, found in many rural areas, that are effective in rocky or steep terrain.

Personal protective equipment (PPE) Clothing and equipment designed to protect fire fighters in conditions that might otherwise result in death or serious injury.

Personnel accountability The ability to account for the location and well-being of one's crew.

Pincer attack (anchor, flank, and pinch attack) A direct fire suppression attack that involves two or more teams of fire fighters establishing anchor points on each side of the fire and working toward the head of the fire until the fire gets "pinched" between them; also known as the anchor, flank, and pinch attack.

Planning section Collects, evaluates, and disseminates information about the incident. This section also manages the planning process and develops the IAP for each of the operational periods. Also, maintains information on the status of equipment and personnel resources, maintains all incident documentation, and prepares the demobilization plan.

Pocket The unburned area between the finger of a fire and the main body of the fire.

Point of ignition (POI) The area where a fire started; also called area of origin.

Pondosa pack A crude backpack made of straps with clasps that allow a fire fighter to carry two to three 100-foot doughnut-rolled hoses.

Portable pump A small pump that can be carried to a water source and used to draft water.

Portable radio A battery-operated, two-way, handheld transceiver.

Portable water tanks Containers used as water reservoirs that engines can use to refill.

Position Task Book (PTB) A document that lists the performance requirements for a position in a format that allows for performance evaluations to determine whether an individual is qualified.

Powered blower A portable tool that uses air to blow leaves and needles away from a structure.

Pressure gradients The differences in atmospheric pressure between two points on a weather map.

Priority traffic Radio message that contains critical information.

Procurement unit Within the finance and administration section and responsible for financial matters involving vendor contracts.

Progressive hoselay Hose connections designed to quickly and continuously move water from the anchor point to the fire. Additional hose lines can be attached without having to shut water off at the pump.

Proportioner Device that adds a predetermined amount of foam concentrate to water to form a foam solution.

Public information officer (PIO) A member of the command staff who acts as the central point of contact for the media.

Pulaski A chopping and trenching tool with one side of the head being a blade used for cutting, and the other side being a flat edge used for grabbing.

Pump-discharge pressure (PDP) The total pressure needed to overcome all friction, appliance, and elevation loss or gain while maintaining adequate nozzle pressure to deliver effective fire streams; calculated with the following equation: PDP = NP + FL.

R

Rapid extraction module support (REMS) unit A specially trained unit that specializes in basic- and paramedic-level medical care and high-angle rescue skills.

Rappel crew A crew of fire fighters who repel down from a hovering helicopter in areas where the helicopter is unable to land.

Rate of spread Speed at which a fire is moving away from the site of origin.

Red card A document that certifies that an individual is qualified to perform the tasks of a specific position on a wildland fire; officially called the Incident Qualification Card.

Relative humidity The ratio of the amount of moisture present in the air compared to the maximum amount of moisture that the air can hold at a given temperature and atmospheric pressure.

Repeater A special base station radio that receives messages and signals on one frequency and then automatically transmits them on a second frequency.

Resource unit Responsible for checking in all incoming resources and personnel on the incident.

Retardant line A control line created by distributing a chemical solution such as foam that slows the forward progress of the fire.

Ring firing A firing technique that involves igniting fire in a circular pattern around specific objects to establish a buffer line against the main wildfire.

Risk management The process of identifying, observing, and acting on potential risks to hinder or eliminate them.

Risk management process A series of steps used to make decisions and observe their outcomes.

Routine traffic Normal radio communication required in response to an incident.

S

Saddles Low topography points between two high topography points.

Safety officer Member of the command staff, reporting to the IC, who is responsible for monitoring and assessing hazardous and unsafe situations and developing measures for assessing personnel safety.

Safety zone An area cleared of flammable materials that is used as refuge when a fire has been determined to be unsafe.

Sandvik A cutting tool with a C-shaped blade used to cut small- and medium-sized brush.

Saw line A fireline constructed by clearing away trees and brush so that remaining fuel can be removed with hand tools.

Scraping section A division of a fire crew responsible for the removal of duff, grass, and smaller brush from the fireline, and completing the handline.

Scratch line A temporary, quickly constructed fireline used as an emergency measure to control the spread of fire.

Scrubbing Agitating foam solution and air within a confined space (usually a hose) to produce tiny, uniform bubbles.

Shovel A tool with a long handled and a tapered blade with both edges sharpened.

Simple hoselay The connection of hoses from a pump to the point-of-discharge.

Simplex channels Push-to-talk, release-to-listen radio systems that use one frequency to transmit and receive all messages; transmissions can occur in either direction but not simultaneously.

Situation unit Collects, processes, organizes, and displays information on the current incident status.

Situational awareness The understanding of what is going on around one's self through observation and communication.

Size-up The process of gathering and evaluating information from a fire to determine a course of action and prepare for response.

Skidgen A vehicle equipped with a plow on the front and a water tank and pump on the back, designed to traverse terrain fire engines cannot. Typically, used for initial attack and mop-up.

Skills crosswalks Specific PTBs used in instances where structural fire fighters want to transition to a wildland fire fighter role to fill in training gaps to reduce the number of training hours needed to obtain qualification.

Slop-over A fire edge that crosses a control line or natural barrier.

Slug flow Discharge of distinct pockets of water and air due to the insufficient mixing of foam concentrate, water, and air in a compressed system.

Smokejumper crew A team of two or more fire fighters who deploy from fixed-wing aircraft by parachute into remote areas for wildfire-suppression efforts.

Smoldering Burning without flame and barely spreading.

Snag A dead or dying tree that is still standing.

Social media Online websites on which people can publicly share personal information, pictures, news, and interests.

Span of control Supervisory ratio of from three to seven individuals, with five individuals to one supervisor being established as optimal.

Spot fire A new fire that starts outside the perimeter of the main fire.

Spotting Occurs when small burning embers and sparks are carried by the wind that land outside the fire perimeter, which starts new fires beyond the zone of direct ignition.

Spot weather forecast A special forecast, issued upon request, that fits the time, topography, and weather of a specific incident.

Staging areas Locations set up at an incident where resources can be placed while awaiting a tactical assignment, on a 3-minute available basis.

Strategy A general plan to meet a set of predetermined objectives.

Strike team Specified combinations of the same kind and type of resources, with common communications and a leader.

String trimmer A long metal rod with a spinning blade or string, used for reducing fuel next to a fireline.

Strip firing A firing technique that involves setting fire to one or more strips of fuel to create indrafts and reduce the intensity of the fire at the control line; also called line firing.

Structure triage A system of classifying structures based on defensibility and prioritizing protection to the least-defensible structures first.

Subsidence Downward or sinking motion of air in the atmosphere.

Subsidence inversions Inversions caused by subsiding air, often resulting in very limited atmospheric mixing conditions.

Supply unit Within the support branch of the logistics section and responsible for ordering equipment and supplies required for incident operations.

Surface fuels Combustible material lying on or near the surface of the ground and a major cause of wildland fire spread. Surface fuels include ground litter, such as leaves, needles, bark, grasses, tree cones, and brush up to 6 ft (1.8 m) in height.

Surfactant Any wetting agent; a formulation that, when added to water in proper amounts, materially reduces the surface tension of the water and increases the penetration and spreading abilities of the water.

Swamper A fire fighter who assists the chainsaw operator.

T

Tactical actions The steps taken after triage, typically utilizing the PACE (primary, alternate, contingency, and emergency) model.

Tactics The specific steps necessary to support and achieve a given strategy.

Task force Any combination of single resources assembled for a particular tactical need, with common communications and a leader.

Technical specialists Personnel who have specialized skills. May be assigned to the planning section, function within a unit, or form a separate unit within the planning section. The most common specialists include field observers, fire effects monitors, FBANs, IMETs, and GIS specialists.

Thermal belt Any area in a mountainous region that typically experiences the least changes in temperature on a daily basis.

Thermal lifting The result of strong heating of the air near the ground. The heated, moisture-laden air rises high enough in the atmosphere to form cumulus clouds.

Time unit Within the finance and administration section and responsible for recording time for incident personnel and hired equipment.

Torching Occurs when tree foliage burns from bottom to top. Torching is intermittent and may be an indicator that fire behavior is increasing. Sometimes, referred to as candling.

Tractor plow A dozer with a plow unit behind it.

Travis packs Hose packs that utilize a configuration of 200 feet of hose for rapid deployment.

True north North according to Earth's axis.

Trunked channels Radio systems that group users to available radio channels as needed through a computerized shared bank of frequencies; also called 700/800 MHz channels.

Type 1 crew A highly experienced, and physically fit crew that can be broken up into smaller squads for constructing firelines and performing complex firing operations.

Type 1 engine A fire apparatus used primarily for fighting urban structural fires.

Type 2 crew A crew that constructs firelines and completes firing operations, with 20 percent of its members having at least one season of experience in fire suppression.

Type 2 engine A fire apparatus that can be used for both structural and wildland firefighting.

Type 2 IA crew A crew with more experience and capability than Type 2 crews, with 60 percent of its members having at least one season of experience in fire suppression, and qualified sawyers and Incident Commander Type 4 or 5 qualified squad bosses.

Type 3 crew A crew with limited to no experience in fire suppression; also known as a camp crew.

Type 3 engine A smaller apparatus with the ability to mobile pump, used typically in wildland areas.

U

United States Geological Survey (USGS) topographic maps Detailed topographic maps produced by the United States Geological Survey.

Universal Transverse Mercator (UTM) A rectangular grid surface designed to make maps simpler to use and avoid the inconvenience of pinpointing locations on curved reference lines.

Unmanned aircraft systems (UASs) Air vehicles without a human pilot on board. Also called unmanned aerial vehicles (UAVs) or drones.

V

Valley winds Daily (diurnal) winds that flow up a valley during the day and down a valley at night.

Vertical fuel arrangement Refers to the different heights of the fuels that are present at a fire.

Virga Precipitation that falls from a cloud but evaporates before reaching the ground.

Vortex A whirling mass of air with a low-pressure area in its center that tends to draw fire and objects, such as firebrands, into the circling action.

W

Water tender Large containers of water, usually equipped with a pump-and-dump valve.

Wet line A control line created by spraying water or a mixture of water and foam or retardant along the ground.

Wet mop-up The use of water or water and soil together to extinguish burning materials.

Wildland Fire Fighter II (NWCG Fire Fighter Type 1) The person at the second level of progression who has demonstrated the depth of knowledge and skills necessary to function as a member of a wildland fire suppression crew under general supervision.

Wildland Fire Fighter I (NWCG Fire Fighter Type 2) The person at the first level of progression who has demonstrated the knowledge and skills necessary to function as a member of a wildland fire suppression crew under direct supervision.

Wildland Fire Incident Management Field Guide A document that contains critical and useful information that is relevant to fire fighters at all levels of responsibility.

Wildland/urban interface (WUI) The area where undeveloped land populated with vegetative fuels meets with manmade structures.

Index

Courtesy of Joe Lowe.

Note: Page numbers followed by *f* and *t* denote figures and tables respectively.

G

M

N

O

P

T